Waves in Metamaterials

Waves in Metamaterials

L. Solymar
Imperial College, London

E. Shamonina
University of Oxford

OXFORD
UNIVERSITY PRESS

Great Clarendon Street, Oxford, OX2 6DP,
United Kingdom

Oxford University Press is a department of the University of Oxford.
It furthers the University's objective of excellence in research, scholarship,
and education by publishing worldwide. Oxford is a registered trade mark of
Oxford University Press in the UK and in certain other countries

First Edition published in 2009
First published in paperback 2014
Impression: 1

Published in the United States of America by Oxford University Press
198 Madison Avenue, New York, NY 10016, United States of America

British Library Cataloguing in Publication Data
Data available

ISBN 978-0-19-921533-1 (hbk.)
ISBN 978-0-19-870501-7 (pbk.)

Printed and bound by Clays Ltd, St Ives plc

To Oscar and Sasha

Preface

Metamaterials is a new subject. The term 'metamaterial' could have been coined at any time in the last hundred years or so (Greek prefixes have always been fashionable) meaning that it is something well beyond ordinary materials, but, apparently, nobody did so. The birth of the term coincides with the birth of the subject. It appears in the first paper (Smith *et al.*, 2000) that showed the existence of an artificial material that can have both negative permeability and permittivity. The roots go back as far as the nineteenth century with a paper by Bose (1898) concerned with the effect of a man-made twisted structure on the polarization of an electromagnetic wave.

So what are metamaterials? There is no consensus. A brief and quite good definition is 'artificial media with unusual electromagnetic properties'. This is good because it captures the spirit of the subject. On the other hand, if we accept this definition then the burden only goes back on the meaning of the word 'unusual'. What is usual to one man/woman might be unusual to another one. And anyway, is the definition general enough? Are we always concerned with artificial media? Not really. Attempts to produce a flat lens with subwavelength resolution rely mostly on silver films, not an artificial material. The unfortunate fact is that no definition exists that would be universally applicable or universally accepted. Thus, the best we can do is to enumerate the topics that we have included in this book and that, at least in our definition, belong to the subject of metamaterials.

Chapter 1 is just a summary of the basic concepts and basic equations that appear later in the book. Chapter 2 is an attempt to give both an introduction and an overview. It is only from Chapter 3 that we start to go into more detail. Chapter 3 is concerned with plasmas and particularly surface plasma waves. Although many in the field might think that plasmas have only marginal relevance to metamaterials we believe that they occupy a central position both in the theoretical formulations and in establishing a physical picture. Growing exponential waves, for example, (often referred to as amplification) make no sense unless we refer to the excitation of surface plasma waves. Chapter 4 is divided into two parts. One is a gallery of various small resonators (we describe more than three dozen) suitable for components in a metamaterial. The second part concentrates on the very popular element known as the split-ring resonator and determines its electrical properties by deriving a differential equation. Chapter 5 discusses subwavelength imaging, which is still a controversial subject. We attempt to give a fair summary of the achievements and of some of the misconceptions. Chapter 6 discusses waveguides, mainly the effect of metamaterials on their cutoff frequency but, in addition, various applications involving metamaterials are also discussed. Chapters 7 and 8 are mostly concerned with our own work on magnetoinductive waves. Chapter 9 is a kind of last-ditch attempt to discuss phenomena that did not fit into earlier treatments. It discusses seven different topics that include nanoparticles and invisibility. In Chapter 10 we look at the subject of metamaterials asking the question how it started and how it became so popular. We have a large number of appendices, most of them serving their traditional purpose; to offer a mathematical derivation that readers might find too time consuming to do themselves or make some comments that are of interest only to a specialized audience.

Having discussed what is included we should also say what is not included. We are aware that we have given very little attention to photonic bandgap materials. Our main argument for this omission is that they are not really part of metamaterials. Both the physics and the mathematical treatments are quite different. The ideal metamaterial contains many elements per free-space wavelength, whereas photonic bandgap materials are traditionally based on Bragg interaction and on a large difference between the dielectric constants. Two further omissions are chiral elements and non-linear effects. We felt that including them would be too much of a diversion from our main aim, to give a fairly unified treatment of the phenomena associated with metamaterials. We made an exception with parametric amplification of magnetoinductive waves which is included in Chapter 8.

After the content it is customary to say what the expected audience is and who in our opinion might benefit from reading the book. We believe it would be perfectly feasible for a research-oriented undergraduate to take this book in hand during his/her last year of study, look at it, turn the pages, read it here and there, draw a few sketches, make a few calculations on the back of an envelope and put it (or not, as the case may be) among those research subjects he/she might be interested to pursue in the future. The attraction for graduate students is to learn about the subject as a whole beyond their narrower speciality. Experts might buy the book for the library of their institution for the benefit of the graduate students but will they read it themselves? A good proportion of the experts are doers and not readers. They may not want to have their ideas tainted by learning about other people's feats. There are many though who are interested in finding out what they might have missed and are willing to delve into a book just in case they come across something new. And there is one more potential market for a book in a fast growing field. Those who feel that their decade (or two?) -old involvement with a particular field no longer leads to tangible results, and are thinking of changing discipline. This book might facilitate the choice for them.

We had no policy on references. It was clear from the beginning that we would not be able to refer to all the papers written about the subject or relevant to the subject. A conservative estimate would make their number exceed 1500. The way we proceeded was to write about a particular topic and choose the references as they came up. Our total number is less than 500. We are bound to have missed important papers. We offer our apologies.

Concerning presentation we have to say a few words about acronyms. We don't like them, although, we must admit, we are responsible for some of them. It is not particularly surprising that in an acronym-mad world many of us in the metamaterial community have been affected by the disease. Just to show how far the disease has spread we enumerate in Appendix A the acronyms we could find after a brief search. The use of acronyms is probably counter-productive. It is supposed to save time. More likely it leads to occasional incomprehension and miscomprehension. We believe authors of technical books should refrain from proliferating them. We shall make a conscious effort not to use any of the new ones invented specifically for the benefit of metamaterials with a few exceptions: SRR for split-ring resonator, SPP for surface plasmon–polariton and MI for magnetoinductive.

We have lots of people to thank for getting to the end of this journey. We were lucky to have plenty of support in our efforts. Our thanks go first of all to Lesha Sydoruk who helped us at every stage of the book by reading every section, finding mistakes and inaccuracies, and making numerous proposals for improvement. We also wish to acknowledge the help we received from our close collaborators, David Edwards, Frank Hesmer, Victor Kalinin, Anya Radkovskaya, Klaus Ringhofer, Misha Shamonin, Chris Stevens, Rich Syms, Zhenya Tatartschuk, Ian Young and Sasha Zhuromskyy. For discussions on a number of subjects we are obliged to P. Belov, N. Engheta, M. Freire, M. Gorkunov, M. Kafesaki, A. F. Koenderink, M. Lapine, R. Marques, S. A. Tretyakov, V. Veselago and M. Wiltshire. Special thanks are due to those whom we asked for original figures and who so warmly responded: K. Aydin, P. Belov, P. Berini, S. R. J. Brueck, W. Cai, C. Caloz, H. Chen, S. A. Cummer, G. V. Eleftheriades, N. Engheta, M. J. Freire, O. Hess, S. Hrabar, M. Kafesaki, A. F. Koenderink, R. Marques, O. J. F. Martin, J. B. Pendry, I. Smolyaninov, D. Schurig, V. M. Shalaev, C. M. Soukoulis, R. R. A. Syms, S. A. Tretyakov, K. L. Tsakmakidis, W. H. Weber, T. Zentgraf, X. Zhang and J. Zhou.

Finally, L.S. would like to thank, and thank profusely, his wife Marianne, who was willing to put up with the long hours he spent on this book in violation of the basic principle that retired people should sit on their laurels and work no longer. E.S. thanks her family for their help, support and encouragement and Sasha for being so generous in letting his mum work and for being such a good boy.

Contents

Basic concepts and basic equations

1

1.1 Introduction

We shall make an attempt in this chapter to introduce most of the basic concepts needed later in the book and in a form upon which we can build in later chapters. Some will be very basic indeed, like fields and potentials, some others will be closer to practical cases, e.g. the incidence of a plane wave upon a dielectric slab. Some will represent useful artifices like electric and magnetic dipoles, leading from there to the introduction of the concept of polarizability: how large are the dipole moments excited by various fields. Boundary conditions and boundary refraction problems must of course be mentioned because they will appear repeatedly in the analysis of metamaterial properties. The Ewald circle construction will then come in as a useful technique. Dispersion is again one of the basic concepts. In any wave phenomenon one must know the relationship between frequency and wave number.[1] Talking about dispersion necessitates the introduction of forward and backward waves. The latter were hardly mentioned in the past, their properties rarely emphasized in studies of electromagnetic waves, but they have lately risen to fame due to their role in metamaterials. Another departure from usual introductions is the prominence of circuit theory, a subject often neglected in physics syllabuses. Very often we shall have to look at phenomena both from the point of view of fields and circuits. Sometimes the two explanations reinforce each other, sometimes we shall resort to only one of them because the other one may be too unwieldy. Acquaintance with Fourier analysis is of course a basic requirement whether one is concerned with the temporal or the spatial regime. In more general terms it means working in reciprocal space but we shall do that only sparingly, mostly in the one-dimensional context.

The subject of metamaterials can be highly mathematical, particularly when radiation effects are included. Our aim, as mentioned in the Preface, is to offer a treatment that can be happily studied by a final-year undergraduate so we shall try to keep it simple. Some mathematics is of course unavoidable. There are Maxwell's equations to start with, which involve a fair amount of vector analysis. So we shall present

[1] Engineers prefer to call it propagation coefficient, a usage we shall sometimes practise when it sounds more appropriate in the context.

the equations but they will not be called upon very often. The wave equation and some simple theory of differential equations is of course necessary. Some tensors will also be mentioned, but only at an elementary level: no fancy co-ordinate transformations, only a description in terms of matrices. There is obviously not enough space to derive all the equations needed. The question is which equations to derive, or at least give hints about the derivation, and which ones to postulate. We hope the compromises chosen will be acceptable.

The aim of this chapter is to serve partly as an introduction to concepts and equations, and partly as a reference library to which the reader can turn when the need arises.

1.2 Newton's equation and electrical conductivity

The most basic equation, historically or otherwise, is the equation of motion postulated by Newton. It comes into electrical studies when a charged particle is affected by an electric field. For an electron of mass m, and charge e, the equation can be written as

$$m\left(\frac{\mathrm{d}v}{\mathrm{d}t}+\frac{v}{\tau}\right)=eE\,, \tag{1.1}$$

where v is the velocity, E is the electric field and t is time. Note that we have added a damping term with a relaxation time, τ. In this book, with very few exceptions, we shall assume temporal variation in the form[2] $\exp(\mathrm{j}\,\omega t)$ where ω is the frequency. With this assumption the velocity may be expressed from eqn (1.1) in the form

$$v=\frac{e}{m}\frac{E}{\mathrm{j}\,\omega+\dfrac{1}{\tau}}\,, \tag{1.2}$$

and the current density as

$$J=Nev=\frac{Ne^2\tau}{m}\frac{E}{1+\mathrm{j}\,\omega\tau}\,, \tag{1.3}$$

which may also be written as

$$J=\sigma E\,, \tag{1.4}$$

where

$$\sigma=\frac{\sigma_0}{1+\mathrm{j}\,\omega\tau}\,,\qquad \sigma_0=\frac{Ne^2\tau}{m}\,, \tag{1.5}$$

and σ_0 is the electrical conductivity. Note that we have abandoned here the time-honoured notation for conductivity that is simply σ instead of σ_0. The reason is that we wish to reserve σ, a complex quantity, for giving the relationship between the current density and the electric field that will be often needed in the form of eqn (1.5).

[2]Unfortunately, the same harmonic time variation is denoted in four different ways in the literature. The exponent may be $-\mathrm{i}\,\omega t$, $\mathrm{i}\,\omega t$, $\mathrm{j}\,\omega t$ and $-\mathrm{j}\,\omega t$. Admittedly, one rarely sees the second and the fourth one but the other two are equally popular, $-\mathrm{i}\,\omega t$ with physicists, and $\mathrm{j}\,\omega t$ with electrical engineers. We shall adopt here $\mathrm{j}\,\omega t$ for the reason that circuit quantities will often appear and then, surely, the reactance of an inductor or a capacitor should be denoted by $\mathrm{j}\,\omega L$ and $1/\mathrm{j}\,\omega C$ and not by $-\mathrm{i}\,\omega L$ and $-1/\mathrm{i}\,\omega C$.

1.3 Maxwell's equations, fields and potentials

Maxwell's equations in the differential form are given below

$$\nabla \times \mathbf{H} = \mathbf{J} + \mathrm{j}\omega\mathbf{D}\,, \tag{1.6}$$

$$\nabla \times \mathbf{E} = -\mathrm{j}\omega\mathbf{B}\,, \tag{1.7}$$

$$\nabla \cdot \mathbf{D} = \rho, \quad \nabla \cdot \mathbf{B} = 0\,, \tag{1.8}$$

$$\mathbf{D} = \varepsilon\mathbf{E}\,, \quad \mathbf{B} = \mu\mathbf{H}\,. \tag{1.9}$$

The electric and magnetic fields may also be expressed in the form

$$\mathbf{E} = -\nabla\varphi - \frac{\partial A}{\partial t} \qquad \text{and} \qquad \mathbf{H} = \nabla \times \mathbf{A}\,. \tag{1.10}$$

In eqns (1.6)–(1.9) the quantities printed with bold letters are vectors, $\mathbf{H}$ is the magnetic field, $\mathbf{B}$ is the magnetic flux density, $\mathbf{D}$ is the electric flux density, also called the dielectric displacement, $\mathbf{J}$ is the current density, ρ is the charge density, $\mathbf{A}$ is the vector potential, φ is the scalar potential, μ and ε are the material constants, permeability and permittivity. ∇ is a differential operator defined as

$$\nabla = \frac{\partial}{\partial x}\mathbf{i}_x + \frac{\partial}{\partial y}\mathbf{i}_y + \frac{\partial}{\partial z}\mathbf{i}_z\,, \tag{1.11}$$

where $\mathbf{i}_x$, $\mathbf{i}_y$ and $\mathbf{i}_z$ are unit vectors in the x, y and z directions.

Two further relationships often referred to in the book, are Faraday's and Ampere's laws, the former relating the temporal derivative of Φ, the magnetic flux, to V, the induced voltage, as

$$V = -\mathrm{j}\omega\Phi\,, \tag{1.12}$$

and the latter relating the line integral of the magnetic field to I, the enclosed current:

$$\int \mathbf{H}\mathrm{d}\mathbf{s} = I\,, \tag{1.13}$$

where $\mathrm{d}\mathbf{s}$ is the line element vector. The third important relationship we shall need is that between the vector potential and the current density. It is of the form[3]

$$\mathbf{A} = \frac{\mu_0}{4\pi}\int \frac{\mathbf{J}}{r}\mathrm{d}\tau\,, \tag{1.14}$$

where $\mathrm{d}\tau$ is the volume element, the integration is over the region in which $\mathbf{J}$ is finite, and r is the distance between the point where the current flows and where the vector potential is evaluated.

Let us now return to the permittivity and permeability that will often be discussed in the book. They can be written as

$$\varepsilon = \varepsilon_0\varepsilon_\mathrm{r} \qquad \text{and} \qquad \mu = \mu_0\mu_\mathrm{r}\,, \tag{1.15}$$

[3]Note that this is valid only in the low-frequency limit that ignores retardation that is the finite velocity of the electromagnetic waves, but this will suffice for most of the book. Retardation will be introduced in Section 8.2.

where ε_r and μ_r are the relative permittivity (also called dielectric constant or dielectric function) and the relative permeability. In the early times when Gaussian units were in fashion ε_0 and μ_0 were taken as unity. With the emergence of SI (System International) units they have the numerical values of

$$\varepsilon_0 = 8.85 \times 10^{-12} \mathrm{A\,s\,V^{-1}\,m^{-1}}, \qquad \mu_0 = 4\pi \times 10^{-7} \mathrm{V\,s\,A^{-1}\,m^{-1}}. \quad (1.16)$$

In vacuum $\varepsilon_r = \mu_r = 1$. In a material, they are different from unity due to the appearance of **P** and **M**, the electric and magnetic polarizations, respectively. In a material containing N particles (atoms, molecules, artificially inserted elements) we can write

$$\mathbf{P} = N\mathbf{p} \qquad \text{and} \qquad \mathbf{M} = N\mathbf{m}, \quad (1.17)$$

where **p** and **m** are the electric and magnetic dipole moments defined as

$$\mathbf{p} = q\mathbf{d} \qquad \text{and} \qquad \mathbf{m} = \mu_0 IS\mathbf{i}_p. \quad (1.18)$$

q is charge and **d** is a vector connecting the negative and positive charges infinitesimally close to each other (Fig. 1.1(a)). The magnetic dipole is equivalent to an infinitesimal loop in which a current I flows. S is the area of the loop. The direction of the magnetic dipole moment, given by the unit vector $\mathbf{i}_p$, is perpendicular to the plane of the loop (Fig. 1.1(b)).

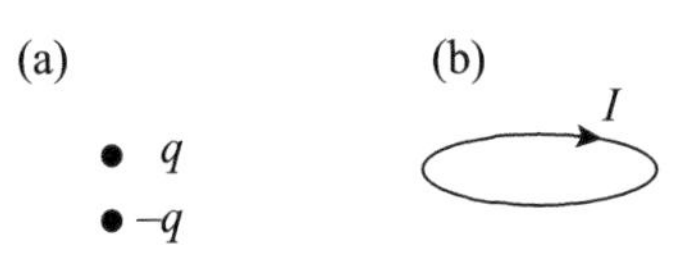

Fig. 1.1 (a) Electric and (b) magnetic dipole

The electric and magnetic dipole moments may already be in the material or may come about as a response to the incident field. The electric and magnetic flux densities may then be written as

$$\mathbf{D} = \varepsilon_0 \mathbf{E} + \mathbf{P} \quad (1.19)$$

and

$$\mathbf{B} = \mu_0 \mathbf{H} + \mathbf{M}. \quad (1.20)$$

The usual problem, both for natural materials and metamaterials, is to find the relationship between the fields and the polarizations and determine from that the values of ε_r and μ_r. In natural materials this is a serious problem because, even in the linear case, we have only a very approximate idea of what is going on at the atomic or molecular level. For man-made materials the difficulties are somewhat easier to overcome because we have a better knowledge of the properties of the elements inserted.

1.4 The wave equation and boundary conditions

The equation we shall need often is the wave equation. It may be obtained from Maxwell's equations in the form

$$\nabla^2\mathbf{E} + k^2\mathbf{E} = 0\,, \tag{1.21}$$

where

$$k^2 = \omega^2\mu\varepsilon\,. \tag{1.22}$$

In Section 1.2 we started with the assumption of harmonic time variation. We add to it now harmonic variation in space as well. The solution of eqn (1.21) is assumed in the form

$$\mathbf{E} = \mathbf{E}_0\,\mathrm{e}^{-\mathrm{j}\,\mathbf{k}\cdot\mathbf{r}}\,, \tag{1.23}$$

where $\mathbf{k}$ is the wave vector that indicates the direction of propagation, $\mathbf{r}$ is a position vector and $\mathbf{E}_0$ is a constant vector perpendicular to $\mathbf{k}$. This is called a plane wave because none of the variables change in a plane perpendicular to $\mathbf{k}$.

The magnitude of the wave vector $k = |\mathbf{k}|$ is known as the wave number by physicists and as the complex propagation coefficient by engineers. Substituting eqn (1.23) into the wave equation and separating the variables we find that a solution exists when the equation

$$k_x^2 + k_y^2 + k_z^2 = k^2 \tag{1.24}$$

is satisfied. Note that k may be complex. Its real and imaginary parts will be denoted by

$$k = \beta - \mathrm{j}\,\alpha\,, \tag{1.25}$$

and they will be called the propagation coefficient, β, and the attenuation coefficient, α. Assuming propagation in the z direction and taking the electric field as a scalar, eqn (1.23) may be written as

$$E = E_0\,\mathrm{e}^{-\mathrm{j}\,\beta z}\,\mathrm{e}^{-\alpha z}\,, \tag{1.26}$$

i.e. the wave declines exponentially in the direction of propagation as may be expected. Although losses are indispensable if we want an accurate answer, very often we can find the main features of a physical phenomenon by neglecting losses. In those cases we shall make no distinction between k and β.

In this plane-wave solution the magnetic field is perpendicular both to the electric field and to the wave vector, and its magnitude is equal to

$$H = \frac{E}{\eta_0}\,, \qquad \eta_0 = \sqrt{\frac{\mu_0}{\varepsilon_0}}\,, \tag{1.27}$$

where η_0 is the free-space impedance equal to 120π ohm. Note that the vectors $\mathbf{E}$, $\mathbf{H}$ and $\mathbf{k}$ are a right-handed set. It has become fashionable to refer to media where this relationship is satisfied as right-handed media in contrast to left-handed media to be discussed in Section 2.11.

The velocity of propagation in free space is equal to c, the velocity of light. If the propagation is in a medium with material constants ε and μ then the velocity of propagation is

$$v = \frac{\omega}{k} = \frac{c}{\sqrt{\mu_\mathrm{r}\varepsilon_\mathrm{r}}}\,, \tag{1.28}$$

and the relationship between the electric and magnetic field becomes

$$H = \frac{E}{\eta} \qquad \text{and} \qquad \eta = \eta_0\sqrt{\frac{\mu_\mathrm{r}}{\varepsilon_\mathrm{r}}} \tag{1.29}$$

is called the impedance of the medium. We may now define the index of refraction as

$$n = \sqrt{\mu_\mathrm{r}\varepsilon_\mathrm{r}}\,, \tag{1.30}$$

which will probably come as a surprise to many in optics because they learned to ignore permeability. In fact, in metamaterials μ_r can be different from unity even in the optical region, therefore eqn (1.30) is the proper definition of the index of refraction. In general, it is a measure of the optical density of the medium. The denser the medium the lower is the velocity of propagation.

When waves are incident from one medium upon another then one would expect some quantities to be continuous across the boundary. These are, as may be found in any textbook on electromagnetism, the tangential components of the electric and magnetic fields and the normal components of the electric and magnetic flux densities.[4] If $\mathbf{H}_1$, $\mathbf{E}_1$ are the fields in medium 1 of material constants μ_1, ε_1, and $\mathbf{H}_2$, $\mathbf{E}_2$, μ_2, ε_2 are the corresponding quantities in medium 2 then the conditions in mathematical form are

[4] This formulation ignores surface currents and surface charges which rarely appear in this book. They can be introduced for surface plasma waves (see Section 3.3.1)

$$H_{1\mathrm{t}} = H_{2\mathrm{t}}\,, \quad E_{1\mathrm{t}} = E_{2\mathrm{t}}\,, \quad \varepsilon_1 E_{1\mathrm{n}} = \varepsilon_2 E_{2\mathrm{n}}\,, \quad \mu_1 H_{1\mathrm{n}} = \mu_2 H_{2\mathrm{n}}\,, \tag{1.31}$$

where subscripts t and n stand for the tangential and normal components. If medium 2 is a metal, it is nearly always (surface plasma waves are again exceptions) taken as having infinite conductivity, which means that the tangential component of the electric field must vanish in medium 2.

1.5 Hollow metal waveguides

Can electromagnetic waves propagate inside a metallic pipe? The experiment to conduct is to raise such a pipe to one of our eyes. If we can see through it that's a clear proof for the propagation of electromagnetic waves. It is true, however, that if we take an ordinary pipe (say a copper pipe used for central heating) its diameter is enormous compared with the wavelength of light, whatever its colour might be. Therefore, we should qualify the previous statement by saying that electromagnetic waves can propagate in a metallic pipe, provided it has a diameter large enough relative to the wavelength. How large should the dimensions be to enable a wave to propagate? For that we need a moderate amount of mathematics. The pipe we shall analyze is a rectangular one as shown in Fig. 1.2. In technical terms it is called a waveguide because it can guide waves.

We shall now assume that only three field components are present, the x component of the electric field[5] and the y and z components of the magnetic field, and we further assume that the electric field is independent of the x co-ordinate. The relevant equations to satisfy are those of Maxwell: eqns (1.6) and (1.7). With the field quantities assumed they may be written as

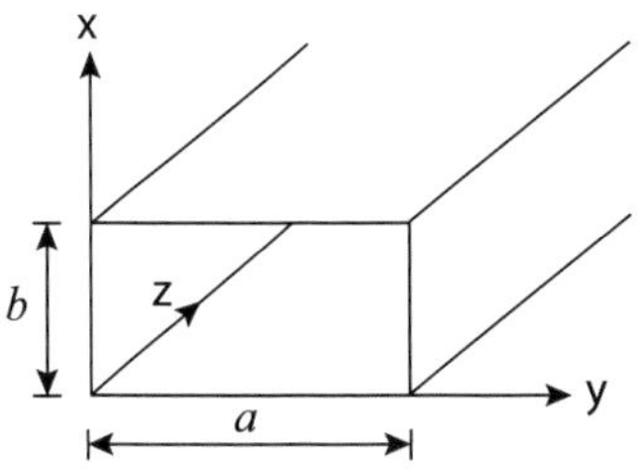

Fig. 1.2 A rectangular metallic waveguide

$$\frac{\partial H_z}{\partial y} - \frac{\partial H_y}{\partial z} = \mathrm{j}\,\omega\varepsilon_0 E_x\,, \tag{1.32}$$

$$\frac{\partial E_x}{\partial z} = -\mathrm{j}\,\omega\mu_0 H_y\,, \tag{1.33}$$

$$\frac{\partial E_x}{\partial y} = \mathrm{j}\,\omega\mu_0 H_z\,. \tag{1.34}$$

Eliminating H_y and H_z from the above equation we end up with the wave equation in the form

$$\frac{\partial^2 E_x}{\partial y^2} + \frac{\partial^2 E_x}{\partial z^2} + k_0^2 E_x = 0\,, \tag{1.35}$$

where $k_0 = \omega\sqrt{\varepsilon_0\mu_0}$. The propagation along the waveguide in the z direction may be described by the function $\exp(-\mathrm{j}\,k_z z)$ as for plane waves. But there is a new feature now. The wave is inside a waveguide. As we learned in the previous section the tangential component of the electric field must vanish at a metal boundary. These boundary conditions are clearly satisfied on the $x = 0$ and $x = b$ surfaces since the chosen electric field is perpendicular to those surfaces. But the E_x component must vanish at $y = 0$ and $y = a$. It would vanish if those boundaries coincided with the nodes of a standing wave. Hence, it is logical to assume that the variation in the y direction will be of the form $\sin(k_y y)$, and if we choose the nearest nodes then we need to take $k_y = \pi/a$. Hence, the solution for the electric field is

$$E_x = E_0 \sin(k_y y)\,\mathrm{e}^{-\mathrm{j}\,k_z z}\,. \tag{1.36}$$

Substituting the above equation into eqn (1.35) we find the relationship

$$k_z^2 = k_0^2 - \left(\frac{\pi}{a}\right)^2. \tag{1.37}$$

Clearly, there is propagation in the z direction when k_z^2 is positive, and exponential decay when k_z^2 is negative. The limiting case is when $k_z = 0$, which occurs when

$$k_0^2 = \left(\frac{2\pi}{\lambda}\right)^2 = \left(\frac{\pi}{a}\right)^2 \quad \text{or} \quad \lambda = 2a\,. \tag{1.38}$$

[5] When the electric field has components only in the transverse plane while propagating in a waveguide, the wave is referred to as a transverse electric or TE wave.

The message is clear. As the size of the waveguide shrinks there is a point at which propagation is no longer possible, and that happens when the waveguide is half-wavelength wide. If $a < \lambda/2$ then no wave can propagate. The waveguide is referred to as a cutoff waveguide.

1.6 Refraction at a boundary: Snell's law and the Ewald circle construction

The next basic concept we shall present is refraction. A wave incident upon a boundary is partly reflected and partly refracted. There is not much point in solving the general case when the fields can have arbitrary polarizations. It is unwieldy and has little practical significance anyway. We shall restrict generality and consider only a two-dimensional situation when there is no variation in the y direction and the wave is incident at an angle θ_1. The plane of incidence is the xz plane. In Fig. 1.3(a) the electric, whereas in Fig. 1.3(b) it is the magnetic polarization that is perpendicular to the plane of incidence. The former is known as a TE and the latter as a TM wave. TE and TM still stand for transverse electric and transverse magnetic but it means now transverse[6] to the plane of incidence.

In both cases the waves incident upon the boundary are partly reflected (at an angle θ_1) and partly refracted (at θ_2) as shown in Fig. 1.3. The aim is to find the reflection and transmission coefficients and the relationship between θ_1 and θ_2.

We shall do the derivation for a TM wave and write the y component of the magnetic field in medium 1 as

$$H_{y1} = A\,\mathrm{e}^{-\mathrm{j}\,(k_{z1}z + k_{x1}x)} + B\,\mathrm{e}^{\mathrm{j}\,(k_{z1}z - k_{x1}x)}\,, \tag{1.39}$$

where A and B are the amplitudes of the incident and reflected waves, and k_{z1} and k_{x1} are the z and x components of the wave vector satisfying the condition

$$k_{z1}^2 + k_{x1}^2 = k_1^2 = \omega^2\mu_1\varepsilon_1\,. \tag{1.40}$$

The electric field will have z and x components as follows from eqn (1.6). For matching we shall need only the tangential component, which is E_{x1}. It may be obtained in the form

$$E_{x1} \quad = \quad -\frac{1}{\mathrm{j}\,\omega\varepsilon_1}\frac{\partial H_{y1}}{\partial z}$$

[6]This terminology is not a fortunate one. For plane waves both the electric and the magnetic fields are in the transverse plane, perpendicular to the direction of propagation. They are called, and should be called, transverse electromagnetic or TEM waves. This is in agreement with the definition in the previous section where a mode in a waveguide is called a TE mode, provided it has only transverse electric fields. Calling waves, as we do in this section and many times later, TE or TM waves depending which field is perpendicular to the plane of incidence can only lead to confusion, but the notation is so widespread in the literature that we reluctantly accept it. There is actually another set of notations, used often by physicists, which is based on the direction of the electric field vector, **E**. It is called *p* polarization when **E** is parallel and *s* polarization when it is perpendicular to the plane of incidence (*s* coming from the German word *senkrecht).*

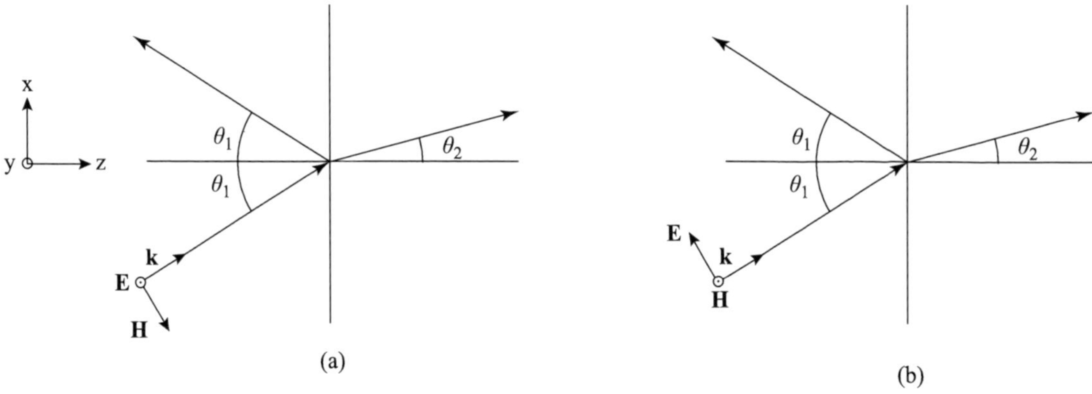

Fig. 1.3 Refraction of a TE (a) and a TM (b) wave

$$= \frac{k_{z1}}{\omega\varepsilon_1}\left[A\,\mathrm{e}^{-\mathrm{j}\,(k_{z1}z + k_{x1}x)} - B\,\mathrm{e}^{\mathrm{j}\,(k_{z1}z - k_{x1}x)}\right]. \tag{1.41}$$

In medium 2 only one wave will propagate, the transmitted wave, because it is assumed to be infinitely wide. Denoting its amplitude by C the tangential components of the electric and magnetic fields in medium 2 are

$$H_{y2} = C\,\mathrm{e}^{-\mathrm{j}\,(k_{z2}z + k_{x2}x)} \tag{1.42}$$

and

$$E_{x2} = \frac{k_{z2}}{\omega\varepsilon_2}C\,\mathrm{e}^{-\mathrm{j}\,(k_{z2}z + k_{x2}x)}. \tag{1.43}$$

According to our boundary conditions (eqn (1.31)) we need to match both field components at the boundary $z = 0$. It is immediately obvious that matching is only possible if the fields vary in the same manner in the x direction, which requires that

$$k_{x1} = k_{x2}, \tag{1.44}$$

meaning that the phase velocity along the x direction must be the same on both sides of the boundary. It follows from the geometry that

$$k_{x1} = k_1 \sin\theta_1 \qquad \text{and} \qquad k_{x2} = k_2 \sin\theta_2, \tag{1.45}$$

whence eqn (1.44) modifies to

$$\sqrt{\mu_1\varepsilon_1}\sin\theta_1 = \sqrt{\mu_2\varepsilon_2}\sin\theta_2, \tag{1.46}$$

which, remembering the definition of eqn (1.30), is known as Snell's law,[7] feared and respected by generations of schoolchildren for at least a couple of centuries.

[7] It was first published by Rene Descartes in his *La Dioptrique* in 1637 but, apparently, a contemporary of his, Villebrord van Roijen Snell, discovered it before him but failed to publish it. In the English language world it is known as Snell's law, in France as *la loi de Descartes*. In many other countries, in a spirit of compromise, it is called the Snell–Descartes law.

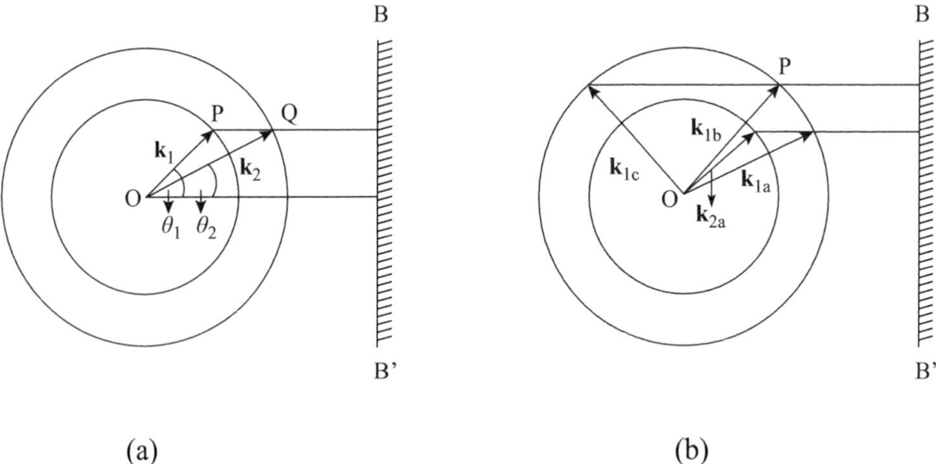

Fig. 1.4 Wave vector diagram at the boundary of two media. (a) $k_1 < k_2$ and (b) $k_1 > k_2$

Further conditions of matching are

$$A + B = C \qquad \text{and} \qquad \frac{k_{z1}}{\varepsilon_1}(A - B) = \frac{k_{z2}}{\varepsilon_2}C\,, \tag{1.47}$$

whence the reflection and transmission coefficients are obtained as

$$R = \frac{B}{A} = \frac{1-\zeta_{\rm e}}{1+\zeta_{\rm e}} \qquad \text{and} \qquad T = \frac{C}{A} = \frac{2}{1+\zeta_{\rm e}}\,, \tag{1.48}$$

where

$$\zeta_{\rm e} = \frac{\varepsilon_1 k_{z2}}{\varepsilon_2 k_{z1}}\,. \tag{1.49}$$

This is a rather unusual form of the well-known reflection and transmission coefficients but convenient for the present book. If instead of a TM wave we assume a TE wave with the incident electric field in the y direction (Fig. 1.3(a)) then the derivation of the reflection and transmission coefficients follows the same pattern and the resulting equations are identical with those of eqn (1.48), but $\zeta_{\rm e}$ needs to be replaced by

$$\zeta_{\rm m} = \frac{\mu_2 k_{z1}}{\mu_1 k_{z2}}\,. \tag{1.50}$$

The condition that the velocity of the wave must be the same on both sides of the boundary leads to a simple construction method for finding the direction of the refracted wave. It is to be recommended not only for its simplicity but also, and mainly, because it gives new insight into the phenomenon, particularly useful when discussing the refraction of backward waves.

The loci of the wave vectors on the two sides of the boundary are represented by concentric circles of radii k_1 and k_2. They show all the

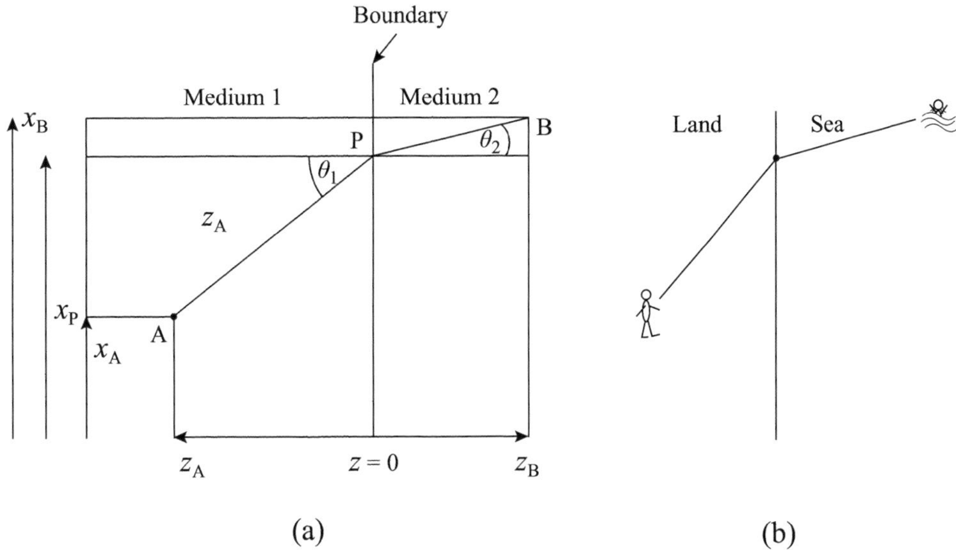

Fig. 1.5 Fermat's principle for the refraction problem. (a) Geometrical scheme. (b) Practical example 'damsel in distress'

possible directions for the wave vectors $\mathbf{k}_1$ and $\mathbf{k}_2$. The geometrical construction is shown in Fig. 1.4(a). We draw the wave vector $\mathbf{k}_1$ corresponding to its angle of incidence, θ_1. On the diagram $\mathbf{k}_1 = \text{OP}$. Now drop a perpendicular from P to the boundary BB′. This line intersects the circle of radius k_2 at Q. Hence $\mathbf{k}_2 = \text{OQ}$ will be the direction of wave propagation in medium 2 since it may be seen to satisfy the condition posed by eqn (1.45). The angle of refraction will be θ_2.

To show the usefulness of the construction let us assume that $k_1 > k_2$. It may now be clearly seen in Fig. 1.4(b) that for an incident wave vector of $\mathbf{k}_{1a}$ a wave can propagate in medium 2 with wave vector $\mathbf{k}_{2a}$, whereas no wave can propagate in medium 2 for the incident wave vector $\mathbf{k}_{1b}$. The perpendicular drawn from P to BB′ does not intersect the inner circle but intersects the outer circle yielding the wave vector $\mathbf{k}_{1c}$, which is that of the reflected wave. It is clearly the case of total internal reflection.

1.7 Fermat's principle

Now we attack the boundary refraction problem from yet another angle using Fermat's principle of minimum time. This will tell us how to get from point A in medium 1 (see Fig. 1.5(a)) to point B in medium 2. The problem is also known as that of the damsel in distress (Fig. 1.5(b)). Assume that medium 1 is land and medium 2 is the sea. A handsome young man (the story comes from the times when sex equality was not uppermost in the minds of storytellers) standing on firm land in point A perceives that a young lady at point B is to be overcome by dangerous waves. He is of a very generous nature and immediately decides to go to the rescue of the lady but there is a problem. To which point on the shore should he run? That surely depends on the relative speed he can muster on land and sea. What he wants is to minimize the time. Mathematically his problem appears as

$$\frac{\mathrm{AP}}{v_{\mathrm{land}}} + \frac{\mathrm{PB}}{v_{\mathrm{sea}}} = \mathrm{Min!} \tag{1.51}$$

With the co-ordinates defined in Fig. 1.5(a) the time needed to reach point B is

$$T = \frac{\sqrt{(x_{\mathrm{P}} - x_{\mathrm{A}})^2 + z_{\mathrm{A}}^2}}{v_{\mathrm{land}}} + \frac{\sqrt{(x_{\mathrm{B}} - x_{\mathrm{P}})^2 + z_{\mathrm{B}}^2}}{v_{\mathrm{sea}}} . \tag{1.52}$$

We find the optimum time with differentiation by the unknown x_P

$$\begin{aligned} \frac{\mathrm{d}T}{\mathrm{d}x_{\mathrm{P}}} = 0 \quad &= \frac{x_{\mathrm{P}} - x_{\mathrm{A}}}{\sqrt{(x_{\mathrm{P}} - x_{\mathrm{A}})^2 + z_{\mathrm{A}}^2}} \frac{1}{v_{\mathrm{land}}} \\ &\quad - \frac{x_{\mathrm{B}} - x_{\mathrm{P}}}{\sqrt{(x_{\mathrm{B}} - x_{\mathrm{P}})^2 + z_{\mathrm{B}}^2}} \frac{1}{v_{\mathrm{sea}}} , \end{aligned} \tag{1.53}$$

which may be seen to reduce to

$$\frac{\sin\theta_1}{v_{\mathrm{land}}} = \frac{\sin\theta_2}{v_{\mathrm{sea}}} . \tag{1.54}$$

If instead of the speed of running and swimming we substitute the speed of an electromagnetic wave in medium 1 as c/n_1 and in medium 2 as c/n_2 then, yet again, we obtain Snell's law as

$$n_1 \sin\theta_1 = n_2 \sin\theta_2 . \tag{1.55}$$

1.8 The optical path and lens design

Next, we ask about the phase of the electromagnetic wave as it travels from A to B via P in Fig. 1.5(a). The phase, as we know, is given by $\mathbf{k} \cdot \mathbf{r}$, which, for the present case, comes to

$$\varphi_{\mathrm{total}} = k_1 \mathrm{AP} + k_2 \mathrm{PB} = \frac{2\pi}{\lambda}(n_1 \mathrm{AP} + n_2 \mathrm{PB}) . \tag{1.56}$$

The total path is AP + PB but that is not the quantity of interest. It is clear from eqn (1.56) that what we need is $n_1\mathrm{AP} + n_2\mathrm{PB}$, which we call the optical path. In general, the total optical path is equal to the geometrical path in each medium multiplied by the corresponding refractive index.

If the optical paths of various trajectories are equal then the phases of the arriving rays must also be equal. If we want to convert a wave emanating from a point to a plane wave then we must make sure that

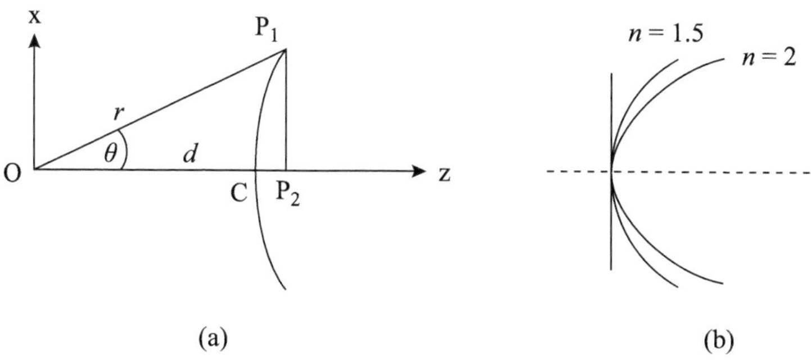

Fig. 1.6 (a) Lens design using the concept of optical path, (b) lens contours

all the optical paths are equal when the plane wavefront is reached. Traditionally it is the job of a lens to realize this conversion. So, let us find the contour of a lens that will do it.

We shall assume a 2D geometry with the rays emanating from point O and work out the optical path of a general ray travelling in the θ direction (see Fig. 1.6(a)) and that of the axial ray when they reach the P_1P_2 wavefront. The former one propagates in free space hence its optical path is OP_1, whereas the latter one propagates in free space from O to C and in the lens from C to P_2. Hence, the equality of optical paths demands

$$\mathrm{OP}_1 = \mathrm{OC} + n\mathrm{CP}_2\,, \tag{1.57}$$

which may be written in polar co-ordinates as

$$r = d + n(r\cos\theta - d)\,, \tag{1.58}$$

whence

$$r = \frac{d(1-n)}{1 - n\cos\theta}\,. \tag{1.59}$$

This may be recognized as a hyperbola in polar co-ordinates. It has an asymptote at

$$\cos\theta = \frac{1}{n}\,. \tag{1.60}$$

Transformation to normalized Cartesian co-ordinates,

$$z = \frac{r}{d}\cos\theta \qquad \text{and} \qquad x = \frac{r}{d}\sin\theta \tag{1.61}$$

yields the more familiar equation of a hyperbola[8]

$$\left(z - \frac{n}{n+1}\right)^2 - \frac{x^2}{n^2-1} = \frac{1}{(n+1)^2}\,. \tag{1.62}$$

For $n = 1.5$ and 2 the contours are plotted in Fig. 1.6(b). As usual, it has a convex shape.

[8] 2D lens contours are generally known as circles. In fact, the above equation of a hyperbola for low angles is approximately the same as that of a circle.

1.9 The effective ε in the presence of a current

We have already mentioned several times the relationship of the current density to other variables like the electric field or the vector potential but the wave solutions presented so far ignored the possible existence of a current. The aim of this section is to take the current into account in a simple way by defining an effective dielectric constant. The way to do this is to rewrite the right-hand side of eqn (1.6) in the form

$$J + \mathrm{j}\,\omega\varepsilon E = \mathrm{j}\,\omega\varepsilon_{\mathrm{eff}} E\,, \tag{1.63}$$

with which $\varepsilon_{\mathrm{eff}}$ is defined. But J is related to E by eqn (1.4), whence we find

$$\varepsilon_{\mathrm{eff}} = \varepsilon + \frac{\sigma_0}{\mathrm{j}\,\omega}\frac{1}{1 + \mathrm{j}\,\omega\tau}\,. \tag{1.64}$$

In the low-frequency limit we find

$$\varepsilon_{\mathrm{eff}} = \varepsilon - \mathrm{j}\,\frac{\sigma_0}{\omega}\,. \tag{1.65}$$

This is the form used for lossy dielectrics. In the high-frequency limit, taking $\varepsilon = \varepsilon_0$,

$$\varepsilon_{\mathrm{eff}} = \varepsilon_0\left(1 - \frac{\omega_{\mathrm{p}}^2}{\omega^2}\right), \tag{1.66}$$

where

$$\omega_{\mathrm{p}}^2 = \frac{Ne^2}{\varepsilon_0 m} \tag{1.67}$$

is known as the plasma frequency. Equation (1.66) is the dielectric constant of an ideal plasma. Interestingly, it is negative below the plasma frequency and positive above the plasma frequency.

Note that eqn (1.22) still applies. We only need to replace ε by $\varepsilon_{\mathrm{eff}}$. When $\varepsilon_{\mathrm{eff}}$ is complex, k will also be complex, implying both propagation and attenuation. When $\varepsilon_{\mathrm{eff}}$ is a negative real number then k is purely imaginary. There is no propagation. The wave declines without any change in the phase.

1.10 Surface waves

We have seen what happens when a wave is incident from one medium upon another one. Let us now explore another aspect of wave propagation related to a boundary between two media, namely that the wave can stick to the boundary. When it does so it is called a surface wave. Let us see the conditions under which these waves may exist.

Assuming a TM wave as in Section 1.6 we have already got the expressions for H_y and E_x on both sides of the boundary. The new feature is that the waves can propagate along the boundary but their amplitudes decline exponentially away from the boundary as shown in Fig. 1.7. This can happen when k_x is sufficiently high so that k_{z1} and k_{z2} are imaginary and must be replaced by $-\mathrm{j}\,\kappa_1$ and $-\mathrm{j}\,\kappa_2$, where κ_1 and κ_2 are real. Hence, the propagation coefficient in the x direction is obtained as

$$k_x^2 = k_1^2 + \kappa_1^2 = k_2^2 + \kappa_2^2 . \tag{1.68}$$

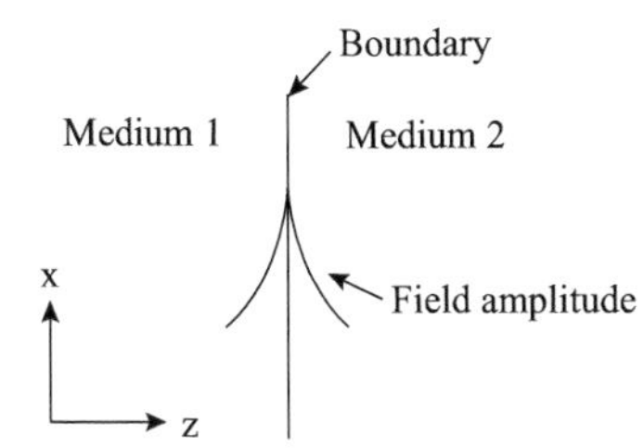

Fig. 1.7 Variation of a surface wave amplitude at a boundary

We are looking for a wave that can exist on the surface without an input.[9] Hence, we can take $A = 0$ but B is taken as finite. It is the amplitude of the wave that declines away from the boundary in the negative z direction. The equations for the magnetic and electric fields are then

[9] In more pretentious language it means that we are looking for eigensolutions.

$$\begin{aligned} H_{y1} &= B\,\mathrm{e}^{\kappa_1 z}\,\mathrm{e}^{-\mathrm{j}\,k_x x}, \\ E_{x1} &= \frac{-\kappa_1}{\mathrm{j}\,\omega\varepsilon_1} B\,\mathrm{e}^{\kappa_1 z}\,\mathrm{e}^{-\mathrm{j}\,k_x x}, \\ E_{z1} &= \frac{-k_x}{\omega\varepsilon_1} B\,\mathrm{e}^{\kappa_1 z}\,\mathrm{e}^{-\mathrm{j}\,k_x x} \end{aligned} \tag{1.69}$$

in medium 1 and

$$\begin{aligned} H_{y2} &= C\,\mathrm{e}^{-\kappa_2 z}\,\mathrm{e}^{-\mathrm{j}\,k_x x}, \\ E_{x2} &= \frac{\kappa_2}{\mathrm{j}\,\omega\varepsilon_2} C\,\mathrm{e}^{-\kappa_2 z}\,\mathrm{e}^{-\mathrm{j}\,k_x x}, \\ E_{z2} &= \frac{-k_x}{\omega\varepsilon_2} C\,\mathrm{e}^{-\kappa_2 z}\,\mathrm{e}^{-\mathrm{j}\,k_x x} \end{aligned} \tag{1.70}$$

in medium 2. In order to satisfy the boundary conditions we need to match H_y and E_x at $z = 0$, which yields[10]

$$B = C \qquad \text{and} \qquad -\frac{\kappa_1}{\varepsilon_1} = \frac{\kappa_2}{\varepsilon_2} . \tag{1.71}$$

[10] Note that the boundary condition $\varepsilon_1 E_{z1} = \varepsilon_2 E_{z2}$ is automatically satisfied, provided eqn (1.71) holds.

Note that we have taken both κ_1 and κ_2 to be positive. Hence, eqn (1.71) can be satisfied only when ε_2 is negative. This is a condition for the existence of a surface wave. It is interesting to note at this stage that eqn (1.71) is equivalent to the condition that

$$\zeta_\mathrm{e} = -1 . \tag{1.72}$$

Another condition to satisfy is given by eqn (1.68). We may then substitute κ_1 and κ_2 from eqn (1.68) into eqn (1.71) and obtain, after some algebraic operations, a relationship between k_x and ω,

$$k_x = \frac{\omega}{c}\sqrt{\frac{\varepsilon_{\mathrm{r}1}\varepsilon_{\mathrm{r}2}}{\varepsilon_{\mathrm{r}1} + \varepsilon_{\mathrm{r}2}}} . \tag{1.73}$$

We want the wave to travel along the surface, hence k_x must be real. With $\varepsilon_{\mathrm{r}2} < 0$ we may then write the condition to satisfy as

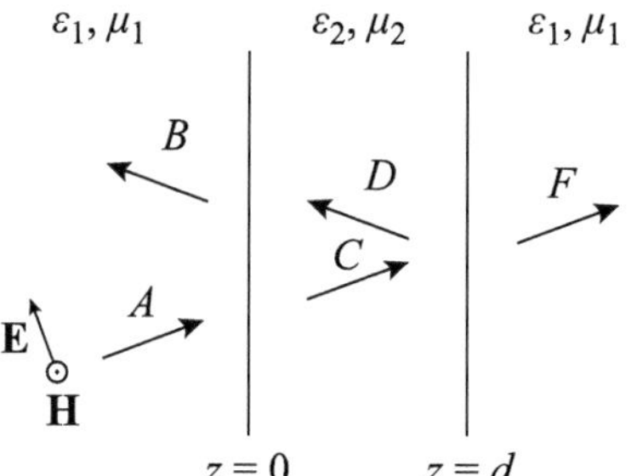

Fig. 1.8 Schematic presentation of wave propagation in three media when a wave is incident from medium 1

$$\varepsilon_{r1}\varepsilon_{r2} < 0 \qquad \text{and} \qquad \varepsilon_{r1} + \varepsilon_{r2} < 0\,. \tag{1.74}$$

This is actually the situation pertaining to an ideal plasma that can have a negative dielectric constant, as shown in the previous section. The waves are usually referred to as surface plasmons, which is really a misnomer because the phenomenon is purely classical. It is as if we would always talk about photons instead of electromagnetic waves. We would of course be right but there is not much point in using quantum-mechanical terminology when the classical would do.

1.11 Plane wave incident upon a slab

In Section 1.4 we investigated the reflection and transmission of a plane wave when incident upon another medium. We shall now add one more boundary, as shown in Fig. 1.8. For simplicity, media 1 and 3 are assumed to have the same material constants and the input wave is a TM wave, i.e. the magnetic field has only a z component.

The major difference from the previous study is that in addition to the transmitted wave there will now be a reflected wave as well in medium 2. The amplitudes, as may be seen in Fig. 1.8, are denoted by A, B, C, D and F. The technique of solution is the same as for the single boundary but now we need to match the tangential components of the electric and magnetic fields at both boundaries, at $z = 0$ and at $z = d$. This leads to four equations from which the unknown amplitudes may be determined. After a number of algebraic operations we find the reflection and transmission coefficients as

$$R = \frac{B}{A} = \frac{2\mathrm{j}\,(1-\zeta_\mathrm{e}^2)\sin(k_{z2}d)}{(1+\zeta_\mathrm{e})^2\,\mathrm{e}^{\mathrm{j}\,k_{z2}d} - (1-\zeta_\mathrm{e})^2\,\mathrm{e}^{-\mathrm{j}\,k_{z2}d}}\,, \tag{1.75}$$

$$T = \frac{F}{A} = \frac{4\zeta_\mathrm{e}}{(1+\zeta_\mathrm{e})^2\,\mathrm{e}^{\mathrm{j}\,k_{z2}d} - (1-\zeta_\mathrm{e})^2\,\mathrm{e}^{-\mathrm{j}\,k_{z2}d}}\,, \tag{1.76}$$

$$\frac{C}{A} = \frac{2(1+\zeta_\mathrm{e})}{(1+\zeta_\mathrm{e})^2\,\mathrm{e}^{\mathrm{j}\,k_{z2}d} - (1-\zeta_\mathrm{e})^2\,\mathrm{e}^{-\mathrm{j}\,k_{z2}d}}\,, \tag{1.77}$$

$$\frac{D}{A} = \frac{2(\zeta_e - 1)}{(1+\zeta_e)^2 \, e^{j k_{z2} d} - (1-\zeta_e)^2 \, e^{-j k_{z2} d}}, \tag{1.78}$$

and, as before, ζ_e should be replaced by ζ_m for an incident TE wave.

1.12 Dipoles

The microscopic electric and magnetic dipole moments, **p** and **m**, were introduced in Section 1.3 for determining the macroscopic electric and magnetic polarizations, **P** and **M**. However, dipoles are useful in many other contexts partly because they can provide good approximate models to describe complicated electromagnetic phenomena but also because they can form the basis of actual devices like the electric dipole antenna.

We shall present here the equations relating the dipole moments to fields in a compact vectorial form that physicists prefer. It might look unfamiliar to electrical engineers:

$$\mathbf{E} = \frac{1}{4\pi\varepsilon_0}\left[(1 + j k_0 r)\frac{3(\mathbf{r}\cdot\mathbf{p})\mathbf{r} - r^2\mathbf{p}}{r^5} + k_0^2 \frac{r^2\mathbf{p} - (\mathbf{r}\cdot\mathbf{p})\mathbf{r}}{r^3}\right] e^{-j k_0 r}, \tag{1.79}$$

with $k_0 = 2\pi/\lambda$. The relationship between the magnetic dipole moment and the magnetic field takes an entirely analogous form: **p** should be replaced by **m** and ε_0 in front of the square bracket by μ_0.

One works usually in spherical co-ordinate systems (z, θ, φ) and the usual assumption is that the dipole moments (whether electric or magnetic) point in the z direction. To find the components of the electric field we need then the unit vector in the r and θ directions:

$$\mathbf{i}_r = \mathbf{i}_x \sin\theta\cos\varphi + \mathbf{i}_y \sin\theta\sin\varphi + \mathbf{i}_z\cos\theta, \tag{1.80}$$

$$\mathbf{i}_\theta = \mathbf{i}_x \cos\theta\cos\varphi + \mathbf{i}_y \cos\theta\sin\varphi - \mathbf{i}_z\sin\theta. \tag{1.81}$$

Substituting eqns (1.80) and (1.81) into eqn (1.79) we obtain the electric-field components in the form

$$E_r = \frac{2p\cos\theta}{4\pi\varepsilon_0 r^3}(1 + j k_0 r)\, e^{-j k_0 r}, \tag{1.82}$$

$$E_\theta = \frac{p\sin\theta}{4\pi\varepsilon_0 r^3}(1 + j k_0 r - k_0^2 r^2)\, e^{-j k_0 r}, \tag{1.83}$$

$$E_\varphi = 0, \tag{1.84}$$

and the corresponding magnetic field is

$$H_\varphi = \frac{j\omega p \sin\theta}{4\pi r^2}(1 + j k_0 r)\, e^{-j k_0 r}. \tag{1.85}$$

The analogous equations for a magnetic dipole are

$$H_r = \frac{m\cos\theta}{2\pi\mu_0 r^3}(1 + \mathrm{j}\,k_0 r)\,\mathrm{e}^{-\mathrm{j}\,k_0 r}, \tag{1.86}$$

$$H_\theta = \frac{m\sin\theta}{4\pi\mu_0 r^3}(1 + \mathrm{j}\,k_0 r - k_0^2 r^2)\,\mathrm{e}^{-\mathrm{j}\,k_0 r}, \tag{1.87}$$

$$H_\varphi = 0, \tag{1.88}$$

and the corresponding electric field is

$$E_\varphi = \frac{\mathrm{j}\,\omega m\sin\theta}{4\pi r^2}(1 + \mathrm{j}\,k_0 r)\,\mathrm{e}^{-\mathrm{j}\,k_0 r}. \tag{1.89}$$

1.13 Poynting vector

The derivation of Poynting's theorem can be found in most textbooks on electromagnetic theory. It relates the movement of electromagnetic power to the temporal variation of stored energy. The theorem is usually presented both in integral and differential form. The essential thing for our purpose is the complex Poynting vector itself,

$$\mathbf{S} = \frac{1}{2}\mathrm{Re}\,(\mathbf{E} \times \mathbf{H}^*), \tag{1.90}$$

where the star denotes complex conjugate. The Poynting vector has the dimensions of power per unit surface; it gives the magnitude and direction of power flow. It is more popular in some branches of electricity than in others. It does not very often appear in metamaterial studies. Two recent books on metamaterials we have looked at do not even have entries for the Poynting vector in the Index. In fact, the Poynting vector played a pivotal role at one of the turning points of materials research. It was shown by Veselago (1968) that for an isotropic material in which both the permittivity and permeability are negative the direction of the Poynting vector is opposite to **k**, the direction of propagation.

The Poynting vector has, however, other significance too in the theory of metamaterials. Once we are concerned with field phenomena on a scale small relative to the wavelength then the best guide to understanding the physics is the streamlines of the Poynting vector. They will tell us where the power originates from, how it propagates and where it is absorbed. Often, it is the only vectorial quantity that can give us guidance because when the electric and magnetic fields are elliptically polarized the field lines can no longer be shown in simple graphical terms.

1.14 Radiation resistance

We have seen in Section 1.11 that the field decay away from the dipole may be described by $1/r^3$, $1/r^2$ and $1/r$ terms. The last one, which has the slowest decay, will be the one that survives a long distance away. It is the term responsible for the power radiated out. How large is that power? For an electric dipole the Poynting vector may be calculated from eqns (1.82) and (1.83) as

$$\mathbf{S} = \frac{1}{2}\mathrm{Re}\,(E_\theta H_\varphi^*) = \frac{1}{2}\eta_0 \left(\frac{kId}{4\pi r}\right)^2 \sin^2\theta\,\mathbf{i}_r\,, \tag{1.91}$$

where the relationship

$$I = \mathrm{j}\,\omega q = \frac{\mathrm{j}\,\omega p}{d} \tag{1.92}$$

has been used. Equation (1.91) gives the power density due to radiation at a point a distance r away from the dipole. The total power can be obtained by integrating the power density over the sphere of radius r. Performing the integration we obtain

$$P = \frac{1}{12\pi}\eta_0 (kId)^2\,. \tag{1.93}$$

We can now introduce the concept of radiation resistance. It is a useful artifice responsible for radiated power. The radiated power may be regarded as that absorbed by the radiation resistance. Hence, its definition comes from the equation

$$\frac{1}{2}R_\mathrm{s}I^2 = P\,. \tag{1.94}$$

From eqns (1.93) and (1.94) we then obtain

$$R_\mathrm{s} = \frac{2\pi}{3}\eta_0 \left(\frac{d}{\lambda}\right)^2\,. \tag{1.95}$$

The radiation resistance of a magnetic dipole may be obtained by similar arguments as

$$R_\mathrm{s} = \frac{\pi}{6}\eta_0 \left(\frac{2\pi r_0}{\lambda}\right)^4\,. \tag{1.96}$$

1.15 Permittivity and permeability tensors

The relationships $\mathbf{B} = \mu\mathbf{H}$ and $\mathbf{D} = \varepsilon\mathbf{E}$ imply that the vectors $\mathbf{B}$ and $\mathbf{H}$ and, similarly, $\mathbf{D}$ and $\mathbf{E}$, point in the same direction because μ and ε are scalar. However, for certain crystals and for certain metamaterials the relationship between the components of $\mathbf{B}$ and $\mathbf{H}$, and $\mathbf{D}$ and $\mathbf{E}$, may be in the form of a matrix. For the magnetic quantities it takes the form

$$\begin{pmatrix} B_x \\ B_y \\ B_z \end{pmatrix} = \mu_0 \begin{pmatrix} \mu_{xx} & \mu_{xy} & \mu_{xz} \\ \mu_{yx} & \mu_{yy} & \mu_{yz} \\ \mu_{zx} & \mu_{zy} & \mu_{zz} \end{pmatrix} \begin{pmatrix} H_x \\ H_y \\ H_z \end{pmatrix}, \tag{1.97}$$

which means that if there is a magnetic field directed along a particular co-ordinate axis it may lead to flux densities that have components along all three axes. Note that the nine components are not independent of each other: the matrix is symmetric. When the permeability is in this form we call it the permeability tensor. Hence, we may write the relationship between **B** and **H** in the form

$$\mathbf{B} = \mu_0 \mu_r \mathbf{H}\,, \tag{1.98}$$

where μ_r denotes now the tensor in eqn (1.97). Analogously, the permittivity may also be a symmetric tensor in both natural materials and in metamaterials.

1.16 Polarizability

Let us take an element and ask the question whether electric or magnetic dipole moments can be induced in it by an electric or magnetic field. Assuming the relationships to be linear we may write them as

$$p = \alpha_e E \qquad \text{and} \qquad m = \alpha_m H\,, \tag{1.99}$$

where α_e and α_m are the electric and magnetic polarizabilities. Equation (1.99) implies that the dipole moments will point in the same direction as the fields that induce them. We can, however, write more general relationships if we allow a field in one direction to induce a dipole moment in another direction. Then, the polarizabilities become tensors and eqn (1.99) may be rewritten as

$$\mathbf{p} = \alpha_e \mathbf{E} \qquad \text{and} \qquad \mathbf{m} = \alpha_m \mathbf{H}\,, \tag{1.100}$$

where α_e and α_m are symmetric tensors. However, this is still not general enough because an electric field might induce a magnetic dipole moment and, vice versa, a magnetic field might induce an electric dipole moment. Acknowledging the possibility of such cross-polarization we write the general relationship as

$$\mathbf{p} = \alpha^{ee}\mathbf{E} + \alpha^{em}\mathbf{H}\,, \tag{1.101}$$

$$\mathbf{m} = \alpha^{me}\mathbf{E} + \alpha^{mm}\mathbf{H}\,, \tag{1.102}$$

where new notations have been introduced. Now the electric and magnetic polarizability tensors are denoted by the superscripts ee and mm, respectively, indicating that the electric field induces an electric dipole and the magnetic field induces a magnetic dipole. The cross-polarizability tensors α^{em} and α^{me} are related to each other (the so-called Onsager relations) as

$$\alpha^{em} = -(\alpha^{me})^{T}\,, \tag{1.103}$$

where the superscript T on a matrix means that it is transposed. The notations are a little clumsy. If, for example, a magnetic field in the z direction induces an electric dipole moment in the y direction then the relationship is written as

$$p_y = \alpha_{yz}^{\text{em}} H_z \,. \tag{1.104}$$

There are also other notations in the literature but they don't seem to be simpler either.

1.17 Working with tensors

The last two sections have been concerned with tensors. They might look a little intimidating to those who come across them for the first time. Physically, all they mean is that the properties of certain materials are different in different directions. Mathematically, it needs a few matrix multiplications, that's all. To show that it is not difficult to manage tensors we shall derive here the wave equation for a case that will be further treated in Chapter 9.

We shall assume a TE wave propagating in the z direction and find a solution for the case when both the permittivity and permeability are diagonal tensors, and the field components are E_y, H_x and H_z. We need just one simple matrix multiplication and we can then write the right-hand sides of Maxwell's equations (1.6) and (1.7) in the form

$$\mathbf{D} = \varepsilon_{yy} E_y \mathbf{i}_y \quad \text{and} \quad \mathbf{B} = \mu_{xx} H_x \mathbf{i}_x + \mu_{yy} H_y \mathbf{i}_y \,. \tag{1.105}$$

The curls on the left-hand sides of eqns (1.6) and (1.7) may be worked out by the usual methods, giving rise to the equations

$$\frac{\partial E_y}{\partial z} = \mathrm{j}\,\omega \mu_0 \mu_{xx} H_x \,, \tag{1.106}$$

$$\frac{\partial E_y}{\partial x} = -\mathrm{j}\,\omega \mu_0 \mu_{zz} H_z \,, \tag{1.107}$$

$$\frac{\partial H_x}{\partial z} - \frac{\partial H_z}{\partial x} = \mathrm{j}\,\omega \varepsilon_0 \varepsilon_{yy} E_y \,, \tag{1.108}$$

from which the wave equation can be derived as

$$\mu_{zz} \frac{\partial^2 E_y}{\partial z^2} + \mu_{xx} \frac{\partial^2 E_y}{\partial x^2} + \frac{\omega^2}{c^2} \varepsilon_{yy} \mu_{xx} \mu_{zz} E_y = 0 \,, \tag{1.109}$$

which is the same kind of second-order differential equation as the wave equation (1.21).

1.18 Dispersion: forward and backward waves

The term dispersion comes from the observation that white light will be decomposed into its constituents when incident on a prism in which the velocity depends on frequency. So, when we plot a dispersion curve it should be, strictly speaking, velocity versus frequency. However, the accepted presentation is frequency against wave number, often called the

ω–k diagram. The dependence of ω on k is then given by the dispersion equation.

Have we already come across dispersion equations? Yes, of course. The simplest one is that of eqn (1.28), $\omega = kv$, where v is the phase velocity. This gives a linear relationship between frequency and wave number, provided the velocity is independent of frequency as it would be in free space. Hence, the $\omega = kc$ curve shows no dispersion, the relationship is called dispersionless. On the other hand, the relationship we have come across for an ideal plasma in Section 1.9 does show dispersion. From eqns (1.22) and (1.66) we obtain, after a moderate amount of algebra,

$$\omega^2 = k^2c^2 + \omega_{\mathrm{p}}^2 . \tag{1.110}$$

It follows from this equation that there is no propagation for $\omega < \omega_{\mathrm{p}}$ and that for large enough ω we again obtain the linear relationship $\omega = kc$.

The phase velocity is the velocity with which a single-frequency wave travels. But of course a single-frequency wave does not carry any information. The velocity of a group of frequencies that do carry information is the group velocity defined as

$$v_{\mathrm{g}} = \frac{\mathrm{d}\omega}{\mathrm{d}k} . \tag{1.111}$$

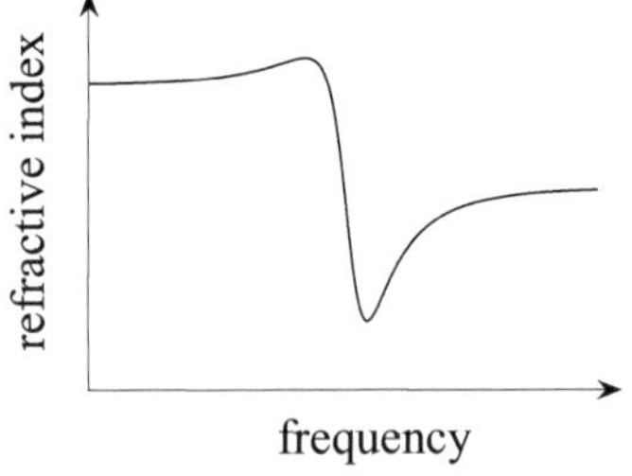

Fig. 1.9 Anomalous dispersion

A wave with positive group velocity is called a forward wave (phase and energy move in the same direction) and one with a negative group velocity is a backward wave. The distinction between forward and backward waves is quite fundamental and particularly important in the theory of metamaterials, as we shall see later in the book.

Under the heading of dispersion it is desirable to mention anomalous dispersion as well. It is a term from the nineteenth century when absorption spectra of various materials were studied in the optical region. The variation of refractive index close to the absorption peak took the shape shown in Fig. 1.9. There is clearly a region where $\mathrm{d}n/\mathrm{d}\omega < 0$ taking its smallest value at the resonant frequency. Does it mean that the corresponding group velocity is negative? Not necessarily. Considering the definition of the refractive index in Section 1.4 as $n = ck/\omega$ we find

$$\frac{\mathrm{d}n}{\mathrm{d}\omega} = \frac{c}{\omega}\left(-\frac{k}{\omega} + \frac{\mathrm{d}k}{\mathrm{d}\omega}\right) , \tag{1.112}$$

whence

$$\frac{\mathrm{d}\omega}{\mathrm{d}k} = \frac{c}{n + \omega\dfrac{\mathrm{d}n}{\mathrm{d}\omega}} , \tag{1.113}$$

i.e. the condition for the group velocity to be negative is not only that $\mathrm{d}n/\mathrm{d}\omega$ must be negative but also the slope of the n versus ω curve must be sufficiently large.

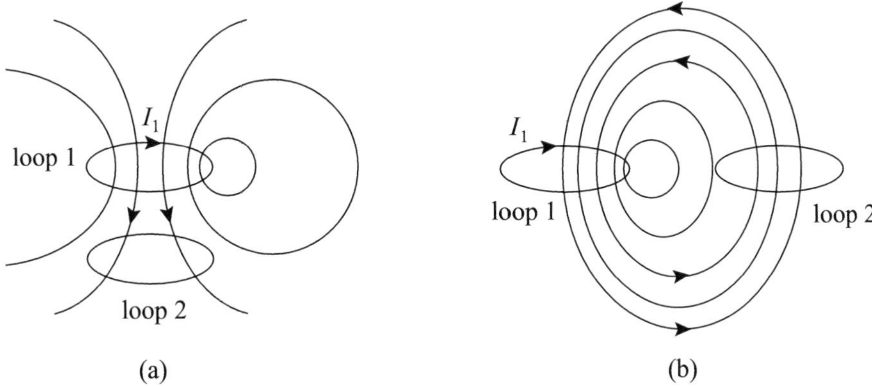

Fig. 1.10 Magnetic coupling for two loops is positive in the axial case (a) and negative in the planar case (b)

1.19 Mutual impedance and mutual inductance

The term 'impedance' has already been used in Section 1.4 when we introduced the free-space impedance, defined as the ratio of the electric to magnetic field in a plane wave. Nevertheless, most people would regard it a circuit concept, related to inductances, capacitances and resistances. The mutual impedance belongs to the same category: it could be claimed both by circuits and fields. Its definition is usually in terms of voltages and currents, which implies that the claim of circuit people might be stronger. Mutual impedance between two elements is defined by the ratio of the voltage in element 2 to the current in element 1 that induced it. It has great significance in the theory of antenna arrays. Our main interest here is in magnetic coupling and, consequently, in mutual inductance. The technique of determining mutual inductance is fairly simple in principle. As an example, we shall go through the steps for two circular loops shown in Fig. 1.10.

Let us assume that a low-frequency current I_1 flows in loop 1 that, true to circuit theory, is constant everywhere in the wire. The corresponding vector potential can be found from eqn (1.14) and the magnetic field from eqn (1.10). Having found the perpendicular magnetic field over the area of loop 2 we can find the total flux threading the loop. The mutual inductance between loops 1 and 2 is then defined by the equation

$$\Phi_2 = M_{21} I_1 \,. \tag{1.114}$$

Note that the mutual inductance can be complex if the distance between the elements becomes comparable with the wavelength. For most of the book we shall be concerned with physical situations in which the mutual inductance is real, but it may still be positive or negative. A simple illustration in Figs. 1.10(a) and (b) shows the difference between the two cases. The mutual inductance is positive if the magnetic field lines

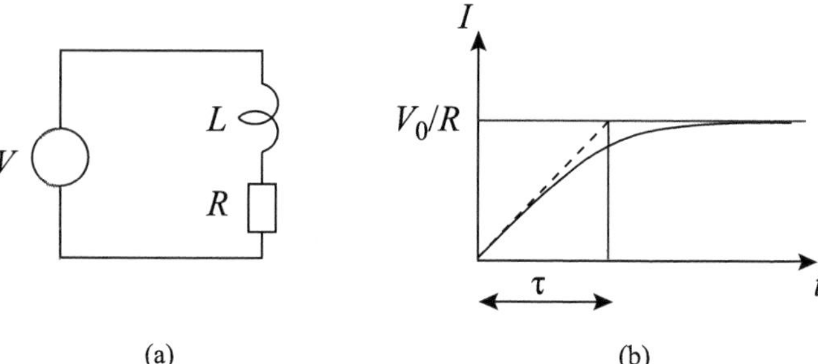

Fig. 1.11 *RL* circuit (a) and the current build-up when a step voltage is applied (b)

cross the two loops in the same direction (we call this the axial configuration), and the mutual inductance is negative (planar configuration) when the magnetic field lines are in the opposite direction.

1.20 Kinetic inductance

Let us first quote the well-known relationship between current and voltage due to the presence of an inductance, L, and a resistance, R, in the circuit shown in Fig. 1.11(a)

$$V = L\frac{\mathrm{d}I}{\mathrm{d}t} + RI\,, \tag{1.115}$$

and the solution for the case when a step voltage is imposed upon the circuit

$$I = \frac{V_0}{R}\left[1 - \exp\left(-\frac{t}{\tau}\right)\right]; \qquad \tau = \frac{L}{R}\,. \tag{1.116}$$

This is shown in Fig. 1.11(b). The current may be seen to be delayed with a time constant τ.

We may now ask the question: is there anything else in an electrical circuit that causes delay? Clearly, in order to have a current, charge carriers must be accelerated, and it takes time to accelerate particles of finite mass. Hence, the current will necessarily lag behind the voltage causing its rise. The basic equation is of course the equation of motion with which we started this chapter. Let us take a piece of conducting material of length l and cross-section S. The voltage and current may then be expressed with the aid of the electric field and electron velocity as

$$V = El \qquad \text{and} \qquad I = NeSv\,, \tag{1.117}$$

which, substituted into eqn (1.1) leads to

$$V = \frac{lm}{Ne^2S}\left(\frac{\mathrm{d}I}{\mathrm{d}t} + \frac{I}{\tau}\right)\,. \tag{1.118}$$

Comparing now eqn (1.115) with the above equation we may define a 'kinetic' inductance[11] and a 'kinetic' resistance with the relationships

[11] It is only a minority of physicists and engineers, mainly those concerned with low-temperature work, who are familiar with the concept of kinetic inductance. As may be seen from the derivation in this section, it is a trivial concept, which, actually, found its way into some undergraduate textbooks, see, e.g., Solymar (1984).

$$L_\mathrm{k} = \frac{lm}{Ne^2S} \qquad \text{and} \qquad R_\mathrm{k} = \frac{lm}{Ne^2S\tau}\,. \tag{1.119}$$

Looking at the expression of the kinetic resistance it becomes clear that it is nothing else but the ordinary resistance. On the other hand, the expression for the kinetic inductance is entirely new. As the cross-section of the conductor gets smaller the kinetic inductance becomes comparable with the magnetic inductance. It could even become the dominant inductance, as for example in circuits containing nanorods. Since the plasma frequency (defined by eqn (1.67)) will often appear in this book we shall rewrite the expression for the kinetic inductance in the form

$$L_\mathrm{k} = \frac{l}{\pi r^2 \varepsilon_0 \omega_\mathrm{p}^2}\,, \tag{1.120}$$

where r is the radius of a conductor of circular cross-section.

1.21 Four-poles: impedance and chain matrices

Four-poles are undeniably circuit quantities but they are also suitable for describing the propagation of waves as will emerge in the next few sections. We shall mostly follow the excellent and instructive treatment of Brillouin (1953).

Let us start with the basics. A four-pole is a black box from which four wires hang out as shown in Fig. 1.12. The input is on the left-hand side and the output is on the right-hand side.We shall use the eminently logical notations V_in and I_in for the input voltage and current, and V_out and I_out for the output voltage and current. A four-pole is characterized by relating two of the above variables to the other two. The most frequently used form of these relations is in terms of impedances, as follows

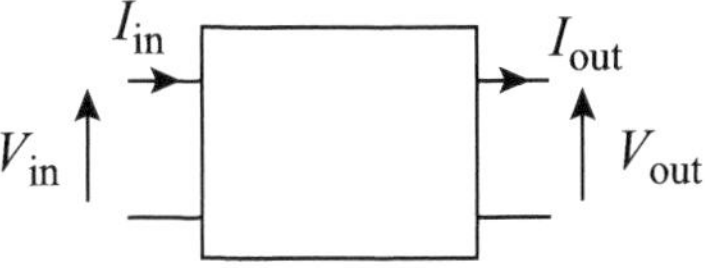

Fig. 1.12 A four-pole

$$\begin{aligned} V_\mathrm{in} &= Z_{11}I_\mathrm{in} - Z_{12}I_\mathrm{out}\,, \\ V_\mathrm{out} &= Z_{21}I_\mathrm{in} - Z_{22}I_\mathrm{out}\,, \end{aligned} \tag{1.121}$$

which may be written in terms of vectors and matrices as

$$\mathbf{V} = Z\mathbf{I}\,, \tag{1.122}$$

where

$$\mathbf{V} = \begin{pmatrix} V_\mathrm{in} \\ V_\mathrm{out} \end{pmatrix}, \quad Z = \begin{pmatrix} Z_{11} & -Z_{12} \\ Z_{21} & -Z_{22} \end{pmatrix}, \quad \mathbf{I} = \begin{pmatrix} I_\mathrm{in} \\ I_\mathrm{out} \end{pmatrix}. \tag{1.123}$$

Due to reciprocity[12] $Z_{12} = Z_{21}$. Note that in the usual representations Z_{12} and Z_{22} have positive signs and the output current points

[12]The principle of reciprocity for a general set of circuit elements consisting of many branches can be formulated as follows: If a voltage generator V, applied to the ith branch, drives a current I in the kth branch, then the same current I is obtained in the ith branch if the voltage is applied to the kth branch. For a proof see, e.g., Simonyi (1963). A consequence of this principle is that for a four-pole described by eqn (1.121) the impedances Z_{12} and Z_{21} must be equal.

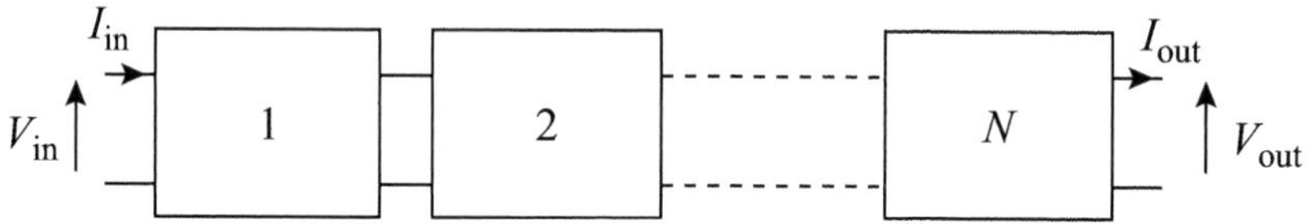

Fig. 1.13 A chain of N four-poles

inwards. They are taken negative in eqn (1.123) to correspond to the case when the output current points outwards, which is more suitable for our purpose because it allows the chain matrix representation that relates the output quantities to the input quantities with the aid of the chain matrix b:

$$\begin{pmatrix} V_{out} \\ I_{out} \end{pmatrix} = \begin{pmatrix} b_{11} & b_{12} \\ b_{21} & b_{22} \end{pmatrix} \begin{pmatrix} V_{in} \\ I_{in} \end{pmatrix} . \tag{1.124}$$

The reciprocity condition may be shown with a little algebra to take the form

$$b_{11}b_{22} - b_{12}b_{21} = 1 . \tag{1.125}$$

The advantage of the representation of eqn (1.124) is that input/output relations for a chain of four-poles (Fig. 1.13) may be simply written as

$$\begin{pmatrix} V_{out}^{(N)} \\ I_{out}^{(N)} \end{pmatrix} = \prod_{l=1}^{N} b^{(i)} \begin{pmatrix} V_{in}^{(1)} \\ I_{in}^{(1)} \end{pmatrix} , \tag{1.126}$$

i.e. the resultant chain matrix is the product of the individual chain matrices.

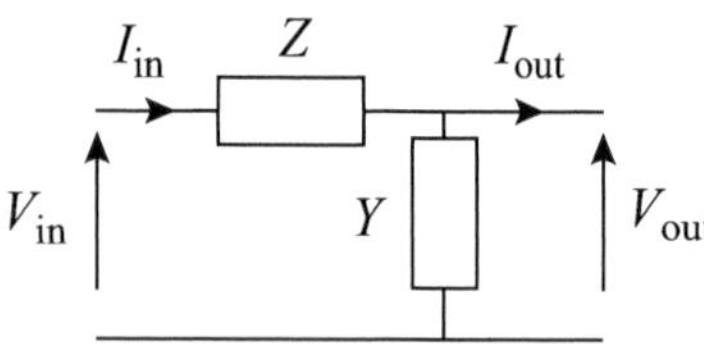

Fig. 1.14 Circuit with a series impedance and a shunt admittance

So far we have been concerned with four-poles characterized by their matrix representation without specifying what the matrix elements are. To get nearer to a practical case we shall choose a simple circuit as shown in Fig. 1.14 with a series impedance Z and a shunt admittance Y.

Kirchhoff's law for the circuit yields the two equations

$$V_{in} = I_{in}Z + V_{out} \qquad \text{and} \qquad V_{out} = \frac{I_{in} - I_{out}}{Y} . \tag{1.127}$$

A little algebraic manipulation then gives the elements of the b matrix[13] as

$$b_{11} = 1, \quad b_{12} = -Z, \quad b_{21} = -Y, \quad b_{22} = 1 + YZ . \tag{1.128}$$

[13]For those in the habit of scrutinizing dimensions it may be disconcerting to look at eqn (1.128) because different elements of the matrix have different dimensions, but that follows from the definition. It is strange when first meeting them; and it takes time to get used to.

1.22 Transmission line equations

We shall reproduce Fig. 1.14 in Fig. 1.15 with the difference that now the circuit represents a dz section of a transmission line and Z_u and Y_u are per unit length impedance and admittance, respectively.

Equation (1.127) is still valid, now yielding

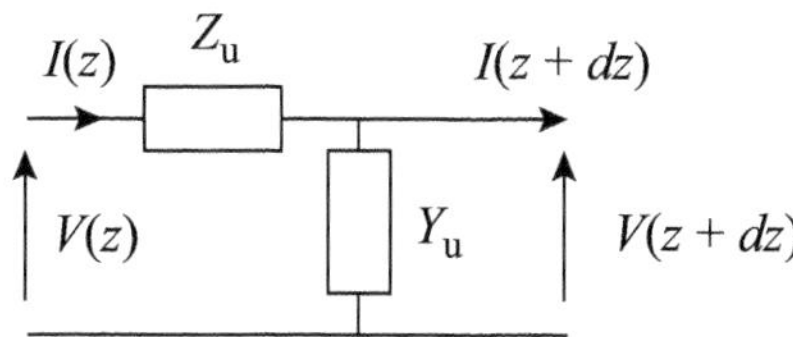

Fig. 1.15 Circuit with a series impedance and a shunt admittance of a dz section of a transmission line

$$V(z + \mathrm{d}z) - V(z) = -Z_\mathrm{u} I(z)\mathrm{d}z \tag{1.129}$$

and

$$I(z + \mathrm{d}z) - I(z) = -Y_\mathrm{u} V(z + \mathrm{d}z)\mathrm{d}z\,, \tag{1.130}$$

whence one can derive the differential equations

$$\frac{\mathrm{d}V}{\mathrm{d}z} = -Z_\mathrm{u} I \qquad \text{and} \qquad \frac{\mathrm{d}I}{\mathrm{d}z} = -Y_\mathrm{u} V\,. \tag{1.131}$$

Differentiating either of the above differential equations and substituting into the other one will yield the wave equation. For the voltage it takes the form

$$\frac{\mathrm{d}^2 V}{\mathrm{d}z^2} + Y_\mathrm{u} Z_\mathrm{u} V = 0\,. \tag{1.132}$$

Comparing the above equation with eqn (1.21) we find that the two are identical if we take for the propagation constant

$$k^2 = Y_\mathrm{u} Z_\mathrm{u}\,. \tag{1.133}$$

Now let us take the values

$$Z_\mathrm{u} = \mathrm{j}\omega L_\mathrm{u} \qquad \text{and} \qquad Y_\mathrm{u} = \mathrm{j}\omega C_\mathrm{u}\,, \tag{1.134}$$

in which case eqn (1.133) reduces to

$$k^2 = \omega^2 L_\mathrm{u} C_\mathrm{u}\,. \tag{1.135}$$

We may now take as an example a two-wire transmission line as shown in Fig. 1.16. When $d \gg r_\mathrm{w}$ (d is the separation of the lines and r_w is the radius of the wire) the per unit inductance and capacitance are (see, e.g., Ramo *et al.* 1965)

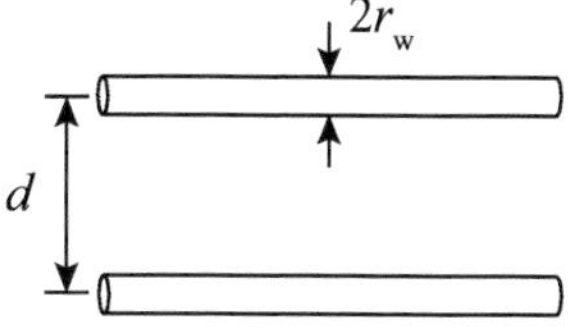

Fig. 1.16 A two-wire transmission line

$$L_\mathrm{u} = \frac{\mu_0}{\pi} \ln\left(\frac{d}{r_\mathrm{w}}\right) \qquad \text{and} \qquad C_\mathrm{u} = \frac{\pi\varepsilon_0}{\ln\left(\dfrac{d}{r_\mathrm{w}}\right)}\,. \tag{1.136}$$

Substituting eqn (1.136) into eqn (1.135) we find

$$k = \frac{\omega}{c}\,. \tag{1.137}$$

This means that the velocity of propagation in a two-wire transmission line is equal to the velocity of light. A wave along a transmission line is of course not a plane wave but Fig. 1.15 and eqn (1.135) are also capable of providing the characteristics of a plane wave if we introduce

$$L_{\mathrm{u}} = \mu_0 \qquad \text{and} \qquad C_{\mathrm{u}} = \varepsilon_0 \,, \tag{1.138}$$

which would lead again, in a different manner, to eqn (1.137).

1.23 Waves on four-poles

Figures 1.14 and 1.15 showed the same kind of simple four-pole and we managed to obtain from that the wave equation in the same form as eqn (1.21). The present section aims at more generality. A four-pole can have a different structure and still support waves.

A characteristic of a wave is that for the same interval the phase of the wave always changes by the same amount. Thus, if we have a chain of four-poles the phase change between the output and input quantities should be the same factor, $\exp(-\mathrm{j}\,ka)$, where a can be regarded as the physical length of a unit. It follows, hence, that

$$V_{\mathrm{out}} = \mathrm{e}^{-\mathrm{j}\,ka} V_{\mathrm{in}} \,, \tag{1.139}$$

and

$$I_{\mathrm{out}} = \mathrm{e}^{-\mathrm{j}\,ka} I_{\mathrm{in}} \,. \tag{1.140}$$

Substituting the above quantities into eqn (1.124) we find that a solution exists when

$$2\cos ka = b_{11} + b_{22} \,. \tag{1.141}$$

This is now an entirely general dispersion equation. If we know the chain matrix of a chain of identical four-poles (in fact, we need only the main diagonal elements) we have immediately the corresponding dispersion relation.

Let us take as an example a circuit with a series impedance Z and a shunt admittance, Y. The main diagonal elements of the corresponding chain matrix are then given by eqn (1.128). Substituting them into eqn (1.141) we find the dispersion relation

$$\cos ka = 1 + \frac{YZ}{2} \,, \tag{1.142}$$

which may be written in the alternative form

$$4\sin^2 \frac{ka}{2} = -YZ \,. \tag{1.143}$$

If $Z = \mathrm{j}\,\omega L$ and $Y = \mathrm{j}\,\omega C$, then from the above two equations we obtain

$$\cos ka = 1 - \frac{1}{2}\omega^2 LC \,, \qquad 4\sin^2 \frac{ka}{2} = \omega^2 LC \,. \tag{1.144}$$

Note that the assumptions were practically the same as in the previous section when discussing the transmission line equations. Nevertheless,

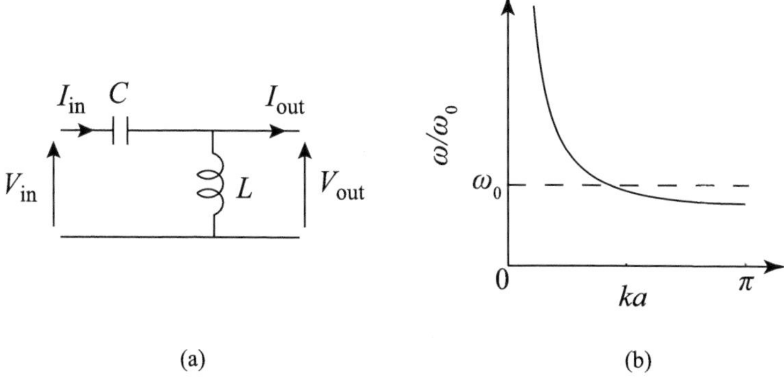

Fig. 1.17 (a) Section of a transmission line with a series capacitance and a shunt inductance and (b) the dispersion curve of a chain of such sections

eqns (1.135) and (1.144) are different. The reason is that in Fig. 1.15 we already assumed that the line is continuous, whereas our chain of four-poles implies a set of discrete elements. We can, of course, still affect the conversion from a discrete to a continuous line by assuming that $ka \ll 1$, in which case eqn (1.144) reduces to

$$(ka)^2 = \omega^2 LC\,, \tag{1.145}$$

which is identical with eqn (1.135) considering that $L_{\text{u}} = L/a$ and $C_{\text{u}} = C/a$.

We have now proven that our last two models may lead to the same result. We should notice, however, that we have now a wave solution for any kind of four-poles. For our second example we shall take the four-pole of Fig. 1.17(a) where

$$Z = \frac{1}{\mathrm{j}\omega C} \qquad \text{and} \qquad Y = \frac{1}{\mathrm{j}\omega L}\,. \tag{1.146}$$

The corresponding dispersion equation is

$$\cos ka = 1 - \frac{1}{2\omega^2 LC}\,. \tag{1.147}$$

plotted in Fig. 1.17(b) for ka in the range 0 to π. The phase velocity, ω/k, is always positive and the group velocity, $\mathrm{d}\omega/\mathrm{d}k$ is always negative. This is our first example of a backward wave.

Our third example is quite a different four-pole, shown in Fig. 1.18(a). We have separated here the mutual inductance, M, and the self-inductance, L, as it leads to simpler mathematics. Its chain matrix can be found quite simply in the form

$$b = \begin{bmatrix} 0 & -\mathrm{j}\omega M \\ \dfrac{1}{-\mathrm{j}\omega M} & -\dfrac{L}{M}\left(1 - \dfrac{\omega_0^2}{\omega^2}\right) \end{bmatrix}, \tag{1.148}$$

where $\omega_0 = 1/\sqrt{LC}$. The dispersion equation derived from the above parameters, solving this time for ω is

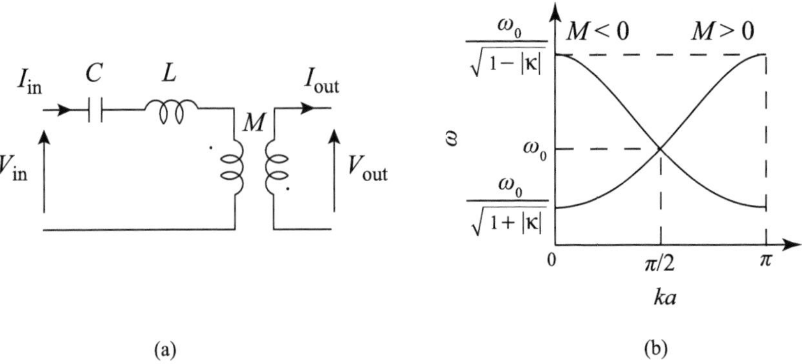

Fig. 1.18 (a) An equivalent circuit for magnetoinductive waves and (b) its dispersion curves for positive and negative coupling

$$\omega = \frac{\omega_0}{\sqrt{1 + \dfrac{2M}{L}\cos(ka)}} \,. \tag{1.149}$$

The corresponding dispersion curves for $2M/L = 0.1$ and -0.1 are shown in Fig. 1.18(b). It may be seen that we have a forward wave for positive M and a backward wave for negative M. Both waves will often appear in this book under the name of magnetoinductive waves.

1.24 Scattering coefficients

The scattering coefficients are defined in terms of wave amplitudes. We are still talking about four-poles, as shown schematically in Fig. 1.12, but instead of voltages and currents the quantities related to each other are the amplitudes of the waves pointing inwards, A_1 at port 1 and A_2 at port 2, and the amplitudes of the waves propagating outwards, B_1 at port 1 and B_2 at port 2. The relationship between them is given by the scattering matrix, S, in the form

$$\begin{pmatrix} B_1 \\ B_2 \end{pmatrix} = \begin{pmatrix} S_{11} & S_{12} \\ S_{21} & S_{22} \end{pmatrix} \begin{pmatrix} A_1 \\ A_2 \end{pmatrix} , \tag{1.150}$$

where, due to reciprocity, $S_{12} = S_{21}$. When there is an incident wave at port 1 and the output waveguide is matched, so that $A_2 = 0$, then

$$B_1 = S_{11}A_1 \qquad \text{and} \qquad B_2 = S_{21}A_1 \,. \tag{1.151}$$

It may be seen from eqn (1.151) that under these conditions S_{11} and S_{21} are the reflection and transmission coefficients. It follows then that for a lossless line

$$|S_{11}|^2 + |S_{22}|^2 = 1 \,. \tag{1.152}$$

Note that in the large majority of measurements, aimed at evaluating device performance, the S_{11} and S_{12} coefficients are measured. These are the values that those more interested in what is going on will compare with theoretical results.

1.25 Fourier transform and the transfer function

Our main interest in this book is in spatial in contrast to temporal variation which is nearly always taken in the form $\exp(\mathrm{j}\,\omega t)$. On the other hand, the spatial information will be of interest in both the actual space domain and in the spatial frequency domain. The relationship between the two is given by the Fourier transform. Let us take a complex two-variable function $g(x, y)$. Its Fourier transform is given by

$$F(g) = G(f_x, f_y) = \int g(x, y)\, \mathrm{e}^{-\mathrm{j}\,2\pi(f_x x + f_y y)} \mathrm{d}x\mathrm{d}y\,. \qquad (1.153)$$

The variables f_x and f_y are spatial frequencies and $G(f_x, f_y)$ is the amplitude of the f_x, f_y pair of spatial frequencies. In intuitive terms we may say that by adding up many sinusoidals of various frequencies and amplitudes the $g(x, y)$ function can be reproduced. Note that the spatial frequencies are related simply to the previously defined wave number as $k_x = 2\pi f_x$ and $k_y = 2\pi f_y$.

The problem usually arising is to find the variation in x and y of some function at the plane z_2 when the function is known at the plane z_1. The space between z_1 and z_2 is filled by some medium (Fig. 1.19). We can then investigate what happens to a particular pair of spatial frequencies of complex amplitude $g(f_x, f_y)$ when it traverses the medium between z_1 and z_2. In a given case this may turn out to be a very difficult problem to solve but it is simple in principle. All that can happen to a particular pair is that at the exit it will have a different amplitude and a different phase. The function that tells us what happens to all the spatial frequencies is called the transfer function $T(f_x, f_y)$. So, how can we find the spatial distribution at the plane z_2?

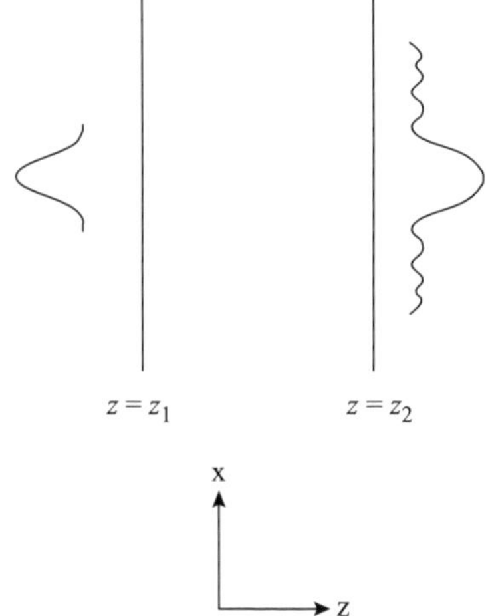

Fig. 1.19 An object at z_1 imaged at z_2

We may denote the function (say the tangential component of the electric field) varying in the plane z_1 as a function of x and y by $g_1(x, y)|_{z=z_1}$. Its spatial harmonics (which together we may call the Fourier spectrum) are given by its Fourier transform $G_1(f_x, f_y)$. When traversing the medium each one of the Fourier components will undergo some change corresponding to the transfer function $T(f_x, f_y)$. Hence, the Fourier spectrum at z_2 is $T(f_x, f_y)G(f_x, f_y)$. But what we are interested in is not the Fourier spectrum but the spatial variation of the field. We can find that by taking the inverse Fourier transform. Hence

$$g(x, y)|_{z=z_2} = \int T(f_x, f_y) G(f_x, f_y)\, \mathrm{e}^{\mathrm{j}\,2\pi(f_x x + f_y y)} \mathrm{d}f_x \mathrm{d}f_y\,. \qquad (1.154)$$

If there are a number of media with different properties between z_1

and z_2 then the transfer function of each one of them must be multiplied together similarly to chain matrices in the previous section.

As an example, let us find the transfer function of a wave in free space propagating at an angle θ_1 relative to the z axis. As we have seen before, the phase varies in the z and x directions as $\exp[-\mathrm{j}\,(k_z z + k_x x)]$ where

$$k_z^2 + k_x^2 = k^2 = \omega^2 \mu\varepsilon \,. \tag{1.155}$$

The period along the x axis is determined by k_x, it is actually equal to $2\pi/k_x$, similarly to the temporal period, which is equal to $T = 2\pi/\omega$. Hence, in analogy, we may call k_x the spatial frequency. Its amplitude at $z = 0$ is equal to 1. When the wave propagates from $z = 0$ to $z = d_1$ its complex amplitude will be $\exp(-\mathrm{j}\,k_z d_1)$ where k_z is related to k_x by eqn (1.155). In this example, clearly, only the phase has changed. The transfer function is

$$T(k_x) = \mathrm{e}^{-\mathrm{j}\,\sqrt{\omega^2\mu_0\varepsilon_0 - k_x^2}\,d_1} \,. \tag{1.156}$$

Our next example is a little more complicated. The space of interest is now from $z = 0$ to $z = d_1 + d_2 + d_3$ as shown in Fig. 1.20. It is free space from $z = 0$ to $z = d_1$ and from $z = d_1 + d_2$ to $z = d_1 + d_2 + d_3$. From $z = d_1$ to $z = d_1 + d_2$ it is a medium with material constants ε, μ. A plane wave is incident at $z = 0$ and we detect it at $z = d_1 + d_2 + d_3$. What will be the total transfer function? We have found the transfer function in free space in the previous example. So, next we must find it for the slab. How can we determine its transfer function? It is a fair amount of work but as it happens we have already done all the work in Section 1.10 but we just did not call it a transfer function. But if we look at eqn (1.76) and realize that k_{z1} and k_{z2} are related to the spatial frequency $k_{x1} = k_{x2} = k_x$ then it can be recognized as a transfer function in disguise. Hence, the total transfer function is

$$T_\mathrm{t}(k_x) = T_1(k_x) T_2(k_x) T_3(k_x) \,, \tag{1.157}$$

and the individual transfer functions are

$$T_1(k_x) = \mathrm{e}^{-\mathrm{j}\,\sqrt{\omega^2\mu_0\varepsilon_0 - k_x^2}\,d_1} \,, \tag{1.158}$$

$$T_2(k_x) = \frac{4\zeta_\mathrm{e}}{(1+\zeta_\mathrm{e})^2\,\mathrm{e}^{\mathrm{j}\,k_{z2}d_2} - (1-\zeta_\mathrm{e})^2\,\mathrm{e}^{-\mathrm{j}\,k_{z2}d_2}} \,, \tag{1.159}$$

$$T_3(k_x) = \mathrm{e}^{-\mathrm{j}\,\sqrt{\omega^2\mu_0\varepsilon_0 - k_x^2}\,d_3} \,. \tag{1.160}$$

Equations (1.158)–(1.160) look complicated but they can simplify much in certain special cases. The possibility of a ‘perfect lens’ does follow from them as will be discussed in Section 2.12.

We have not so far asked any questions about the range of spatial frequencies. From the analogy with temporal frequency we may assert that it can take any value from zero to infinity. Obviously the smaller

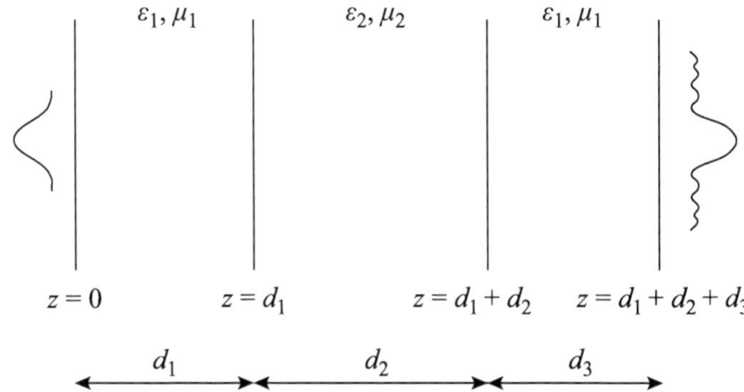

Fig. 1.20 Imaging in a three-media configuration

the details of a spatial function the higher the spatial frequencies that can reproduce those details. Note, however, that eqn (1.155) must be satisfied. When the spatial frequency is high enough to satisfy the inequality

$$k_x > \omega\sqrt{\mu\varepsilon}\,, \tag{1.161}$$

then, according to eqn (1.155), k_z, the wave number responsible for propagation in the z direction, becomes imaginary. The wave declines exponentially, as we saw for surface waves in Section 1.9, but the expressions derived are still valid.

A bird's-eye view of metamaterials

2

2.1 Introduction

The aim of this chapter is to provide an introduction to the subject. It is called a bird's-eye view because it looks at the subject of metamaterials, admittedly superficially, from above, pausing at certain views that the authors like to share with the reader and making some relevant comments without going into very much detail. Physical concepts, most of them already introduced in Chapter 1, will be further developed. The mathematics will be kept to an absolute minimum. It will mostly be simple algebra.

Who might benefit from this survey? Well, the authors have certainly benefited from writing it because they were forced to pronounce judgment on the relative simplicity of the various parts of the subject and had to decide on where to start, how to proceed and what to include. For a beginner, who first comes into contact with the subject, there might be too many new concepts. Even then it is hoped that some concepts will be picked up, stored in the brain, and will later act as catalysts facilitating the absorption of further information. For a research student who has already acquired some familiarity with the fundamentals it might serve as a reinforcement of existing knowledge. For unbelievers who question the correctness of all new ideas wherever they come from, until properly checked, this chapter might offer new things to worry about. For those who are thoroughly familiar with all the basic tenets of modern research in metamaterials this might still be suitable for bedtime reading.

This chapter makes no claim to rigour, which will be sacrificed at the altar of simplicity. We shall not be concerned with priorities either. When we just want to get through a large number of different concepts it is not worth pausing and telling the reader who did what and when and it is particularly difficult to reconcile rival claims to priority. We shall therefore quote relatively few references, but that would not, of course, apply to other chapters.

2.2 Natural and artificial materials

We all know what natural materials are. They are made up by lots and lots of small elements like atoms and molecules. Some of these materials

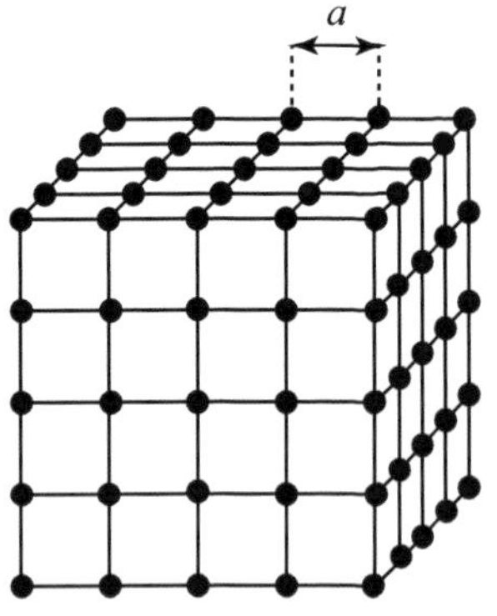

Fig. 2.1 Cubic lattice

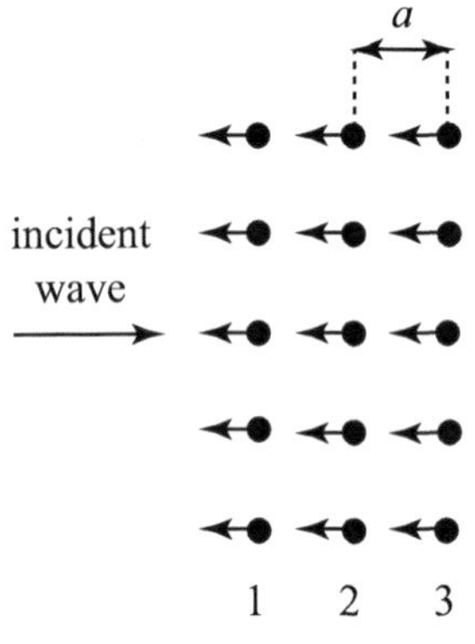

Fig. 2.2 Electromagnetic wave incident normally at a lattice

are amorphous, meaning that all those elements are heaped upon each other in a random manner, others are crystalline, which means that they arrange themselves into some regular periodical pattern.

Our main interest is in the interplay of waves and materials restricted to classical physics. The key parameter is a/λ, where a is the distance between elements in the material and λ is the free-space wavelength. For simplicity, let us assume that the elements arrange themselves in a regular cubic lattice, the same in all three directions, as may be seen in Fig. 2.1. We may now look at two cases: the wavelength is comparable with a or much larger than a. In the first case the Bragg effect comes into play. The simplest example is shown schematically in Fig. 2.2. An electromagnetic wave may be seen to be incident perpendicularly at a lattice. The wave propagating then from row 1 to row 2 will cover a path a. The part of the wave that is reflected by row 2 will have covered an additional distance a when arriving back at row 1. When a happens to be equal to one half of a wavelength then the waves reflected by all the rows will have the same phase and will reinforce each other. If there are many rows, and there are indeed many of them in a crystal, then most of the incident power may be reflected. This effect is at the basis of X-ray and electron diffraction in crystals.

When the wavelength is much larger than the lattice period then no such dramatic effect occurs, but it is nonetheless significant. There may not be major reflection or diffraction but the electromagnetic wave is still considerably affected when it enters a material. We may then ignore the details and pretend that there is no discrete structure: the material is homogeneous and continuous. The aim is then to find some effective parameters like electric permittivity and magnetic permeability. This is known as the effective-medium approximation. Summarizing, there is the Bragg effect, when the distance between the elements is comparable with the wavelength, and there is effective-medium response when that distance is much less than the wavelength.

Now let's think of artificial materials in which atoms and molecules are replaced by macroscopic, man-made, elements. Let's not worry for the moment how the elements remain in their allotted space. That may not be always obvious but we can safely assume that we have complete freedom in choosing both the elements and the distance between them. Now, all dimensions are bigger than in natural materials but the division into the above two categories is still valid. When the separation between the elements is comparable with the wavelength we have the Bragg effect, and when the separation is much smaller than the wavelength we can resort to effective-medium theory. In the former case we talk about photonic bandgap materials and in the latter case about metamaterials. Can we have a better definition of metamaterials? Not easily. There is broad agreement on what the subject is about but not about all the details. It would need a fairly long description accompanied by a number of examples to be more precise. We shall give here two definitions in current use.

1. Metamaterials are engineered composites that exhibit superior prop-

erties not found in nature and not observed in the constituent materials.
2. A metamaterial is an artificial material in which the electromagnetic properties, as represented by the permittivity and permeability, can be controlled. It is made up of periodic arrays of metallic resonant elements. Both the size of the element and the unit cell are small relative to the wavelength.

Definition 1 is too general, whereas definition 2 is not general enough: neither of them mention applications and do not even pay homage to negative refraction. We shall make no attempt here to give a more comprehensive definition. Perhaps definition 2 could be made more general by adding that control, among other things, means that it is possible to achieve simultaneously negative permittivity and negative permeability at the same frequency, which will then lead to negative-index media and to negative refraction. It is not easy to find a definition that would satisfy everybody. For a discussion of the difficulties of a proper definition and for many other ideas on the subject see Sihvola (2007).

We have talked about natural and artificial materials and their relationship to transverse electromagnetic waves. But electromagnetic waves are not the only ones that should be considered. In a crystal, for example, the atoms and molecules may move relative to each other. They cannot move far away because there are some restoring forces. One of the manifestations of these motions and of the forces opposing them is the emergence of acoustic waves.

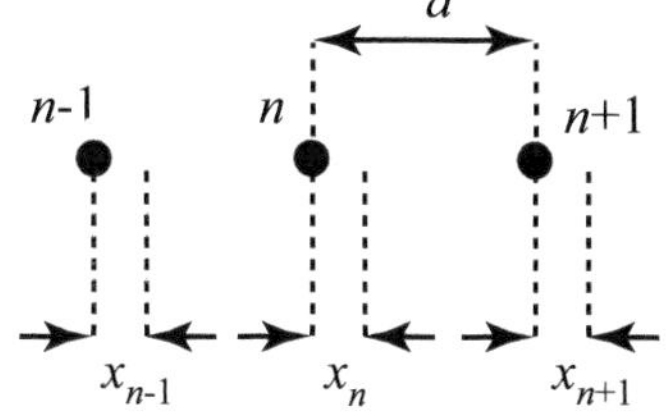

Fig. 2.3 1D chain of atoms showing displacement from the quiescent position

Let us take a one-dimensional chain of atoms in which the elements are at a distance a from each other at rest. Figure 2.3 shows schematically the positions of three of the atoms at x_{n-1}, x_n and x_{n+1}. Three elements are usually sufficient when we can get away with an approximation that takes into account only nearest neighbours (see, e.g., Brillouin 1953; Dekker 1965). Note that all the displacements are in the longitudinal direction, which makes the problem conceptually simpler. How can we work out the net force? If $x_{n+1} > x_n$ then there will be a force on the atom x_n wanting to move it to the right. Conversely, if $x_n > x_{n-1}$ then there will be a force to the left. A proper mathematical formulation using Newton's equation (eqn (1.1)) followed by the assumption of a wave solution will yield the dispersion equation for acoustic waves (Fig. 2.4). It shows the relation between frequency, ω, and wave number, k. The uppermost frequency ω_a at which acoustic waves can propagate occurs at $ka = \pi$ or $\lambda_a = a/2$, where λ_a is the acoustic wavelength. At frequencies above ω_a acoustic waves of the kind we investigated cannot propagate. The band up to ω_a is the pass band and above it, where the wave cannot propagate, is the stop band.

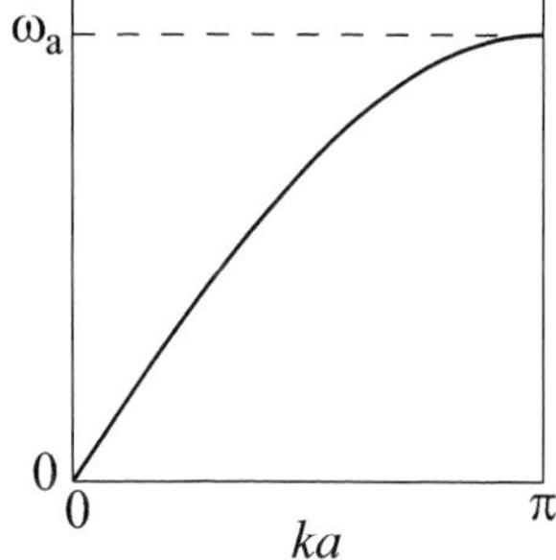

Fig. 2.4 Dispersion curve of acoustic waves

To give another example plasma waves may also propagate in a natural material. Take sodium for example. It is a metal in which numerous electrons float in a pool and are compensated by positive ions. In the simplest case all the electrons undergo simple harmonic motion in the direction of their propagation. If the frequency is high enough, above the plasma frequency somewhere in the ultraviolet, these electrons can move quite freely. We should also mention spin waves in magnetic materials in which the direction of the magnetic moment changes from element to element.

What we wish to say is that our attention should not be restricted to the effect of a material upon the propagation of transverse electromagnetic waves. There can be waves in a natural material due to the interaction between the particles, and these waves may not exhibit any electrical or magnetic phenomena as in our example of acoustic waves.

Fig. 2.5 Capacitively loaded loop

Now back to metamaterials. Can there be waves on the elements that are unrelated to transverse electromagnetic waves? There can be. In fact, a considerable part of this book (Chapters 7 and 8) will be concerned with one of these waves that we called magnetoinductive (MI) waves (Shamonina *et al.*, 2002*a*). They have already been introduced in Section 1.23 as the waves propagating on the chain of four-poles shown in Fig. 1.18. The simplest elements that will propagate this wave are capacitively loaded metallic loops shown schematically in Fig. 2.5.

The resonant frequency of the elements, which can be simply regarded as LC circuits, is

$$\omega_0 = \frac{1}{\sqrt{LC}}, \tag{2.1}$$

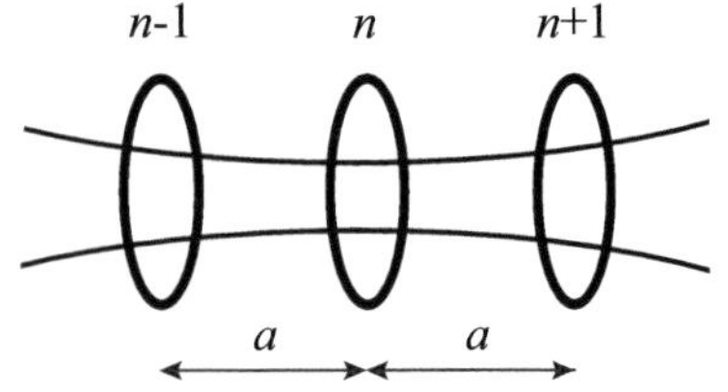

Fig. 2.6 Magnetic nearest-neighbour interactions in a metamaterial chain of identical elements

where L is the inductance of the loop and C is the capacitance of the loading capacitor. Considering again nearest-neighbour interactions[1] we show three such elements separated by a distance a from each other in Fig. 2.6.

It may be seen that the magnetic field created by element n will also thread elements $n-1$ and $n+1$. Clearly, there is some magnetic coupling between the elements. Can it lead to waves? It can. The mathematical formulation may be based on Kirchhoff's voltage equation. It says that the total voltage in a closed circuit must be zero. When we consider only three currents, I_{n-1}, I_n and I_{n+1}, then the total voltage in circuit n will have three contributions: (i) the self-voltage, equal to $I_n Z_0$ (where Z_0 is the self-impedance of the element), (ii) the voltage induced by element $n-1$ that is equal to $\mathrm{j}\omega M I_{n-1}$ (where M is the mutual inductance between elements $n-1$ and n) and (iii) a contribution from element $n+1$ that is $\mathrm{j}\omega M I_{n+1}$. Hence, the relevant equation is

$$Z_0 I_n + \mathrm{j}\omega M(I_{n-1} + I_{n+1}) = 0. \tag{2.2}$$

Next, assume a wave solution in the form

$$I_n = I_0\,\mathrm{e}^{-\mathrm{j}nka}. \tag{2.3}$$

We obtain the dispersion equation for MI waves as

[1]This is now the second time in this section that we mention nearest-neighbour interaction. It is not something we would readily associate with waves. It makes no sense for water waves and even less for electromagnetic waves in vacuum. There are no neighbours in vacuum. On the other hand, it would not be difficult to develop a physical picture of wave propagation based on nearest neighbours if we give a little thought to it. Imagine, for example, a large number of houses next to each other in a street in which nearest neighbours can talk to each other over the fence. An interesting piece of news could certainly reach the last house in the street by propagating along the row of houses via nearest-neighbour interaction.

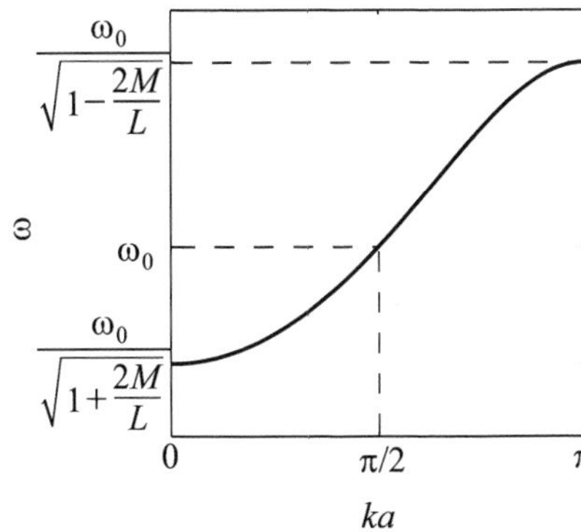

Fig. 2.7 Dispersion of magnetoinductive waves with positive magnetic coupling

$$L\left(1-\frac{\omega_0^2}{\omega^2}\right)+2M\cos(ka)=0, \tag{2.4}$$

where we took into account that

$$Z_0=\mathrm{j}\omega L+\frac{1}{\mathrm{j}\omega C}. \tag{2.5}$$

Note that eqn (2.4) is identical with eqn (1.149). The corresponding dispersion equations have already been plotted in Fig. 1.18. It is replotted in Fig. 2.7 for $2M/L>0$ for a more pervasive examination.

It shows some similarity to the dispersion curve of acoustic waves at least in the sense that the group velocity, $\mathrm{d}\omega/\mathrm{d}k$, is always positive and at the band edge the group velocity is zero, as it is for all waves on discrete structures. Note also that there is a lower cutoff frequency below which the MI wave cannot propagate. The pass band is within the range

$$\frac{\omega_0}{\sqrt{1+\dfrac{2M}{L}}}<\omega<\frac{\omega_0}{\sqrt{1-\dfrac{2M}{L}}}. \tag{2.6}$$

There is no reason of course that the coupling between the metamaterial elements has to be magnetic. It can be electric. Well before the advent of metamaterials an experiment, shown in Fig. 2.8(a), was performed by Shefer (1963), using a set of metallic rods (Fig. 2.8(b)). One of the horns is a transmitter of microwaves, the other horn is a receiver, and a wave travels along the rods from one horn to the other horn due to electric coupling. Typical dimensions in the experiments were $l=12$ mm, $d=1$ mm and a distance between the rods of 5 mm. They found good transmission between the horn antennas at around the frequency of 1.2 GHz. A more recent example of wave propagation by rods on a substrate (Hohenau *et al.*, 2005), at a frequency five orders of magnitude higher (360 THz), is shown in Fig. 2.8(c), where the element dimensions are 800 nm × 80 nm × 50 nm. The distance between the elements is 320 nm.

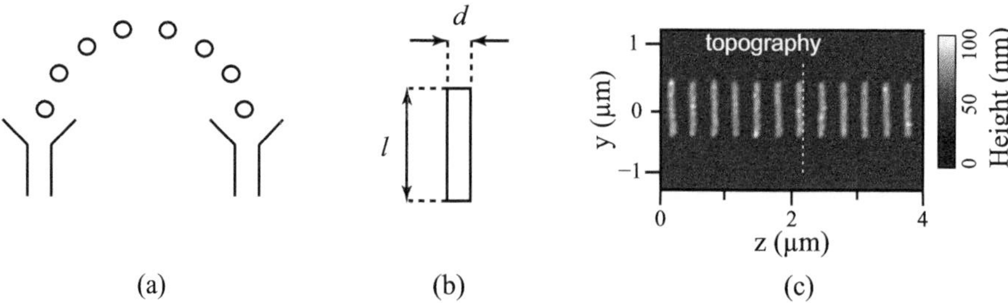

Fig. 2.8 Electric coupling on a chain of rods. (a) Schematic of a setup and (b) dimensions of the rod used in the experiment by Shefer (1963). Copyright © 1963 IEEE. (c) Topography of the setup used by Hohenau *et al.* (2005). Copyright © 2005 EDP Sciences

2.3 Determination of the effective permittivity/dielectric constant in a natural material

The principles are quite simple. They are related to the interdependence of three vectors: the electric field, **E**, the electric flux density, **D**, and the electric polarization, **P**. In free space, in SI (Systeme International) units this can be written as

$$\mathbf{D} = \varepsilon_0 \mathbf{E} + \mathbf{P}\,, \tag{2.7}$$

as has already been given in eqns (1.9) and (1.19).

The central question is the relationship between **E** and **P** in a material in which electric dipoles appear in response to an electric field. Let us consider an element inside a cubic material at the centre of a rectangular co-ordinate system $(0, 0, 0)$ and apply an electric field, $\mathbf{E}_{\text{ext}}$, to the material parallel to the z co-ordinate. What will be the electric field at our chosen element? One's first thought is that the electric field will be equal there to the external field. This would indeed be the case if the electric field would not cause the positive and negative charges to separate. But the charges do separate. Each dipole will then contribute to the electric field at $(0, 0, 0)$. We shall call the field there $\mathbf{E}_{\text{loc}}$, the local field, which is equal to

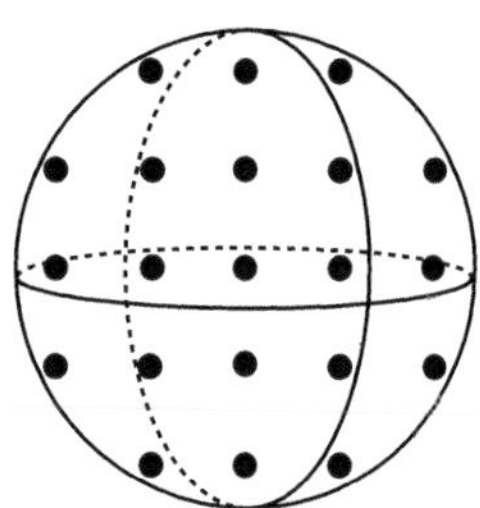

Fig. 2.9 A sphere within a cubic lattice used for averaging of macroscopic parameters

$$\mathbf{E}_{\text{loc}} = \mathbf{E}_{\text{ext}} + \mathbf{E}_{\text{dipole}}\,. \tag{2.8}$$

The usual technique for determining $\mathbf{E}_{\text{dipole}}$ is available in practically every textbook on solid state physics (see, e.g., Kittel 1953). One takes a sphere (see Fig. 2.9) inside the material that is large microscopically in the sense that it contains many elements but small macroscopically, meaning that it is small relative to the electromagnetic wavelength. So the material is divided into two regions: inside the sphere and outside the sphere.

Since the radius of the sphere is macroscopically small we can use the static formulae for the electric field of a dipole, the $1/r^3$ terms of eqns (1.82) and (1.83). When we sum up the contributions at the point $(0, 0, 0)$ in a cubic lattice from every element inside the sphere it yields zero electric field as shown in Appendix B. So, after all, the electric field is equal there to the external field. Is it? Not really, because this was the contribution of the elements inside the sphere. We still need to consider the effect of the elements outside the sphere. Now we can no longer say that the effect will be zero because for elements far away we also need to take into account the slowly declining radiation field. It is a tremendously difficult problem. For a recent analysis see Belov and Simovski (2005*a*). Here, we shall disregard all those difficulties and follow the time-honoured method of accounting for the effect of the elements far away by charges induced on the surface of the sphere. The electric field at the centre may then be determined by a simple integration from the charges far away as equal to $\mathbf{P}/3\varepsilon_0$, which gives for the local field

$$\mathbf{E}_{\mathrm{loc}} = \mathbf{E}_{\mathrm{ext}} + \frac{\mathbf{P}}{3\varepsilon_0} . \tag{2.9}$$

But the polarization $\mathbf{P}$ at $(0, 0, 0)$ is proportional to the local electric field there

$$\mathbf{P} = N\alpha_{\mathrm{e}}\mathbf{E}_{\mathrm{loc}} , \tag{2.10}$$

where α_{e} is the atomic/molecular polarizability discussed in Section 1.16. It tells us how effective the local electric field is in producing an electric polarization. From eqns (2.9) and (2.10) we obtain

$$\mathbf{P} = \mathbf{E}_{\mathrm{ext}} \frac{N\alpha_{\mathrm{e}}}{1 - \dfrac{N\alpha_{\mathrm{e}}}{3\varepsilon_0}} . \tag{2.11}$$

We may now rely on the definition of $\varepsilon_{\mathrm{eff}}$ as

$$\mathbf{D} = \varepsilon_{\mathrm{eff}}\mathbf{E}_{\mathrm{ext}} = \varepsilon_0\mathbf{E}_{\mathrm{ext}} + \mathbf{P} , \tag{2.12}$$

to find

$$\varepsilon_{\mathrm{eff}} = \varepsilon_0 + \frac{N\alpha_{\mathrm{e}}}{1 - \dfrac{N\alpha_{\mathrm{e}}}{3\varepsilon_0}} . \tag{2.13}$$

We may also define the relative permittivity (or relative dielectric constant) as

$$\varepsilon_{\mathrm{r}} = \frac{\varepsilon_{\mathrm{eff}}}{\varepsilon_0} = 1 + \frac{N\alpha_{\mathrm{e}}/\varepsilon_0}{1 - \dfrac{N\alpha_{\mathrm{e}}}{3\varepsilon_0}} . \tag{2.14}$$

Alternatively, we can solve eqn (2.14) for $N\alpha_{\mathrm{e}}$, which may be expressed as

$$N\alpha_{\mathrm{e}} = 3\varepsilon_0 \frac{\varepsilon_{\mathrm{eff}} - \varepsilon_0}{\varepsilon_{\mathrm{eff}} + 2\varepsilon_0} = 3\varepsilon_0 \frac{\varepsilon_{\mathrm{r}} - 1}{\varepsilon_{\mathrm{r}} + 2} . \tag{2.15}$$

This is the well-known Clausius–Mossotti equation. The merit of eqns (2.13)–(2.15) is that they relate the polarizability to the effective dielectric constant.

Our interest in this section is in the electronic contribution to the dielectric constant. For natural materials there are other contributions too, e.g. orientational polarization, but they do not occur for metamaterials, and our aim is to focus upon such effects that occur both in natural and in metamaterials.

The above analysis has been done for finding the effective permittivity. The method is equally valid for finding the effective permeability of a natural material, and, as we shall see later, of a metamaterial too. The only difference is then that the electric polarizability needs to be replaced by the magnetic polarizability.

2.4 Effective plasma frequency of a wire medium

The plasma frequency of a natural material and its relationship to the dielectric constant was given in Section 1.9. It was also made clear in that section that there is no propagation when the dielectric constant is negative. Plasmas in a solid are well behaved. They are confined within the material. On the other hand, plasmas in general are difficult to control. In order to study their properties related to the re-entry problem[2] plasma simulation was a fashionable topic some time ago. One such simulation was by a lattice of metallic wires as shown in Fig. 2.10(a). It was shown by Rotman (1962) that they behave as a plasma with a much reduced plasma frequency.

[2]One of the interesting questions studied was the effect of rocket exhaust upon the radiation of re-entry vehicle antennas.

For determining the plasma frequency we shall use a very simple model that gives a good approximation. The calculation is based on the relationship between current and electric field within a cubic unit cell of side a. Note that a is then the distance between the elements and, also, a is the length of the rod within the unit cell. Let us first find the current in a thin piece of wire of length a and radius r_{w}. An incident electric field of E parallel to the wire will yield a current according to Ohm's law equal to

$$I = \frac{Ea}{Z_{\mathrm{w}}}, \tag{2.16}$$

where

$$Z_{\mathrm{w}} = R_{\mathrm{w}} + \mathrm{j}\omega L_{\mathrm{w}} \tag{2.17}$$

is the impedance of the wire. Next, we shall find the average current density in the unit cell that has an area of a^2. It is

$$J_{\mathrm{av}} = \frac{E}{(R_{\mathrm{w}} + \mathrm{j}\omega L_{\mathrm{w}})a}. \tag{2.18}$$

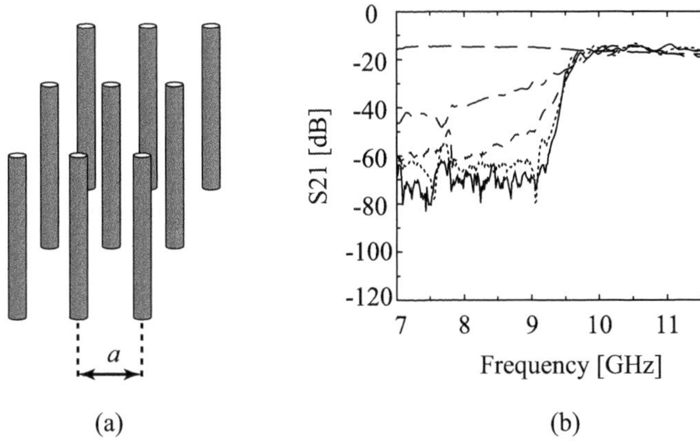

Fig. 2.10 (a) Schematic presentation of a rodded medium. (b) Scattering coefficient S_{21} for samples ($r_{\mathrm{w}} = 30\,\mu$m, $a = 6$ mm) with an increasing number of layers: 5 layers (dot-dashed line), 10 layers (dashed line), 15 layers (dotted line) and 20 layers (solid line). The reference without a sample is also shown (long-dashed line). From Gay-Balmaz *et al.* (2002). Copyright © 2002 American Institute of Physics

Having found the relationship between the electric field and the current density we can follow the method outlined in Section 1.9 to find the effective dielectric constant as

$$\varepsilon_{\mathrm{r}} = 1 + \frac{1}{\mathrm{j}\,\omega}\varepsilon_0(R_{\mathrm{w}} + \mathrm{j}\,\omega L_{\mathrm{w}})a\,. \tag{2.19}$$

Defining now

$$\omega_{\mathrm{p}}^2 = \frac{1}{\varepsilon_0 a L_{\mathrm{w}}}\,, \tag{2.20}$$

with ω_{p} being an effective plasma frequency,[3] we may rewrite eqn (2.19) as

$$\varepsilon_{\mathrm{r}} = 1 + \frac{\omega_{\mathrm{p\,eff}}^2}{\omega^2 - \dfrac{\mathrm{j}\,\omega}{\tau_{\mathrm{w}}}}\,, \tag{2.21}$$

where losses are characterized by the time constant

$$\tau_{\mathrm{w}} = \frac{L_{\mathrm{w}}}{R_{\mathrm{w}}}\,. \tag{2.22}$$

The expressions for the resistance and inductance (see, e.g., Grover 1981) are

[3]It is always an advantage to look at a phenomenon from different angles. The effective plasma frequency in eqn (2.20) is written by Pendry *et al.* (1996) in the form

$$\omega_{\mathrm{p}}^2 = \frac{e^2 N_{\mathrm{eff}}}{\varepsilon_0 m_{\mathrm{eff}}}\,.$$

It is the same as eqn (1.67) but instead of the actual electron density, N, and actual electron mass, m, the physical quantities appearing in the above equation are N_{eff}, the average electron density in the unit cell, and m_{eff}, the effective mass of the electron defined so that eqn (2.20) should be satisfied. According to Pendry *et al.* the reduction of plasma frequency may be interpreted as due partly to a decrease in the effective electron density and partly to an increase in the effective mass, and that increase may amount to four orders of magnitude. It is an interesting proposal but there is some inconsistency in it. If the mass of the electron has increased reducing thereby the plasma frequency then the increased mass must have also reduced the conductivity. A decrease in conductivity by four orders of magnitude would make the losses enormous and there is no evidence of that. For a different criticism of the introduction of an effective mass see also Pokrovsky and Efros (2002*c*).

$$R_{\mathrm{w}} = \frac{a}{\pi r_{\mathrm{w}}^2 \sigma_0} \qquad \text{and} \qquad L_{\mathrm{w}} = \frac{\mu_0 a}{2\pi}\left[\ln\left(\frac{2a}{r_{\mathrm{w}}}\right) - \frac{3}{4}\right]. \tag{2.23}$$

As an example, let us take $a = 6$ mm, $r_{\mathrm{w}} = 0.03$ mm, and $\sigma_0 = 5.8 \times 10^7$ S/m for copper. The resultant plasma frequency may be calculated from eqns (2.20) and (2.23) to be 8.73 GHz, well in the microwave region. For copper at room temperature the time constant is $\tau_{\mathrm{w}} = 2.24 \times 10^{-8}$ s, which makes the factor $\omega\tau_{\mathrm{w}} = 1230$. For these parameters it is large enough to ignore losses.

For a detailed experimental study see Gay-Balmaz *et al.* (2002). With the above choice of $r_{\mathrm{w}} = 0.03$ mm and $a = 6$ mm the transmission (S_{21}) through N layers (each layer consisted of 39 wires clamped in a groove in a substrate) as a function of frequency is plotted in Fig. 2.10(b). The transition occurs at about 9.2 GHz, which is in fairly good agreement with the theoretical value of 8.73 GHz calculated above. The long dashed line in Fig. 2.10(b) is the reference level measured when no wires are present. The curves with dot-dash, dashed, dotted and solid lines correspond to $N = 5$, 10, 15 and 20 layers. It may be seen that 5 layers already cause considerable attenuation but in order to have a sharp transition 15 layers are needed. Increasing the number of layers to 20 makes hardly any difference.[4]

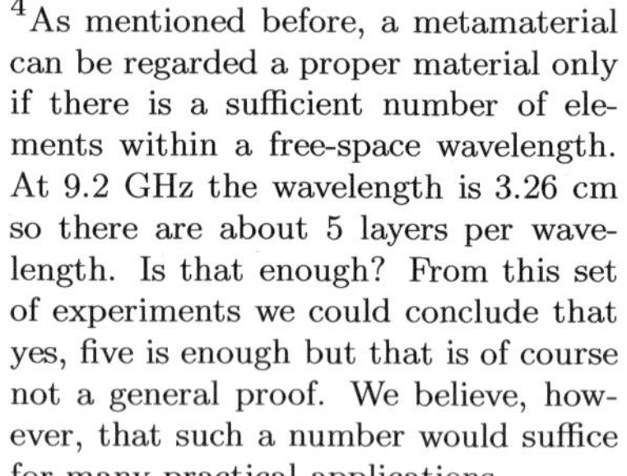
[4]As mentioned before, a metamaterial can be regarded a proper material only if there is a sufficient number of elements within a free-space wavelength. At 9.2 GHz the wavelength is 3.26 cm so there are about 5 layers per wavelength. Is that enough? From this set of experiments we could conclude that yes, five is enough but that is of course not a general proof. We believe, however, that such a number would suffice for many practical applications.

2.5 Resonant elements for metamaterials

The wire elements in the previous section are not resonant. They are useful because they can provide negative dielectric constant at frequencies below the effective plasma frequency, which can be adjusted by choosing the thickness of the wire and the density of the wire mesh. However, most metamaterial elements are resonant and then the problem arises how to make them small. It is not trivial to satisfy the requirement for the elements to be resonant and at the same time to be small relative to the wavelength.

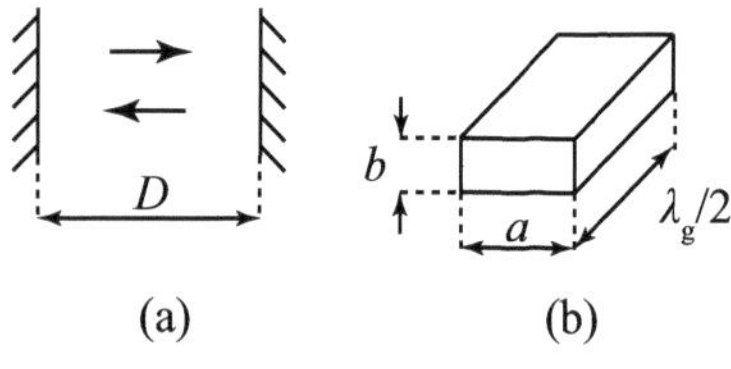

Fig. 2.11 (a) Fabry–Perot resonator. (b) Microwave resonator

When we think of an electromagnetic resonator the one first coming to mind is probably the Fabry–Perot resonator used in lasers. It consists of two parallel mirrors a distance D apart (Fig. 2.11(a)). Resonance occurs when D is equal to an integral (very large) number of wavelengths. It is then easy to imagine a wave trapped between the two mirrors just bouncing back and forth between them. A microwave resonator (Fig. 2.11(b)) with dimensions a, b and λ_{g} may be as small as half a guide wavelength, which still enables the waves to bounce between the metal walls but it is still far too big. If we want a resonator small relative to the wavelength that can be easily realized by lumped circuit elements then all we need is an inductance L and a capacitance C. With a lumped inductance and a lumped capacitance the size of the resonant circuit can be very small relative to the wavelength. Those circuit elements, however, will not do because they cannot easily couple to electric or magnetic fields. One could, however, use a lumped capacitor that can be very compact, and for the inductance a loop that can couple to a magnetic field whether it comes from an incident electromagnetic wave or from currents flowing in neighbouring elements.

Are there other resonant elements in the same category? There are plenty of them starting, probably, with the re-entrant cavity used in klystron amplifiers going back more than half a century (see, e.g., Hansen 1939). We shall present a gallery of small resonators in Chapter 4. For the time being we shall concentrate on one resonator, a member of the family of split-ring resonators (SRRs), which has become very popular in the last decade (Pendry *et al.*, 1999). It consists of two concentric split rings with gaps on opposite sides. Two realizations with small pipes and in printed circuit form are shown in Figs. 2.12(a) and (b). The third one (Fig. 2.12(c)) is the so-called complementary split-ring resonator where metal replaces air and vice versa (see Baena *et al.* 2005*a*).

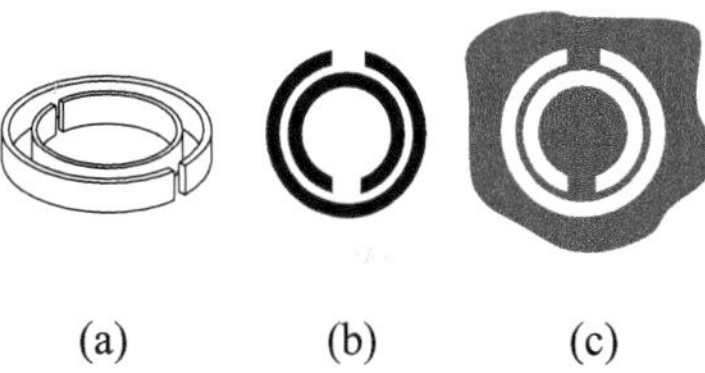

Fig. 2.12 Split-ring resonators (a) as pipes, (b) in printed circuit form, and (c) as a complementary variety

At first sight the physical phenomena governing the operation of a SRR are quite complicated. Each ring has a self-inductance, there is a mutual inductance between them, there is a capacitance between the rings and there are gap capacitances at the splits. If one wants to take into account all these factors then it is difficult indeed to determine its properties. It turns out, however, that a simplified physical picture can lead to an excellent approximation (Marques *et al.*, 2002*c*). First, ignore the gap capacitances on the basis that they are small and they are unlikely to have a major influence on the flow of currents. Secondly, ignore the mutual inductance. In the third place, take the self-inductance equal to the average self-inductance of the two rings. In the fourth place, consider the two inter-ring capacitances between the splits as being connected in series. We may then put these assumptions into mathematical form. Take the average radius of the SRR to be equal to r_0, the average inductance of the two rings equal to L and the inter-ring capacitance per unit length equal to C_{pu}. Then, the capacitance of a half-ring is equal to

$$C_{\mathrm{half-ring}} = \pi r_0 C_{\mathrm{pu}}\,, \tag{2.24}$$

and the total capacitance of the DSDR is equal to

$$C = \frac{1}{2} C_{\mathrm{half-ring}} = \frac{1}{2}\pi r_0 C_{\mathrm{pu}}\,, \tag{2.25}$$

whence the resonant frequency is

$$\omega_0 = \frac{2}{\sqrt{\pi r_0 L C_{\mathrm{pu}}}}\,. \tag{2.26}$$

2.6 Loading the transmission line

As we have seen in Section 1.21, a four-pole can be regarded as a basic unit of a transmission line and it can even represent a plane wave propagating in free space. We have also discussed the loading of the transmission line based on Brillouin's (1953) book.[5] Can we use that

[5] In fact, the benefits of loading a transmission line have been known well before Brillouin's work. A patent filed by Pupin in 1900 stipulated that the transmission of telephone signals can be much improved if the line is periodically loaded by inductances. The patent was bought by American Telephone and Telegraph for \$ 185,000 + royalties. By 1911 this technique enabled AT&T to extend their telephone lines from New York to Denver.

approach in our quest to unravel the properties of metamaterials? Well, we can argue that in the presence of our metamaterial elements the plane wave is not free to propagate. In other words it can only propagate in a jerky manner: it bumps regularly into obstacles presented by the metamaterial elements. It propagates a little, then it meets an element by which it is affected, it propagates further, it meets the next element by which it is again affected, and so on, and so on. We call this the loading of the transmission line or of free space. In order to proceed we need to know two things about the element: what is its circuit equivalent, and how is it coupled to the plane wave?

2.6.1 By resonant magnetic elements in the form of LC circuits

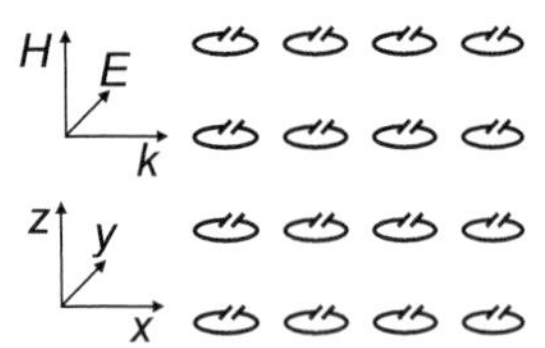

Fig. 2.13 Cubic lattice of capacitively loaded loops

Let us assume a plane wave incident perpendicularly upon a cubic lattice of capacitively loaded loops as shown in Fig. 2.13. The loops are in the xy plane. The propagation direction of the plane wave is along the x axis. The polarization of the magnetic field is in the z direction so that it can interact with the loops. The coupling between the magnetic field and the resonant loop can be represented by a mutual inductance M'.

An approximate value of that can be obtained from the following considerations (Syms *et al.*, 2005*b*). The voltage induced in the resonant element can be expressed as $\mathrm{j}\omega M' I_\mathrm{t}$, where I_t is the current flowing in the transmission line. The same voltage can be expressed by field quantities as $\mathrm{j}\omega\pi r_0^2\mu_0 H_\mathrm{t}$, where H_t is the magnetic field of the electromagnetic wave. Noting further that for the unit cell $H_\mathrm{t} = I_\mathrm{t}/a$ we find

$$M' = \frac{\pi r_0^2 \mu_0}{a} \,. \tag{2.27}$$

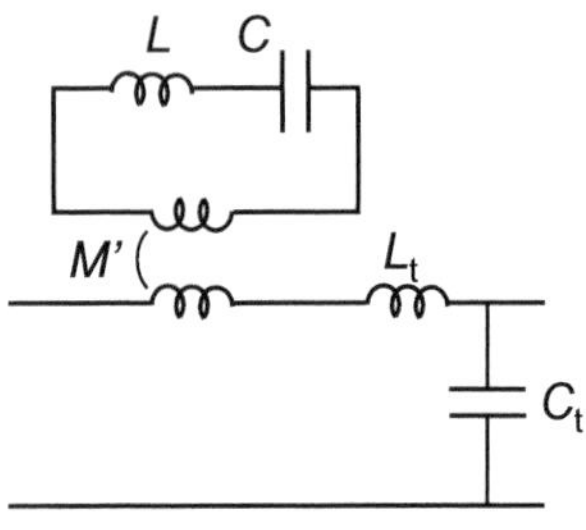

Fig. 2.14 The four-pole equivalent of a transmission line loaded by a resonant loop

The equivalent circuit of the loaded transmission line in terms of four-poles may be seen in Fig. 2.14.

We can find the dispersion equation from there by the method outlined previously. First, we need to find the chain matrix of the 'load', i.e. that of the LC circuit coupled by M' to the transmission line. A little algebra yields

$$\begin{aligned} b_{11}^\mathrm{L} &= 1\,, \quad b_{21}^\mathrm{L} = 0\,, \quad b_{22}^\mathrm{L} = 1\,, \\ b_{12}^\mathrm{L} &= \frac{\mathrm{j}\omega M'^2}{L}\frac{1}{1-\dfrac{\omega_0^2}{\omega^2}}\,, \quad \omega_0^2 = \frac{1}{LC}\,. \end{aligned} \tag{2.28}$$

The elements of the chain matrix for the $L_\mathrm{t}C_\mathrm{t}$ transmission line are given by eqn (1.128). We then need to multiply the chain matrix of the load with the chain matrix of the transmission line, and then take the main diagonal elements to obtain the dispersion equation from eqn (1.141). After a fair amount of algebra we find

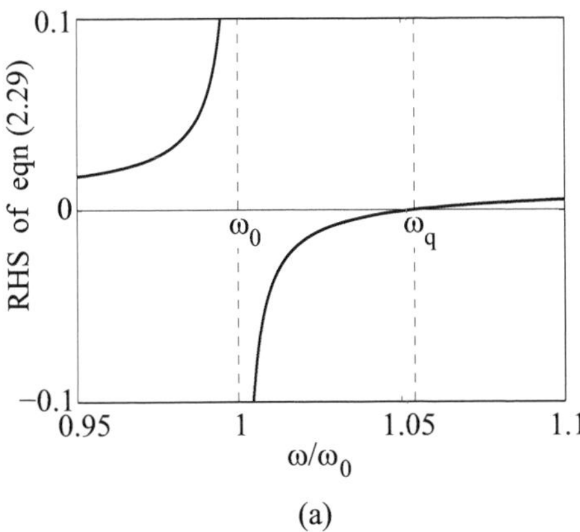

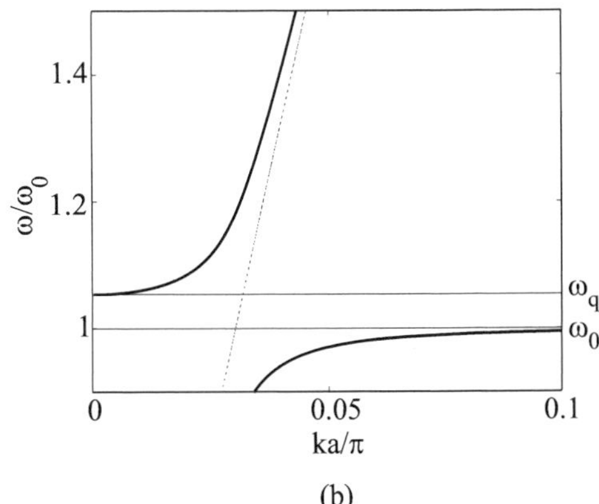

Fig. 2.15 Loading by rings. (a) Plot of the RHS of eqn (2.29) against ω/ω_0 and (b) dispersion curve

$$4\sin^2\frac{ka}{2} = (1-q^2)\frac{\omega^2}{\omega_t^2}\frac{\omega^2-\omega_q^2}{\omega^2-\omega_0^2}, \tag{2.29}$$

where

$$q^2 = \frac{M'^2}{LL_t}, \quad \omega_t^2 = \frac{1}{L_tC_t} \quad \text{and} \quad \omega_q^2 = \frac{\omega_0^2}{1-q^2}. \tag{2.30}$$

The left-hand side can vary only between 0 and 4, corresponding to $ka = 0$ and $ka = \pi$. Hence, there is solution of eqn (2.29) only for those values of the right-hand side that vary within the same limits. For values of $q^2 = 0.1$, and $\omega_0/\omega_t = 0.1$ we plot the right-hand side as a function of ω/ω_0 showing the range for which a solution exists. It may be seen (Fig. 2.15(a)) that with good approximation there is solution between 0 and ω_0 and again above ω_q. There is no solution for ω between ω_0 and ω_q. The corresponding dispersion curve is shown in Fig. 2.15(b). We can say that there are pass bands between 0 and ω_0 and above ω_q up to a frequency comparable with ω_t, and there is a stop band between ω_0 and ω_q.

If we want to regard the medium made up by these resonant magnetic elements as a continuous one then we can replace $\sin(ka/2)$ by $ka/2$. Equation (2.29) then takes the form

$$k^2 = (1-q^2)\frac{\omega^2}{c^2}\frac{\omega^2-\omega_q^2}{\omega^2-\omega_0^2}. \tag{2.31}$$

The main, and obvious, change is that in the continuous limit the separation of the elements, a, no longer appears in the dispersion equation. It is also clear now that the changes from pass band to stop band and vice versa occur exactly at the frequencies ω_0 and ω_q. It may also be seen that the frequency range extends to infinity.

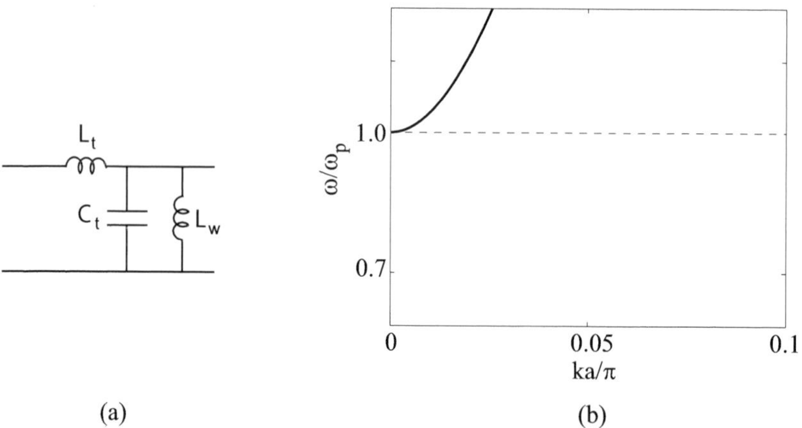

Fig. 2.16 Loading by rods. (a) Four-pole equivalent and (b) dispersion curve

2.6.2 By metallic rods

A thin, short metallic rod of radius r and length l can be characterized by a self-inductance L_w. Its value has already been given in eqn (2.23). Such a rod may interact with the electric field of an input electromagnetic wave. It will shunt it. Hence, the 'load' is here an inductance across the transmission line. The loaded transmission line is shown in Fig. 2.16(a). We may now follow the method of the previous subsection for finding the dispersion equation by multiplying the chain matrices of load and transmission line but it is much simpler just to look at the four-pole of Fig. 2.16(a) and find the elements of the chain matrix from there. The general relationship was already given by eqn (1.128). All we need to notice is that Y is now equal to the parallel combination of the capacitance C_t and the inductance L_w hence

$$Y = \mathrm{j}\omega C_\mathrm{t} + \frac{1}{\mathrm{j}\omega L_\mathrm{w}} = \mathrm{j}\omega C_\mathrm{t}\left(1 - \frac{\omega_\mathrm{p}^2}{\omega^2}\right)\,; \qquad \omega_\mathrm{p} = \frac{1}{\sqrt{L_\mathrm{w} C_\mathrm{t}}}\,. \tag{2.32}$$

With $Z = \mathrm{j}\omega L_\mathrm{t}$ remaining unchanged we can find the main diagonal elements of this chain matrix, and from that the dispersion equation in the form

$$4\sin^2\frac{ka}{2} = \frac{\omega^2 - \omega_\mathrm{p}^2}{\omega_\mathrm{t}^2}\,. \tag{2.33}$$

Clearly there is no solution when $\omega < \omega_\mathrm{p}$ but a solution exists up to the frequency, $\omega = \sqrt{4\omega_\mathrm{t}^2 + \omega_\mathrm{p}^2}$. The corresponding dispersion curve is shown in Fig. 2.16(b).

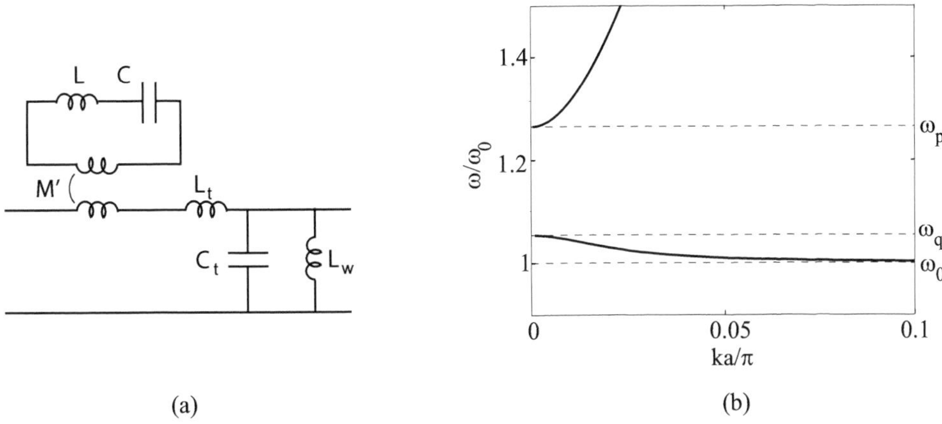

Fig. 2.17 Loading by resonant loops and rods. (a) Four-pole equivalent. (b) Dispersion curve

2.6.3 By a combination of resonant magnetic elements and metallic rods

We shall now combine into one circuit both the magnetic element and the metallic rod as shown in Fig. 2.17(a). The technique for finding the elements of the chain matrix is still the same. The chain matrix of the magnetic load should be multiplied with the chain matrix of the rod-loaded transmission line. The resultant equation is

$$4\sin^2\frac{ka}{2} = (1-q^2)\frac{\omega^2-\omega_{\mathrm{p}}^2}{\omega_{\mathrm{t}}^2}\frac{\omega^2-\omega_{\mathrm{q}}^2}{\omega^2-\omega_0^2}\,. \tag{2.34}$$

The above equation is very similar to eqn (2.29). The only difference is that in the first term ω^2 should be replaced by $\omega^2-\omega_{\mathrm{p}}^2$. What kind of dispersion characteristics could we expect from eqn (2.34)? It will depend on the relative values of the three characteristic frequencies, ω_{p}, ω_0, and ω_{q}. We know that ω_{q} is larger than ω_0 but ω_{p} could be anywhere depending on the parameters of the rods chosen. Let us choose it to be larger than ω_{q}. Then, for $\omega<\omega_0$ we have a stop band, for $\omega_0<\omega<\omega_{\mathrm{q}}$ we have a pass band that happens to be a backward-wave region, for $\omega_{\mathrm{q}}<\omega<\omega_{\mathrm{p}}$ we have a stop band, and finally for $\omega>\omega_{\mathrm{p}}$ we have again a pass band. The dispersion curve for $\omega_{\mathrm{p}}=1.2\,\omega_{\mathrm{q}}$ is plotted in Fig. 2.17(b).

The appearance of pass and stop bands is summarized in the diagram of Fig. 2.18 for the three cases: Rods alone, resonant magnetic elements alone, rods and magnetic elements combined. Let us just concentrate our attention on the frequency region between ω_0 and ω_{q} in the three cases. In case 1 that frequency region is a stop band. In case 2 it is again a stop band, but in case 3 it is a pass band. This is a significant conclusion. For rods alone or for magnetic elements alone it is a stop band, but when the two are combined the stop band turns into a pass band. The rules are quite easy and quite interesting. Two pass bands

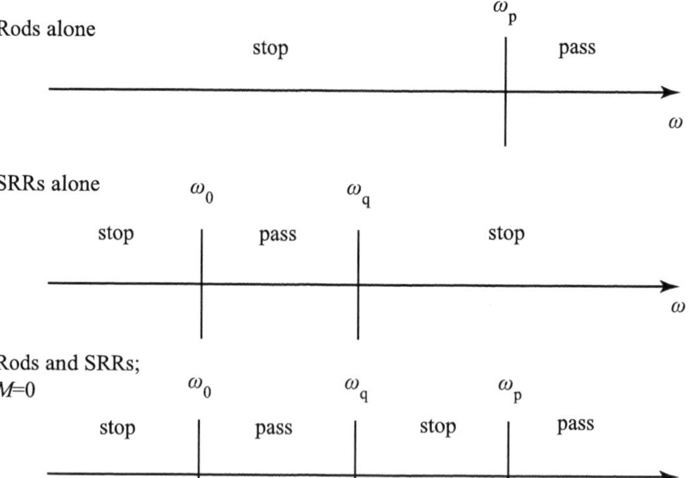

Fig. 2.18 Pass and stop bands for rods, rings and a combination of both. From Syms *et al.* (2005*a*). Copyright © 2005 American Institute of Physics

make a pass band, a pass band and a stop band make a stop band, and two stop bands make a pass band. According to these rules we can assign +1 to a pass band and −1 to a stop band. Then $1 \times 1 = 1$, $1 \times (-1) = -1$, $(-1) \times 1 = -1$ and $(-1) \times (-1) = 1$. These are in fact the same rules that will arise later when discussing negative-index materials.

2.7 Polarizability of a current-carrying resonant loop: radiation damping

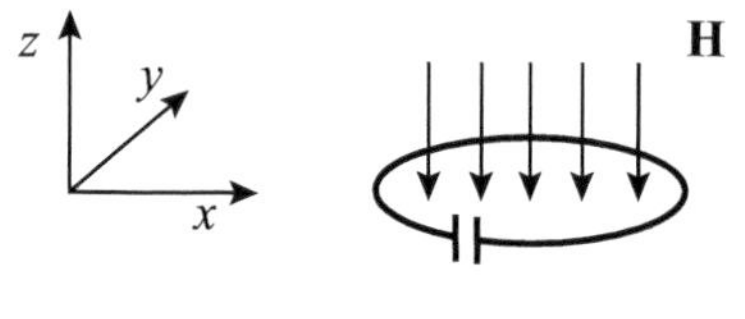

Fig. 2.19 Resonant loop in a magnetic field

We shall now find the magnetic polarizability in the simple case of a loop. More precisely we shall look for the relationship between the z component of a spatially constant magnetic field and the induced magnetic moment when the loop is in the xy plane (Fig. 2.19).

It is a simple problem because the magnetic field does not change over the area of the loop and we also know from the geometry that the induced magnetic dipole moment will have a z component only. We can find it by first determining the flux threading the loop, equal to $\mu_0 SH$, then the voltage excited in the loop as being $-\mathrm{j}\omega\mu_0 SH$, whence the current is

$$I = -\frac{\mathrm{j}\omega\mu_0 SH}{Z}, \tag{2.35}$$

where Z is the impedance of the resonant element,

$$Z = \mathrm{j}\omega L + \frac{1}{\mathrm{j}\omega C} + R, \tag{2.36}$$

and losses in the form of a resistance have also been taken into account. The induced magnetic moment is then (see eqn (1.18))

$$m = \mu_0 S I = -\frac{\mathrm{j}\,\omega \mu_0^2 S^2 H}{Z}, \tag{2.37}$$

whence according to the definition of eqn (1.100)

$$\alpha_{\mathrm{m}} = -\frac{\mathrm{j}\,\omega \mu_0^2 S^2}{Z}. \tag{2.38}$$

If we want to think in more general terms then we can regard the above polarizability as the $\alpha_{zz}^{\mathrm{mm}}$ component of the polarizability tensor relating the z component of the magnetic dipole moment to the magnetic field applied in the z direction.

Note that in the absence of losses α_{m} is real. It becomes complex when R is added to the impedance. However, this is not the only source of loss. Power can turn not only into heat but it can also be lost by radiation. In Section 1.14 we asked the question whether we can take into account the radiated power by assigning to the element a resistance that we called the radiation resistance. The derivation was done for an electric dipole but the value of the radiation resistance for a magnetic dipole was also quoted (see eqn (1.96)). A small loop is of course equivalent to a magnetic dipole, hence we can take care of the radiation loss by adding

$$R_{\mathrm{s}} = \frac{\pi}{6}\eta_0 \left(\frac{2\pi r_0}{\lambda}\right)^4 \tag{2.39}$$

to the ohmic resistance. For most metamaterials we do not need to consider the radiation resistance because all dimensions are small relative to the wavelength. But when the length of the line is larger than the free-space wavelength, as will be the case in Section 8.2, then radiation effects must be taken into account, which we can do by adding the radiation resistance to the ohmic resistance. Under these conditions it is preferable to work in terms of the inverse of magnetic polarizability, which will then take the form

$$\frac{1}{\alpha_{\mathrm{m}}} = \left(\frac{1}{\alpha_{\mathrm{m}}}\right)_{\text{lossless}} + \left(\frac{1}{\alpha_{\mathrm{m}}}\right)_{\text{ohmic loss}} + \left(\frac{1}{\alpha_{\mathrm{m}}}\right)_{\text{radiation loss}}, \tag{2.40}$$

where

$$\left(\frac{1}{\alpha_{\mathrm{m}}}\right)_{\text{lossless}} = -\frac{f_{\mathrm{r}} L}{\mu_0^2 S^2}; \qquad f_{\mathrm{r}} = 1 - \frac{\omega_0^2}{\omega^2}, \tag{2.41}$$

$$\left(\frac{1}{\alpha_{\mathrm{m}}}\right)_{\text{ohmic loss}} = -\mathrm{j}\frac{L}{Q\mu_0^2 S^2}; \qquad Q = \frac{\omega L}{R}, \tag{2.42}$$

where Q is the quality factor. For the third term a little algebra will yield

$$\left(\frac{1}{\alpha_{\mathrm{m}}}\right)_{\text{radiation loss}} = \mathrm{j}\frac{k_0^3}{6\pi\mu_0}. \tag{2.43}$$

The above expression is called radiation damping by physicists. It is the physicists' measure of the radiation due to a magnetic dipole. Antenna engineers don't like it and don't use it. It is one of the examples when physicists and engineers look differently at the same phenomenon.

For completeness, we shall give below the expression for the radiation damping associated with an electric dipole, which can be determined in an analogous manner,

$$\left(\frac{1}{\alpha_{\mathrm{e}}}\right)_{\text{radiation loss}} = \mathrm{j}\,\frac{k_0^3}{6\pi\varepsilon_0}\,. \tag{2.44}$$

2.8 Effective permeability

We have already found effective permittivities in Section 2.3. We shall now make an attempt to find the effective permeability in some simple cases. In our first attempt we make the assumption that the local field is equal to the externally applied field, i.e. we completely disregard the effect of all the other elements. We assume now a cubic lattice of loops as shown in Fig. 2.13. The magnetic field of the incident plane wave is in the z direction and the loops are in the xy plane. Then, the magnetization due to the effect of the incident field upon the elements is

$$M_{\mathrm{m}} = Nm = N\alpha_{\mathrm{m}}H\,. \tag{2.45}$$

We shall now find the effective relative permittivity of the medium perpendicular to the plane of the element (according to the notations of Section 1.16 this is the μ_{zz} component of the permeability tensor) from the definition

$$\mu_{\mathrm{r}} = \frac{B}{\mu_0 H} = \frac{\mu_0 H + M_{\mathrm{m}}}{\mu_0 H} = 1 + \frac{M_{\mathrm{m}}}{\mu_0 H}\,. \tag{2.46}$$

With the aid of eqn (2.45), and the definition of magnetic polarizability, eqn (2.38), we find

$$\mu_{\mathrm{r}} = 1 - \frac{\mu_0 N S^2}{L\left(f_{\mathrm{r}} - \dfrac{\mathrm{j}}{Q}\right)} \tag{2.47}$$

or

$$\mu_{\mathrm{r}} = 1 - \frac{F}{f_{\mathrm{r}} - \dfrac{\mathrm{j}}{Q}}\,, \tag{2.48}$$

which is often regarded as the standard form in the literature. F is defined as

$$F = \frac{\mu_0 N S^2}{L}\,. \tag{2.49}$$

Another form of eqn (2.47) for the lossless case may be obtained with a little algebra as

$$\mu_r = (1 - F)\frac{\omega^2 - \omega_F^2}{\omega^2 - \omega_0^2}, \qquad (2.50)$$

where $\omega_F = \omega_0/\sqrt{1-F}$. It may be seen from eqn (2.50) that in the absence of losses there is a pole at ω_0 and a zero at ω_F. The variation of μ_r with frequency is plotted schematically in Fig. 2.20.

The interesting thing is that between the pole and the zero the permeability is negative. How wide is the range of negative permeability? It can be easily calculated when F is small. Then,

$$\Delta\omega_{\text{neg.perm.}} = \omega_0 \left(\frac{1}{\sqrt{1-F}} - 1\right) \simeq \omega_0 \frac{F}{2}. \qquad (2.51)$$

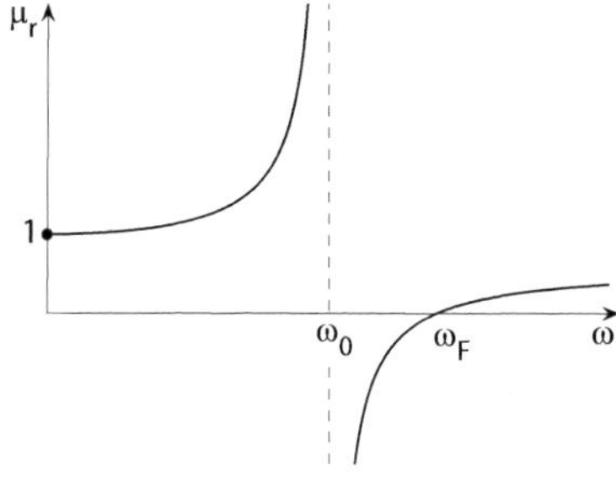

Fig. 2.20 Frequency variation of permeability

In our second model we shall take into account that all the other elements will also contribute to the flux at element n. The total flux is then due to the applied field plus the flux provided by all the other elements,

$$\Phi = \mu_0 S H + I \sum M_{nn'}, \qquad (2.52)$$

where $M_{nn'}$ is the mutual inductance between element n and n' and the current I is assumed to be the same in all the elements. The corresponding current must then satisfy the equation

$$I = \frac{-\mathrm{j}\omega}{Z}\left(\mu_0 S H + I \sum M_{nn'}\right), \qquad (2.53)$$

whence we may determine the current, from the current the magnetic dipole density, and from that the effective permeability. We obtain finally

$$\mu_r = 1 - \frac{F}{f_r + \Delta f_r - \dfrac{\mathrm{j}}{Q}}, \qquad (2.54)$$

where

$$\Delta f_r = \frac{1}{L}\sum M_{nn'}. \qquad (2.55)$$

The difference between our first and second model is that a new term, Δf_r, enters the denominator that involves all the mutual inductances. How can we find that term for a cubic lattice? We have met this problem in a somewhat different form in Section 2.3. There, we were concerned with the total electric field at an element due to all other elements. That sum was shown to be zero in Appendix B. The same applies here. If the total magnetic field is zero then the mutual inductance must also be zero. For other lattice configurations the additional term represents a small shift in the position of the pole where the effective permeability tends to infinity.

Our third model for finding the effective permeability is based on eqn (2.31) that gives the dispersion equation for loading by magnetic elements in the continuous limit. Clearly, the dispersion equation and the effective permeability are closely connected to each other. For a plane wave propagating in a medium with a relative permeability μ_r the relationship between ω and k (see Section 1.4) is

$$k = \frac{\omega\sqrt{\mu_r}}{c}. \tag{2.56}$$

The corresponding relationship in the continuous approximation between ω and k for a magnetic element yields

$$\mu_r = (1 - q^2)\frac{\omega^2 - \omega_q^2}{\omega^2 - \omega_0^2}, \tag{2.57}$$

which is of the same form as eqn (2.50) but q^2 appears instead of F. The two expressions would be exactly the same if F were to agree with q^2. In fact, if we look at eqns (2.27), (2.30) and (2.49) we shall find that they do agree. Hence, our third model gives the same result as the first one.

Our fourth model is the oldest of them all, the one that has been applied to finding the material constants for natural materials for well over a century, the Clausius–Mossotti model. We derived it in Section 2.3 for the effective permittivity, but the same equation is valid of course in the magnetic case. We only need to substitute magnetic polarization for electric polarization in eqn (2.14). We find for the lossless case

$$\mu_r = \frac{\omega^2\left(1 - \dfrac{2F}{3}\right) - \omega_0^2}{\omega^2\left(1 + \dfrac{F}{3}\right) - \omega_0^2}. \tag{2.58}$$

It may be seen that the shape of the μ_r curve has hardly changed but the region of negative permeability has shifted towards lower frequencies. The pole has moved from ω_0 to $\omega_0/\sqrt{1+F/3}$ and the zero from $\omega_0/\sqrt{1-F}$ to $\omega_0/\sqrt{1+2F/3}$. It follows from the above relations that for small values of F the width of the negative permeability region has not changed. It is $\omega_0 F/2$.

The fifth model is that of Gorkunov *et al.* (2002) who essentially repeat the derivation of the Clausius–Mossotti equation but include the effect of mutual inductances. Then, as we may guess from our second model the additional term Δf_r appears. To be exact we need to add Δf_r both to $2F/3$ and to $F/3$ in eqn (2.58).

Losses. Although losses were taken into account in deriving eqn (2.48) we disregarded them later for simplicity. In the general case when μ_r is complex we use the notation

$$\mu_r = \mu' - \mathrm{j}\,\mu''. \tag{2.59}$$

This is plotted in Fig. 2.21 in the vicinity of the resonant frequency for $Q = 100$, 1000 and 10 000 from eqn (2.48) with $F = 0.1$. As may be expected, the real part of the permeability no longer tends to infinity and the maximum μ' achievable is reduced to 2, 3.6 and 6.1 for the three values of the quality factor. The imaginary part of μ_r can be quite large. Its maximum may be larger than that of the real part.

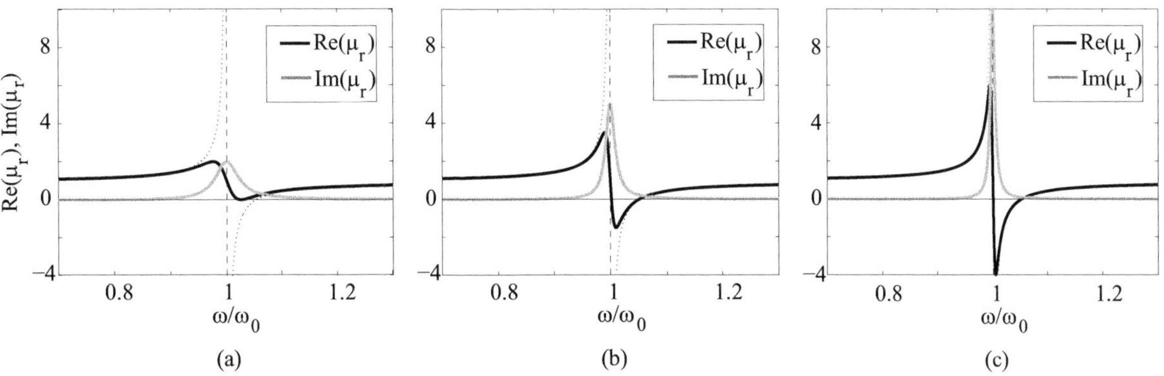

Fig. 2.21 Frequency variation of the real and imaginary parts of permeability for (a) $Q = 100$, (b) $Q = 1000$, (c) $Q = 10\,000$

2.9 Dispersion equation of magnetoinductive waves derived in terms of dipole interactions

We shall now look again at the linear array of loops shown schematically in Fig. 2.6, but assume this time that there is an infinite number of elements, and derive once more the dispersion equation in a different manner. It is more general in the sense that the interactions are not limited to nearest neighbours but less general in another sense that our loops will be replaced by magnetic dipoles, true only if the loops are small enough and two loops are not too close to each other. This is a technique used by theoretical physicists who like to start with simple concepts so that they can later add to them lots of complications of their own liking.

From the definition of the magnetic moment and magnetic polarizability, as applied to element 0, we may write

$$m_0 = \alpha_{\rm m} H\,, \tag{2.60}$$

where H comes from all the other dipoles. The element positioned at a distance na away for $n > 0$ will give a contribution

$$(H_n)_{\rm r} = m_n f(n)\,, \tag{2.61}$$

where

$$f(n) = \frac{1 + {\rm j}\, k_0 na}{2\pi\mu_0 n^3 a^3}\, {\rm e}^{-{\rm j}\, k_0 na}\,. \tag{2.62}$$

Hence, the contribution of all the elements from $n = 1$ to ∞ will amount to

$$H_{\rm total} = H(n > 0) + H(n < 0) = \sum (m_n + m_{-n})\, f(n)\,. \tag{2.63}$$

Next comes the wave assumption

$$m_n = m_0 \, \mathrm{e}^{-\mathrm{j}nka}, \tag{2.64}$$

which, substituted into eqn (2.63), leads to the equation

$$\frac{1}{\alpha_\mathrm{m}} = \frac{1}{\pi\mu_0 a^3}\sum_{n=1}^{\infty}\cos(kna)\frac{(1+\mathrm{j}\,k_0 na)\,\mathrm{e}^{-jk_0 na}}{n^3}. \tag{2.65}$$

The right-hand side of eqn (2.65) is known as the interaction function. We shall denote it by I_F. Hence, the general form of the dispersion equation is

$$\alpha_\mathrm{m} I_\mathrm{F} = 1. \tag{2.66}$$

In what sense is eqn (2.66) a dispersion equation? In the same sense as the previous dispersion equations. I_F depends on ω and k and the polarizability depends on ω, hence eqn (2.66) relates ω and k to each other.

This is now the third time that we derive a dispersion equation for a set of resonant loops. The derivation in Section 1.23 was based purely on circuit concepts, whereas the one in Section 2.2 was based on nearest-neighbour coupling and Kirchhoff's law. It was also assumed there that all dimensions are small relative to the free-space wavelength. Hence, if we want eqn (2.66) to reduce to eqn (2.4) we have to take the lossless case, assume that $k_0 a \ll 1$ and secondly that the infinite sum should be reduced to $n = 1$. We then obtain

$$L\left(1-\frac{\omega_0^2}{\omega^2}\right) + \mu_0\frac{\pi r_0^4}{a^3}\cos ka = 0. \tag{2.67}$$

Now we are nearly there. Remember that in the derivation used in the present section we assumed magnetic dipoles instead of loops and we found the magnetic field due to a loop in the dipole approximation. Thus, if we further note that in this approximation the mutual inductance between two loops a distance a apart in the axial configuration is[6]

$$M = \frac{\mu_0\pi r_0^4}{2a^3}, \tag{2.69}$$

then eqns (2.67) and (2.4) may be seen to be identical.

[6] For nearest neighbours and small wavelength the radial magnetic field takes the form

$$H = \frac{m}{2\pi\mu_0 a^3}. \tag{2.68}$$

Considering further that $m = \mu_0\pi r_0^2 I$, that the flux Φ is $\mu_0 H\pi r_0^2$ and the mutual inductance is defined as $M = \Phi/I$ we can find M in the form of eqn (2.69).

2.10 Backward waves and negative refraction

Backward waves, as mentioned before, have phase and group velocities in opposite directions. Does it make a difference whether the waves are forward or backward when it comes to refraction at a boundary? Let's start with two isotropic media 1 and 2, both of which can support

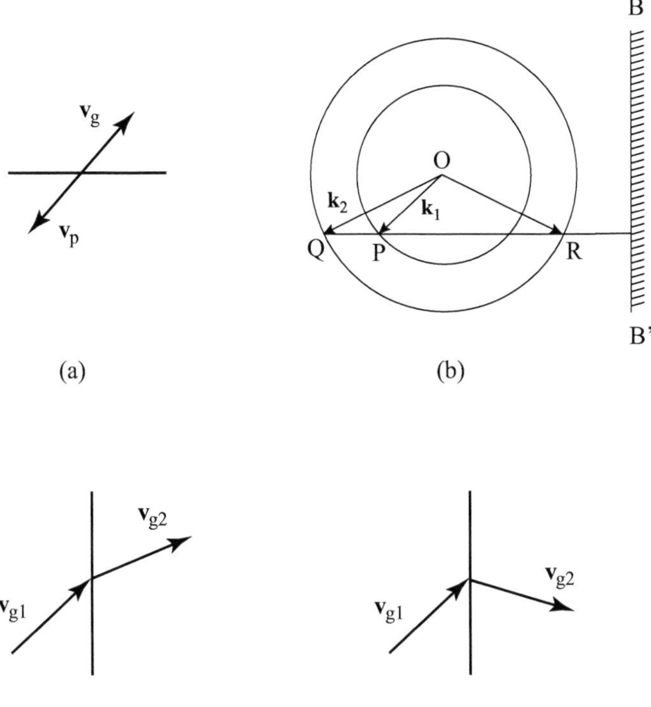

Fig. 2.22 Refraction problem. (a) In a backward wave phase and group velocities are opposed to each other. (b) Ewald circle for two media with $k_2 > k_1$. (c) Refraction of a backward wave into a backward wave. (d) Refraction of a backward wave into a forward wave

backward waves at a particular frequency, and assume that the wave number k_2 is larger than k_1. If wave 1 is incident upon the boundary then its group velocity must be in the first quarter, and, consequently, its phase velocity will be in the third quarter, as shown in Fig. 2.22(a). Next, we draw the Ewald circle with radii k_1 and k_2 in Fig. 2.22(b) in the same way as in Section 1.6. The wave vector of the input wave is $k_1 = \text{OP}$. If we want to satisfy the boundary condition that the phase velocities along the boundaries (denoted by BB$'$ in Fig. 2.22(b)) match then the possible wave vectors are either OQ or OR. However, the group velocity of the wave in medium 2 must point away from the boundary, i.e. it must be in the first quarter. Hence, the only solution is $k_2 = \text{OQ}$. If we look at the direction of the group velocities at the boundary (Fig. 2.22(c)) then we can see that nothing interesting has happened. The refraction of backward waves is of the same kind as the refraction of forward waves.

What about media that support different types of waves, say the wave in medium 1 is a backward wave but the wave in medium 2 is a forward wave. If medium 1 is the backward-wave medium then the Ewald circle construction of Fig. 2.22(b) is still valid: OP is still the wave vector in medium 1 and OQ and OR are still the two possibilities for k_2 that satisfy the boundary condition. The difference is that the wave in medium 2 is a forward wave therefore the group velocity is in the same direction as OR and that is now the only solution. The group velocities for this case are shown in Fig. 2.22(d). Now there is something interesting! In contrast to the usual case, taught in all science courses, the refraction is in the negative direction. In current terminology we talk about negative refraction.

The possibility of negative refraction has been known for a very long time (see the historical review in Chapter 10) but it became a subject of intense study only quite recently after Smith *et al.* (2000) discovered Veselago's paper on negative refractive index.

2.11 Negative-index materials

2.11.1 Do they exist?

The possibility of negative-index materials was broached by Veselago in a paper written in Russian in 1967 and published in English in 1968. It lay dormant for many years, until Smith *et al.* (2000) discovered it.

We have seen (eqn (1.30)) earlier the expression for the refractive index in terms of the relative permittivity and permeability,

$$n = \sqrt{\varepsilon_r \mu_r}\,. \tag{2.70}$$

When both μ_r and ε_r are positive then there is no problem, everything is familiar apart from the cry of those experienced in optical phenomena who will very likely ask what μ_r is doing in that equation. That objection can be easily overcome by asserting that the relative permeability may indeed be different from unity for metamaterials. What happens when only ε is negative? That is well known. That happens in plasmas, we talked about this in Sections 1.9 and 2.4. But Veselago asked a more daring question: What happens when both μ_r and ε_r are negative? He discusses the possible responses as quoted below:

'The situation can be interpreted in various ways. First, we may admit that the properties of a substance are actually not affected by a simultaneous change of the signs of ε and μ. Second, it might be that for ε and μ to be simultaneously negative contradicts some fundamental law of nature, and therefore no substance with $\varepsilon < 0$ and $\mu < 0$ can exist. Finally, it could be admitted that substances with negative ε and μ have some properties different from those of substances with positive ε and μ.'

Veselago then puts forward the third explanation. He goes on to show the consequences of negative material constants straight from Maxwell's equations. Assuming a plane wave propagating in a medium with material constants ε and μ in the form $\exp(-\mathrm{j}\,\mathbf{k}\cdot\mathbf{r})$ eqns (1.6) and (1.7) take the form

$$\mathbf{k} \times \mathbf{H} = \omega\varepsilon\mathbf{E} \qquad \text{and} \qquad \mathbf{k} \times \mathbf{E} = -\omega\mu\mathbf{H}\,. \tag{2.71}$$

It may be seen from the above equations that it makes a difference whether the material constants are both positive or both negative. In the former case the vectors $\mathbf{E}$, $\mathbf{H}$ and $\mathbf{k}$ constitute a right-handed set, whereas for negative ε and μ we have a left-handed set. The wave vector **k** tells us the direction of the phase velocity, the Poynting vector tells us the direction of the group velocity. If the two are in opposite directions we have a backward-wave material with all that implies. Thus, negative refraction at the boundary of two materials, one having positive material constants and the other negative ones, follows immediately. But there is an alternative explanation. We may argue that the square root in eqn (2.70) can be positive or negative. It is sensible to take it positive when the material constants are both positive and take it negative when both material constants are negative. But that will have an influence on Snell's law (eqn (1.46)),

$$n_1 \sin\theta_1 = n_2 \sin\theta_2 \,. \tag{2.72}$$

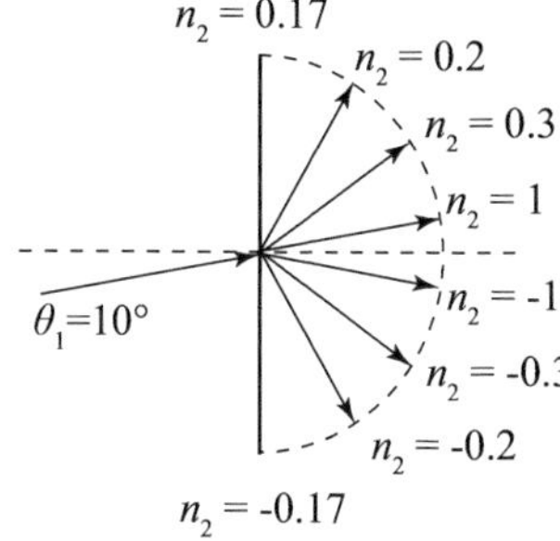

Fig. 2.23 Refraction at a boundary for various values of n_2

Let us take now medium 1 as free space, $n_1 = 1$, and see the direction of the refracted wave (Fig. 2.23) when medium 2 has a large range of refraction indices, which may be smaller than 1 and may take any negative value. The angle of refraction is 90° when $n_2 = \sin\theta_1$ (if n_2 is even smaller then total internal reflection occurs in medium 1). As n_2 increases from this value below unity up to infinity the refracted angle declines from 90° to 0°. Note that the angle of refraction is the same for $n_2 = -\infty$ as for $n_2 = \infty$. Now, as n increases from minus infinity to $-\sin\theta_1$ the angle of refraction declines from 0° to −90°. If n_2 is between $-\sin\theta_1$ and 0 then there is again total internal reflection. Clearly, negative n_2 implies negative refraction. It may be worth reiterating at this stage that we got negative refraction in two different ways. In Section 2.10 from the properties of backward waves and in the present section from the concept of negative refractive index.

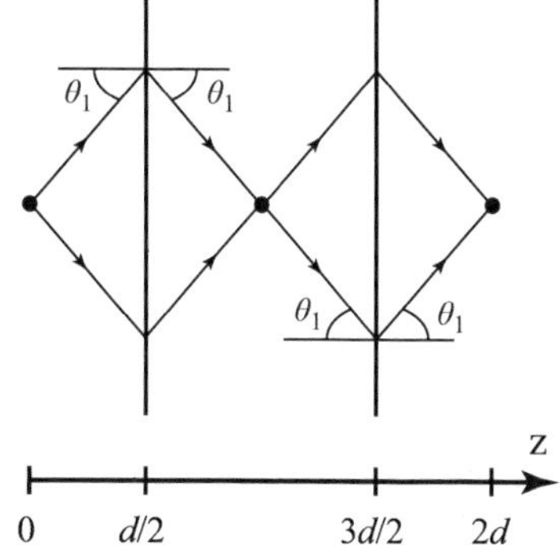

Fig. 2.24 Negative refraction for $n = -1$ (Veselago's flat lens)

The most striking example of what we can do with a negative-index material is Veselago's flat lens. For $n = -1$ the angle of refraction is equal to the negative angle of incidence hence all the rays emanating from a line source will be refocused inside the material and brought to another focus outside the material as shown in Fig. 2.24. The lens thickness is d. A point source at a distance $d/2$ in front of the lens is reproduced at a distance $d/2$ behind the lens.

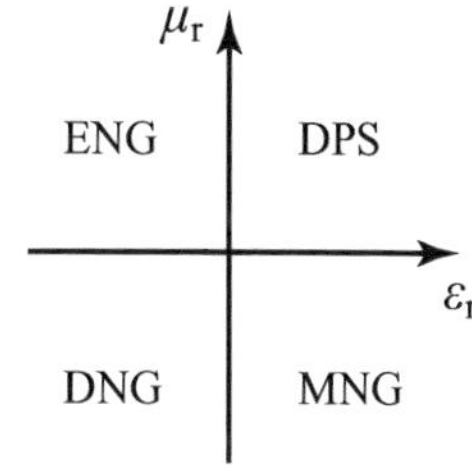

Fig. 2.25 Terminology for materials with various signs of permittivity and permeability after Engheta *et al.* (2005)

2.11.2 Terminology

Having realized that negative material constants may be interpreted as having negative refractive index, and that the consequence is a left-handed **E**, **H**, **k** relationship, Veselago called these materials left-handed. In the many papers that followed Smith *et al.*'s (2000) rediscovery quite a number accepted this terminology and referred to these materials as left-handed media (LHM) and to those with positive material constants as right-handed media or (RHM). There is, however, a vocal minority unhappy with this description. They argue that left-handedness and right-handedness had been terms widely used before in chiral materials referring to the direction of chirality.

A different terminology related to signs of permittivity and permeability was proposed by Engheta *et al.* (2005) and may be seen in Fig. 2.25. Materials with both constants positive are double-positive (DPS) materials, with only ε negative they are ENG materials, with only ε positive they are MNG materials and with both of them negative they are double-negative (DNG) materials. Also, ENG and MNG materials are sometimes referred to as SNG or single-negative materials.

Other terms used are NRI for negative refractive index media, NIM for negative-index media. The one we prefer is BW (Lindell *et al.*, 2001) meaning backward-wave media, which is a generic term. All negative-index media are backward-wave media, but there are plenty of backward-wave media that, mainly because they are one-dimensional structures, do not qualify as negative-index media.

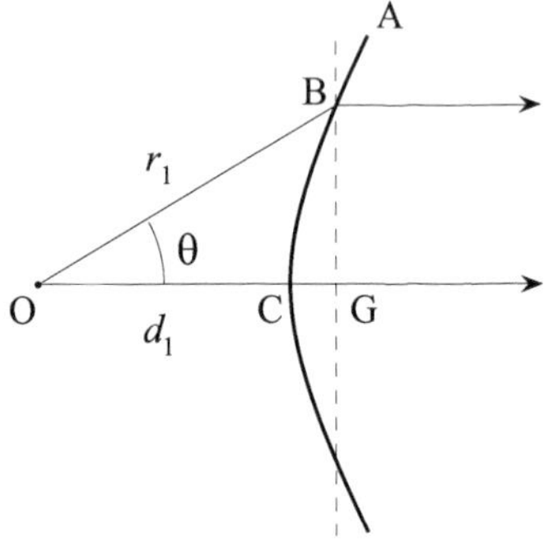

Fig. 2.26 Ray picture for a lens

2.11.3 Negative-index lenses

The possibility of negative-index materials can lead to the design of new families of lenses. The design procedure is the same as in Section 1.8 and the same equation applies. We shall reproduce Fig. 1.6(a) in Fig. 2.26 and eqn (1.62) in the equation below in normalized Cartesian coordinates:

$$\left(x - \frac{n}{n+1}\right)^2 - \frac{y^2}{n^2-1} = \frac{1}{(n+1)^2}\,. \tag{2.73}$$

Let us now see what the contours are if n can take any value from $n = -\infty$ to ∞. When n is very large the hyperbola is close to the straight line $x = d$ (see Fig. 2.27(a) for $n = 10$). For $n = 1.5$ the hyperbola is shown in Fig. 2.27(b). As $n \to 1$ the hyperbola exists for smaller and smaller angular regions (see Fig. 2.27(c) for $n = 1.05$) degenerating eventually into the $y = 0$ line. What happens when n becomes less than unity? It means that the phase velocity of the wave is larger than the velocity of light. That is fine, that does not violate any physical principle. In fact, such a lens was realized in the 1960s by Kock (1964) using hollow metallic waveguides in which the phase velocity can be arbitrarily high. But let us continue our quest. As n becomes less than unity there is no longer an asymptote, the hyperbola turns into an ellipse, as may be seen in Fig. 2.27(d) for $n = 0.95$, the contour still being quite close to the $y = 0$ line. As n declines further towards zero the curvature of the ellipse declines and the contour tends towards a circle (see Fig. 2.27(e) for $n = 0.01$). As n becomes negative nothing radical happens. The contour for $n = -0.01$ is hardly different from that at $n = 0.01$, only the ellipse becomes elongated in the y direction. At $n = -1$ the contour switches back to a hyperbola (see Fig. 2.27(f)) but the asymptote occurs at $\theta > \pi/2$, so it is irrelevant for the shape of the contour. As n decreases further towards $-\infty$ the hyperbola becomes flatter (see Fig. 2.27(g) for $n = -10$) finally reaching the $x = 1$ line, this time from the left. As we have seen in Fig. 2.23 it makes no difference whether the index of refraction is positive infinity or negative infinity.

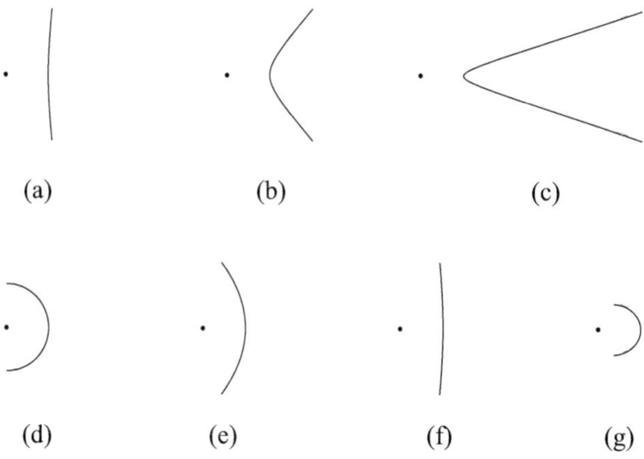

Fig. 2.27 Contours of negative-index lenses. $n = 10,\ 1.5,\ 1.05,\ 0.95,\ 0.01,\ -1,\ -10$(a)–(g)

With the advent of negative-index materials new possibilities were born for the design of lenses. In fact, such lenses may have superior properties (smaller aberrations) to positive-index lenses, as pointed out by Schurig and Smith (2004).

2.11.4 The flat-lens family

We have seen in the previous section that the contour of a lens may tend towards a straight line as the index of refraction tends to infinity. In the limit that would be a flat lens but it would be of little use. If nothing else high refractive index leads to high reflection, which makes it unsuitable for practical applications. However, a flat lens working on somewhat different principles was proposed by Veselago (1968). He showed that for a particular geometry a flat lens with $n = -1$ will bring a diverging beam to a converging beam, as shown in Fig. 2.24. The new feature was the presence of an internal focus. Note that the optical path from the external focus to the internal focus is zero.

We may now ask the question whether this lens is unique, whether it is the only example of a lens with an internal focus? There is no reason in principle why one should not be able to realize an internal focus with a positive refractive index but the lens would be very thick. If we want a reasonably thin lens we need to design another set of lenses that this time will have an internal focal point.

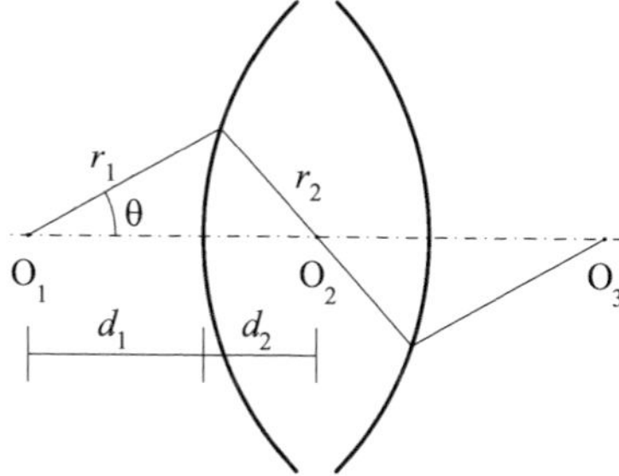

Fig. 2.28 Ray picture of a lens having an internal focus

The variation of the required contour with angle can be found with the aid of the sketch in Fig. 2.28 where both contours are shown. Due to symmetry, it is sufficient to look at the problem between the points O_1 and O_2 representing the two foci. The equality of optical paths demands that

$$r_1 + nr_2 = d_1 + nd_2\,. \tag{2.74}$$

A second relationship between r_1 and r_2 follows from the geometry

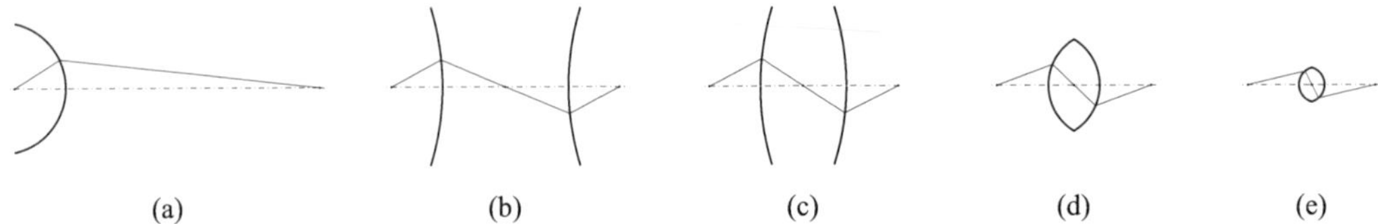

Fig. 2.29 Contours of lenses with internal foci. $n = -0.2, -0.8, -1.2, -2, -4$ (a)–(e)

$$r_2^2 = r_1^2 + (d_1 + d_2)^2 - 2r_1(d_1 + d_2)\cos\theta\,. \tag{2.75}$$

If we want to have the same type of lens as Veselago's flat lens then we should assume that the optical path from focus to focus is zero, i.e.

$$d_1 + nd_2 = 0\,. \tag{2.76}$$

Then, the equation in polar co-ordinates reduces to

$$(n+1)\left(\frac{r_1}{d_1}\right)^2 - 2n\cos\theta\left(\frac{r_1}{d_1}\right) + (n-1) = 0\,. \tag{2.77}$$

It may again be converted into normalized Cartesian co-ordinates yielding the equation

$$\left(x - \frac{n}{n+1}\right)^2 + y^2 = \frac{1}{(n+1)^2}\,, \tag{2.78}$$

which is clearly that of a circle centred at $(n/(n+1), 0)$. In the limit of $n \to -1$ it becomes a straight line, i.e. we obtain the flat lens as a special case.

Next, let us find the contours for various values of n. Clearly, the condition of zero optical path posed by eqn (2.76) can only be satisfied if $n < 0$. Thus, the range of interest is for n between zero and $-\infty$. When n is small and negative the radius of the circle is close to unity, but the internal focus is quite far away from the front surface because $d_2 = -d_1/n$. The contour and the focusing mechanism are shown in Fig. 2.29(a) for $n = -0.2$. The internal focus is there but the lens is far too thick. As n tends towards -1 the lens flattens, as may be seen in Fig. 2.29(b) for $n = -0.8$. At $n = -1$ the lens is flat. At this point the centre of the circle switches from minus infinity to plus infinity. As n decreases below $n = -1$ the radius of the circle decreases, as may be seen in Fig. 2.29(c) for $n = -1.2$. For even lower values, $n = -2$ and -4 the full lens is shown in Figs. 2.29(d) and (e). It may be seen that the lens gets smaller as n declines. As n tends towards minus infinity the lens becomes a circle but its radius declines to zero.

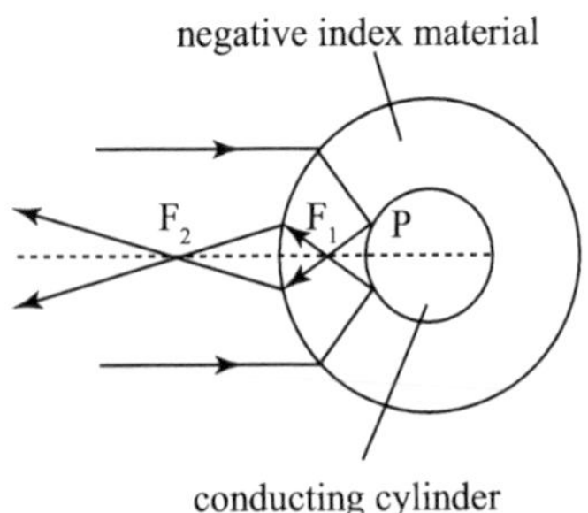

Fig. 2.30 A reflector that brings a parallel beam to a focus. From Lagarkov and Kissel (2001). Copyright © 2001 Springer Science + Business Media

An interesting variation on this theme is the reflector of Lagarkov and Kissel (2001) also in a 2D geometry. The reflector consists of a layer of negative-index material upon a conducting cylinder as shown in Fig. 2.30. An incident parallel beam refracts in the negative direction reaching the conducting (and therefore reflecting) cylinder at P from which it is reflected, going through the focal point F_1 inside the material and after a further negative refraction on the air/negative-index material boundary it comes to a second focus at F_2. A possible application of this reflector is for a microwave antenna. Inverting the path of propagation a line source positioned at F_2 would produce a parallel beam.

We have looked in this section at another family of lenses made possible by the existence of negative-refractive-index materials. It gives new options for designing lenses.

2.11.5 Experimental results and numerical simulations

It took 33 years to confirm experimentally Veselago's theory but then it started an avalanche. The credit goes to the group at the University of California, San Diego headed by David Smith. They were the first to set out to find a negative-index material, they performed the crucial experiment and interpreted their results in terms of negative permittivity and negative permeability. We have seen that it is possible to have a material, consisting of metallic rods, which gives negative ε, and it is also possible to produce negative μ by SRRs. The idea of Smith *et al.* (2000) was simply to superimpose in one structure the two different kinds of elements. A schematic representation of the unit cell is shown in Fig. 2.31.

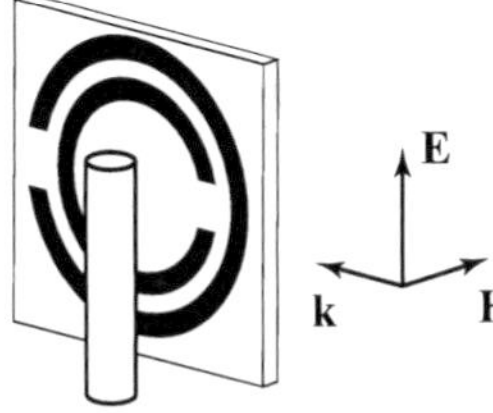

Fig. 2.31 Unit cell consisting of a SRR and of a metallic rod. From Smith *et al.* (2000). Copyright © 2000 by the American Physical Society

The experiments were performed with the wave confined between parallel metal plates and incident perpendicularly upon arrays of those unit cells. In order to couple to the SRRs the magnetic field had to be perpendicular to the plane of the SRR, and in order to couple to the rods the electric field had to be parallel with the rods, and of course the direction of propagation was perpendicular to the electric and magnetic fields as shown in Fig. 2.31.

The crucial experimental results are shown in Fig. 2.32. If only the SRRs are present there is a stop band between the frequencies of (roughly) 4.7 and 5.2 GHz. In the stop band the attenuation increases from 2 dB to about 35 dB. When the rods are also included then the stop band turns into a pass band. The attenuation declines from 50 dB to about 32 dB. Note that the attenuation is very high even in the pass band because there is considerable power absorption by the rods. However, the stop band turning into a pass band proves that a material with both material constants negative can propagate electromagnetic waves.

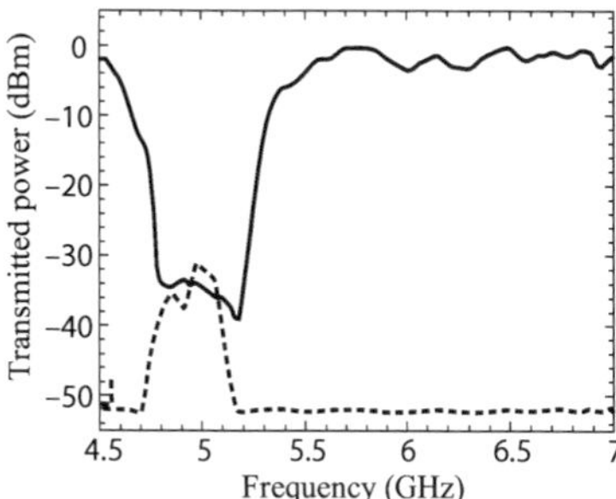

Fig. 2.32 Transmission as a function of frequency for SRRs (solid line) and for SRRs + rods (dotted line). From Smith *et al.* (2000). Copyright © 2000 by the American Physical Society

We shall present here another set of experimental results by Li *et al.* (2003). The negative-index material consisted of a similar combination of SRRs and rods with the difference that there were two rods in the unit cell. The sample was placed at the focal point of a lens-compensated horn antenna with a similar antenna as the receiver. The experimental setup is shown in Fig. 2.33(a) and the results in Fig. 2.33(b). This time, the S-parameters were measured (see Section 1.24 for their definition), i.e. the reflection in the transmitter and the transmission in the receiver. We find again that a pass band appears in the frequency range (from

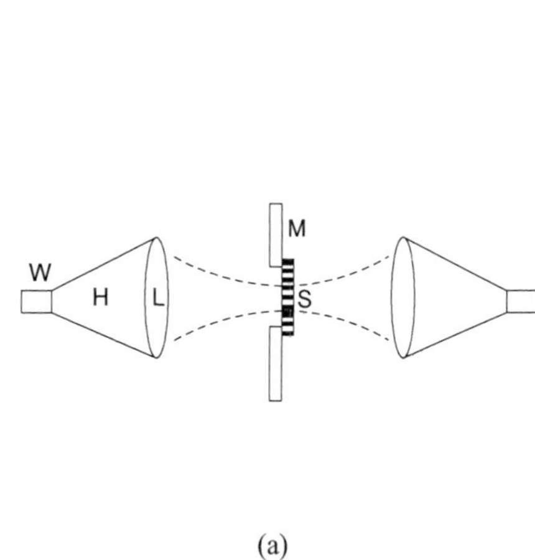

(a)

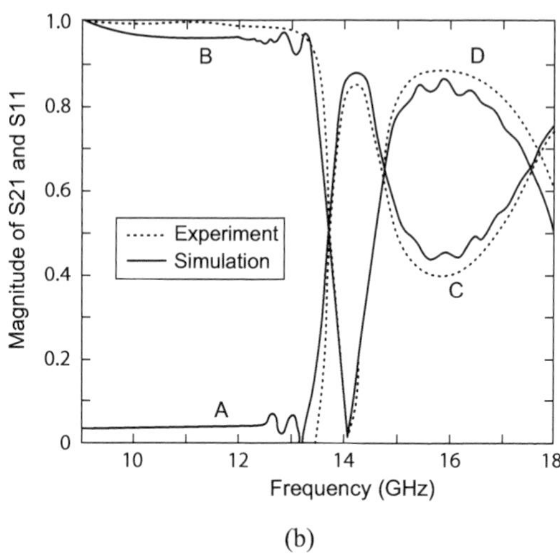

(b)

Fig. 2.33 (a) A negative-index material between two lens-compensated horns. (b) Experimental results (solid lines) for transmission (A) and reflection (B). Numerical simulations (dotted lines) for transmission (C) and reflection (D). From Li *et al.* (2003). Copyright © 2003 American Institute of Physics

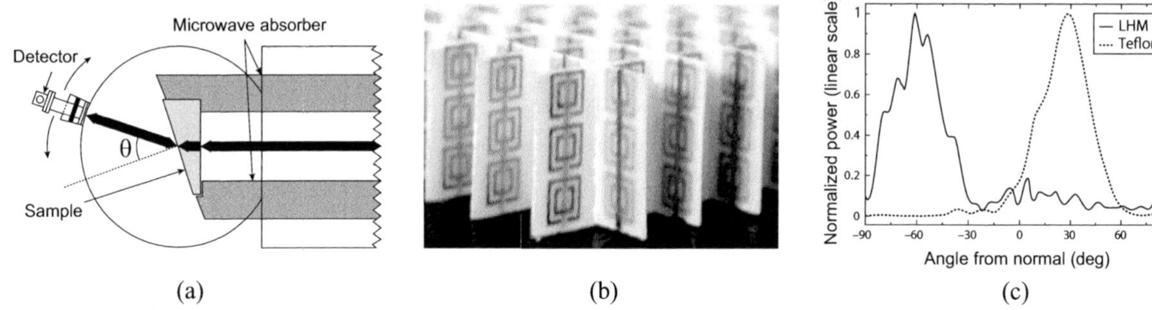

(a) (b) (c)

Fig. 2.34 (a) Experimental setup for measuring refraction, (b) Left-handed material realized by SRRs and rods, (c) measured results. From Shelby *et al.* (2001*a*). Copyright © 2001 AAAS

13.2 to 14.2 GHz) where both material parameters are negative. Curves A and B show the measured values of S_{21} and S_{11}, whereas curves C and D are derived from numerical simulation. The package used was HFSS,[7] a frequency domain Maxwell's equations solver. The sample was simulated by a unit cell with periodic boundary conditions. The agreement may be seen to be extremely good.

[7]High Frequency Structure Simulation, Ansoft Inc., Pittsburgh, 2002

The crucial experiment concerning negative refraction was done by Shelby *et al.* (2001*a*). The sample made up by a combination of SRRs and rods and both produced on printed circuit boards is shown in Fig. 2.34(a). The experimental setup with which they measured the angle of refraction may be seen in Fig. 2.34(b). The arrays of SRRs and rods were arranged in the shape of a prism. Microwaves at a frequency of 10.5 GHz were confined vertically by parallel metal plates and laterally by absorbers. They were incident upon the back of the prism as shown

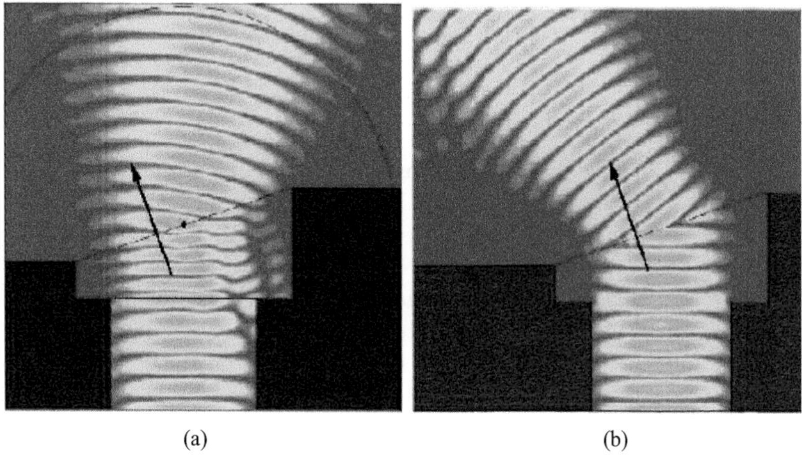

Fig. 2.35 Numerical simulations of the refraction of a finite beam by a wedge, (a) $\varepsilon_r = 2.2$, $\mu_r = 1$, (b) $\varepsilon_r = -1$, $\mu_r = -1$. From Kolinko and Smith (2003). Copyright © 2003 Optical Society of America. For coloured version see plate section

by the thick black lines. After propagating through the prism the waves were refracted at an angle of $-61°$ corresponding to an effective refractive index of -2.7. As a control experiment the prism was replaced by another prism of the same shape but made of Teflon, which has a refractive index of 1.4. It duly refracted in the positive direction with an angle corresponding to $n = 1.4$. The measured beam profiles for Teflon (dotted lines) and for the composite material (solid lines) are shown in Fig. 2.34(c) normalized to unity. The actual measured peaks are of course quite different because the composite material has much higher losses.

We shall show here simulation of negative refraction by Kolinko and Smith (2003). A finite beam is incident on a wedge-shaped material that in the first case (Fig. 2.35(a)) has material parameters $\varepsilon_r = 2.2$, $\mu_r = 1$, and in the second case (Fig. 2.35(b)) $\varepsilon_r = -1$ and $\mu_r = -1$. Similarly to the experimental results shown in Fig. 2.34(c) there is positive refraction in the first case and negative refraction in the second case. The numerical package used was HFSS. The finite elements were tetrahedrons. For convergence, as many as 100 000 elements were required.

2.11.6 Derivation of material parameters from reflection and transmission coefficients

We shall now be concerned with the inverse problem. Can the values of ε and μ be deduced from simple measurements, in particular when the S_{11} and S_{21} scattering coefficients are measured for perpendicular incidence? A simple argument suggests that it can be done. The unknowns are two complex quantities, so they can be determined from the two complex quantities measured. The analytical forms for the transmission and reflection coefficients are given by eqns (1.76) and (1.75). Changing the notation from exponential to trigonometric ones we then obtain for perpendicular incidence

$$\frac{1}{T'} = \frac{1}{T}\,\mathrm{e}^{\mathrm{j}\,kd} = \left[\cos(nkd) + \frac{\mathrm{j}}{2}\left(\eta + \frac{1}{\eta}\right)\sin(nkd)\right], \tag{2.79}$$

where η is the impedance of the medium. We also find that

$$\frac{R}{T'} = \mathrm{j}\,\frac{1}{2}\left(\eta - \frac{1}{\eta}\right)\sin(nkd)\,. \tag{2.80}$$

Instead of ε and μ we have here the intermediate variables $n = \sqrt{\varepsilon\mu}$ and $\eta = \sqrt{\mu/\varepsilon}$. The technique is then to express $\cos(nkd)$ with the aid of R and T' and then find ε and μ from that. The expressions can be found in the paper by Smith *et al.* (2000). The main problem is that $\cos(nkd)$ is multivalued, which makes the problem a little unwieldy, but soluble. The authors were capable of recovering both negative ε and μ, although their meaning for an insufficient number of elements per wavelength is controversial. For a discussion see Appendix C.

2.12 The perfect lens

2.12.1 Does it exist?

It is an old subject, going back at least for half a century (Toraldo di Francia, 1953), how one can beat the classical limit of resolution. There are essentially two approaches: one uses the far field and relies on changing the field distribution in the aperture of the lens, the second one is based on near fields. The first attempt at high-resolution near-field imaging was made by Ash and Nicholls (1972).[8] The idea was to make use of the field leaking out of a microwave cavity through a small hole. If an object with a structure somewhat larger than the hole is scanned in front of the hole then the resonant frequency of the cavity depends on the relative position of the object. By monitoring the resonant frequency it turned out to be possible to obtain information about the structure with a resolution of $\lambda/60$, close to that of the size of the hole. Their work initiated the whole new field of scanning near-field optical microscopy.

[8]There is actually a much earlier paper recently discovered, in which Synge (1928) proposed to improve resolution in a similar manner. His aim was to look at a stained biological specimen, ground and planed so that its surface does not diverge from a plane by more than 10 nm. A further requirement is an opaque plate polished to similar planeness and in which there is a hole of 10 nm diameter. The object can be scanned by illuminating the hole with a high intensity light source and moving the plate in small steps. He envisaged that these steps might be as small as 10 nm.

An entirely new idea of near-field imaging came with a proposal by Pendry (2000). He calculated that the flat lens of Veselago (which required a refractive index of minus unity) will be able to image an object with infinite resolution, provided $\varepsilon_{\mathrm{r}} = -1$ and $\mu_{\mathrm{r}} = -1$. The positions of the object, lens and image correspond to those of Veselago shown in Fig. 2.24.

Infinite resolution means perfect imaging. Perfect imaging means that every single detail of the object is reproduced in the image (including both propagating and evanescent components). In terms of a spatial frequency (discussed in Section 1.25) it means that the spatial frequency spectrum of the image (including both propagating and evanescent components) will be identical with the spatial frequency spectrum of the object. In terms of a transfer function it means that the transfer function is flat. Entirely flat. It is the same for every spatial frequency component. Is that possible? Not really. A limit will be set, if by nothing else, then by the period of the negative-index material (Haldane, 2002). If we can make metamaterial elements of the size of 100 nm and if the distance between them is also 100 nm then there would be a chance of making a lens with a resolution approaching 100 nm. At the time of writing it does not seem to be likely that such a lens could be made and such resolution could be obtained, but it is possible in principle. Another chance is obtained with a material in which only the dielectric constant is negative. That will not yield a flat transfer function but it would be flat enough for many purposes, and it would have the great advantage that natural materials (e.g. silver) with that property exist. The period in that material will be of the order of one tenth of a nanometer, thus, at least on that account, the resolution could be extremely high. We shall discuss the details of this mechanism in Chapter 5. In the present section we shall show only a few representative examples.

2.12.2 The ideal situation, $\varepsilon_{\mathrm{r}} = -1$ and $\mu_{\mathrm{r}} = -1$

Let us first look at the situation when the refractive index is exactly minus unity in the slab. We found the field distribution in the slab in Section 1.11 and talked about the spatial frequency spectrum in Section 1.25. We shall need both for the analysis. The geometry of the lens (the Veselago geometry) is plotted in Fig. 2.24. The object is located at $z = 0$, the flat lens is between $z = d/2$ and $z = 3d/2$, and the image plane is at $z = 2d$. In order to find the total transfer function we need to find their value in all three intervals separately. But that has already been done. Equations (1.158)–(1.160) provide the three transfer functions. To apply those to the geometry of Fig. 2.24 we only need to substitute $d_2 = d$ and $d_1 = d_3 = d/2$ into those equations. Assuming further an incident TM wave we still have to find the value of ζ_{e} from eqn (1.49). Note that it is necessary to be a little more specific than for the Veselago lens. For that to work it was sufficient to take $n = -1$. For the perfect lens we need both ε_{r} and μ_{r} to be equal to -1.

We shall now take a particular spatial frequency k_x and work out the total transfer function. Note that in order to satisfy the boundary conditions the x component of the wave vector is the same in all three media, so that $k_x = k_{x1} = k_{x2} = k_{x3}$. The propagation coefficient in the z direction, k_z, is given in medium 1 ($\varepsilon_1 = \varepsilon_0$, $\mu_1 = \mu_0$) as

$$k_{z1} = \sqrt{\frac{\omega^2}{c^2} - k_x^2}\,. \tag{2.81}$$

In medium 2 ($\varepsilon_2 = -\varepsilon_0$, $\mu_2 = -\mu_0$) we find

$$k_{z2} = \sqrt{\frac{\omega^2}{c^2} - k_x^2}\,. \tag{2.82}$$

Since medium 3 is the same as medium 1 it follows that $k_{z3} = k_{z1}$.

Thus, the values of k_z are identical in all three media. As a result, from eqn (1.49)

$$\zeta_e = -1. \tag{2.83}$$

Substituting these values into eqns (1.158)–(1.160) we find

$$T_1 = e^{-\frac{j k_z d}{2}}, \quad T_2 = e^{j k_z d}, \quad T_3 = e^{-\frac{j k_z d}{2}}, \tag{2.84}$$

and, consequently,

$$T_1 T_2 T_3 = 1\,. \tag{2.85}$$

The total transfer function is flat, but not only flat it is actually unity for all values of the spatial frequency for which there is propagation.

The physics is quite simple. The amplitudes are always the same because we neglected losses. The phase goes forward in medium 1 by $k_z d/2$, it goes backward in medium 2 by $k_z d$ (remember, it is a backward wave) and it goes forward again in medium 3 by $k_z d/2$. The total phase change is zero. Thus, the transfer function is flat for all propagating waves. This is remarkable but it is still far from the perfect lens. A perfect lens should reproduce not only those spatial frequencies for which there is propagation but also those for which the waves are evanescent. This occurs when the spatial frequency is larger than k_0, i.e. the details of the object are smaller than the free-space wavelength. This is when we talk about subwavelength imaging.

The waves are evanescent in all three media when $k_x > k_0$. Then k_{z1} and k_{z2} become imaginary (as in Section 1.10) and should be replaced by $-j\,\kappa_1$ and $-j\,\kappa_2$. Then, ζ_e takes the form

$$\zeta_e = \frac{\varepsilon_{r2}\kappa_2}{\varepsilon_{r2}\kappa_1}\,. \tag{2.86}$$

We can then still use the condition $\zeta_e = -1$, in which case eqn (2.84) modifies to

$$T_1 = e^{-\frac{\kappa_1 d}{2}}, \quad T_2 = e^{\kappa_1 d}, \quad T_3 = e^{-\frac{\kappa_1 d}{2}}\,. \tag{2.87}$$

Remarkably, once more,

$$T_1 T_2 T_3 = 1\,. \tag{2.88}$$

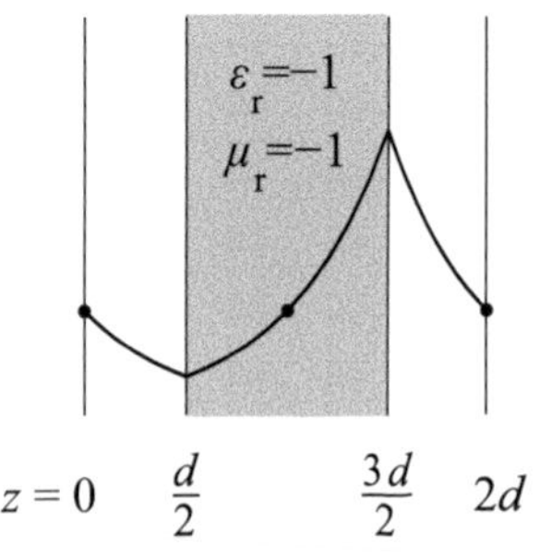

Fig. 2.36 Evolution of an evanescent component

The variation of H_y is plotted in Fig. 2.36. It may be seen that the magnetic field is the same at $z = 2d$ as at $z = 0$, and that applies to every single spatial frequency. The conclusion is that the transfer function is unity whether the wave is propagating or evanescent. However large k_x is, a slab having the material constants $\varepsilon_{r2} = -1$ and $\mu_{r2} = -1$ will perfectly reproduce the object.

It is easy to understand the physics of phase cancellation. It is also easy to understand (one might even say it's trivial) that an evanescent

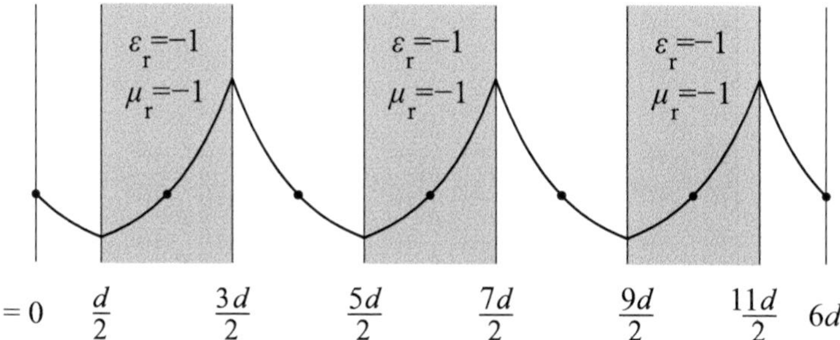

Fig. 2.37 Multilayer superlens

wave declines exponentially in media 1 and 3. What is difficult to understand is how one can have a growing wave in medium 2. The usual expression in the literature is that 'the wave is amplified'. There is no doubt that the amplitude of the wave increases within the slab in the $+z$ direction but 'amplification' is too strong a word. It makes one think that the amount of power in the beam continuously increases, which is impossible because there is no source of power anywhere. So what is the explanation?

In general, under evanescent conditions, in the slab there will be a forward propagating wave that declines in the $+z$ direction and a reflected wave that declines in the $-z$ direction. For $\zeta_e = -1$ it turns out that $C = 0$ (see eqn (1.77)), i.e. the wave that is supposed to decline in the direction of the group velocity is not present. The only wave present in the slab is the one that has amplitude D and declines from the boundary at $z = 3d/2$ in the $-z$ direction. And, of course, the wave declines away from that boundary in the positive z direction as well. Where did we see waves declining in both directions from a boundary? It was in Section 1.10 where we introduced the concept of a surface wave. The role of the wave growing in the z direction is to excite the surface wave at the rear boundary.

The problem is of course to realize a material with $\varepsilon_r = -1$ and $\mu_r = -1$ and to use sufficiently small elements to realize significant subwavelength imaging. This has not been done so far and the chances do not seem to be very good at the time of writing that it will be done in the near future. However, there are natural materials for which ε_r is negative, silver and gold were examples given before.

We have so far disregarded losses. In their presence, as may be expected, the transfer function cannot be flat for all values of k_x. There must be a cutoff, as discussed in detail in Chapter 5.

2.12.3 The periodic solution

Looking at Fig. 2.36 it is not difficult to come to the idea (proposed by Shamonina *et al.* (2001)) that a periodic solution must exist. If we create an array by placing further identical slabs at periods of d (see Fig. 2.37) the total transfer function will still be unity. The situation is not quite the same when $\mu_r = 1$ and only $\varepsilon_r = -1$ but it is still true that a multilayer lens gives a much better image than a single-layer lens of the same total thickness. This is of great practical significance, as will be further discussed in Section 5.5.

2.12.4 The electrostatic limit: does it exist?

In the electrostatic limit $k_x \gg k_0$, hence eqns (2.81) and (2.82) reduce to

$$k_{z1} = k_{z2} = k_{z3} = -\mathrm{j}\, k_x\,. \tag{2.89}$$

When $\varepsilon_\mathrm{r} = -1$ it follows then that $\zeta_\mathrm{e} = -1$ and the derivation in Section 2.12.2 yields a flat transfer function, provided the incident wave is TM polarized. Is this correct? Does it apply to large values of k_x? One could argue that as k_x increases the approximation given by eqn (2.89) becomes better and better. This is true, but the trouble is elsewhere. Let us look at the transfer function of the slab. The denominator of eqn (1.76) is of the form

$$(1+\zeta_\mathrm{e})^2\, \mathrm{e}^{\mathrm{j}\, k_{z2} d} - (1-\zeta_\mathrm{e})^2\, \mathrm{e}^{-\mathrm{j}\, k_{z2} d}\,. \tag{2.90}$$

If ζ_e is exactly equal to -1 then the first term in the above expression is zero and we do get the flat transfer function. Alas, the small deviation of $|k_x|$ from $|k_{z2}|$ that leads to ζ_e being slightly different from -1 is responsible for the large difference in the value of this term because the accompanying exponential factor, $\exp(k_x d)$, may become very large. Hence, both terms need to be taken into account and the transfer function is no longer flat, as was pointed out by Shamonina *et al.* (2001). At certain values of k_x the transfer function suddenly declines. For further details and curves see Chapter 5.

So, does the electrostatic limit exist? It does not. It is not true that the transfer function can be flat when only ε_r is equal to -1. However, as shown in Chapter 5, the approximation works in a number of well-defined cases up to a certain value of k_x.

2.12.5 Far field versus near field: Veselago's lens versus Pendry's lens

It is easy to describe the operation of Veselago's lens. All one needs to know is geometrical optics and the laws of refraction and reflection. For $n = -1$ Snell's law tells us that a wave incident at an angle θ will be refracted at an angle of $-\theta$. The corresponding ray picture has already been shown in Fig. 2.24. All dimensions are large relative to the wavelength.

Pendry's lens also has $n = -1$ but it is also important what the material parameters are. Both the relative permittivity and the relative permeability must be equal to -1. This has the further advantage that the impedance of the medium $\eta_\mathrm{m} = \eta_0\sqrt{\mu_\mathrm{r}/\varepsilon_\mathrm{r}}$ is equal to the impedance of free space η_0. Consequently, Pendry's lens is matched. None of the

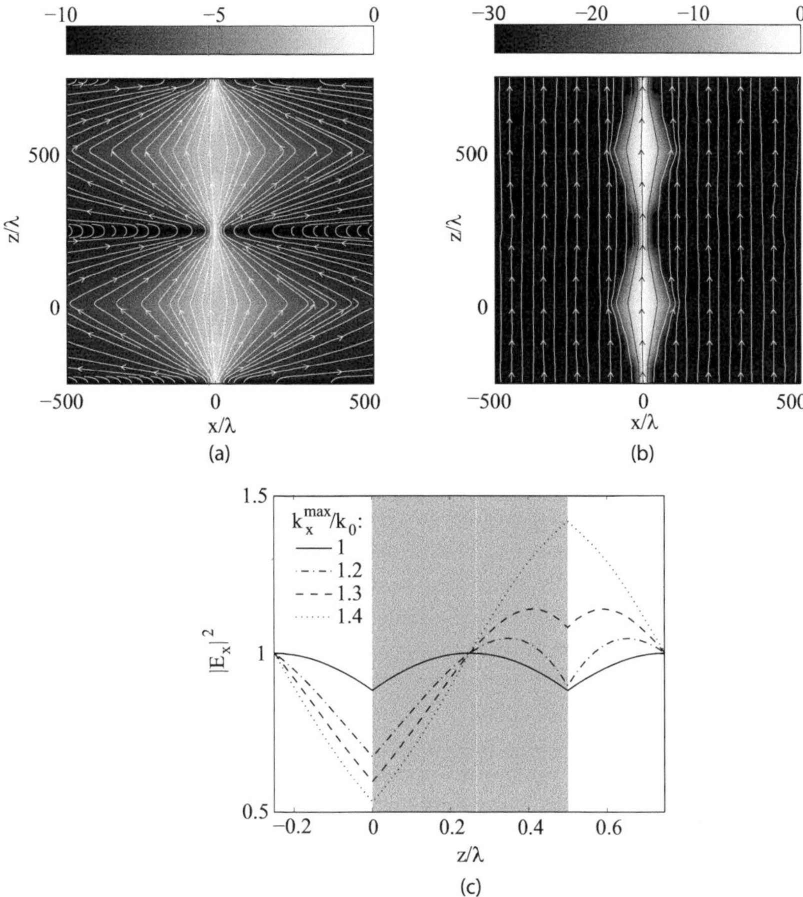

Fig. 2.38 (a), (b) Poynting vector streamlines for a flat lens with $d/\lambda = 500$ taking only propagating components into account. Object half-width is (a) one wavelength and (b) five wavelengths. (c) Evolution of the field across the lens with $d/\lambda = 0.5$ and for various number of evanescent components included. Object half-width is 0.1 λ

incident waves are reflected. However, this is only a minor difference between the two lenses. After all, Veselago's lens could also have both material parameters equal to -1, which would still lead to $n = -1$. The major difference is that Veselago's lens relies on geometrical optics, and hence on the far field, whereas Pendry's lens is designed for subwavelength imaging, and must therefore be concerned with the near field.

Let us see a few examples. We shall take the object in the form of a Gaussian $f(x) = \exp(-x^2/\tau^2)$ and assume to start with that the object size (half-width of the Gaussian) is equal to one wavelength and the width of the lens is $d = 500\,\lambda$. The object and image planes are, as usual, at a distance of $d/2$ in front of and behind the lens. We could of course use rays to illustrate the operation of the lens but let us use instead the streamlines of the Poynting vector to see how the power moves from object to image. The streamline picture (Fig. 2.38(a)) is exactly the same as the ray picture that we, and everyone else, showed before. The magnitude of the Poynting vector, presented by the gray scale, declines in the lateral direction, as it should. Note that when the amplitude is below the maximum by a factor of 10^{-10} the scale becomes completely black. What about evanescent waves? Surely when they travel a distance of 250 wavelengths they are entirely negligible. For a value of k_x just above the propagating spectrum (say, $1.0001k_0$) the value of $\exp(-\kappa_1 d/2)$ comes to 10^{-682}, very small indeed.

Next, we shall choose the half-width of the object five times as large. The streamlines of the Poynting vector still look the same (see Fig. 2.38(b)) but they are more concentrated because a bigger object gives a narrower beamwidth. It may be clearly seen in both Figs. 2.38(a) and (b) that there is a nice internal focus.

Next, let us choose a subwavelength object of half-width 0.1λ and a lens of $d = 0.5\,\lambda$ and plot in Fig. 2.38(c) $|E_x|^2$ along the z axis. When we consider propagating waves only $0 < k_x < k_0$ (solid line) there is a maximum (a rather gentle maximum) in the middle of the slab corresponding to an internal focus. Let us include now in the spectrum evanescent components in the range $k_0 < k_x < 1.2k_0$ (dashed-dotted line) and in the range $k_0 < k_x < 1.3k_0$ (dashed line). The maximum may be seen to shift towards the rear surface. When the upper limit of k_x is chosen up to a value of $1.4\ k_0$ (dotted line) then the maximum is clearly at the rear surface. The internal focus disappears, but without influencing the image. In all four cases the image invariably appears in the image plane. If we include more and more evanescent components the maximum at the rear surface rises rapidly but the position of the image is not affected.

Our tentative conclusion is that geometrical optics is valid in the far-field lens and we can see a corresponding internal focus, whereas the internal focus disappears in the near-field lens as more and more evanescent waves are added. Is this true? No, it is not. The weak point in the argument was when we said above about the evanescent waves that 'Surely when they travel a distance of 250 wavelengths they are entirely negligible'. Yes, they are very small indeed but we forgot to say that when those evanescent waves enter the lens they will grow by a factor $\exp(\kappa_2 d)$. Thus, the net growth is enormous.

The growth of the incident evanescent waves is exactly what makes Pendry's lens a perfect lens but the same applies whether the lens is in the far field or in the near field. In principle, there is no difference between the two lenses. The simple calculations above highlight the problem with the lossless perfect lens. It cannot possibly exist. However low are the amplitudes of the evanescent waves incident upon the lens they will grow in the lens to astronomical figures. The remedy is to include losses. In the presence of losses the growth in the lens is checked and near-perfect resolution might be achievable. We shall say a lot more about this in Chapter 5 concerned with subwavelength imaging.

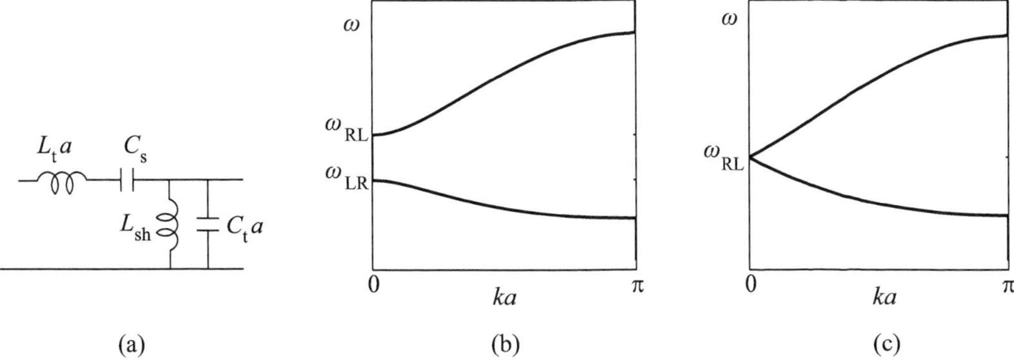

Fig. 2.39 Circuit analogue of a left-handed medium (a). Its dispersion curve when $\omega_{RL} \neq \omega_{LR}$ (b) and $\omega_{RL} = \omega_{LR}$ (c)

2.13 Circuits revisited

Section 2.6 was devoted to the study of loaded transmission lines. The main aim was to show in circuit terms how particular loads can lead to stop bands and pass bands, anticipating the introduction of positive and negative material constants. Take, for example, Fig. 2.14 where a resonant LC circuit is coupled by a mutual inductance M' to a transmission line. It is a realistic, though approximate, description of what happens when a plane wave is incident on a lattice of SRRs. Similarly, Fig. 2.17(a) is a good description in terms of loaded transmission lines of a lattice of combined SRRs and metallic rods. The primary aim was to find circuit equivalents of actual physical configurations.

However, once it is conceived that left-handed media can be modelled by circuits propagating backward waves there is no reason any longer for the circuits to imitate the physical phenomena occurring in left-handed media. If a simple non-resonant loading circuit can produce the same type of dispersion characteristics that would be equally good in building an analogue of a left-handed medium. In fact, it would be better because there would be more flexibility in choosing the parameters. Such a simple circuit in the form of Fig. 2.39(a) was proposed by Eleftheriades *et al.* (2003). Here, L_t and C_t are the inductance and capacitance of the transmission line per unit length, a is the length of the transmission line section, L_{sh} is the shunt inductance and C_s is the series capacitance with which the transmission line is loaded.[9]

An identical circuit, but with a different reasoning, was proposed by Sanada *et al.* (2004). They start with the statement that a left-handed material may be represented by a series capacitor, C_L, and a shunt inductor, L_L (for circuit and dispersion curve see Fig. 1.17). But in any practical realization, they argue, there will be a series inductance L_R, associated with the capacitor, and a shunt capacitance C_R, associated with the inductor. It is clear from the notations that L_R and C_R on their own represent a right-handed line and C_L and L_L a left-handed line. The authors called this general line a composite right/left-handed

[9] It is rather confusing that the same notations L and C mean sometimes inductances and capacitances proper and some other time inductances and capacitances per unit length. Alas, these notations are so widespread in the literature that the only remedy is to call attention to them each time they appear.

transmission line.

The two models are of course identical, provided we change the notations to

$$C_{\mathrm{L}} = C_{\mathrm{s}}\,, \qquad L_{\mathrm{L}} = L_{\mathrm{sh}}\,, \qquad C_{\mathrm{R}} = C_{\mathrm{t}}a\,, \qquad L_{\mathrm{R}} = L_{\mathrm{t}}a\,. \tag{2.91}$$

The dispersion equation may be obtained with the aid of eqn (1.142) by noting that

$$Z = \mathrm{j}\left(\omega L_{\mathrm{R}} - \frac{1}{\omega C_{\mathrm{L}}}\right) \qquad \text{and} \qquad Y = \mathrm{j}\left(\omega C_{\mathrm{R}} - \frac{1}{\omega L_{\mathrm{L}}}\right), \tag{2.92}$$

yielding

$$4\sin^2\frac{ka}{2} = -ZY = \frac{(\omega^2 - \omega_{\mathrm{RL}}^2)(\omega^2 - \omega_{\mathrm{LR}}^2)}{\omega^2\omega_{\mathrm{RR}}^2}\,, \tag{2.93}$$

where

$$\omega_{\mathrm{RR}} = \frac{1}{\sqrt{C_{\mathrm{R}}L_{\mathrm{R}}}}\,; \qquad \omega_{\mathrm{RL}} = \frac{1}{\sqrt{C_{\mathrm{R}}L_{\mathrm{L}}}}\,; \qquad \omega_{\mathrm{LR}} = \frac{1}{\sqrt{C_{\mathrm{L}}L_{\mathrm{R}}}}\,. \tag{2.94}$$

The corresponding dispersion curve is plotted schematically in Fig. 2.39(b). There are two branches: the upper one is a forward wave and the lower one is a backward wave. An interesting special case arises when $\omega_{\mathrm{RL}} = \omega_{\mathrm{LR}}$, i.e. the series resonant circuit and the shunt resonant circuit have the same resonant frequencies. In that case the upper and lower curves have a common point at $ka = 0$, $\omega = \omega_{\mathrm{RL}}$ (see Fig. 2.39(c)) where the group velocities are non-zero.

For some practical realizations of the equivalent circuits in the microwave region see Chapter 6.

Plasmon–polaritons

3

3.1 Introduction

Plasmas is an old and respectable subject. It was started by Langmuir[1] in the 1920s. Its properties have been studied ever since but not always with the same vigour. Topics rose in response to new applications, and declined on reaching saturation or realizing that the chances of the envisaged solution are fast receding. The development of radio broadcasting led to the discovery of the Earth's ionosphere that reflects radio waves and is responsible for reception of radio signals when the transmitter is over the horizon. Hannes Alfven's theory of magnetohydrodynamics (1940) that treated plasma as a conducting fluid has been employed to investigate sunspots, solar flares, the solar wind, star formation and other topics in astrophysics. The interest in thermonuclear fusion came in the wake of the hydrogen bomb. At one time it was believed that fusion, assisted by plasmas, is round the corner. The belief turned out to be not well founded. Later, the invention of lasers opened the new chapter of laser plasma physics with some promise of fusion but that has not been realized either. High-energy physicists hope to use plasma acceleration techniques to dramatically reduce the size and cost of particle accelerators. Among the applications that have come off are microscopy and sensing (see, e.g., Yeatman 1996; Homola 2003; Barnes 2006).

We have enumerated only a fraction of plasma phenomena that could be discussed. The subject is not only old and respectable: it is also very big. Our interest is limited to plasmas on surfaces, the kind that was presented very briefly in Section 1.10. They are important for metamaterial applications that belong to two categories. One is their role in the subwavelength manipulation of images. If $\mu_\mathrm{r} = 1$ and $\varepsilon_\mathrm{r} = -1$ then the plasma resonances will limit the range of spatial frequencies for which the transfer function is approximately flat. This was already briefly discussed in Section 2.12.4. The second relationship with metamaterials is via elements that may exhibit plasma resonances at high enough frequencies. They are mainly nanosized rods and rings and their combinations, reminiscent of electric dipole and loop antennas operating at radio frequencies.

There is, however, a third category that hardly exists at the moment. Nanoscale geometry, for example, allows us to change plasma properties and understanding of this mechanism would allow us to design metamaterial elements with tailored properties. This is just one example that may or may not turn out to be feasible but there are many others. We

[1] A transparent liquid that remains when blood is cleared of various corpuscles was named plasma (after the Greek word that means 'formed, jelly-like') by a Czech medical scientist Purkinje. The Nobel prize-winning American chemist Irving Langmuir first used this term to describe an ionized gas in 1927: Langmuir was reminded of the way blood plasma carries red and white corpuscles by the way an electrified fluid carries electrons and ions. This analogy is, however, slightly odd: blood plasma is blood that is free from blood corpuscles; electric plasma would lose its properties if we were to get rid of free charges in it.

believe that there is such a wealth of plasma phenomena available that lots of new metamaterial applications are bound to come. And that we regard as an equally important motive for delving into the multifarious physical phenomena.

The simplest plasma wave is a density wave of mobile electrons in the background of immobile positive charges. They interact with one another via Coulomb forces. The responsiveness of a plasma is a direct consequence of the mobility of its constituents, and occurs only when the charged particles that comprise it are relatively free. Examples of systems that show plasma-like behaviour include gas discharges, electrons and holes in semiconductors, and, what is of particular interest for us, free electrons in metals. The dispersion of a plasma wave is quite an unusual one (see, e.g., Solymar and Walsh 2004),

$$\omega(k) = \omega_{\mathrm{p}} = \text{const}\,. \tag{3.1}$$

ω, being a constant independent of k, means that only a single value of the frequency, the plasma frequency, is allowed, while the wave number can be arbitrary. Does this equation, which gives a straight horizontal line on an ω–k diagram, show a dispersive behaviour? The answer is yes.[2] Of course not all charged-particle systems show such behaviour. An ionic crystal, e.g. sodium chloride, is made up of positively and negatively charged Na^+ and Cl^- ions. The ions are so tightly bound to one another by electrostatic forces that they cannot move about freely; this system supports acoustic rather than plasma waves. And yet both plasma waves and acoustic waves show remarkable similarities in their ability to interact with electromagnetic waves, giving rise to hybrid modes. And that brings us to the point where we have to say something about terminology. As an electromagnetic wave propagates through a polarizable medium, the polarization it induces modifies the wave and the electromagnetic wave becomes coupled to the induced polarization. We could call this a hybrid mode because its properties depend both on those of the electromagnetic wave and also of the medium. The problem is then that apart from vacuum we nearly always have hybrid modes because the properties of the medium can very rarely be completely disregarded. The argument may then be used that we call them hybrid modes only when the effect of the medium is significant. This is apparently what happened to acoustic and plasma waves. When they significantly alter the properties of electromagnetic waves they acquire the postfix 'polariton'. And if that does not sound sufficiently pretentious their name is also upgraded. They change from the classical[3] 'plasma waves' and 'acoustic waves' to 'plasmons' and 'phonons' ending up eventually as plasmon–polaritons and phonon–polaritons. To make things even more complicated we shall have to distinguish between bulk plasmon–polaritons that propagate inside a homogeneous medium, and surface plasmon–polaritons that stick to a surface. Most physicists refer to them nowadays by the acronym SPP.

We shall first look at bulk plasmon–polaritons in Section 3.2 on the basis of the Drude model, and then find the conditions for the existence

[2] If we plot a dispersion curve and we find that the frequency is independent of the wave number then it is tempting to call it a non-dispersive wave—and some authors indeed do that. On the other hand, if we remember the original aim of the dispersion curves to show the change of wave velocity with frequency then it is quite obvious that a horizontal line is very dispersive. The phase velocity ω/k varies strongly as a function of k. The only dispersionless curve is a straight line that goes through the origin.

[3] Admittedly classical physics fails when the propagation length of the mode and the wavelength are comparable with atomic distances (Yang *et al.*, 1991) but we are very far from that limit.

of surface plasmon–polaritons along a single interface in Section 3.3. We shall discuss there the dispersion characteristics with particular reference to losses and also the behaviour of the Poynting vector, how it can predict whether a wave is of the forward or of the backward variety. In Section 3.4 a more complicated situation is investigated, the plasma properties of a metal slab embedded in a dielectric. There are then two brief sections, Sections 3.5 and 3.6, the former on metal–dielectric–metal slabs, and the latter on one-dimensional structures like a metal sheet of infinite length but finite width. In Section 3.7 we return to a specifically metamaterial theme: the surface modes that may exist for arbitrary values of the permittivity and permeability.

3.2 Bulk polaritons. The Drude model

We have actually encountered bulk plasmon–polaritons, without calling them so, in Section 1.9. There, we discussed the dielectric function of a lossless isotropic medium in the presence of a current,

$$\varepsilon_{\mathrm{r}} = 1 - \frac{\omega_{\mathrm{p}}^2}{\omega^2}, \tag{3.2}$$

known as the Drude model for a free-electron gas. In the presence of losses the above expression modifies to

$$\varepsilon_{\mathrm{r}} = 1 - \frac{\omega_{\mathrm{p}}^2}{\omega\left(\omega - j\gamma_{\mathrm{p}}\right)}, \tag{3.3}$$

with the damping constant, $\gamma_{\mathrm{p}} = 1/\tau$, vanishing if the collision time, τ, becomes infinite. In most metals, the plasma frequency is in the ultraviolet,[4] making them shiny in the visible range as all the incident light is reflected. In doped semiconductors, the plasma frequency is usually in the infra-red.

[4] Taking, for instance, the electron density in a typical metal as $N_{\mathrm{e}} = 6 \times 10^{28}$ m^{-3} we can calculate that $\omega_{\mathrm{p}} = 2\pi \times 2.2 \times 10^{15}\ \mathrm{s}^{-1}$, which corresponds to the wavelength $\lambda_{\mathrm{p}} = 2\pi c/\omega_{\mathrm{p}} = 136$ nm, which lies in the ultraviolet.

The asymptotic and limiting cases for eqn (3.2) are

$$\begin{cases} \varepsilon_{\mathrm{r}} \to -\infty & \text{as} \quad \omega \to 0 \\ \varepsilon_{\mathrm{r}}(\omega_{\mathrm{p}}) = 0 & \\ \varepsilon_{\mathrm{r}} \to 1 & \text{as} \quad \omega \to \infty \end{cases}. \tag{3.4}$$

What are the implications of eqn (3.2) for electromagnetic wave propagation in such a medium? An electromagnetic wave of the form $\exp\left[\mathrm{j}\left(\omega t - \mathbf{k}\cdot\mathbf{r}\right)\right]$ no longer propagates without dispersion. Its wave number has to satisfy the condition

$$k = \frac{\omega}{c}\sqrt{\varepsilon_{\mathrm{r}}\mu_{\mathrm{r}}} = \frac{\omega}{c}\sqrt{1 - \frac{\omega_{\mathrm{p}}^2}{\omega^2}} = k_0\sqrt{1 - \frac{\omega_{\mathrm{p}}^2}{\omega^2}}, \tag{3.5}$$

resulting in a dispersion equation

$$\omega^2 k_0^2 = \omega^2 k^2 + \omega_{\mathrm{p}}^2 k_0^2. \tag{3.6}$$

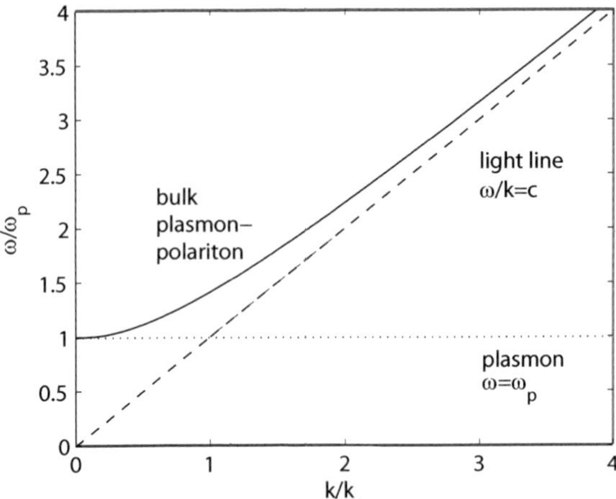

Fig. 3.1 Dispersion of bulk plasmon–polaritons, of plasmons and of the electromagnetic wave in vacuum

It is plotted in Fig. 3.1 as a function of k/k_{p} (where $k_{\mathrm{p}} = \omega_{\mathrm{p}}/c$), together with the 'light line' $k = \omega/c$ of the electromagnetic wave in vacuum and with the horizontal dispersion line $\omega = \omega_{\mathrm{p}}$.

The dispersion curve in Fig. 3.1 may be seen as resulting from the crossing of the bulk plasmon line $\omega = \omega_{\mathrm{p}}$ and the light line $k = \omega/c$. A possible way of interpreting the dispersion equation (3.6) is that there is no longer a pure electromagnetic and a pure plasma wave. Instead, the plasma wave interacts strongly with light, resulting in our bulk plasmon–polariton. The asymptotic and limiting cases for the dispersion are

$$\begin{cases} \omega \to \omega_{\mathrm{p}} & \text{as} \quad k \to 0 \\ \omega \to kc & \text{as} \quad k \to \infty \end{cases} . \tag{3.7}$$

At high frequencies the bulk mode lies close to the light line and is light-like; at low wave numbers it moves away from the light line until the cut-off plasma frequency is reached. Bulk plasmon–polaritons of frequency below the plasma frequency cannot propagate[5] in plasma because its free electrons screen the electric field of the light. Bulk plasmon–polaritons of frequency above the plasma frequency can propagate through plasma because the electrons cannot respond fast enough to screen it.

[5]There is a stop band for $\omega < \omega_{\mathrm{p}}$. Note a general feature (Tilley, 1988) that surface modes can be found within the bulk mode stop bands. It is in this range that our other type of plasmon–polariton, the SPP, can propagate.

Since many of the properties of plasmon–polaritons are similar to those of phonon–polaritons it may be worthwhile to include here a brief note on the latter. The interaction is then between the electromagnetic wave and the optical branch (optical phonons) of the acoustic wave in an ionic crystal. There, the dielectric function has a form

$$\varepsilon_{\mathrm{ph}} = \varepsilon_\infty + \frac{\omega_{\mathrm{T}}^2}{\omega_{\mathrm{T}}^2 - \omega^2}(\varepsilon_0 - \varepsilon_\infty) = \varepsilon_\infty \left(1 + \frac{\omega_{\mathrm{L}}^2 - \omega_{\mathrm{T}}^2}{\omega_{\mathrm{T}}^2 - \omega^2}\right) , \tag{3.8}$$

with ω_{T} the so-called TO (transverse optical) phonon frequency and $\omega_{\mathrm{L}} = \omega_{\mathrm{T}}\sqrt{\varepsilon_0/\varepsilon_\infty}$ the LO (longitudinal optical) phonon frequency (see, e.g., Kittel 1963).[6] The asymptotic and limiting cases here are

[6]Note that ε_0 in eqn (3.8) is not the free-space permittivity but the value of the dielectric constant at $\omega = 0$.

$$\begin{cases} \varepsilon_{\rm ph} \to \pm\infty & \text{as} \quad \omega \to \omega_{\rm T} \\ \varepsilon_{\rm ph}(\omega_{\rm L}) = 0 & \\ \varepsilon_{\rm ph} \to \varepsilon_\infty & \text{as} \quad \omega \to \infty \end{cases} . \tag{3.9}$$

The dispersion equation is then $k = \omega\sqrt{\varepsilon_{\rm ph}\mu}$. The frequency at which $k = 0$ is known as the *Reststrahl* frequency (Fox, 2001). Note further that in the frequency range $\omega_{\rm T} < \omega < \omega_{\rm L}$ the dielectric constant is negative[7] and the wave vector is imaginary, meaning a stop band for the bulk phonon–polariton. It may be instructive to note that formally the plasmon–polariton dispersion equation (3.5) is a special case of the phonon–polariton dispersion equation with $\varepsilon_\infty = 1$, $\omega_{\rm L} = \omega_{\rm p}$ and $\omega_{\rm T} = 0$.

[7] An example is SiC, a polar material. It was used for subwavelength imaging in the negative dielectric constant region (see Section 5.4.6) by Korobkin *et al.* 2006*a*; Korobkin *et al.* 2006*b*.

3.3 Surface plasmon–polaritons. Semi-infinite case, TM polarization

The electromagnetic field of SPPs at a dielectric–metal interface is obtained from the solution of Maxwell's equations in each medium, and the associated boundary conditions. The latter express the continuity of the tangential components of the electric and magnetic fields across the interface. A further obvious condition is that the fields must decline away from the boundary and vanish infinitely far away.

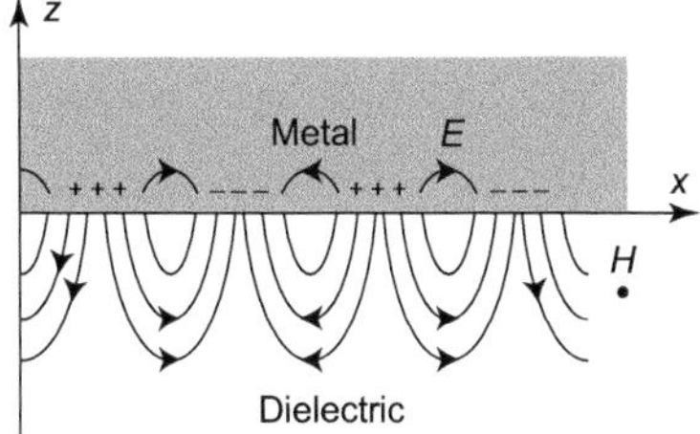

Fig. 3.2 SPP mode. A sketch of charge and electric-field distributions at the metal–dielectric boundary

3.3.1 Dispersion. Surface plasmon wavelength

We consider now a plane interface $z = 0$ between two semi-infinite media, the first one being vacuum, air or a dielectric with the permittivity, ε_1, taken here as frequency-independent; the second one being a metal with the permittivity, ε_2, described by the Drude model (eqn (3.2)). We shall first look at the case of TM polarization. Recalling the results from Section 1.10 the conditions for the existence of a surface mode propagating along the surface in the x direction[8] are given as

$$\varepsilon_1\varepsilon_2 < 0 \qquad \text{and} \qquad \varepsilon_1 + \varepsilon_2 < 0\,. \tag{3.10}$$

[8] For $\varepsilon_{\rm r1} + \varepsilon_{\rm r2} < 0$, $\mu_{\rm r1} = \mu_{\rm r2} = 1$ a TM surface mode with components E_x, E_z and H_y and propagating in the x direction is the only option. Explicit calculations show that there is no propagation for this mode in the y direction, nor is there a surface TE mode carrying H_x, H_z and E_y components.

The first of these conditions, requiring that the dielectric constant is changing sign at the boundary, has a simple meaning (Barnes, 2006). A surface mode involves charges accumulated at the boundary between the metal and the dielectric. The corresponding electric field that is built up would originate on positive charges and terminate on negative charges and do so, in fact, in both medium 1 and medium 2. This is shown schematically in Fig. 3.2. That means that in the presence of surface charges the components of the electric field normal to the surface must have opposite signs.

A boundary condition for the normal component of the displacement field is[9]

$$\varepsilon_1 E_{1z} = \varepsilon_2 E_{2z}\,, \tag{3.11}$$

[9] We have just described the role of surface charges and yet, when it comes to the boundary conditions, we ignore the surface charge. This apparent contradiction is resolved in Appendix D.

which is another way of saying that for a surface mode to exist dielectric constants in the two media must have different signs. The second condition of eqn (3.10) follows from eqn (1.73), which is given below

$$k_x = k_0\sqrt{\frac{\varepsilon_{\mathrm{r}1}\varepsilon_{\mathrm{r}2}}{\varepsilon_{\mathrm{r}1}+\varepsilon_{\mathrm{r}2}}}\,. \tag{3.12}$$

It ensures that k_x is real so that the mode can propagate along the surface. It is a rather innocent-looking equation but it contains a large amount of information about the properties of SPPs at a single interface. Since $\varepsilon_{\mathrm{r}2}$ is a function of frequency, eqn (3.12) is the dispersion equation, i.e. it provides the relationship between the frequency and the wave number. Inserting eqn (3.2) from the lossless Drude model into eqn (3.12) we find

$$k_x^2\left[(1+\varepsilon_{\mathrm{r}1})k_0^2 - k_{\mathrm{p}}^2\right] = k_0^2\varepsilon_{\mathrm{r}1}(k_0^2 - k_{\mathrm{p}}^2)\,, \tag{3.13}$$

where

$$k_{\mathrm{p}} = \frac{\omega_{\mathrm{p}}}{c} = \frac{2\pi}{\lambda_{\mathrm{p}}}\,. \tag{3.14}$$

Equation (3.13) may then be solved for k_0 to obtain the dispersion equation in terms of ω. It looks rather complicated so we shall give here the solution only for $\varepsilon_{\mathrm{r}1} = 1$

$$\omega^2 = \frac{\omega_{\mathrm{p}}^2}{2} + c^2k_x^2 \pm \sqrt{\frac{\omega_{\mathrm{p}}^4}{4} + c^4k_x^4}\,, \tag{3.15}$$

which is not too long. Taking the negative sign[10] we shall plot in Figs. 3.3(a) and (b) the dispersion curves at an interface metal–vacuum ($\varepsilon_{\mathrm{r}1} = 1$) and metal–glass ($\varepsilon_{\mathrm{r}1} = 2.25$).

[10]The positive sign in eqn (3.15) also has physical meaning and yields the Brewster wave discussed in Appendix E.

The dispersion curve shows that at low frequencies the surface mode lies close to the light line and is light-like. As frequency increases, the mode moves away from the light line, gradually approaching an asymptotic limit. This occurs when the permittivities in the two media are of the same magnitude but opposite sign, thus producing a pole in the dispersion equation (3.12) (Barnes, 2006). The asymptotes are

$$\begin{cases} k_x \to \frac{\omega}{c}\sqrt{\varepsilon_{\mathrm{r}1}} & \text{as} \quad \omega \to 0 \\ k_x \to \infty & \text{as} \quad \omega \to \omega_{\mathrm{s}} \end{cases}\,, \tag{3.16}$$

where

$$\omega_{\mathrm{s}} = \frac{\omega_{\mathrm{p}}}{\sqrt{1+\varepsilon_{\mathrm{r}1}}} \tag{3.17}$$

is the upper cutoff frequency for the surface mode. For vacuum it is

$$\omega_{\mathrm{s}} = \frac{\omega_{\mathrm{p}}}{\sqrt{2}}\,. \tag{3.18}$$

Medium 2 with negative permittivity is often called the surface-active medium, while medium 1 with positive permittivity is called the surface-passive medium. The reason is quite simple. The terms express the fact

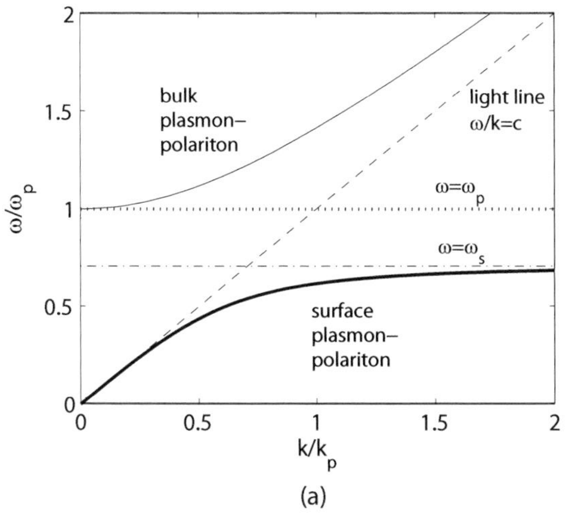

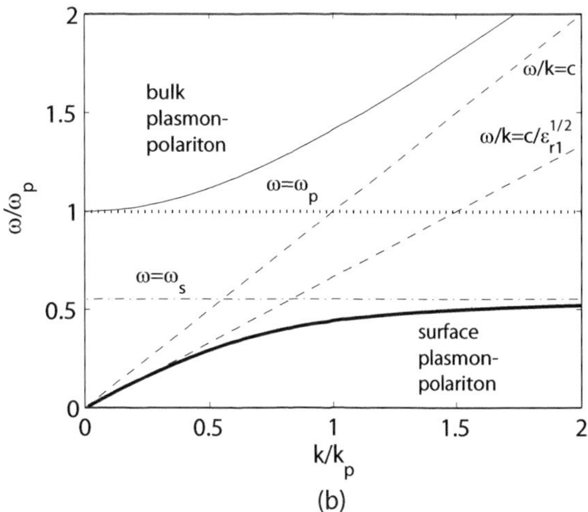

Fig. 3.3 Dispersion of SPPs (solid line) at the boundary vacuum–metal (a) and at the boundary glass–metal (b). Also shown is the dispersion curve of the bulk mode with $k_y = k_z = 0$ (thin solid line), the light line in vacuum and in the dielectric (dashed lines), the dispersions of bulk (dotted line) and surface plasmons (dashed-dotted line)

that it is charges in medium 2 that lead to electric-field oscillations at the boundary (Abeles, 1986; Tilley, 1988).

Note that the SPP dispersion lies in a region to the right of the light lines in medium 1, where the k_x values are larger than the wave number of the propagating electromagnetic wave

$$k_x > \frac{\omega}{c}\sqrt{\varepsilon_{r1}}\,, \tag{3.19}$$

and therefore this mode is non-radiative, or bound (Burke *et al.*, 1986). This means that this wave has no propagating component in the z direction; z components of the k vector are purely imaginary. There are two important consequences. First, the surface mode on such a flat interface cannot decay by emitting a photon, and, conversely, it cannot be excited directly by an incident plane wave. An incident plane wave can never have a wave vector parallel to the interface large enough to couple to this surface mode. In order to couple to the SPP, both k and ω must match. In engineering terms one could say that both frequency and velocity must agree, whereas physicists would talk about conservation of energy and momentum. One may use a prism-coupling scheme based on the method of attenuated total reflection, which would provide an evanescent component with imaginary k_z and sufficiently large k_x. Best known are the Kretschmann and the Otto configurations (Kretschmann and Raether, 1968; Otto, 1968). Another consequence is that this mode is capable of interacting with the evanescent part of a spectrum from a small subwavelength object, providing an important mechanism for subwavelength imaging lying at the heart of the concept of the 'perfect lens' (see Chapter 5 for a detailed analysis).

According to the inequality (3.19) the SPP wavelength is always smaller

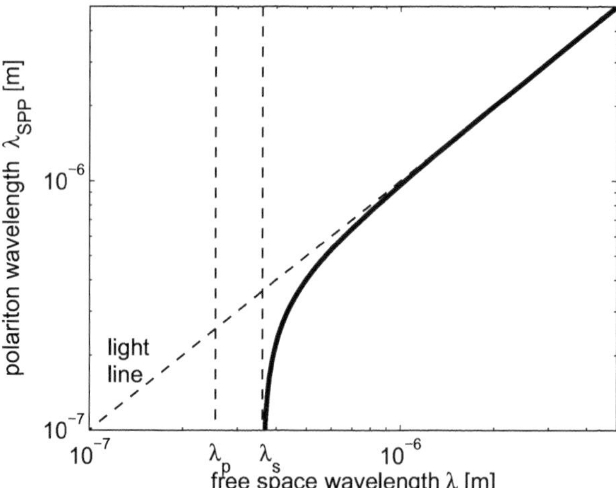

Fig. 3.4 Dispersion curve for the metal–vacuum interface

than the free-space wavelength. It would be desirable at this stage to abandon the normalized quantities ω/ω_p and k/k_p and find the actual values of the various characteristics of the waves, starting with the plasma frequency. For silver, the value of the plasma frequency was determined experimentally by McAlister and Stern (1963) from the dip in the transmission of electromagnetic waves through a thin film. They measured the plasma frequency in terms of electron volt units, giving a value of 3.8 eV, which in SI units corresponds to 918 THz or a wavelength of 326 nm. According to Pendry (2000) the plasma frequency is approximately 9 eV, corresponding to 2183 THz, a big difference. The main problem is of course that the Drude model, assumed so far is not valid for frequencies close to the plasma frequency. When $\omega = \omega_p$ the dielectric constant is not equal to 1 but to a value of ε_0 (a horrendous deviation from the meaning of ε_0 in the SI system) due to interband transitions. If we follow Pendry and take its value as 5.7 then ε_r is equal to -1 at a wavelength of 359 nm. When coming to imaging by a silver lens then this is the wavelength range at which successful imaging was obtained (see, e.g., Melville and Blaikie 2005). Can we save the Drude model without contradicting experimental results? We can do it by choosing the plasma frequency so that the Drude relation remains approximately true[11] in the critical region around $\varepsilon_r = -1$. This choice led us to $f_p = 1200$ THz. We can now plot (Fig. 3.4) the wavelength of the SPP mode $\lambda_{SPP} = 2\pi/k_x$ versus free-space wavelength $\lambda = 2\pi c/\omega$. As the free space wavelength approaches the asymptotic value $\lambda_s = 2\pi c/\omega_s$ the SPP wavelength becomes much shorter than the free-space wavelength, opening up the possibility for subwavelength manipulation of the fields. For larger values of the free-space wavelength, λ_{SPP} is only marginally smaller than λ.

[11]A technique often used (see, e.g., Gray and Kupka 2003).

In view of the dispersion equations derived we should say here a few more words about terminology. The logical thing would be to call the waves surface plasmon–polaritons for small k_x when the dispersion curve is close to that of electromagnetic waves in free space and it is some kind

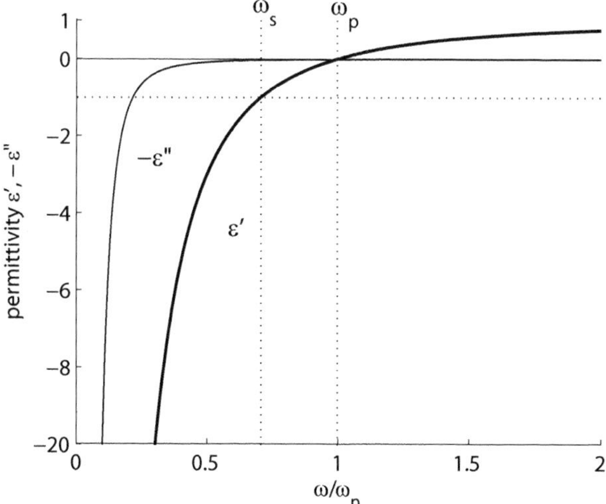

Fig. 3.5 The real and imaginary parts of the dielectric function for a lossy metal according to the Drude model ($\gamma_p/\omega_p = 0.01$)

of hybrid wave, but call them surface plasma waves for large k_x. Alas, logic is not an influential factor when it comes to terminology. We shall accept the majority view, accept the term surface plasmon–polariton, and keep on referring to them as SPPs. This may also be the place where we can mention the electrostatic limit, which means obtaining the dispersion equation by solving Laplace's equation instead of the full apparatus of Maxwell's equations. This is permissible for large k_x. For more details see Appendix F.

3.3.2 Effect of losses. Propagation length

If we take losses of medium 2 into account, we should expect that a surface wave bound to the interface would propagate along the surface a finite distance, until its energy is dissipated as heat in the metal.

In eqn (3.12), in the presence of losses, the right-hand side is a complex quantity, and so is the left-hand side, meaning that k_x will have an imaginary part as well. We need to write

$$k_x = k'_x - \mathrm{j}\, k''_x\,, \tag{3.20}$$

and the complex relative permittivity of the metal as

$$\varepsilon_{r2} = \varepsilon' - \mathrm{j}\,\varepsilon''\,. \tag{3.21}$$

We obtain for its real and imaginary parts (Boardman, 1982; Berini, 2000*a*)

$$\varepsilon' = 1 - \frac{\omega_p^2}{\omega^2 + \gamma_p^2}\,, \qquad \varepsilon'' = \frac{\omega_p^2 \gamma_p}{\omega\left(\omega^2 + \gamma_p^2\right)}\,. \tag{3.22}$$

The corresponding values of k'_x and k''_x may be determined from eqn (3.12) by substituting into it the complex value of the dielectric

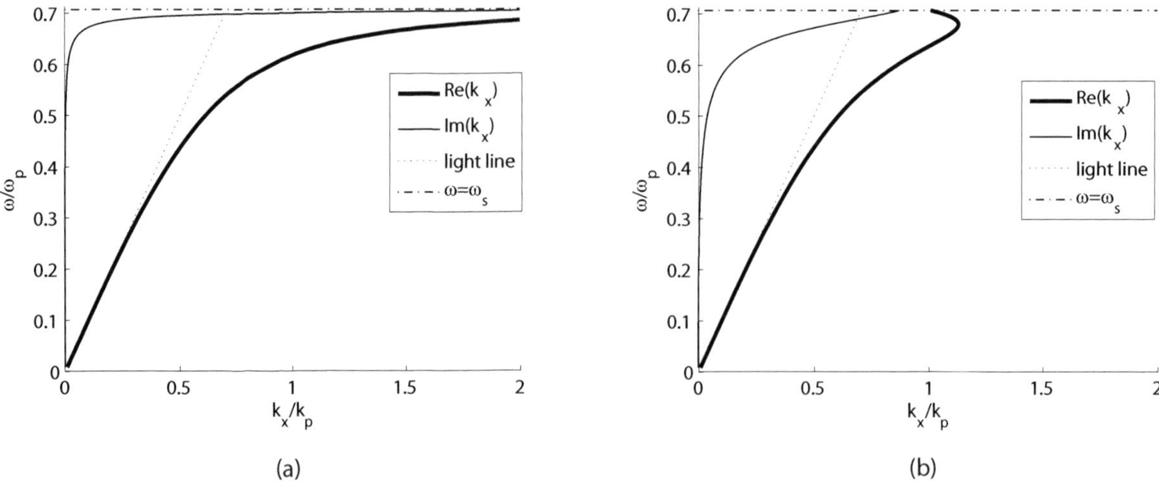

Fig. 3.6 Dispersion for lossy metal. $\gamma_p/\omega_p = 0.01$ (a) and 0.1 (b)

constant. The real and imaginary parts of ε_{r2} are plotted in Fig. 3.5 as a function of frequency for $\gamma_p = 0.01\omega_p$.

The dispersion curves in the presence of losses may now be calculated with the aid of eqn (3.12). They are plotted in Figs. 3.6(a) and (b). The normalized horizontal co-ordinate is $k'_x c/\omega_p = k'_x/k_p$ for the propagation coefficient (solid lines) and $k''_x c/\omega_p$ for the attenuation. For the lossless case $\gamma_p = 0$ there is no attenuation in the pass band between 0 and ω_s. For $\gamma_p/\omega_p = 0.01$ the attenuation becomes significant just below ω_s but it declines sharply as the frequency decreases (Fig. 3.6(a)). The same is true for $\gamma_p/\omega_p = 0.1$ but the effect is more substantial (see Fig. 3.6(b)). All the information about propagation and attenuation is properly contained in Figs. 3.6(a) and (b) but this is not the way most people in the art of plasmas like to talk about attenuation. They prefer to use the measure of propagation length, the distance the SPP travels before its intensity is reduced by a factor of e. The relationship between the propagation length and the attenuation coefficient is $L_{\text{SPP}} = 1/2k''_x$ (the factor 2 is there because propagation length refers to intensity, whereas we have used k''_x for the attenuation of the wave amplitude). For this reason we also plot the propagation length in Fig. 3.7 for the same set of loss parameters. It may be seen that in the visible region (with our choice of ω_p) it falls roughly between $0.2\omega_p$ and $0.4\omega_p$. The largest value of the propagation length for $\gamma_p = 0.01\,\omega_p$ is about 10 μm, a very small length. On the other hand, it is still large relative to both the free-space wavelength and the SPP wavelength. So there would be about 20 wavelengths available for manipulation of information and that might be enough for practical application.

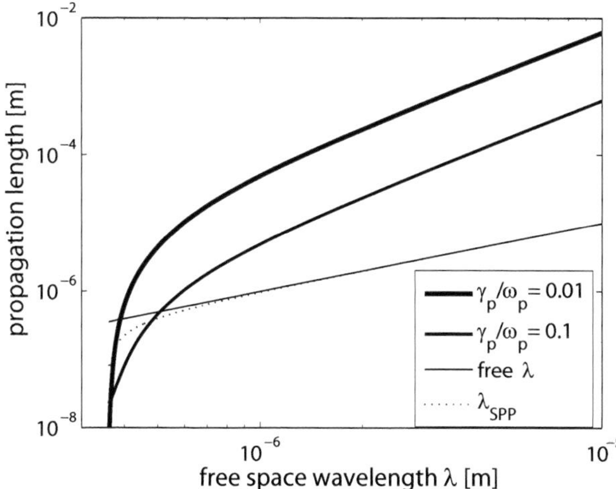

Fig. 3.7 Propagation length. $f_{\mathrm{p}} = \omega_{\mathrm{p}}/2\pi = 1200$ THz

3.3.3 Penetration depth

The penetration depth into the dielectric, δ_{d}, or into the metal, δ_{m}, is the distance in the z direction from the surface over which the intensity of the field is reduced by a factor of e. It follows immediately that the penetration depth is inversely proportional to the imaginary part of the z component of the wave vector,

$$\delta_{\mathrm{d}} = \frac{1}{2k''_{z1}}, \qquad \delta_{\mathrm{m}} = \frac{1}{2k''_{z1}}. \tag{3.23}$$

In Section 1.10 we have already calculated the values of the normal components of the k vector in both media. The relationship is

$$k_{z1} = \sqrt{\varepsilon_{\mathrm{r1}}\mu_{\mathrm{r1}}k_0^2 - k_x^2}\,, \qquad k_{z2} = \sqrt{\varepsilon_{\mathrm{r2}}\mu_{\mathrm{r2}}k_0^2 - k_x^2}\,. \tag{3.24}$$

Disregarding losses we have seen already that the normal components are purely imaginary, which means that the field amplitudes decay away from the surface,

$$k_{z1} = -\mathrm{j}\,\kappa_1\,, \qquad k_{z2} = -\mathrm{j}\,\kappa_2\,. \tag{3.25}$$

A number of important conclusions may be drawn. For the lossless case it may be shown that

$$\frac{\delta_{\mathrm{m}}}{\delta_{\mathrm{d}}} \simeq \left|\frac{\varepsilon_{\mathrm{r1}}}{\varepsilon_{\mathrm{r2}}}\right|. \tag{3.26}$$

Since in the frequency range of surface plasmon–polaritons $|\varepsilon_{\mathrm{r2}}| \geq \varepsilon_{\mathrm{r1}}$ (see eqns (3.10)), this means that the penetration depth in the surface active medium, metal, is always smaller than in the dielectric. At low frequencies, $\omega \ll \omega_{\mathrm{s}}$, the penetration depth in the metal is much smaller than in the dielectric, whereas in the high-frequency limit with $\omega \to \omega_{\mathrm{s}}$ the penetration depths in the two media become comparable. Another

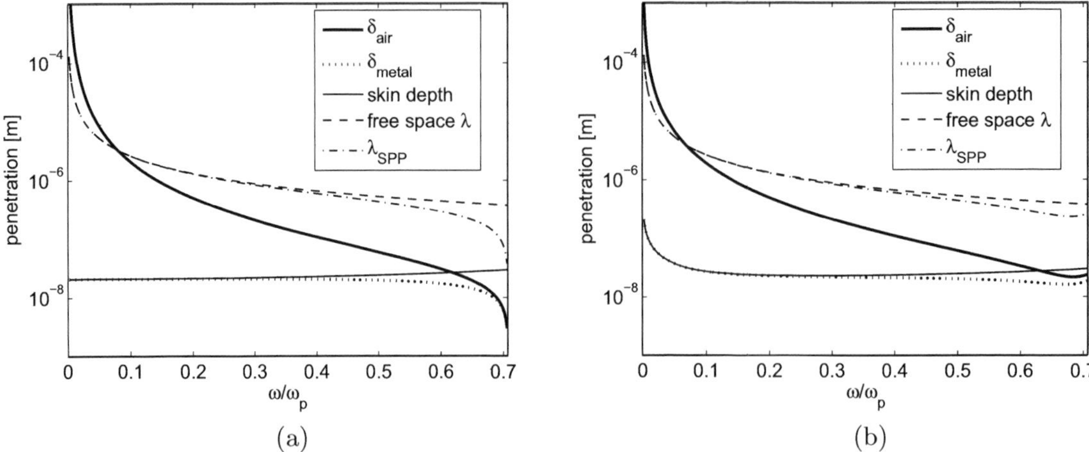

Fig. 3.8 Penetration depth for (a) $\gamma_{\mathrm{p}} = 0$, (b) $\gamma_{\mathrm{p}} = 0.1\omega_{\mathrm{p}}$

interesting point is their very different asymptotic behaviour at low frequency,

$$\delta_{\mathrm{m}} \simeq \frac{\lambda_{\mathrm{p}}}{4\pi}, \qquad \delta_{\mathrm{d}} \simeq \frac{\lambda^2}{4\pi\lambda_{\mathrm{p}}}. \tag{3.27}$$

The penetration depth in the metal in the low-frequency limit is practically frequency-independent, whereas the penetration depth in the dielectric is strongly frequency-dependent and can be quite large in the infra-red (Abeles, 1986).

Barnes *et al.* (2003) in their review pointed out the importance for the SPP-based nanocircuitry of these three characteristic length scales, the propagation length, the penetration depth into the dielectric and into the metal. The propagation length of the SPP mode, L_{SPP}, is usually determined by the loss in the metal. For a low-loss metal, for example, silver, at a wavelength of 500 nm it is as large as 20 mm. The propagation length sets the upper size limit for any photonic circuit based on SPPs. The penetration depth into the dielectric material, δ_{d}, is typically of the order of one half the wavelength of light involved and dictates the maximum height of any individual features, and thus components, that might be used to control SPPs. The ratio of $L_{\mathrm{SPP}}/\delta_{\mathrm{d}}$ thus gives one a measure of the number of SPP-based components that may be integrated together. The penetration depth into the metal, δ_{m}, determines the minimum feature size that can be used; this is between one and two orders of magnitude smaller than the wavelength involved, thus highlighting the need for good control of fabrication at the nanometer scale.

Another important point is that the penetration depth into the metal gives an idea of what thickness is required to allow coupling between the modes on the two surfaces of a thin metallic film. As shown later in Fig. 3.17(b) for a certain set of parameters, and as will be discussed in much

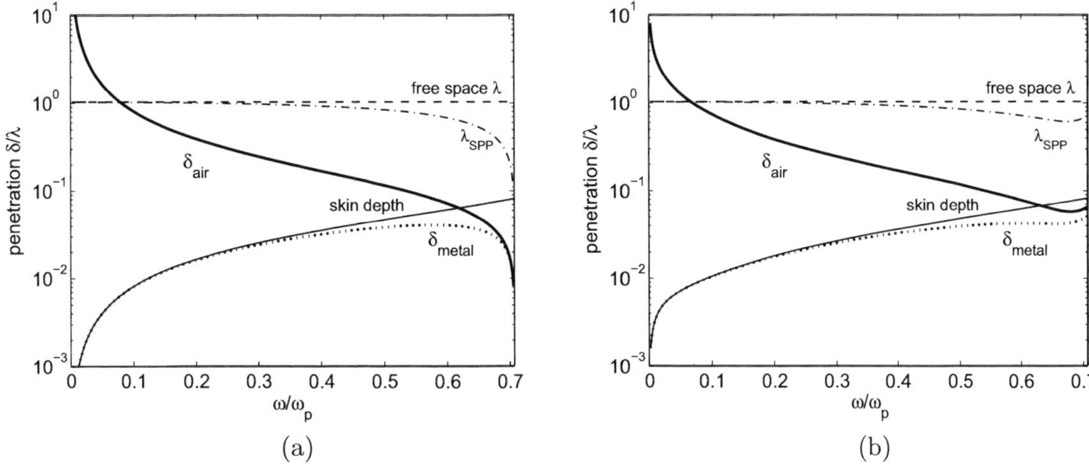

Fig. 3.9 Penetration depth normalized to the free-space wavelength for (a) $\gamma_{\mathrm{p}} = 0$, (b) $\gamma_{\mathrm{p}} = 0.1\omega_{\mathrm{p}}$

more detail in Chapter 5, a film of 60 nm thickness at a wavelength around 360 nm may be regarded as too thick for an effective coupling between the two surfaces. The subwavelength imaging mechanism relies on this coupling and therefore very thin films are essential there.

The penetration depth into the dielectric and into the metal are plotted in Figs. 3.8(a) and (b) as a function of $\omega/\omega_{\mathrm{p}}$ for the lossless case, and for $\gamma_{\mathrm{p}} = 0.1\omega_{\mathrm{p}}$ within the range of the SPPs from 0 to ω_{s}. The free-space wavelength and the SPP wavelength are plotted as well for comparison. The penetration depth in the metal exhibits a remarkably constant value for a wide range of frequencies and there is not much difference whether losses are present or not. There is though a difference at very low frequencies where the penetration depth increases with decreasing frequency. At this stage it is interesting to compare the penetration depth with the skin depth (see, e.g., Solymar 1984), well known from undergraduate studies. Both the penetration depth and the skin depth stand for the same thing: how far the field will penetrate into the metal. The difference between them is that the usual measure of the skin depth is for perpendicular incidence, implying a $k_x = 0$, whereas k_x is finite in the SPP case. As may be seen in Fig. 3.8, the deviation of the penetration depth from the skin depth is quite small. The curves of Fig. 3.8 are replotted in Fig. 3.9 but now normalized to the free-space wavelength.

The penetration depth into the dielectric is lower than λ_{SPP} at higher frequencies and larger than λ_{SPP} at lower frequencies. That increase, relative to the free-space wavelength, arises because at lower frequencies the metal is a better conductor and the dispersion of the SPP is closer to the light line and is less confined to the surface.

It is interesting to note that the penetration depth into the dielectric for a localized mode associated with nanoparticles can become much

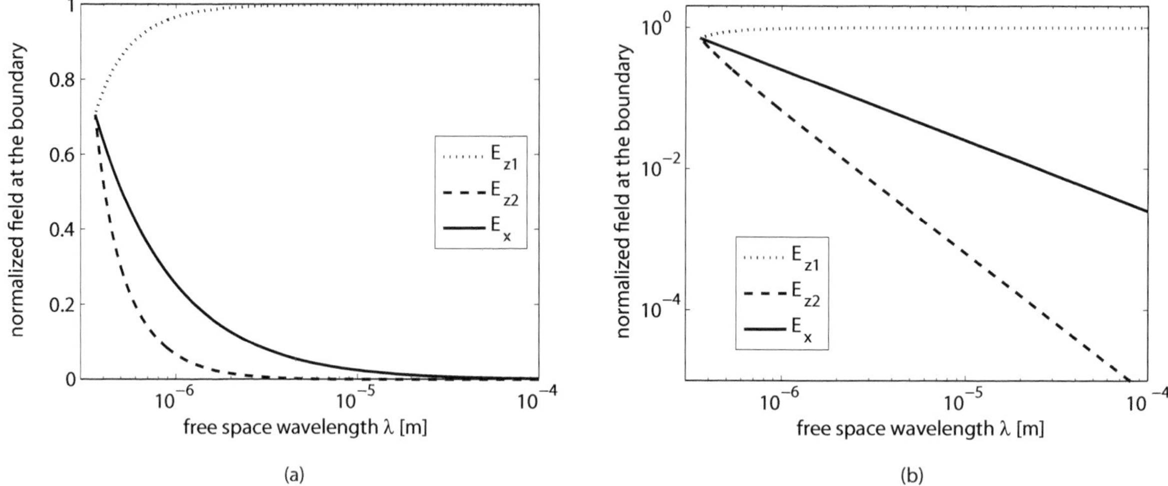

Fig. 3.10 Electric-field components at the metal–air interface

smaller, of the order of 10 nm (Whitney *et al.*, 2005; Barnes, 2006). This is attributed to the divergence of the electric field around the tips of such structures. The larger the localization of the field near the metal surface, the larger is the field enhancement (Barnes, 1998). This enhanced field makes SPP modes sensitive to changes at the surface and opens up possibilities to use SPP modes for sensor applications (Barnes, 2006; Homola, 2003).

We can conclude from looking at Figs. 3.8 and 3.9 that as the free-space wavelength decreases both the propagation length and the penetration depth into the dielectric decrease. As ω approaches ω_s, the wave penetrates to equal extents into both the dielectric and the metal, and as this happens the propagation length declines as well. So the correlation between confinement and attenuation may be roughly expressed as follows: the smaller the fraction of the wave in the metal, the smaller is the attenuation. We will return to this question when discussing the case of two interfaces.

3.3.4 Field distributions in the lossless case

Next, we shall look at field distributions at the boundary. As seen before there is only one component of the magnetic field, H_y, along the y axis, while the electric field has two components, E_x and E_z in the sagittal plane that is defined by the normal to the surface (the z axis) and the direction of propagation (the x axis).

The field in the z direction is evanescent, reflecting the bound non-radiative nature of the SPP. No power is transferred in the normal direction away from the surface. The H_y and E_x components are continuous at the boundary, whereas the E_z component changes sign and can have a large discontinuity in its magnitude, depending on the ratio $\varepsilon_{r2}/\varepsilon_{r1}$.

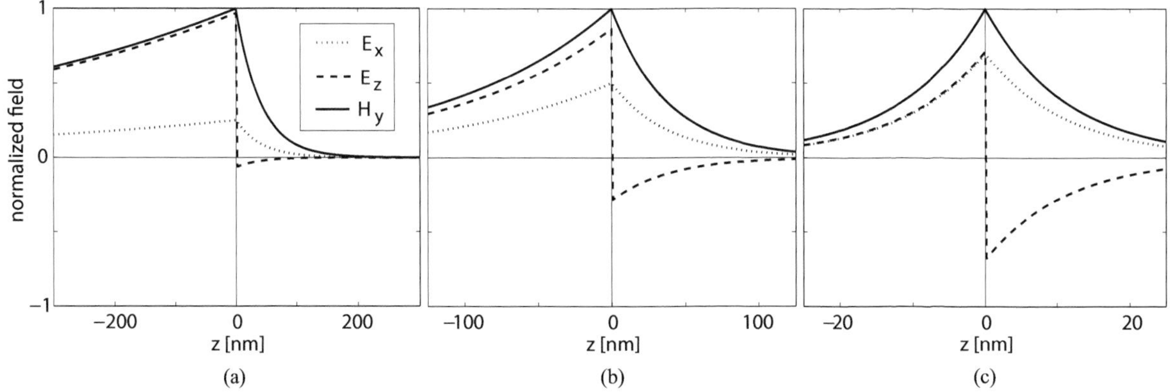

Fig. 3.11 Field amplitudes as a function of z for (a) $\omega/\omega_\mathrm{p} = 0.25$, (b) $\omega/\omega_\mathrm{p} = 0.5$, (c) $\omega/\omega_\mathrm{p} = 0.7$

The ratio of the transverse and longitudinal components of the electric field in the dielectric medium may be obtained from eqns (1.69) and (3.12) as

$$\frac{E_{z1}}{E_x} = \mathrm{j}\sqrt{\frac{-\varepsilon_{\mathrm{r}2}}{\varepsilon_{\mathrm{r}1}}} = \mathrm{j}\sqrt{\frac{\omega_\mathrm{p}^2 - \omega^2}{\varepsilon_{\mathrm{r}1}\omega^2}} . \tag{3.28}$$

The transverse component is dominant in the electric field of SPPs with small wave numbers and low frequencies (light-like case) for which the dispersion curve is close to the dielectric light line. When $k_x \gg k_0$ then the transverse and longitudinal components are comparable. They are equal when $k_x \to \infty$. The two components of the electric field at the boundary as a function of frequency, normalized to its maximum, are plotted in Fig. 3.10(a). They are replotted in Fig. 3.10(b) using a logarithmic scale in order to show how much smaller E_x is for most of the range.

The decay of the fields is faster in the metal than in the dielectric as follows from $|\kappa_2|$ being always larger than $|\kappa_1|$. In Figs. 3.11(a)–(c) we show the decay of E_z, $|E_x|$ and H_y as a function of z in both media for $\omega/\omega_\mathrm{p} = 0.25$, 0.5 and 0.7. It may be seen that as ω tends towards $\omega_\mathrm{p}/\sqrt{2}$ the decay in air becomes relatively stronger and the x and z components of the electric field become equal. The absolute decay strongly increases as ω increases, which may be seen by realizing that the scales in the three figures are different.

The amplitude of the magnetic field in the xz plane is shown in Figs. 3.12(a)–(c) by the gray-scale contrast for $\omega/\omega_\mathrm{p} = 0.25$, 0.5 and 0.7. The magnetic field may be seen to be more confined at $\omega/\omega_\mathrm{p} = 0.7$. Notice again the change of scale. The SPP wavelength is strongly reduced as ω approaches ω_s. The electric field lines are also plotted in the same figure.

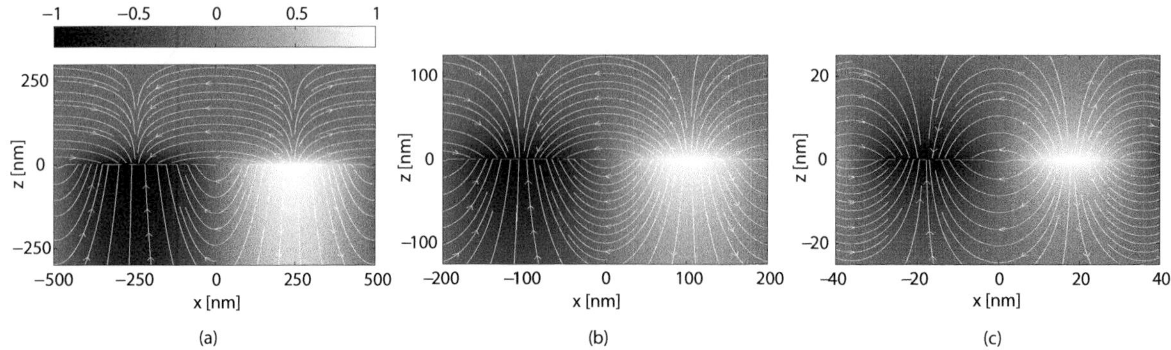

Fig. 3.12 Magnetic-field (contour plot) and electric-field (field lines) distribution at the metal–air boundary for (a) $\omega/\omega_{\mathrm{p}} = 0.25$, (b) $\omega/\omega_{\mathrm{p}} = 0.5$, (c) $\omega/\omega_{\mathrm{p}} = 0.7$

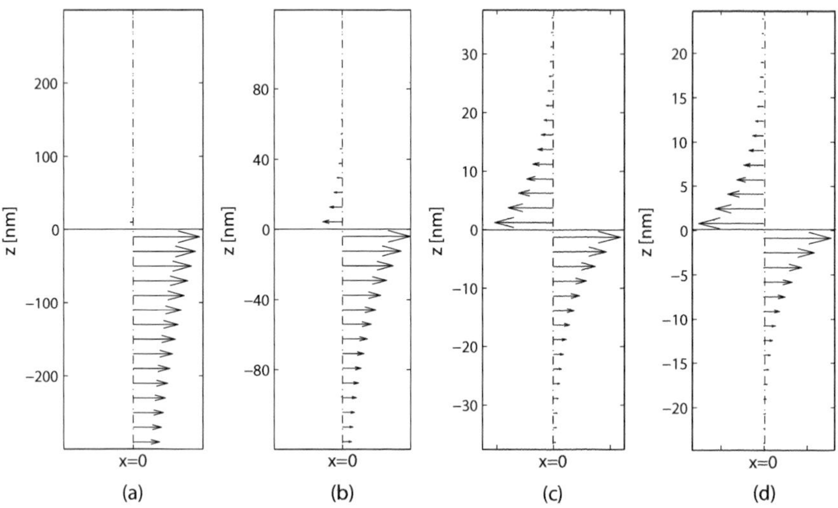

Fig. 3.13 Poynting vector in the vicinity of the metal–air interface for (a) $\omega/\omega_{\mathrm{p}} = 0.25$, (b) $\omega/\omega_{\mathrm{p}} = 0.5$, (c) $\omega/\omega_{\mathrm{p}} = 0.685$, (d) $\omega/\omega_{\mathrm{p}} = 0.7$. Lossless case

3.3.5 Poynting vector: lossless and lossy

Knowing all field components we can easily calculate the magnitude and direction of the energy flow in the two media using the definition for the Poynting vector (Section 1.13). Without losses, there can be no Poynting vector component across the interface. Along the interface the x component of the Poynting vector may be written as

$$S_x = -\frac{1}{2}\mathrm{Re}\left(E_z H_y^*\right) . \tag{3.29}$$

With the aid of eqns (1.69) and (1.70) it takes the form

$$S_{x1} = \frac{k_x}{2\omega\varepsilon_{\mathrm{r1}}}|B|^2 \,\mathrm{e}^{\,2\kappa_1 z} \qquad \text{and} \qquad S_{x2} = \frac{k_x}{2\omega\varepsilon_{\mathrm{r2}}}|B|^2 \,\mathrm{e}^{\,-2\kappa_2 z} , \tag{3.30}$$

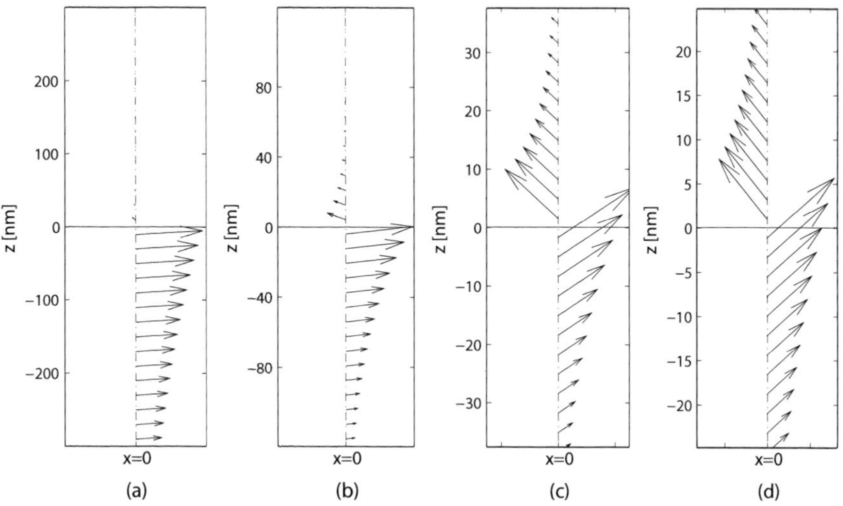

Fig. 3.14 Poynting vector in the vicinity of the metal–air interface for (a) $\omega/\omega_\mathrm{p} = 0.25$, (b) $\omega/\omega_\mathrm{p} = 0.5$, (c) $\omega/\omega_\mathrm{p} = 0.685$, (d) $\omega/\omega_\mathrm{p} = 0.7$. Lossy case

where S_{x1} and S_{x2} are the x components of the Poynting vector in the dielectric and in the metal, respectively. Taking $\varepsilon_\mathrm{r1} = 1$ the distribution of the Poynting vector is shown in Figs. 3.13(a)–(d) both in air and in the metal for $\omega/\omega_\mathrm{p} = 0.25$, 0.5, 0.685 and 0.7. It is interesting to note that practically all the power is flowing in air at $\omega/\omega_\mathrm{p} = 0.25$. The corresponding wave vector is close to the light line. As ω/ω_p increases some of the power flows in the metal, oddly enough in the opposite direction. In the vicinity of the asymptote at $\omega/\omega_\mathrm{p} = 0.7$ the flow of power flowing in the positive and negative directions is nearly the same. There is very little net power flowing. If we want to find the net power of the SPP moving in the positive x direction we need to integrate S_{x1} for z from $-\infty$ to zero, and S_{x2} from $z = 0$ to $z = \infty$. The integration can be easily performed. The final result after some algebra can be found as

$$\text{Net power} = |B|^2 \frac{k_x}{4\omega\kappa_1} \frac{\varepsilon_\mathrm{r2}^2 - \varepsilon_\mathrm{r1}^2}{\varepsilon_\mathrm{r1}^2 \varepsilon_\mathrm{r2}^2} . \tag{3.31}$$

When $\varepsilon_\mathrm{r1} = |\varepsilon_\mathrm{r2}|$ there is no net power flow at all. Otherwise, the net power is in the forward direction. We have a forward wave as follows anyway from the dispersion curve.

Why is there no Poynting vector component in the z direction? Both E_x and H_y are non-zero hence the normal component of the Poynting vector

$$S_z = \frac{1}{2} \mathrm{Re}\left(E_x H_y^*\right) \tag{3.32}$$

should exist. It does not for the reason that E_x and H_y are 90 degrees out of phase, and their product has zero real part.[12] However, in the presence of losses ($\gamma_\mathrm{p} = 0.01\omega_\mathrm{p}$) S_z must be finite. The Poynting vector

[12]The same picture is known to occur in the case of total internal reflection.

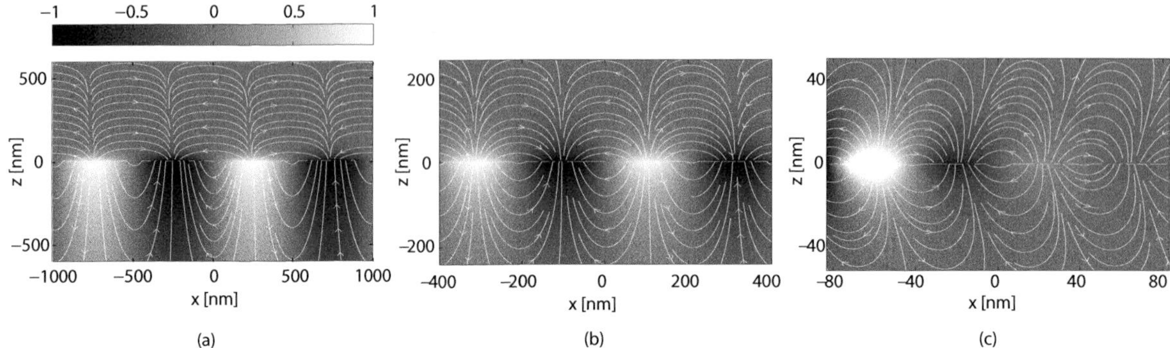

Fig. 3.15 Magnetic-field (contour plot) and electric-field (streamlines) distribution at the metal–air boundary. Lossy case

must cross the boundary between the dielectric and the metal in order to replenish the absorption occurring in the metal. This is shown in Figs. 3.14(a)–(d) for the same set of frequencies. An obvious consequence of losses is that the wave attenuates as it propagates in the x direction. Figures 3.15(a) and (b) show again (as in Fig. 3.12) the contour plots of H_y in the xz plane, and the field lines of the electric field. The strong decline of the magnetic field at $\omega = 0.7\omega_\mathrm{p}$ may be clearly seen. There is a bright spot between $x = -80$ nm and -40 nm but one period away the intensity has so much declined that the maximum is hardly visible. Note also that the field lines, just as the streamlines of the Poynting vector, are slanted.

3.4 Surface plasmon–polaritons on a slab: TM polarization

So far we have had a single interface between two semi-infinite media. Now we consider a metallic slab of dielectric constant ε_2 and thickness d in the z direction (see Fig. 3.16), and infinitely large in the other two directions. It is sandwiched between two homogeneous isotropic semi-infinite media of dielectric constant ε_1. If the two surfaces are far away from each other then each one is unaware of the existence of the other one and no new phenomena occur. However, if the slab is sufficiently thin then the SPP at one of the surfaces 'feels' the existence of the SPP at the other surface. A 'repulsion of levels' occurs, and the dispersion curves for the SPP localized at each of the two interfaces become split due to the interaction of these waves (Zayats *et al.*, 2005).

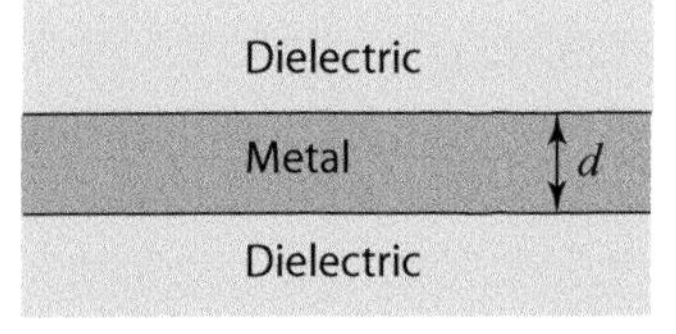

Fig. 3.16 Metallic slab surrounded by two identical semi-infinite dielectrics

3.4.1 The dispersion equation

We have already considered the oblique incidence of a TM wave upon a slab in Section 1.11 and have already looked upon the transfer function with 'perfect imaging' in mind. Our interest now is to derive the disper-

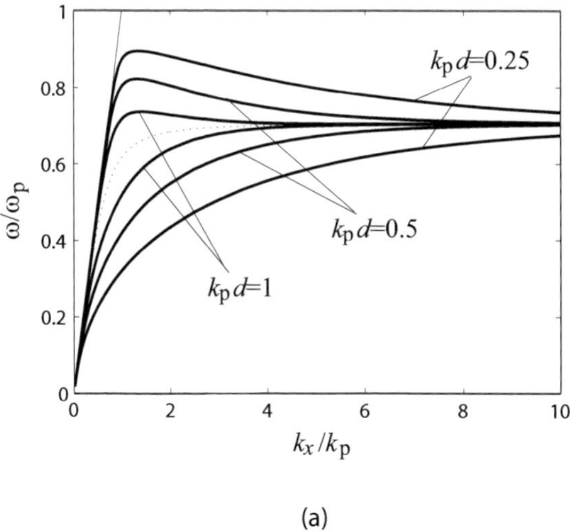

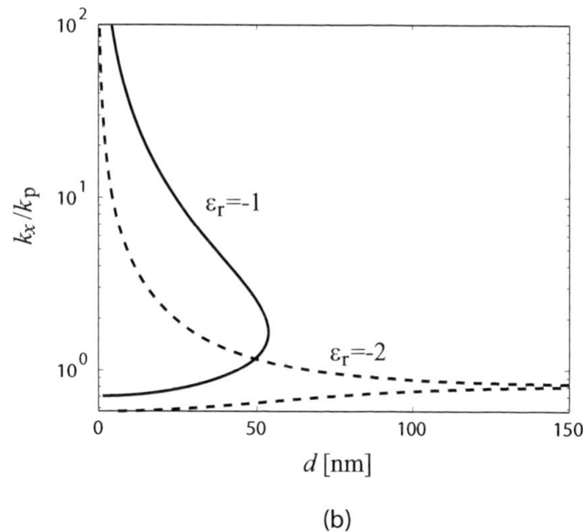

Fig. 3.17 (a) Dispersion curves for thin slabs for $k_{\mathrm{p}}d = 0.25$, 0.5 and 1. (b) k_x versus thickness for two different values of ω corresponding to $\varepsilon_{\mathrm{r}} = -1$ and -2

sion equation. Like in the single-interface case we wish to find a solution with non-zero components H_y, E_x and E_z. The relationship between the input wave with amplitude A and the output wave with amplitude F is given by eqn (1.76). We have an eigensolution when there is an output without an input. This occurs when the denominator of eqn (1.76) is zero,[13]

$$(1+\zeta_{\mathrm{e}})^2 - \mathrm{e}^{-2\kappa_2 d}(1-\zeta_{\mathrm{e}})^2 = 0\,, \tag{3.33}$$

where $\zeta_{\mathrm{e}} = \kappa_2\varepsilon_1/(\kappa_1\varepsilon_2)$. The two solutions are

$$(1+\zeta_{\mathrm{e}}) = \pm\,\mathrm{e}^{-\kappa_2 d}(1-\zeta_{\mathrm{e}})\,, \tag{3.34}$$

representing two modes guided by the metallic slab (in which the wave, now aware of both surfaces, travels with one single velocity), one symmetric and one antisymmetric with respect to their field distributions. As far as the dispersion curve is concerned it may be expected that the unperturbed dispersion curve (i.e. the one for the single interface) will split. The two branches of the dispersion equation given by eqn (3.34) can be expressed in more concise form as

$$\zeta_{\mathrm{e}} = -\tanh\frac{\kappa_2 d}{2}\,, \tag{3.35}$$

and

$$\zeta_{\mathrm{e}} = -\coth\frac{\kappa_2 d}{2}\,. \tag{3.36}$$

These transcendental equations can be easily solved numerically. As shown above there are two solutions, first reported by Oliner and Tamir (1962). We shall denote the upper branch of the solution by $\omega^{(+)}$ and

[13]The presence of a pole in expressions for field amplitudes is a typical feature of an eigenmode of a system. An alternative derivation, based on Maxwell's equations, is given in Appendix G.

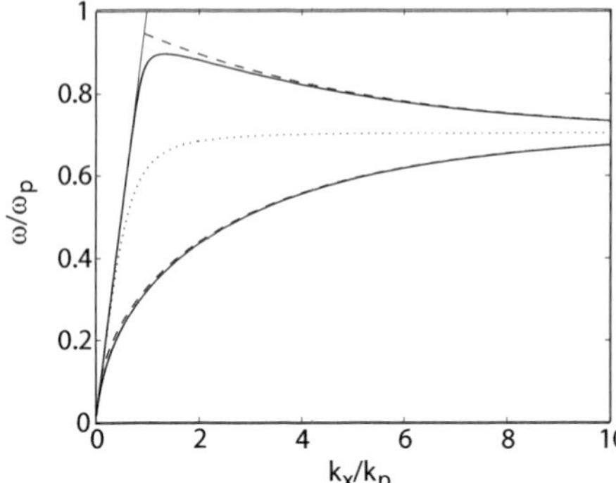

Fig. 3.18 Electrostatic approximation and full solution for a slab. $k_{\mathrm{p}}d = 0.25$

the lower branch by $\omega^{(-)}$. In Fig. 3.17(a) we plot the upper and lower branches for $\varepsilon_{\mathrm{r}1} = 1$ and for three different thicknesses ($k_{\mathrm{p}}d = 0.25$, 0.5 and 1) together with the solution for the single interface ($d \to \infty$) that is plotted with the dotted line. The meaning of the adjectives 'upper' and 'lower' make sense because they are above and below the dotted line. For large k_x both curves tend towards ω_{s}, the same limit as for the single interface. For small k_x both types of dispersion curves tend to the light line. We also find, not unexpectedly, that the split between the modes becomes smaller as the slab becomes thicker. The propagation coefficient is plotted in Fig. 3.17(b) as a function of slab thickness for two different values of ω corresponding to $\varepsilon_{\mathrm{r}} = -1$ and -2. It may be seen that for a large enough thickness only one mode may exist, or even none. The two surfaces are then uncoupled.[14]

[14]It makes good sense that the surface modes that exist on a single interface at a given ω will split when the two surfaces are coupled. It is, however, less obvious that in the presence of coupling the so far prohibited territory between ω_{s} and ω_{p} may also be inhabited by a new mode, the upper branch of the dispersion curve. Physical considerations would suggest that such a mode is permissible because the basic condition $\varepsilon_{\mathrm{r}2} < 0$, is still valid, allowing charges to accumulate at the surface.

It is always nice to find analytical approximations. They give a sense of security. We know what is going on. The first one presented is the electrostatic approximation that we have already come across in Section 2.12.4. It is described in Appendix F. For the present case we find the two branches of the dispersion equation in the form

$$\omega^2 = \frac{\omega_{\mathrm{p}}^2}{2}\left(1 \pm \mathrm{e}^{-k_x d}\right). \tag{3.37}$$

For thin enough slabs the electrostatic approximation turns out to be extremely good for the lower branch, as may be seen in Fig. 3.18. It is good too for the upper branch until it reaches the light line but then it fails miserably. It cannot possibly follow the light line. This is of course not surprising. When the wave velocity is close to the velocity of light then nobody would expect an electrostatic approximation to work. Interestingly, the approximation for the lower branch is still quite good in the vicinity of the light line, although for sufficiently low values of k_x the approximation is bound to fail. How the approximation deteriorates for the lower branch as well as for low values of k_x may be seen in Figs. 3.19(a)–(c) for three different thicknesses. The approximate curve is denoted by dashed lines. It may be seen that firstly the approximation crosses the light line and secondly that the approximation becomes worse as the thickness increases.

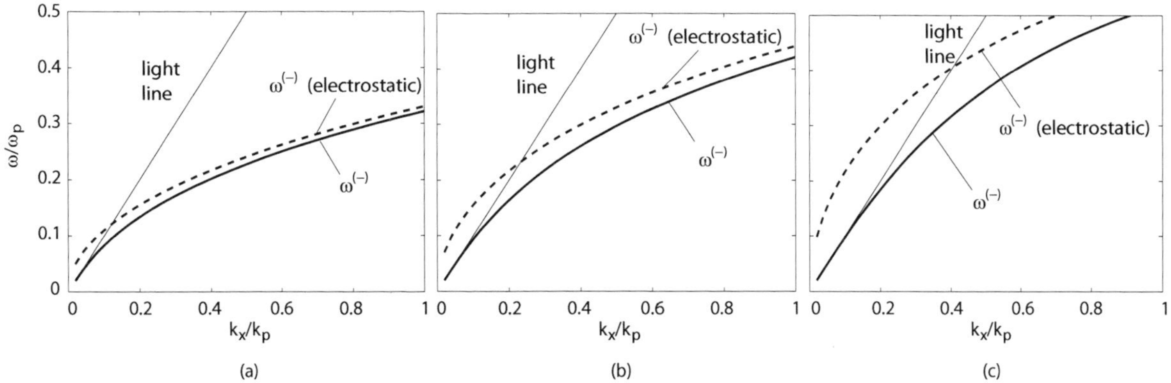

Fig. 3.19 (a) Electrostatic approximation and full solution for a slab. $k_{\mathrm{p}}d = 0.25,\ 0.5,\ 1$ (a)–(c)

Two other approximations for small k_x were first given by Economou (1969). The derivation is given below. For small enough k_x we may expect the solution to be close to the light line, i.e. we may assume

$$k_x^2 = k_0^2(1 + \delta^2)\,, \tag{3.38}$$

where $\delta^2 \ll 1$. It follows then that

$$\kappa_2 = k_{\mathrm{p}} \qquad \text{and} \qquad \kappa_1 = \delta k_0\,. \tag{3.39}$$

We may then determine δ by substituting eqn (3.39) into eqn (3.35) yielding

$$\delta = \frac{\omega}{\omega_{\mathrm{p}}} \coth\left(\frac{k_{\mathrm{p}}d}{2}\right), \tag{3.40}$$

which leads to the approximate dispersion equation

$$k_x^2 = k_0^2\left[1 + \frac{\omega^2}{\omega_{\mathrm{p}}^2}\coth^2\left(\frac{k_{\mathrm{p}}d}{2}\right)\right]. \tag{3.41}$$

If $\delta \to \infty$, i.e. we are back to the single interface, then the approximate curve is

$$k_x^2 = \frac{\omega^2}{c^2} + \frac{\omega^4}{c^2\omega_{\mathrm{p}}^2}\,. \tag{3.42}$$

Note that eqn (3.41) would further simplify when $k_{\mathrm{p}}d/2 \ll 1$ to

$$k_x^2 = \frac{\omega^2}{c^2} + \frac{4}{d^2}\frac{\omega^4}{\omega_{\mathrm{p}}^4}\,. \tag{3.43}$$

It can be immediately seen that the smaller d is the farther is the dispersion equation from the light line for a given ω. Next, we shall look for an approximation for the upper branch. We shall assume again that

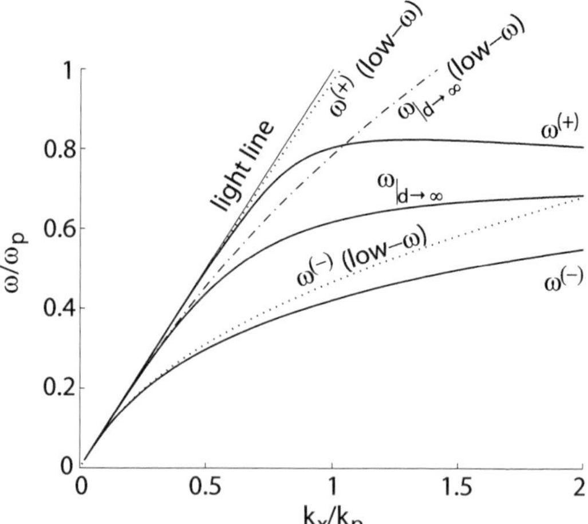

Fig. 3.20 Full solution and various approximations for a slab. $k_\mathrm{p}d = 0.5$

k_x is small and close to the light line so that we can start again with eqn (3.38). Then, using the same technique we find

$$\delta = \frac{\omega}{\omega_\mathrm{p}} \tanh\left(\frac{k_\mathrm{p}d}{2}\right), \tag{3.44}$$

leading to

$$k_x^2 = k_0^2 \left[1 + \frac{\omega^2}{\omega_\mathrm{p}^2} \tanh^2\left(\frac{k_\mathrm{p}d}{2}\right)\right], \tag{3.45}$$

or after expanding the tanh function to

$$k_x^2 = \frac{\omega^2}{c^2} + \frac{\omega^4 d^2}{4c^4}. \tag{3.46}$$

It may now be seen that the smaller d is the nearer is this branch of the dispersion equation to the light line.

Using the above-derived approximations for $\omega^{(-)}$ and $\omega^{(+)}$ we plot in Fig. 3.20 the approximate and exact curves for $k_\mathrm{p}d = 0.5$. The approximation is very good for small k_x up to about $\omega/\omega_\mathrm{p} = 0.35$ and deteriorates afterwards.

So far we have looked at the lossless case but of course attenuation is equally important if we have practical applications in mind. We need to find not only the ω versus k'_x curve but also ω versus k''_x. How will k''_x enter our equations? It will be through κ_2, which in the presence of losses will take the form

$$\kappa_2^2 = (k'_x - \mathrm{j}\,k''_x)^2 + k_0^2(\varepsilon' - \mathrm{j}\,\varepsilon''). \tag{3.47}$$

We need to substitute κ_2 from above into eqns (3.33) and (3.34) taking good care that ζ_e depends both on κ_2 and on $\varepsilon_{\mathrm{r}2}$. We are not aware of any analytical approximations for this case. We need to resort to numerical solutions.[15]

[15] We used an iterative technique first assuming the value of k'_x from the lossless solution and finding k''_x, and then for this value of k''_x finding the corresponding value of k'_x, etc.

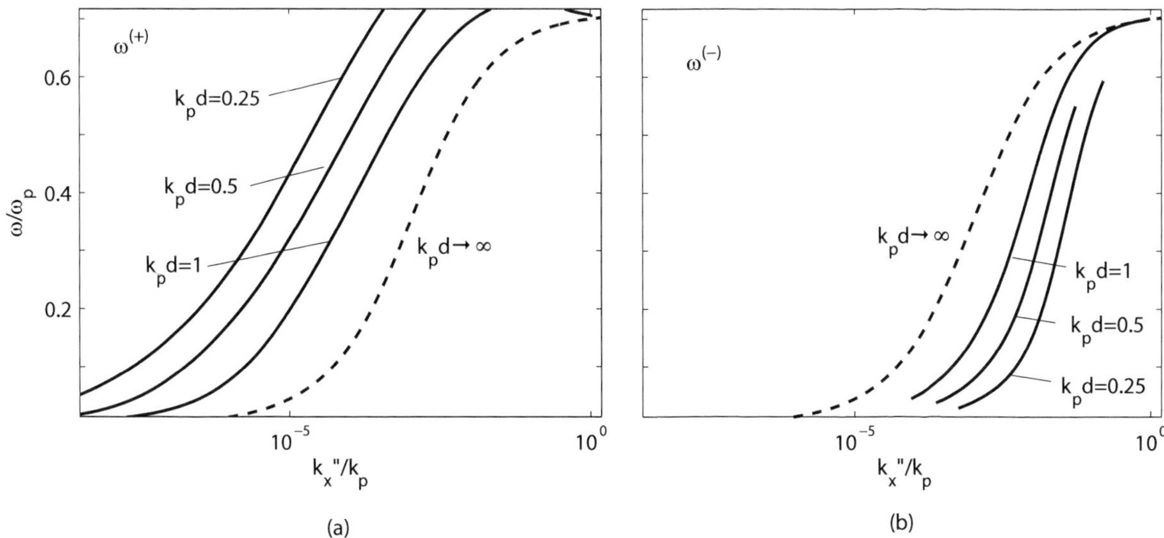

Fig. 3.21 Attenuation coefficient versus frequency for $\omega^{(+)}$-branch (a) and $\omega^{(-)}$-branch (b)

We shall not replot the frequency against normalized propagation constant curve—which remains practically the same. The frequency against normalized attenuation coefficient is plotted in Figs. 3.21(a) and (b) for the $\omega^{(+)}$ and the $\omega^{(-)}$ branches, respectively, for a range of normalized slab thicknesses. It may be seen that the waves in the upper branch are much less attenuated than the waves in the lower branch. In Fig. 3.22 we shall return to the concept of propagation length. It is plotted against the free-space wavelength for the $\omega^{(+)}$ branch. It may be clearly seen that as the thickness declines, the propagation length increases. The reason why attenuation is lower for the upper branch is far from being obvious. We shall show first the field distributions and then discuss in quite some detail the physical reasons why the upper branch deserves the epithet of 'long range'.

3.4.2 Field distributions

Let us now see the field distributions at one of these thicknesses ($k_{\mathrm{p}}d = 0.25$) for $\omega/\omega_{\mathrm{p}} = 0.3$, 0.5 and 0.7. They are plotted in Figs. 3.23(a)–(c). Let us first look at the symmetry. For the upper mode E_z and H_y (and, consequently, S_x) are symmetric as a function of z, whereas E_x is antisymmetric. For the lower mode E_z and H_y are antisymmetric (consequently, S_x is symmetric) and E_x is symmetric. It is also true, as we have already known for the single interface, that the decay away from the interfaces becomes stronger as $\omega/\omega_{\mathrm{p}}$ increases. It also follows from the figures that S_x is positive in the dielectric and flows in the opposite direction in the metal. As ω increases towards ω_{s} the net power flow tends to zero. We need to emphasize that for a given ω up to ω_{s} we had two different values of k_x belonging to two different branches of

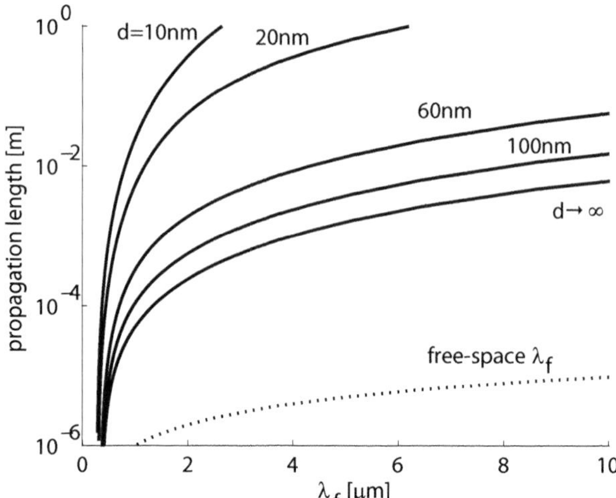

Fig. 3.22 Propagation length versus free-space wavelength

the dispersion curve. Above ω_s the situation is different. Although for a given value of ω there are still two values of k_x, they happen to belong to the same branch, to the upper branch. The field distribution corresponding to $\omega/\omega_p = 0.735$ is shown in Fig. 3.23(d). The symmetries are now the same at both values of k_x: the fields E_z, H_y and the Poynting vector S_x are symmetric and E_x is antisymmetric. As we may also expect, the decay of the fields away from the metal is steeper for the higher k_x. A good look at S_x will show (and numerical integration will prove it) that the total power carried is now in the opposite direction. The wave is a backward wave.

As we have seen, a field distribution is either symmetric or antisymmetric. So terminology should be easy: those that have symmetric distribution should be called symmetric modes and denoted by s, and those with antisymmetric distribution could be called antisymmetric (or asymmetric) modes and denoted by a. The problem is which field component should we refer to? Is it E_z, the field component normal to the direction of propagation, or E_x the field component in the direction of propagation? Opinions have been divided. Economou (1969) refers to the lower branch, $\omega^{(-)}$, as that with symmetric oscillation and to the upper branch as that with antisymmetric oscillation. For him it is the charge distribution, and the effect of the tangential component of the electric field on it, that is of primary importance. Burke *et al.* (1986) call the upper branch symmetric and the lower branch antisymmetric. According to Welford (1988), in the upper branch the two surface waves propagate π out of phase and hence should be referred to as the antisymmetric mode. In the branch with the lower ω the surface waves propagate in phase and should therefore be referred to as the symmetric mode. Burke *et al.* (1986) attributed the conflict in notations to the different background of the authors. The terminology in which the antisymmetric mode has a zero in its transverse electric field inside the film comes from integrated optics and is just opposite to the solid-state version, which

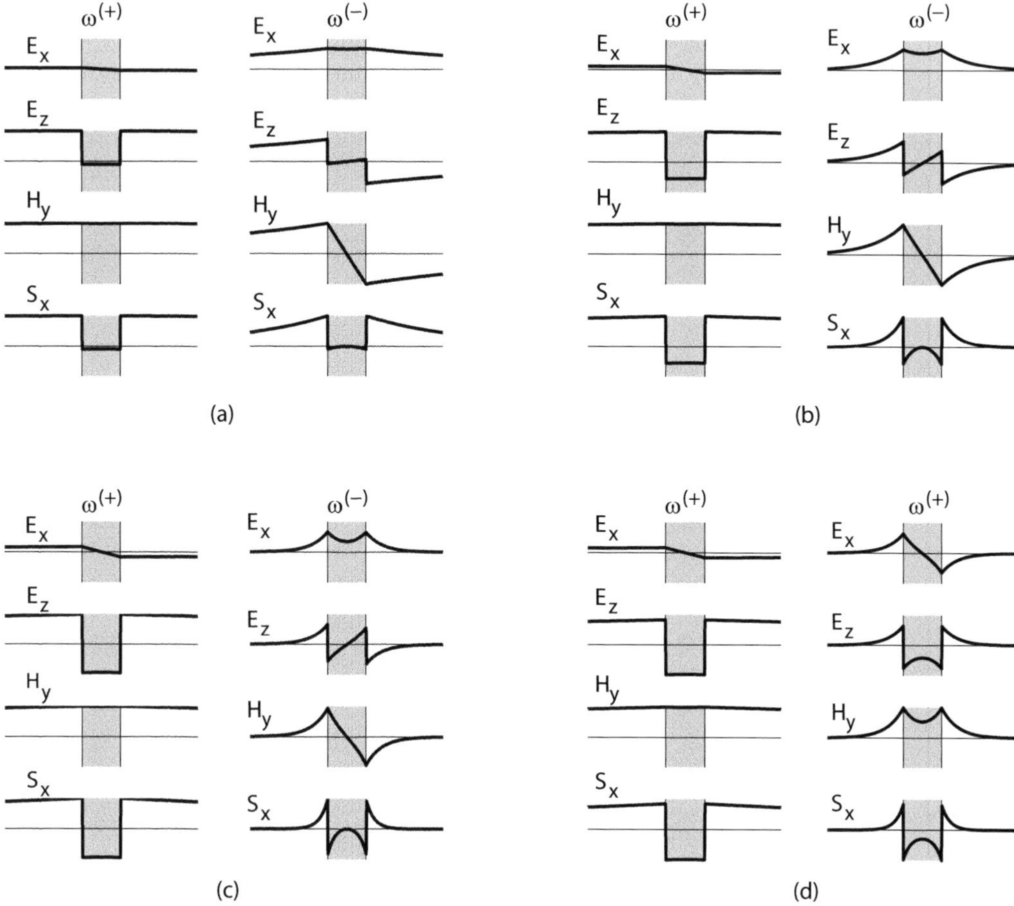

Fig. 3.23 Field distributions of $\omega^{(+)}$ and $\omega^{(-)}$-branches for four typical frequencies (a)–(d) $\omega/\omega_p = 0.3,\ 0.5,\ 0.7,\ 0.735$

deals with the symmetries of the charge distribution and therefore with the longitudinal components of the electric field. By now it has been largely accepted[16] that the mode in which the normal component of the electric field is symmetric is denoted by s and that in which the normal component is antisymmetric is called a. In addition, when the waves are bound the modes are referred to as s_b and a_b and when they are leaky[17] as s_l and a_l.

If the metal is lossless the modes are lossless. If the metal is lossy the modes are lossy, so much is clear. But which one is lossier? The calculations show that the s_b mode is less lossy, meriting the description of 'long-range surface plasmon'. We shall show curves later but first let us try to give a physical explanation.

One explanation may be based on the fact that the $\omega^{(+)}$-mode is nearer to the light line. If it is nearer it resembles more an electromagnetic wave. Electromagnetic waves travel in vacuum with no attenuation, hence the upper branch will be the one that has the lower attenuation.

[16] Maier *et al.* (2005) use a notation somewhat at variance with the usual notions of symmetry and antisymmetry. They refer to the $\omega^{(+)}$-mode as 'antisymmetric (s_b) mode' and to the $\omega^{(-)}$-mode as the 'symmetric (a_b) mode'.

[17] Leaky waves, as the name implies, leak out power, which is good for antennas but not for guiding the wave. Here, we are concerned only with bound modes so, with regret, we shall not discuss the properties of leaky waves.

Mansuripur *et al.* (2007) offers a somewhat more complicated explanation for the loss mechanism. With the even mode, the field component E_z has the same sign on the boundaries of the slab; therefore at a given point along x, the electrical charges on the upper and lower surfaces have opposite signs. Inside the metallic slab, the field component E_z—reduced by a factor $\varepsilon_{r2}/\varepsilon_{r1}$ relative to the E_z immediately outside—helps move the charges back and forth between the two surfaces. The slab being thin, the transport distance is short; hence the corresponding electrical current is small. In contrast, the charges of the odd mode have the same sign on opposite sides of the slab. Consequently, positive and negative charges must move in the x direction during each period of oscillation. The travel distance is now of the order of the SPP wavelength, which is greater than the slab thickness. Therefore, the current densities of the odd mode are relatively large, leading to correspondingly large losses.

A fairly similar reasoning is due to Zayats *et al.* (2005) giving the E_x component as being responsible for loss. For the $\omega^{(+)}$ mode E_x vanishes at the midplane of the slab, while it is a maximum at that plane for the $\omega^{(-)}$ mode. The mode with the smaller fraction of the electric field inside the metal causes less dissipation.

Barnes (2006) explains the loss mechanism as follows. The $\omega^{(+)}$ mode propagates in such a way that less of the power is carried in the metal than in the case of a single-interface mode. This reduces the effect of loss. Propagation lengths of the order of centimeters may be achieved. There is though a price to pay for such long propagation: the mode is not localized at the surface, it extends further into the dielectric, resembling the transverse plane electromagnetic wave.

Maybe the best explanation is based on the penetration depth in the metal. It may be seen in Figs. 3.23(a)–(c) that for a given frequency the antisymmetric mode at the higher k_x has a fast decay away from the metal surface. The symmetric mode has a slower decay, hence more field is in the dielectric, consequently, it is less lossy.

Another consideration is confinement of the fields. The more they are confined, the larger they are and that's a good thing for applications in which high fields are required. On the other hand, as the arguments above have shown, confinement leads to higher losses. Hence, some compromise between the two must always be made. For a discussion of the trade-off see Berini (2006).

3.4.3 Asymmetric structures

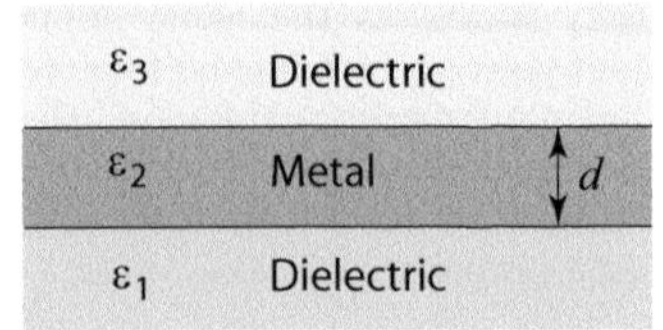

Fig. 3.24 Metallic slab surrounded by different dielectric media

We use here the terms symmetric and asymmetric in yet another sense, referring this time to structures. A symmetric structure is when the dielectrics on the two sides of the metal slab are identical and it is asymmetric when the dielectrics are different (see Fig. 3.24).

The asymmetric structure was investigated in some of the early papers (Sarid, 1981; Wendler and Haupt, 1986; Burke *et al.*, 1986; Yang *et al.*, 1991) but received much less attention than the symmetric case. If there

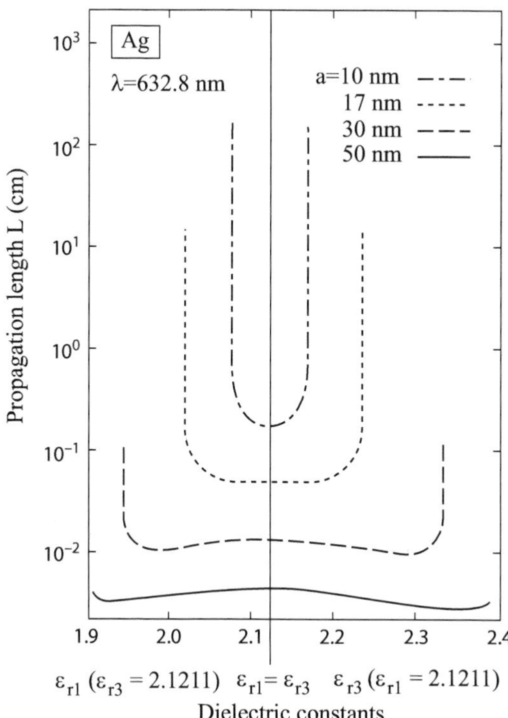

Fig. 3.25 Propagation length of the long-range SPP vs. dielectric constants ε_{r1} of the superstrate and ε_{r3} of the substrate for different thicknesses a of the Ag layer. From Wendler and Haupt (1986). Copyright © 1986 American Institute of Physics

is only a small difference in the dielectric constant then we can still talk about quasi-symmetric and quasi-antisymmetric modes but we find that the field confinement is very asymmetric. The quasi-symmetric mode is localized at the boundary between the metal and the dielectric with the lower dielectric constant, whereas the quasi-antisymmetric mode is localized at the higher dielectric constant boundary. The quasi-symmetric mode inherits the mantle of the s_b mode: it is still the long-range mode, it still has a lower attenuation and, as may be expected, the attenuation decreases as the slab thickness decreases. The new phenomenon is that the quasi-symmetric mode has a cutoff thickness below which no propagation is possible. This is bound to occur (Berini, 2001) because the s_b mode no longer has the chance quietly to convert into the TEM electromagnetic wave as $d \to 0$ because the different dielectrics on the two sides do not permit it. The quasi-antisymmetric mode remains the short-range mode, its attenuation increases with decreasing slab thickness, it has not got a cutoff thickness.

For detailed calculations of the propagation length in asymmetric structures see Wendler and Haupt (1986). They investigate theoretically thin silver slabs at the He-Ne wavelength of 632.8 nm with a dielectric constant of $\varepsilon_{r2} = -18 - \mathrm{j}\,0.47$ obtained from Johnson and Christy (1972). They first assume that both dielectrics have identical dielectric constants of $\varepsilon_{r1} = \varepsilon_{r3} = 2.1211$, then change the dielectric constant of one of them within a range of about 20%. They find that the propagation length first remains unchanged and then suddenly increases. Their results, quite spectacular, are shown in Fig. 3.25.

In a further study Zervas (1991) showed that long-range SPPs are supported even by highly asymmetric configurations.

3.5 Metal–dielectric–metal and periodic structures

It is now an easy exercise to derive dispersion for the surface modes for a thin dielectric layer with two semi-infinite metallic media adjacent to it. The dispersion equations are very similar to those of eqns (3.35) and (3.36) and can be derived in an analogous manner. In the final result the only difference is that κ_2 is being replaced by κ_1 in the arguments of the tanh and coth functions, i.e. the equations are given as

$$\zeta_{\rm e} = -\tanh\frac{\kappa_1 d}{2}\,, \tag{3.48}$$

and

$$\zeta_{\rm e} = -\coth\frac{\kappa_1 d}{2}\,. \tag{3.49}$$

As before, we shall try to find some approximations. For the upper branch $\omega^{(+)}$ we can take the argument of the coth function small and then without further approximations we find

$$k_x^2 = \left(\frac{2}{d}\right)^2\left[1-\left(\frac{\omega}{\omega_{\rm p}}\right)^2\right]\left[\left(\frac{\omega_{\rm p}}{\omega}\right)^4-\left(\frac{\omega_{\rm p}}{\omega}\right)^2-\left(\frac{k_{\rm p}d}{2}\right)^2\right]. \tag{3.50}$$

The frequency at which $k_x = 0$ may be obtained from the above equation as

$$\frac{\omega}{\omega_{\rm p}} = 1-\frac{1}{8}(k_{\rm p}d)^2\,, \tag{3.51}$$

i.e. the dispersion curve cuts the vertical axis just below the bulk plasma frequency.

We may again resort to the electrostatic approximation that turns out to be identical with that (eqn (3.36)) already derived for the dielectric–metal–dielectric structure.

The dispersion curves calculated from eqns (3.48) and (3.49) are plotted in Figs. 3.26(a) and (b) together with the electrostatic approximation and that given by eqn (3.50) for $k_{\rm p}d = 0.25$ and 0.5. The first thing to notice is that the $\omega^{(+)}$ branch crosses the light line. The wave can propagate at phase velocities higher than the velocity of light. The $\omega^{(-)}$ branch behaves as it did for the dielectric–metal–dielectric structure. It tends to $\omega_{\rm s}$ for $k_x \to \infty$ and to the light line as $k_x \to 0$. The electrostatic approximation is quite good for the $\omega^{(-)}$ mode and even better for the $\omega^{(+)}$ mode. The low k_x approximation, not surprisingly, deteriorates as k_x increases.

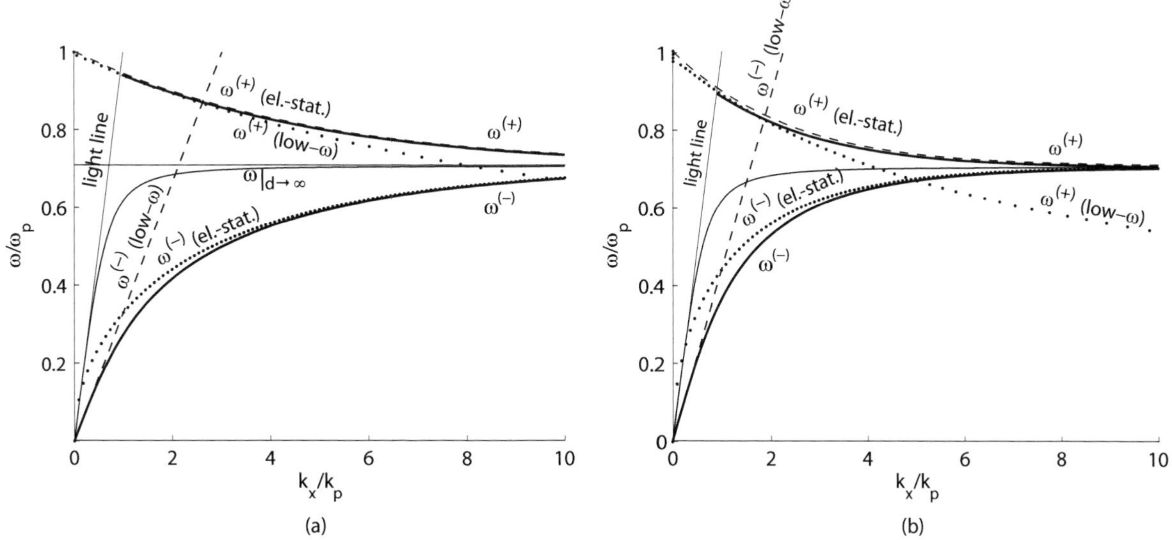

Fig. 3.26 Dispersion curves for a metal–dielectric–metal structure. Full solutions and approximation for (a) $k_{\mathrm{p}}d = 0.25$ and (b) $k_{\mathrm{p}}d = 0.5$

Economou (1969) goes on investigating two and more metal slabs between dielectrics, including the generalization to a periodic structure (see Fig. 3.27(a)). More and more layers in the structure lead to multiple splitting, i.e. to the appearance of more and more dispersion curves with the final result that both the $\omega^{(+)}$ and the $\omega^{(-)}$ branches widen into bands, as shown schematically in Fig. 3.27(b).

3.6 One-dimensional confinement: shells and stripes

All the problems we have investigated so far have been concerned with two-dimensional confinement of the surface waves, to a single surface or to multiple surfaces. Clearly, for any practical application we need a one-dimensional structure to take information from point A to point B. Such a structure was analyzed by Al-Bader and Imtar (1992). They considered a cylindrical metallic shell of inner radius a, and thickness t embedded in a dielectric or in two different dielectrics. They set up the differential equations for the fields of a TM mode and solved them numerically for silver at the He-Ne wavelength of 633 nm. They presented their results in terms of a mode index $n_{\mathrm{e}} = n'_{\mathrm{e}} - \mathrm{j}\, n''_{\mathrm{e}}$ where the complex propagation coefficient is given as $k = k_0 n_{\mathrm{e}}$. Their results were similar to those found for planar structures to which they reduced in the limit of infinite radius. They were also interested in the evolution of the TM_{01} fibre mode into surface plasmons, the model they considered was a dielectric cylinder embedded in metal. They found that the surface

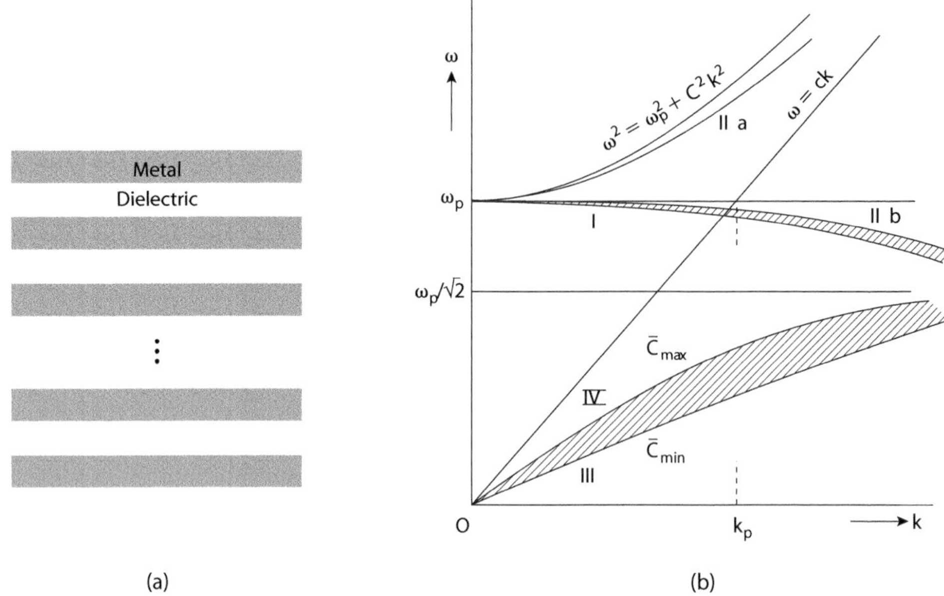

Fig. 3.27 Multilayer metal–dielectric structure. (a) Sketch of the configuration. (b) Dispersion relations. From Economou (1969). Copyright © 1969 by the American Physical Society

mode appeared for sufficiently large radius.

The physics for cylindrical structures is quite similar to that of planar structures. However, a host of new phenomena appear when the infinite slab is turned into the one-dimensional metal stripe of width w and thickness t, as shown in Fig. 3.28(a). They have received considerable attention lately (Berini, 1999; Berini, 2000*a*; Berini, 2000*b*; Berini, 2001; Berini, 2006; Al-Bader, 2004) due to their promise of signal processing in the visible and infra-red region. They solved the relevant partial differential equations subject to the boundary conditions by the method of lines. The main difference they found relative to the 2D case was the variety of field distributions that could occur. Simple TM modes no longer exist. All six field components must be present in all modes. The question then arises as to what nomenclature to use to identify the modes. The obvious start (Berini, 1999; Berini, 2000*a*) is with the $\mathrm{s_b}$ and $\mathrm{a_b}$ modes of 2D structures because the symmetry of the 1D structure will again permit symmetric and antisymmetric modes, in fact there are two symmetry axes in the x and y directions of Fig. 3.28(a). When $w/t \gg 1$ then the main transverse electric field is E_z whose symmetry with respect to the y and z axes is reflected in the use of the subscripts ss, sa, as, aa. A further superscript is then used to track the number of extrema observed in the spatial distribution of E_z along the y axis, and a second superscript would describe the extrema in the z direction (likely to exist but they have not been found so far). They are all bound modes, hence the subscript b is also added as for the 2D case. The new feature is that the higher-order modes have a cutoff width below

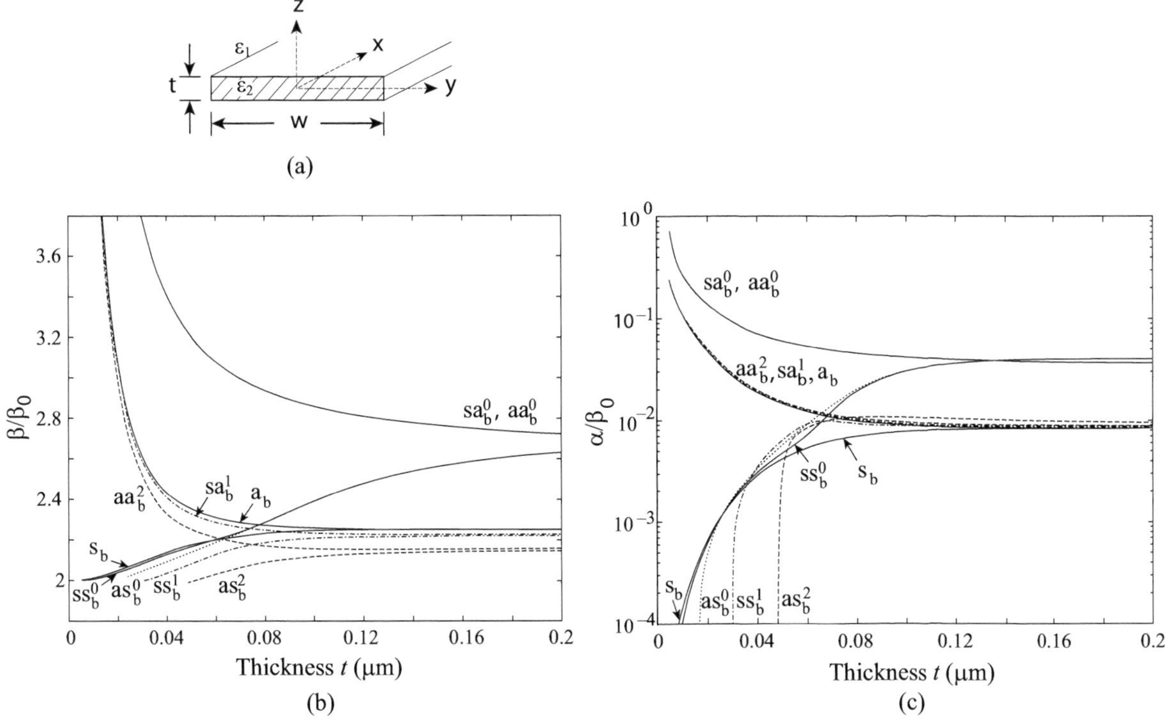

Fig. 3.28 Metallic stripe. (a) Sketch of the geometry. (b), (c) Dispersion with thickness of the first six modes. Also shown the a_b and s_b modes for $w \to \infty$ for comparision. From Berini (2000*b*). Copyright © 2000 by the American Physical Society

which they cannot propagate and some of the modes also have a cutoff thickness.

For a stripe width of $w = 0.5\,\mu$m the normalized propagation and attenuation coefficients are plotted in Figs. 3.28(b) and (c) for the first six modes as a function of stripe thickness. The s_b and a_b modes of the 2D structure are also shown for comparison. As may be expected, the antisymmetric modes have higher attenuation than the symmetric modes. The most relevant conclusion is that the 1D stripe geometry is in no way worse than the amply analyzed 2D geometry. In fact, it may be much better, at least in theory. The attenuation of two of the higher modes may be seen to decrease rapidly before they reach cutoff. Presumably, it happens at the expense of confinement. The ss_b^0 mode has probably more practical significance. If we compare it with the s_b mode at the lowest thicknesses shown, below 20 nm, we find that the attenuation of the stripe mode is very considerably below that of the 2D planar mode, and the attenuation can be further reduced by decreasing the stripe width.

Stripes having different dielectrics for substrates and superstrates have also been investigated by Berini (2001). Small differences in the upper and lower dielectric constants have been found to lead to larger differences in propagation properties. There has been no reply given as yet to the question posed by the very large propagation lengths (see Fig. 3.25) obtained theoretically by Wendler and Haupt (1986), whether they can be realized in stripe form.

3.7 SPP for arbitrary ε and μ

Looking at the surface modes at interfaces of two media we have so far assumed that one of the media is a dielectric with $\mu = 1$ and ε being positive and frequency-independent and that the other one is a metal.

The condition of existence of a surface TM mode at a single metal–dielectric interface was given as

$$\varepsilon_{\mathrm{r2}}/\varepsilon_{\mathrm{r1}} < -1\,, \tag{3.52}$$

which, assuming the Drude model for a metal, is satisfied below the surface plasma frequency,

$$\omega < \frac{\omega_{\mathrm{p}}}{\sqrt{1+\varepsilon_{\mathrm{r1}}}}\,. \tag{3.53}$$

In the case of a metal slab, the SPP modes excited at the two surfaces interact, the dispersion splits into two, one below and another one above the unperturbed dispersion curve. The resulting condition for the existence of surface modes is much more relaxed. In the limit of an infinitely thin slab it is

$$\varepsilon_{\mathrm{r2}}/\varepsilon_{\mathrm{r1}} < 0, \tag{3.54}$$

meaning that

$$\mathrm{sign}(\varepsilon_{\mathrm{r2}}) \neq \mathrm{sign}(\varepsilon_{\mathrm{r1}})\,, \tag{3.55}$$

and the upper SPP branch can reach up to the bulk plasma frequency

$$\omega < \omega_{\mathrm{p}}\,. \tag{3.56}$$

3.7.1 SPP dispersion equation for a single interface

We will now generalize these results to the case when ε and μ can take arbitrary values. At this stage we are not concerned with the question of how or whether these values can be realized. For an early treatment see Ruppin (2000). The derivations to follow are based on the work of Darmanyan *et al.* (2003). The mathematical problems of solving the field equations subject to the boundary conditions are not particularly difficult. There is, however, a problem with terminology. Allowing now the possibility of negative permeability what should we call them? Some people talk about magnetic plasmons,[18] which gives the opportunity to call them electric and magnetic surface plasmon–polaritons but that is quite a mouthful and it is unlikely that it would catch on. So we just decided to stick to the original name and in spite of the changes in permeability we shall still call them surface plasmon–polaritons.

[18] The analogy is not very good because there are no magnetic charges.

Having done the derivation for the case when the permittivity could take negative values it is an easy task to derive the dispersion equations for the surface waves in the general case. The approach is the same. The results are though more far reaching and will have significant implications for the imaging mechanism of the 'perfect' lens to be discussed in Chapter 5.

We disregard losses for simplicity. They can be included later if needed. Looking for surface-wave solutions declining exponentially we require the z components of the k vector to be purely imaginary so that both

$$\kappa_1 = \sqrt{k_x^2 - \varepsilon_{\mathrm{r}1}\mu_{\mathrm{r}1}k_0^2} \tag{3.57}$$

and

$$\kappa_2 = \sqrt{k_x^2 - \varepsilon_{\mathrm{r}2}\mu_{\mathrm{r}2}k_0^2} \tag{3.58}$$

are to be positive and real. The dispersion equation is formally the same as in the case of a metal. It may be obtained from the poles of the field solutions. For the TM polarization it is

$$\zeta_{\mathrm{e}} + 1 = 0 \qquad \text{(TM case)} \tag{3.59}$$

or

$$\frac{\kappa_2}{\kappa_1}\frac{\varepsilon_{\mathrm{r}1}}{\varepsilon_{\mathrm{r}2}} = -1 \qquad \text{(TM case)}, \tag{3.60}$$

whereas for the TE polarization we obtain

$$\zeta_{\mathrm{m}} + 1 = 0 \qquad \text{(TE case)} \tag{3.61}$$

or

$$\frac{\kappa_1}{\kappa_2}\frac{\mu_{\mathrm{r}2}}{\mu_{\mathrm{r}1}} = -1 \qquad \text{(TE case)}. \tag{3.62}$$

Inserting eqns (3.57) and (3.58) into eqn (3.60) we obtain for the TM case

$$(k_x^2 - \varepsilon_{\mathrm{r}2}\mu_{\mathrm{r}2}k_0^2)\varepsilon_{\mathrm{r}1}^2 = (k_x^2 - \varepsilon_{\mathrm{r}1}\mu_{\mathrm{r}1}k_0^2)\varepsilon_{\mathrm{r}2}^2 \qquad \text{(TM case)}, \tag{3.63}$$

leading to the dispersion equation

$$k_x^2 = k_0^2 \frac{\varepsilon_{\mathrm{r}1}\varepsilon_{\mathrm{r}2}}{\varepsilon_{\mathrm{r}1} + \varepsilon_{\mathrm{r}2}} \frac{\mu_{\mathrm{r}1}\varepsilon_{\mathrm{r}2} - \mu_{\mathrm{r}2}\varepsilon_{\mathrm{r}1}}{\varepsilon_{\mathrm{r}2} - \varepsilon_{\mathrm{r}1}} \qquad \text{(TM case)}. \tag{3.64}$$

Similarly for the TE case, inserting eqns (3.57) and (3.58) into eqn (3.62) we obtain the dispersion equation

$$k_x^2 = k_0^2 \frac{\mu_{\mathrm{r}1}\mu_{\mathrm{r}2}}{\mu_{\mathrm{r}1} + \mu_{\mathrm{r}2}} \frac{\varepsilon_{\mathrm{r}1}\mu_{\mathrm{r}2} - \varepsilon_{\mathrm{r}2}\mu_{\mathrm{r}1}}{\mu_{\mathrm{r}2} - \mu_{\mathrm{r}1}} \qquad \text{(TE case)}. \tag{3.65}$$

For the z components of the k vector in the TM case we obtain by using eqns (3.57), (3.58) and (3.64)

$$\kappa_1^2 = k_0^2 \varepsilon_{r1}^2 \frac{\varepsilon_{r1}\varepsilon_{r2}}{\varepsilon_{r1}+\varepsilon_{r2}} \frac{\mu_{r1}\varepsilon_{r1}-\mu_{r2}\varepsilon_{r2}}{\varepsilon_{r2}^2-\varepsilon_{r1}^2} \qquad \text{(TM case)} \tag{3.66}$$

and

$$\kappa_2^2 = \kappa_1^2 \varepsilon^2 \qquad \text{(TM case)}, \tag{3.67}$$

and in the TE case by using eqns (3.57), (3.58) and (3.65)

$$\kappa_1^2 = k_0^2 \mu_{r1}^2 \frac{\mu_{r1}\mu_{r2}}{\mu_{r1}+\mu_{r2}} \frac{\varepsilon_{r1}\mu_{r1}-\varepsilon_{r2}\mu_{r2}}{\mu_{r2}^2-\mu_{r1}^2} \qquad \text{(TE case)} \tag{3.68}$$

and

$$\kappa_2^2 = \kappa_1^2 \mu^2 \qquad \text{(TE case)}, \tag{3.69}$$

where $\varepsilon = \varepsilon_{r2}/\varepsilon_{r1}$ and $\mu = \mu_{r2}/\mu_{r1}$.

3.7.2 Domains of existence of SPPs for a single interface

It follows immediately from eqn (3.60) that, as both κ_1 and κ_2 are to be real and positive, for TM modes to exist the condition

$$\varepsilon_{r2}\varepsilon_{r1} < 0 \qquad \text{(TM case)} \tag{3.70}$$

must be satisfied. For TE modes, as follows from eqn (3.62) the condition is

$$\mu_{r2}\mu_{r1} < 0 \qquad \text{(TE case)}. \tag{3.71}$$

The result for the TM polarization makes good sense. We are talking here about the electric field with its E_z component changing sign across the boundary, a condition compatible with accumulation of charges at the surface, as we discussed earlier for metals. It is important to note that this condition holds for any value of μ: surface TM modes exist only if ε changes sign at the boundary.

The same argument could be applied for the TE polarization, but it is somewhat difficult to provide a simple physical interpretation of the requirement that μ_1 and μ_2 must be of different sign for a TE surface mode to exist. The reason why it is difficult to picture what is going on is that in the case of negative permeability we cannot appeal to our common-sense expectations. Negative-ε materials have received much more attention in the past and the condition $\varepsilon_{r1}\varepsilon_{r2} < 0$ for the existence of TM surface modes has been known ever since Fano's article over sixty years ago. It is quite different with negative-μ materials. All such materials, constructed so far, have negative permeability within a frequency band and not below a certain frequency, as it happens with negative-permittivity materials. Also, we cannot think of surface charges. Instead we must consider surface currents, a less familiar concept, and it is particularly difficult to think about it when the structure is discrete.

Returning to the dispersion equations and taking into account that for TM polarization $\varepsilon_{r2} < 0$, eqns (3.64) and (3.66) can be rewritten as

$$k_x^2 = k_0^2 \varepsilon_{r1} \mu_{r1} \frac{|\varepsilon|(|\varepsilon| + \mu)}{\varepsilon^2 - 1} \qquad \text{(TM case)} \qquad (3.72)$$

and

$$\kappa_1^2 = k_0^2 \varepsilon_{r1} \mu_{r1} \frac{1 + |\varepsilon|\mu}{\varepsilon^2 - 1} \qquad \text{(TM case)}. \qquad (3.73)$$

Similarly, taking into account that for TE polarization $\mu_{r2} < 0$, eqns (3.65) and (3.67) can be rewritten as

$$k_x^2 = k_0^2 \varepsilon_{r1} \mu_{r1} \frac{|\mu|(|\mu| + \varepsilon)}{\mu^2 - 1} \qquad \text{(TE case)} \qquad (3.74)$$

and

$$\kappa_1^2 = k_0^2 \varepsilon_{r1} \mu_{r1} \frac{1 + |\mu|\varepsilon}{\mu^2 - 1} \qquad \text{(TE case)}. \qquad (3.75)$$

The TM and TE surface modes of eqns (3.72) and (3.74) exist if k_x^2, κ_1^2 and κ_2^2 are real and positive. In the case of TM polarization, it follows from eqn (3.73) that κ_1^2 is positive if either

$$|\varepsilon| > 1 \qquad \text{and} \qquad \mu > -\frac{1}{|\varepsilon|} \qquad \text{(TM case)} \qquad (3.76)$$

or

$$|\varepsilon| < 1 \qquad \text{and} \qquad \mu < -\frac{1}{|\varepsilon|} \qquad \text{(TM case)}. \qquad (3.77)$$

Similarly, for TE polarization, the condition $\kappa_1^2 > 0$ in eqn (3.75) is fulfilled if either

$$|\mu| > 1 \qquad \text{and} \qquad \varepsilon > -\frac{1}{|\mu|} \qquad \text{(TE case)} \qquad (3.78)$$

or

$$|\mu| < 1 \qquad \text{and} \qquad \varepsilon < -\frac{1}{|\mu|} \qquad \text{(TE case)}. \qquad (3.79)$$

If κ_1^2 is positive then κ_2^2 is positive as well, see eqns (3.67) and (3.69). Similar considerations for the positiveness of k_x^2 (eqns (3.64) and (3.74)), do not provide any further constraints for the existence of the surface modes. The domains of existence of the TM mode and of the TE mode are plotted in Fig. 3.29 in the plane (ε, μ).

The horizontal dashed line $\mu = 1$ corresponds to the case of the second medium being a metal with a TM surface mode for $\varepsilon < -1$ ($\varepsilon_{r2} < -\varepsilon_{r1}$) being the only possibility.

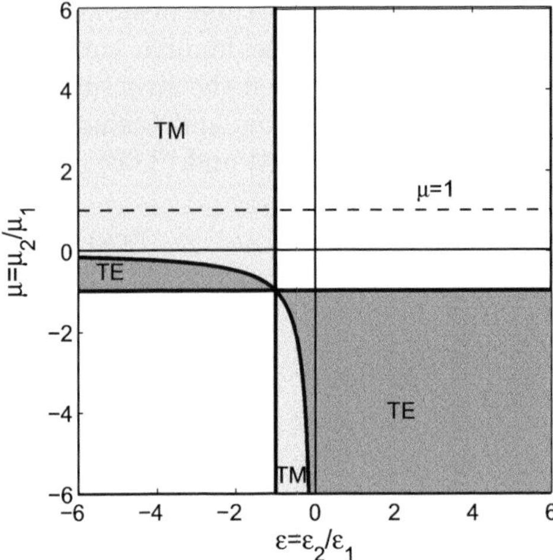

Fig. 3.29 Domains of TM and TE modes for an interface dielectric–metamaterial with arbitrary values of permittivity and permeability

It is worth mentioning that there can be no surface modes at the boundary of two left-handed media with $\varepsilon_{\mathrm{r}1}$, $\varepsilon_{\mathrm{r}2}$, $\mu_{\mathrm{r}1}$, $\mu_{\mathrm{r}2}$ all being negative (Darmanyan *et al.*, 2003) (this corresponds to the first quadrangle of the plane ε, μ).

It is important to note that if, in a certain frequency range, in medium 2 both $\varepsilon_{\mathrm{r}2}$ and $\mu_{\mathrm{r}2}$ are negative, i.e. if medium 2 is a left-handed one then a transverse electromagnetic (bulk) wave can propagate with the dispersion equation

$$k = \frac{\omega}{c}\sqrt{\varepsilon_{\mathrm{r}2}\mu_{\mathrm{r}2}}\,. \tag{3.80}$$

For a surface mode (with fields exponentially decaying away from the interface) to exist, its wave vector must be larger than those of the bulk modes in either media, resulting in the condition

$$k_x > \max\left\{\left(k_0 = \frac{\omega}{c}\sqrt{\varepsilon_{\mathrm{r}1}\mu_{\mathrm{r}1}}\right), \left(k = \frac{\omega}{c}\sqrt{\varepsilon_{\mathrm{r}2}\mu_{\mathrm{r}2}}\right)\right\}\,. \tag{3.81}$$

This is a new feature that we have not met in the case of a metal where surface modes were only found in the stop bands of the bulk modes. In the case of a left-handed metamaterial with both ε_2 and μ_2 being negative a bulk and a surface mode can coexist at the same frequency. If their dispersion curves $\omega(k)$ intersect, only those parts of the surface mode solutions that lie to the right of the bulk modes are meaningful.

3.7.3 SPP at a single interface to a metamaterial: various scenarios

The results of the previous section are general and can be used to look at possible SPP modes for any values of ε and μ. There is an obvious symmetry regarding ε and μ on one side and TM and TE modes on the other side. It can be easily seen that the results for TM and TE SPP

Table 3.1 Numerical values for the frequencies at which surface modes start or stop. $F = 0.56$

$\frac{\omega_0}{\omega_p}$	$\frac{\omega\|_{\varepsilon\mu=1}}{\omega_p}$	$\frac{\omega\|_{\mu=-1}}{\omega_p}$	$\frac{\omega\|_{\mu=0}}{\omega_p}$	(i)TM forw.	(ii)TE backw.	(ii)TM backw.	(iii)TM forw.	(iii)TE forw.
0.4	0.52	0.47	0.6	yes	yes	—	yes	—
0.6	0.71	0.71	0.9	yes	—	—	—	—
0.8	0.86	0.94	1.2	yes	—	yes	—	yes
1.0	—	1.2	1.5	yes	—	—	—	yes

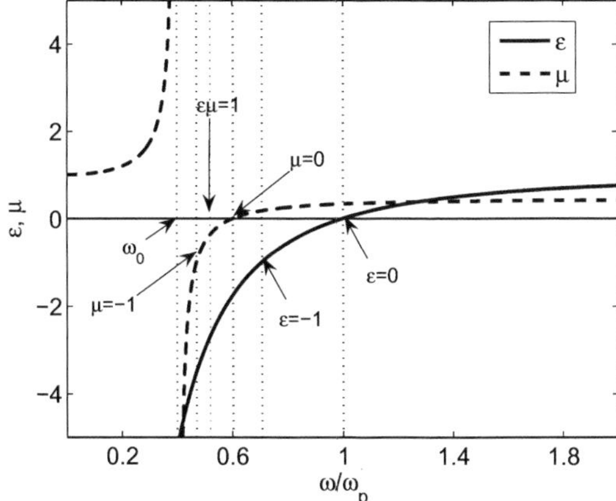

Fig. 3.30 Frequency dependence of permittivity and permeability for $F = 0.56$, $\omega_0 = 0.4\,\omega_p$

modes transform into each other by the mutual substitution $\varepsilon \leftrightarrow \mu$. As we will see in this section the symmetry is broken when ε_{r2} and μ_{r2} have different types of frequency dependence.

Here, we will look at the dispersion curves of the possible surface modes assuming that the dielectric constant of medium 2 follows the Drude model,

$$\varepsilon_{r2} = 1 - \frac{\omega_p^2}{\omega^2}, \tag{3.82}$$

and that the permeability has a resonant behaviour that can be described as (see Section 2.8)

$$\mu_{r2} = 1 - \frac{F\omega^2}{\omega^2 - \omega_0^2}. \tag{3.83}$$

These forms of ε_{r2} and μ_{r2} were frequently used in describing the response of metamaterials comprising rods (Pendry *et al.*, 1996) and SRRs (Pendry *et al.*, 1999). The choice of the parameters, of the plasma frequency, ω_p, of the magnetic resonance frequency ω_0 and of the filling factor, F, has a crucial effect on the behaviour of the metamaterial, in

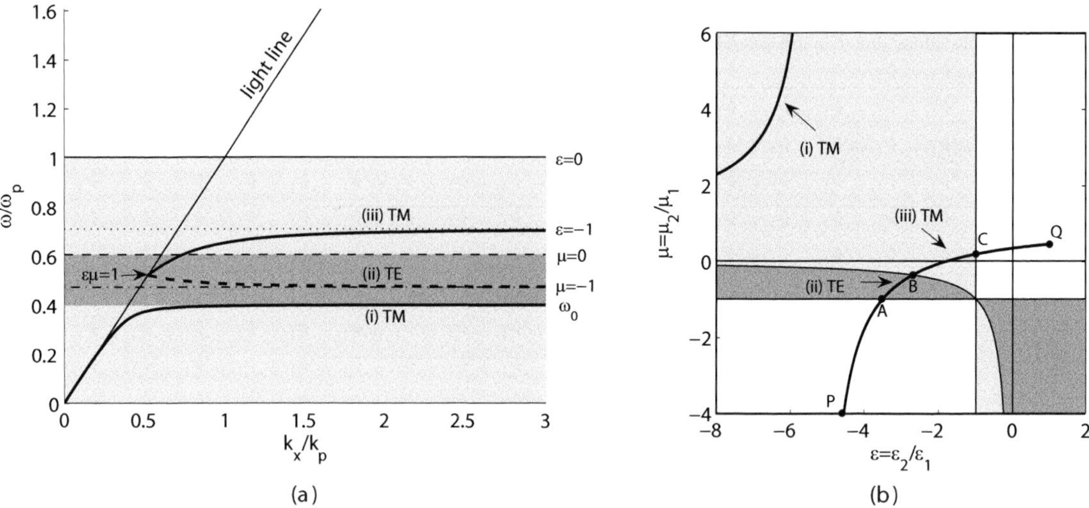

Fig. 3.31 SPP dispersion for single interface. $F = 0.56$, $\omega_0 = 0.4\,\omega_{\rm p}$. (a) ω–k_x diagram. (b) μ–ε diagram

general, and on the character of the surface modes, in particular. Let us recall the most important asymptotes and limits for permeability and permittivity,

$$\varepsilon_{\rm r2}: \begin{cases} \varepsilon_{\rm r2} \to 1 & \text{for} \quad \omega \to \infty \\ \varepsilon_{\rm r2} > 0 & \text{for} \quad \omega > \omega_{\rm p} \\ \varepsilon_{\rm r2} < 0 & \text{for} \quad \omega < \omega_{\rm p} \\ \varepsilon_{\rm r2} = -1 & \text{for} \quad \omega = \dfrac{\omega_{\rm p}}{\sqrt{2}} \\ \varepsilon_{\rm r2} \to -\infty & \text{for} \quad \omega \to 0 \end{cases} \tag{3.84}$$

and

$$\mu_{\rm r2}: \begin{cases} \mu_{\rm r2} \to 1 - F & \text{for} \quad \omega \to \infty \\ \mu_{\rm r2} > 0 & \text{for} \quad \omega > \dfrac{\omega_0}{\sqrt{1-F}} \\ \mu_{\rm r2} < 0 & \text{for} \quad \omega_0 < \omega < \dfrac{\omega_0}{\sqrt{1-F}} \\ \mu_{\rm r2} = -1 & \text{for} \quad \omega = \dfrac{\omega_0\sqrt{2}}{\sqrt{2-F}} \\ \mu_{\rm r2} \to \mp\infty & \text{for} \quad \omega \to \omega_0 \pm 0 \\ \mu_{\rm r2} > 0 & \text{for} \quad \omega < \omega_0 \\ \mu_{\rm r2} \to 1 & \text{for} \quad \omega \to 0 \end{cases} . \tag{3.85}$$

We will now give examples of four quite different scenarios for SPP modes using typical sets of parameters. In all examples we keep $F = 0.56$ and choose for $\omega_0/\omega_{\rm p}$ values of 0.4, 0.6, 0.8 and 1. In all four cases, the frequency at which $\varepsilon_{\rm r} = -1$ is the surface plasmon frequency, $\omega_{\rm p}/\sqrt{2} \simeq 0.71\omega_{\rm p}$ and the frequency at which $\varepsilon_{\rm r} = 0$ is of course the plasma frequency, $\omega_{\rm p}$. Table 3.1 gives numerical values for the frequencies of interest from eqn (3.85) in the four cases.

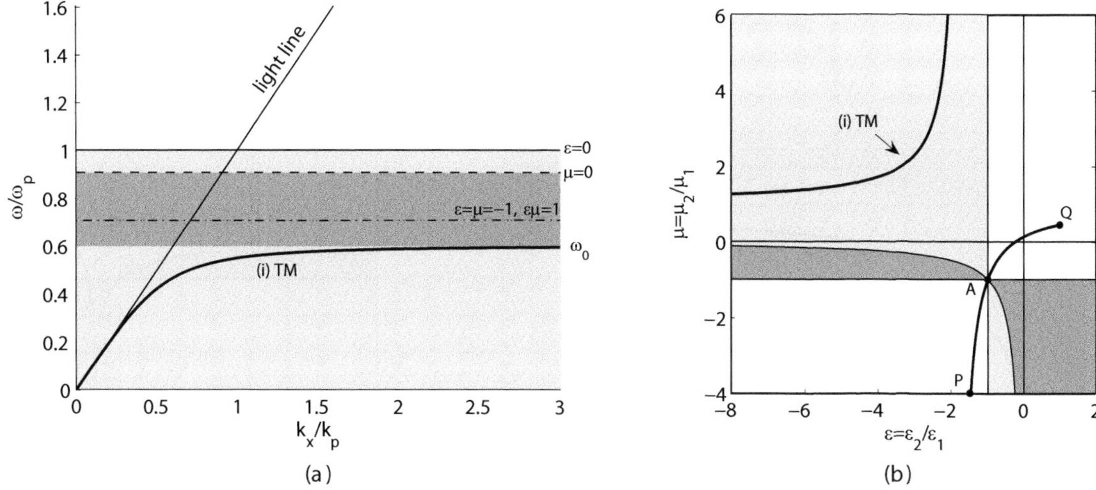

Fig. 3.32 SPP dispersion for single interface. $F = 0.56$, $\omega_0 = 0.6\,\omega_\mathrm{p}$. (a) ω–k_x diagram. (b) μ–ε diagram

Case (a) $\omega_0/\omega_\mathrm{p} = 0.4$. The variations of $\varepsilon(\omega)$ and $\mu(\omega)$ as a function of frequency are shown in Fig. 3.30. The question we are asking now is what kind of surface modes may exist for this set of parameters. For simplicity we shall take $\varepsilon_{\mathrm{r}1} = \mu_{\mathrm{r}1} = 1$, although the equations are valid for the case of a general lossless dielectric. It can be shown from eqns (3.76)–(3.79) that there are three SPP modes, as may be seen in Fig. 3.31(a). The first one is a TM mode that exists in the range $0 < \omega < \omega_0$. Thus, the presence of the magnetic resonance limits the range of the TM mode. Instead of moving up to ω_s it stops now at ω_0. In other words, the upper frequency limit occurs where $\mu_{\mathrm{r}2}$ turns negative instead of $\varepsilon_{\mathrm{r}2} = -1$. The second mode is a TE one that starts at the light line at a value of ω where $\varepsilon_{\mathrm{r}2}\mu_{\mathrm{r}2} = 1$ and tends asymptotically to the value of ω where $\mu_{\mathrm{r}2} = -1$. It is a backward wave. The third mode is again a TM one. It starts at the same point as the TE mode but it moves upwards to tend asymptotically to the $\varepsilon_{\mathrm{r}2} = -1$ line. Not shown in the figure is the bulk mode that propagates for the range of frequencies for which both $\varepsilon_{\mathrm{r}2}$ and $\mu_{\mathrm{r}2}$ are negative.

Let us now return to our diagram of Fig. 3.29 showing the areas in the $\varepsilon\mu$ plane in which surface modes are possible. Each of our three modes may be represented by a curve in this plane replotted in Fig. 3.31(b). For the lower TM mode μ is positive and ε is negative. As ω varies from 0 to ω_0 the curve denoted by (i) is described. As ω tends to ω_0 we know that the permeability tends to infinity and then suddenly changes to minus infinity. Up to $\mu = \infty$ there is a TM mode, but when the permeability switches to $-\infty$ the surface mode is no longer there. As the frequency increases further the permeability now increases from minus infinity. At a certain value of the frequency it reaches the point P and then traverses a number of regions between P and Q. In the PA region, as may be seen from the shading, there cannot be surface modes.

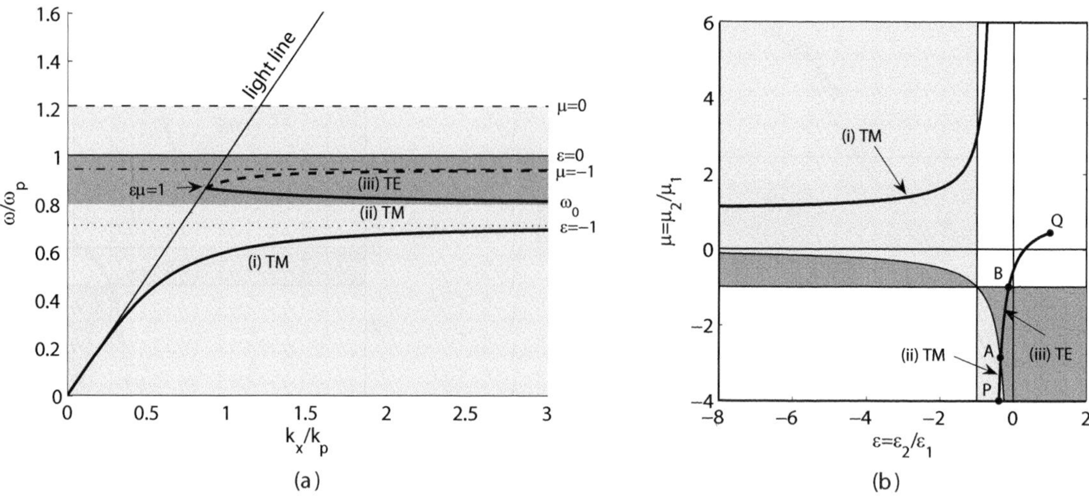

Fig. 3.33 SPP dispersion for single interface. $F = 0.56$, $\omega_0 = 0.8\,\omega_\mathrm{p}$. (a) ω–k_x diagram. (b) $\mu - \varepsilon$ diagram

The TE region is entered at A and this mode exists until the frequency is high enough to reach B. This is the mode we have denoted by (ii). Increase ω again and we move from B to C, which is in the TM mode region corresponding to mode (iii). The curve continues beyond C as ω increases but no further surface modes are possible. It is interesting to note that this novel presentation in terms of geometrical loci in the ε,μ plane is not only capable of accounting for all the modes but also makes it clear how the various surface modes arise.

Next, let us consider what happens when ω_0 increases, i.e. the magnetic resonance moves up in frequency. It may be shown that both the TE mode and the upper TM mode we have seen in Fig. 3.31(b) become flatter and then disappear altogether. For case (b), when $\omega_0/\omega_\mathrm{p} = 0.6$ is taken, only the lower TM mode survives, as shown in the dispersion curve of Fig. 3.32(a). The geometrical locus of this TM mode in the $\varepsilon\mu$ plane resembles the one in Fig. 3.31(b), but has shifted to the right, as illustrated in Fig. 3.32(b). What happened to the other two modes? Could we get our answer from the geometrical loci? What happened is that the PQ curve also moved to the right. At this particular value of $\omega_0/\omega_\mathrm{p} = 0.6$ it crosses the $\varepsilon = -1$, $\mu = -1$ point at A without ever crossing the area in which either surface mode exists.

When ω_0 increases further there is an inversion of the upper TE and TM modes. First comes a backward TM mode and further up (although starting again at the same point) a forward TE mode. These changes are indicated by the next set of curves taken for $\omega_0/\omega_\mathrm{p} = 0.8$ (Fig. 3.33). Note that $\omega_0 > \omega_\mathrm{s}$, hence the magnetic resonance does not affect the lower TM mode. Its range is, as in the case when μ is frequency-independent, between $\omega = 0$ and $\omega = \omega_\mathrm{s}$. The dispersion curves of the three modes are plotted in Fig. 3.33(a) and the corresponding curves in the ε,μ plane in Fig. 3.33(b). It may be clearly seen that in the interval

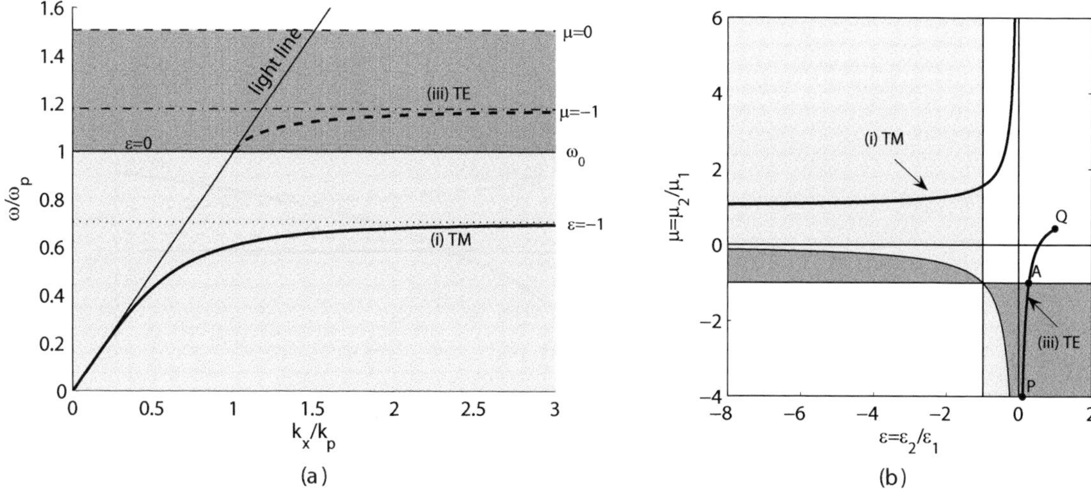

Fig. 3.34 SPP dispersion for single interface. $F = 0.56$, $\omega_0 = \omega_\mathrm{p}$. (a) ω–k_x diagram. (b) μ–ε diagram

PA we have a TM mode, between A and B a TE mode, and no surface mode beyond B.

In case (d) we have chosen $\omega_0/\omega_\mathrm{p} = 1$. There is now no region for which μ and ε are simultaneously negative. The lower TM mode, as may be expected, remains unchanged. The upper TM mode has disappeared and only one TE mode remains in the frequency region where the permeability is negative. The dispersion curve is shown in Fig. 3.34(a) and the geometrical locus of the modes in Fig. 3.34(b). The region PA is now responsible for the upper TE mode and no surface mode exists beyond A.

We may summarize at this point the rules governing the appearance of the upper TM and TE modes. The upper TM mode exists in the frequency range between the values that give $\varepsilon\mu = 1$ and that giving $\varepsilon = -1$. Depending on which one is higher for the particular value of $\omega_0/\omega_\mathrm{p}$ it is a forward wave or a backward wave. When those two limits take identical values the upper TM mode disappears. The upper TE mode originates at the same point as the upper TM mode and tends asymptotically to the line $\mu = -1$. Depending again on the relative positions of these two points the TE wave is a backward wave or a forward wave or it just vanishes when it changes from one into the other one.

Concerning the number of modes for a given value of $\omega_0/\omega_\mathrm{p}$ they vary between one and three. Specifically, there is only one mode for case (b), two modes for case (d), and three modes for cases (a) and (d). Case (a) has been considered by Ruppin (2001). Case (b) includes the perfect lens condition $\varepsilon = \mu = -1$ at $\omega = \omega_\mathrm{s}$.

Note that the total number of modes is dictated by the chosen form for the frequency dependence of ε and μ that result in various ways of crossing the regions in which surface modes are allowed. A different

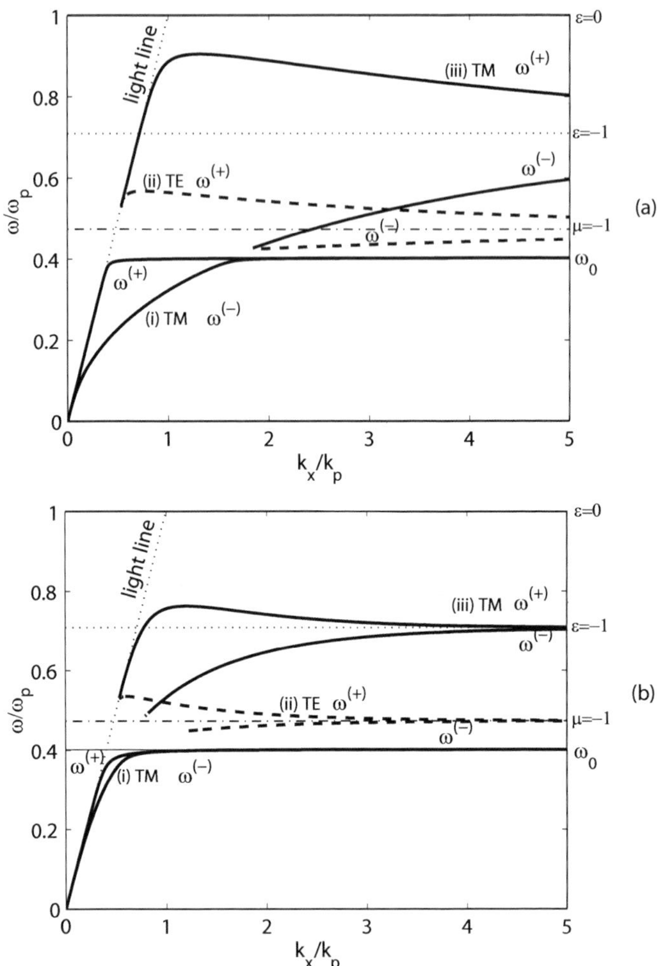

Fig. 3.35 SPP dispersion for a slab. $F = 0.56$, $\omega_0 = 0.4\,\omega_\mathrm{p}$. $k_\mathrm{p}d = 0.25$ (a) and 1 (b)

frequency dependence for ε and μ (e.g. with multiple resonances) can result in a smaller or larger number of modes. The results of Section 3.7.2 summarized in Fig. 3.29 would still apply.

A number of authors looked at different possibilities for the dependence of ε and μ on frequency. The results reported by Ruppin 2000; Darmanyan *et al.* 2003 can be reproduced with our model.

3.7.4 SPP modes for a slab of a metamaterial

The story is the same again as for the epsilon-negative-only medium. If the slab is sufficiently thin the modes on each side are coupled to each other. As a consequence, the number of branches in the dispersion equation of a slab will double in comparison to the single interface. Each mode splits into a symmetric and an antisymmetric branch. We take here only two examples, case (a) and case (b) of the previous section, i.e. for $\omega_0 = 0.4\omega_\mathrm{p}$ and $\omega_0 = 0.6\omega_\mathrm{p}$. The resulting dispersion curves for

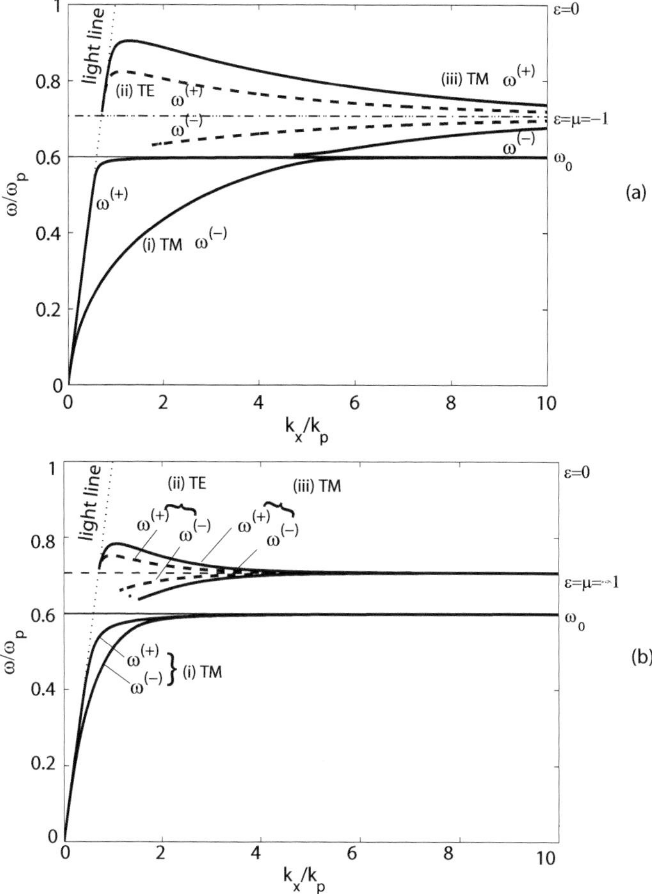

Fig. 3.36 SPP dispersion for a slab. $F = 0.56$, $\omega_0 = 0.6\,\omega_\mathrm{p}$. $k_\mathrm{p}d = 0.25$ (a) and 1 (b)

$k_\mathrm{p}d = 0.25$ and 1.0 are shown in Figs. 3.35(a) and (b) and Figs. 3.36(a) and (b). Comparing them to the unperturbed solutions for the single interface shows some striking features in case (b). Apart from the low-ω TM, which splits into two, two further modes, a backward and a forward one, appear that asymptotically approach the miraculous frequency at which $\varepsilon_\mathrm{r} = \mu_\mathrm{r} = -1$. The higher k_x is, the closer are the two modes (one above and one below) to the single interface curve. Due to the different character of the frequency dependence of ε and μ, the split between the TE modes looks different from the split between the TM modes (it is actually larger for TM). We will return to this case later when discussing subwavelength imaging properties of the perfect lens.

The asymmetric structure when the two media surrounding the slab are different were first considered by Ruppin (2001) and later generalized by Tsakmakidis *et al.* (2006).

We finish this section by mentioning an idea due to Oulton *et al.* (2008). The authors note that, due to losses and stringent fabrication requirements, practical SPP waveguides have not succeeded in producing field confinement beyond that of dielectric waveguides. The new idea is to form a hybrid waveguide between a cylindrical dielectric nanowire and a metallic plate. In that case much of the power will propagate in the space between the nanowire and the plate, reducing thereby the losses, and confinement will be determined by the distance between them, which can be made very small by available semiconductor fabrication techniques. For further comments on this waveguide see Maier (2008).

Small resonators

4

4.1 Introduction

The kind of resonators we need for metamaterials has already been discussed in Section 2.5. They have to be small relative to the wavelength and at the same time should be accessible to electric and/or magnetic fields. The elements that were proved useful for showing the existence of negative material parameters and negative refraction (Smith *et al.*, 2000; Shelby *et al.*, 2001*a*) were metallic rods and split-ring resonators. The combination was new. It is true, however, that those elements and all those introduced later were the products of a long line of development. This development is of interest on its own account because it tells us how the practical need of the manipulation of RF and microwave signals led to a variety of approaches that still give inspiration when new applications arise or simply when one wishes to improve an earlier design.

The aim of this chapter is to present most of the small resonators proposed up to the time this manuscript was sent to the publisher. It provides a fairly comprehensive list. Obviously, there is no space to go into much detail but we shall make an attempt to emphasize the main features of the design, and of course, in many cases it is sufficient just to look at the figures to appreciate how the resonance comes about and what function the resonator might fulfil. The order of presentation will be chronological at first. In Section 4.2 we shall introduce small resonators, starting with the invention of the re-entrant cavity in 1939. For decades afterwards the main application of small resonators was for providing crucial components in microwave tubes, for detecting nuclear magnetic and electron spin resonance, and for microwave filters. The emphasis changed around 1999–2000. Since then, most of the research on small resonators has been directed towards realizing negative material parameters: negative permittivity and negative permeability. Those new developments will be presented in Section 4.3, including the attempts to reach higher and higher frequencies. These two sections will provide an extensive list, a kind of catalogue, with relatively little discussion and without mathematical models, but showing some experimental results as well. In Section 4.4 the mathematical model of Shamonin *et al.* 2004; Shamonin *et al.* 2005 is introduced. In contrast to other analyses, the equivalent circuit of the split-ring resonator is presented there in terms of distributed instead of the usual lumped circuits. It is shown there that the current and voltage distributions may be obtained

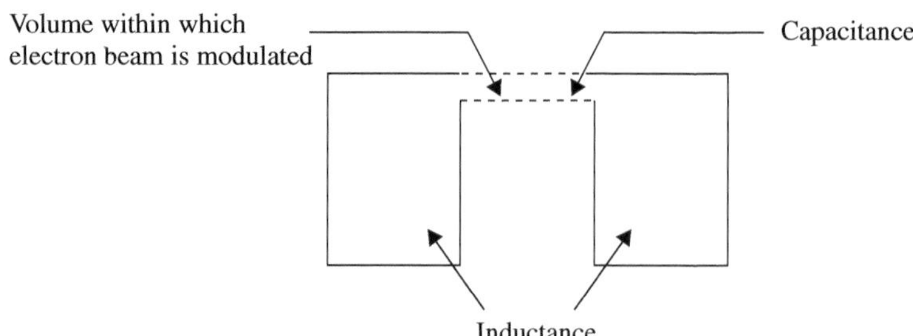

Fig. 4.1 Re-entrant cavity

by solving a second-order differential equation, and the solution will also yield the resonant frequency. Theoretical predictions of the resonant frequency will be compared with the experimental results of Radkovskaya *et al.* (2005) and also with those obtained by a numerical package. In Section 4.5 we have a brief look at the neglected topic of higher-order resonances. We shall compare some of the predictions of the distributed circuit model with those of simulations.

4.2 Early designs: a historical review

The need for small resonators arose at the design of the first microwave tubes. In a klystron, physical space was needed for modulating the electron beam[1] but at the same time the resonator had to be small (tubes need to be small) and it had to be resonant for using efficiently the electromagnetic power. The main requirement of the design was to ensure a space in which a longitudinal electric field could interact with a drifting electron beam. The resonator designed (Hansen, 1939; Hansen and Richtmeyer, 1939; Whinnery and Jamieson, 1944), called a re-entrant cavity, is shown in Fig. 4.1. The resonance is now not based on the bouncing of waves between reflectors but comes about because parts of the single metallic structure can be identified as an inductance and as a capacitance, respectively. The capacitance is given by the opposing metallic grids and the inductance is provided by the rest of the cavity. Clearly, the resonant frequency can be lowered by making the opposing metal surfaces closer to each other and increasing thereby the capacitance. A recent realization of a re-entrant cavity is that of White *et al.* (2005) who designed a tunable resonator in the GHz region having a diameter of about $\lambda/10$.

[1] There had to be an interaction space where the field affected the motion of the electrons and vice versa when the electrons delivered power to the field.

The classic example of a small resonator is that of the cavity magnetron that played such prominent role in Second World War radar. For an account by the inventors see Boot and Randall (1976). The magnetron cavity (three adjacent ones are shown in Fig. 4.2(a)) is in a sense the ancestor of all small resonators we use today: (i) it is an open structure, (ii) it has a gap through which it can interact with external electric fields, (iii) it is small relative to the wavelength, (iv) its resonant frequency is determined by the capacitance and inductance of the resonator. The capacitance is only partly due to the gap; there are

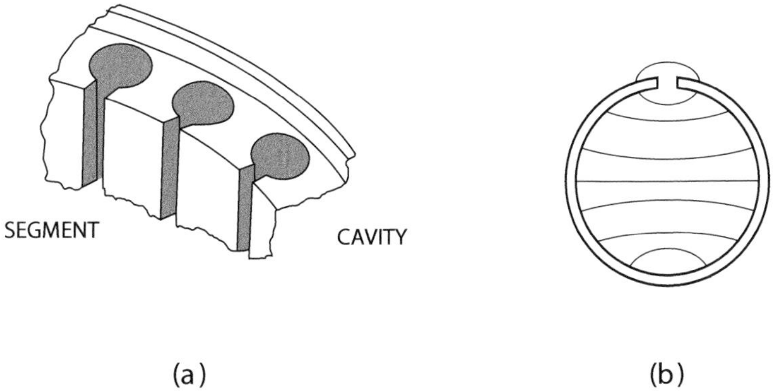

Fig. 4.2 (a) Three adjacent magnetron cavities. (b) Electric-field lines.

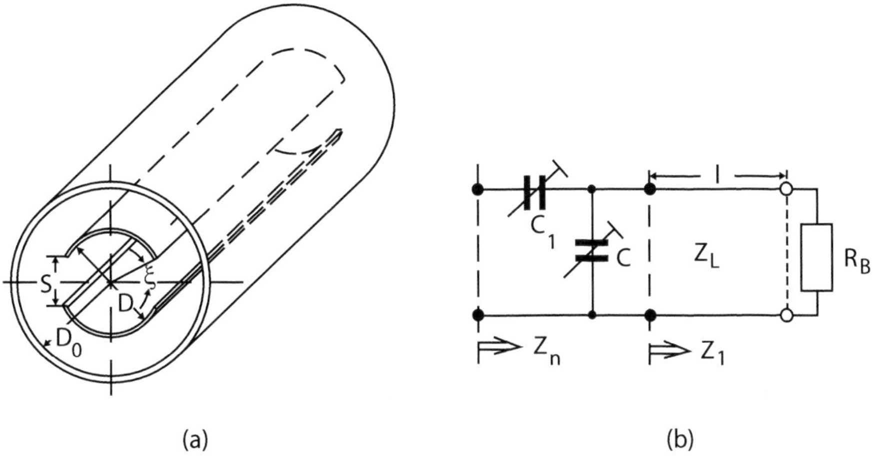

Fig. 4.3 Shielded symmetrical slotted-tube line (a). Matching and tuning of the slotted tube resonator (b). From Schneider and Dullenkopf (1977). Copyright © 1977 American Institute of Physics

contributions from other parts of the resonator as well as follows from the field pattern of Fig. 4.2(b). The inductance is due to the horizontal current flowing in the cylinder.

Our first two examples were on microwave tubes. Interestingly, the need for small resonators also arose in the detection of nuclear magnetic resonance as the resonant frequency moved towards the hundreds of MHz region. A resonator with bouncing waves would have been too large, whereas the traditionally used solenoid plus lumped capacitor no longer worked satisfactorily. The solution proposed by Schneider and Dullenkopf (1977) is shown schematically in Fig. 4.3(a). It is a cylindrical structure called the slotted-tube resonator. As may be seen, it bears a strong resemblance to the magnetron cavity. The main difference is that the inner tube is split so that the upper and lower parts of the inner tube can serve as a transmission line similarly to a strip line. The outer cylinder provides a shield. Combining these field concepts with circuit considerations the authors used a matching capacitance C_1

(a) (b)

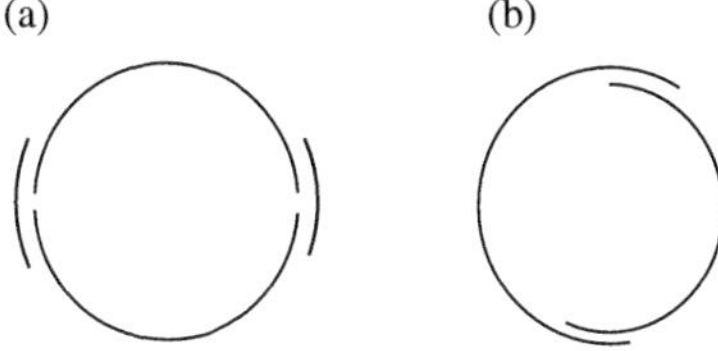

Fig. 4.4 Loop gap resonators

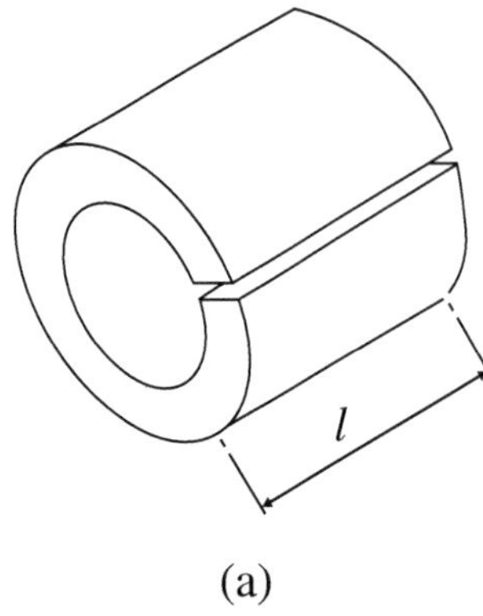

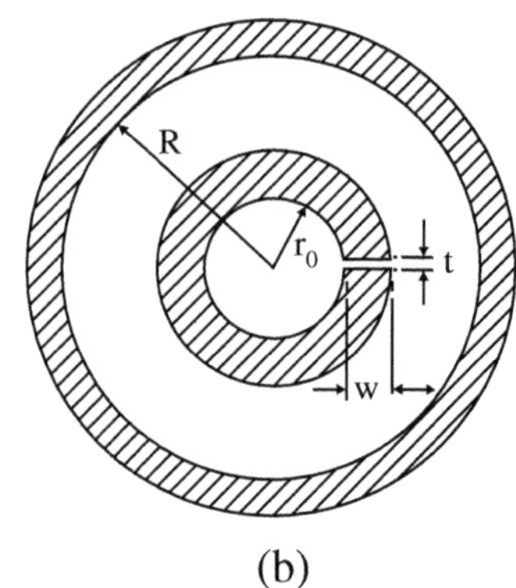

Fig. 4.5 Split-ring resonator. (a) the simplest version and (b) the double-ring version. From Hardy and Whitehead (1981). Copyright © 1981 American Institute of Physics

and a capacitance C to reduce the length of line for resonance as shown in Fig. 4.3(b). The same type of resonator for electron spin resonance measurements in the range of 1–10 GHz was proposed by Mehring and Freysoldt (1980). Tuning was achieved by inserting a dielectric. A very similar design by Froncisz and Hyde (1982) was called a loop gap resonator and the same terminology was used by Pfenninger *et al.* (1988) and Ghim *et al.* (1996). The essential features of their resonators are shown in Figs. 4.4(a) and (b). That of Pfenninger *et al.* (1988), the so-called bridged loop gap resonator, has an additional capacitance due to two surfaces opposite the two gaps. In the design of Ghim *et al.* (1996) two circles of slightly different radii overlap, producing the capacitance needed for the resonance.

A variation on the slotted-tube resonator is the split-ring resonator[2] of Hardy and Whitehead (1981) shown in Figs. 4.5(a) and (b). Now the metallic tube is split on one side only. The split tube is a resonator on its own, conceptually exactly the same as the magnetron cavity but the geometry of the gap is different, the contribution of the gap to the capacitance is higher. The outer cylinder is for confining the magnetic field in the annular region. Then, the magnetic flux in the outer region can be taken to be equal to that in the inner region providing the relationship

[2]This is the first time that the term split-ring resonator is mentioned in the literature. It consists of a double ring but only the inner ring is split, whereas in the split-ring resonator of Pendry *et al.* (1999), so popular nowadays, both rings are split. In the terminology of Shamonin *et al.* 2004; Shamonin *et al.* 2005 the resonator of Hardy and Whitehead (1981) is called a singly split double ring.

$$BS_2 = -S_1 B_0 \tag{4.1}$$

holds, where

$$S_1 = \pi r_0^2, \qquad S_2 = \pi[R^2 - (r_0 + w)^2]\,. \tag{4.2}$$

w is the thickness of the inner tube and R and r_0 are the inner radii of the outer and inner tubes, respectively. Then, from the equality of magnetic and electric energy they determine the resonant frequency as

$$\omega_0 = \sqrt{1 + \frac{S_1}{S_2}}\sqrt{\frac{t}{\pi} w}\frac{c}{r_0}\,, \tag{4.3}$$

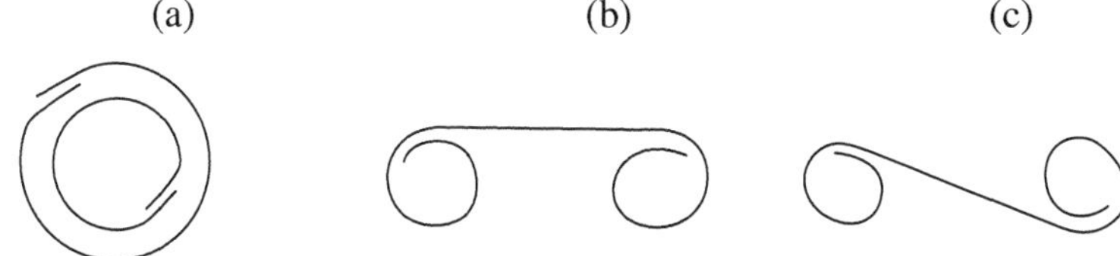

Fig. 4.6 Double circular elements. From Kostin and Shevchenko 1993; Kostin and Shevchenko 1994

where t is the gap width. They found the above expression to be accurate within about 20%. For an extension of the frequency range of the split-ring resonator to 4 GHz see Momo *et al.* (1983).

The aims of Kostin and Shevchenko 1993; Kostin and Shevchenko 1994 were somewhat different. They wanted to realize an artificial medium using metallic elements that may yield a high value of the imaginary part of the permeability. They wanted high magnetic losses and at the same time they wished to minimize the amount of metal, which made them use thin films. Their first design was a non-resonant metallic loop. Later ones were resonant solutions combining in slightly different ways capacitances and inductances, as shown in Figs. 4.6(a)–(c).

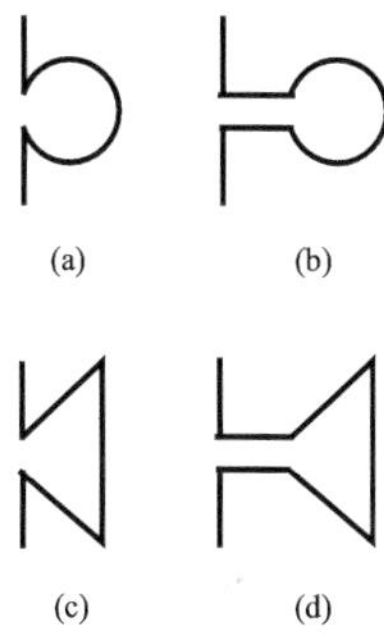

Fig. 4.7 Ω-particles. From Engheta *et al.* (2002)

An element combining an electric and a magnetic dipole was proposed by Saadoun and Engheta (1992). The corresponding media were called by the authors omega or pseudochiral media. The reason for the first designation is obvious, the elements are shaped in the form of the Greek letter, Ω, as may be seen in Fig. 4.7(a) (some related elements are shown in Figs. 4.7(b)–(d)). The second description is due to the fact that although the element possesses no handedness, it couples to each other electric and magnetic fields with a phase difference of 90 degrees. The general relationship is formulated[3] by the authors as

$$\mathbf{D} = \varepsilon \mathbf{E} + \Omega_{\mathrm{em}} \mathbf{B} \qquad \text{and} \qquad \mathbf{H} = \frac{1}{\mu}\mathbf{B} + \Omega_{\mathrm{me}} \mathbf{E}\,, \tag{4.4}$$

where ε, μ, Ω_{em} and Ω_{me} are the permittivity, permeability and coupling tensors, respectively. ε and μ are diagonal tensors, whereas

$$\Omega_{\mathrm{em}} = \mathrm{j}\,\Omega_{yz} \qquad \text{and} \qquad \Omega_{\mathrm{me}} = \mathrm{j}\,\Omega_{zy}\,. \tag{4.5}$$

A generalization of the element by Saadoun and Engheta (1994) to one including a small transmission line between the stems and the loop is shown in Fig. 4.7(b). The effective permittivity and permeability of the medium, as calculated by them, are shown in Figs. 4.8(a) and (b). The remarkable fact is that both materials constants turn negative at the same frequency and remain negative in a certain frequency band. This is actually the medium Veselago (1968) was looking for but neither Saadoun and Engheta nor anybody else who came across their paper was interested at the time to make use of it. Saadoun and Engheta had actually some doubts whether their circuit model was valid at such high frequencies. Some nine years later an analytical study by Simovski and He (2003) confirmed that both material parameters can be negative in a certain frequency band. Experimental proof that a material containing such elements can exhibit negative refraction was provided by Ran *et al.*

[3]This is a somewhat different form of the tensorial relationships given in Sections 1.15 and 1.16.

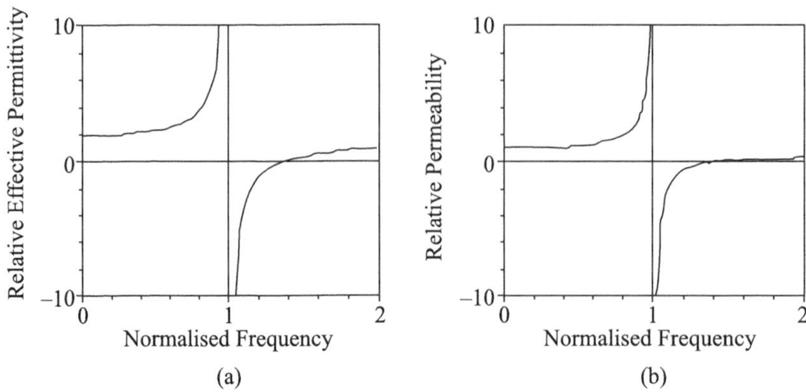

Fig. 4.8 Effective permittivity and permeability of an Ω-particle medium. From Saadoun and Engheta (1994)

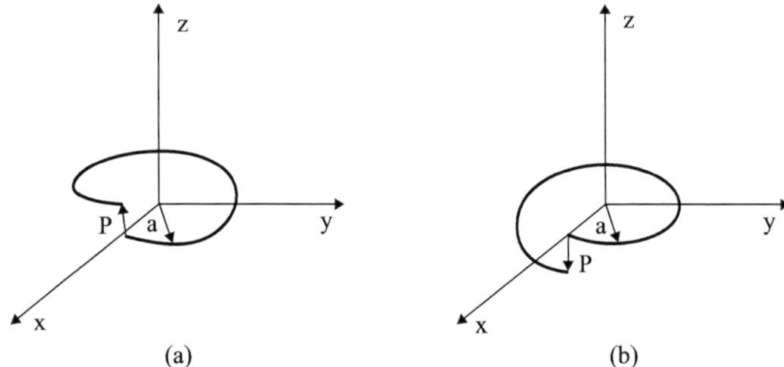

Fig. 4.9 One-turn helices. (a) Right-handed helix. (b) Left-handed helix. From Bahr and Clausing (1994). Copyright © 1994 IEEE

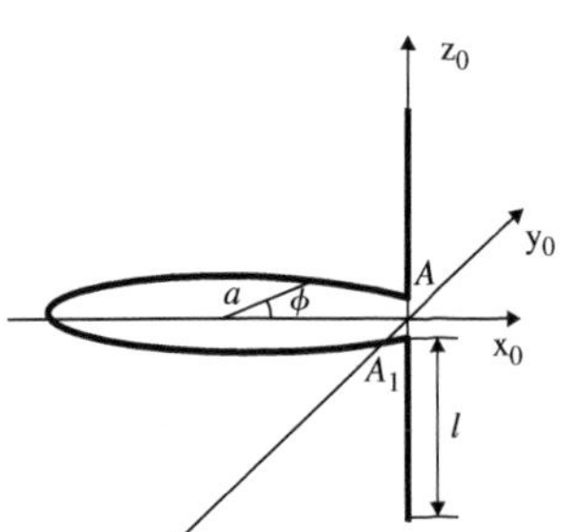

Fig. 4.10 Geometry of a chiral particle. From Tretyakov *et al.* (1996). Copyright © 1996 IEEE

(2004). It is a rather curious fact that the omega particle, in spite of its obvious suitability as a metamaterial element, has been eclipsed by the SRR plus rod combination.

A chiral element with actual handedness was analyzed by Bahr and Clausing (1994). The one-turn helix (Figs. 4.9(a) and (b)) was either right-handed or left-handed. A further chiral element similar to that of Saadoun and Engheta (1994) was investigated by Tretyakov *et al.* (1996). The analysis was based on the available theory of both linear and loop antennas (King, 1969). Their basic element is shown in Fig. 4.10. It consists again of a combination of an electric and a magnetic dipole but this time the perpendicular to the loop is in the direction of the electric dipole. The excitation of dipoles (electric and magnetic) in terms of fields (electric and magnetic) mediated by polarizability tensors is summarized in Table 4.1. It shows which tensors play a role for particular excitations. The terminology is the same as that presented in Section 1.16.

Let us take as an example a plane wave incident upon such an element with electric-field polarization in the x direction and magnetic-field polarization in the z direction. The corresponding tensor elements are $\alpha_{zz}^{\mathrm{mm}}$, $\alpha_{zz}^{\mathrm{em}}$, $\alpha_{yz}^{\mathrm{me}}$, and $\alpha_{xx}^{\mathrm{ee}}$. The corresponding physics is quite straightforward. A magnetic field in the z direction sets up a current flowing in the loop that gives rise to a magnetic moment in the z direction and

Table 4.1 Effective dipole moments for different exciting field directions. From Tretyakov *et al.* (1996). Copyright © 1996 IEEE

Field orientation	Effective dipole moments	Polarizability components
E	$\mathbf{p}_z$, $\mathbf{p}_y$, $\mathbf{m}$	α_{zz}^{ee}, α_{yz}^{ee}, α_{zz}^{me}
H	$\mathbf{p}_z$, $\mathbf{p}_y$, $\mathbf{m}$	α_{zz}^{em}, α_{yz}^{em}, α_{zz}^{mm}
E	$\mathbf{p}_z$, $\mathbf{p}_y$, $\mathbf{m}$	α_{yy}^{ee}, α_{zy}^{ee}, α_{zy}^{me}
E ⊗	⊗ $\mathbf{p}_x$	α_{xx}^{ee}

to an electric dipole moment in the y direction. But the same current flows also in the electric dipole and sets up therefore an electric dipole moment in the z direction. Finally, the electric field in the x direction sets up an electric dipole moment in the same direction. Note that for α_{yz}^{me} and α_{xx}^{ee} to be finite the current in the loop must vary as a function of the angle, i.e. the current in the loop cannot be uniform, which means that the loop cannot be infinitesimally small.

Another version of the magnetron cavity realized in thin-film form was used by Hong and Lancaster 1996*a*; Hong and Lancaster 1996*b*; Hong and Lancaster 1999; Hong 2000 in designing microwave filters. Two of their typical filter configurations may be seen in Figs. 4.11(a) and (b), showing a four-pole and a six-pole design, respectively. In the same spirit, Hong and Lancaster (1998) produced yet another microwave resonator for filter applications that was U-shaped, as shown in Fig. 4.11(c). They called them hairpin resonators.

The elements that have been discussed so far were designed either for some specific purpose (e.g. for a filter or for the measurement of nuclear magnetic resonance) or as potential elements in some, not very well specified, artificial material. A wide variety of resonators had already been designed and measured. Any new resonator was bound to be some modified version of an existing one.

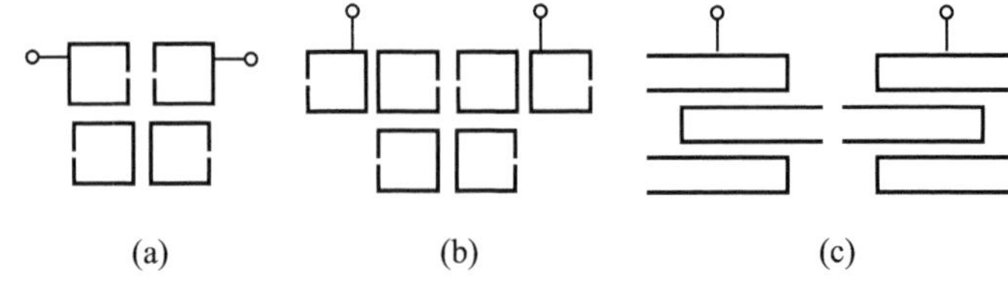

Fig. 4.11 Microwave filters in thin-film form. (a) and (b) From Hong and Lancaster (1996*a*). Copyright © 1996 IEEE. (c) From Hong and Lancaster (1998). Copyright © 1998 IEEE

4.3 A roll-call of resonators

As we have seen before the term 'split-ring resonator' was coined by Hardy and Whitehead (1981) referring to a design in which the inner ring is split, and the same name was retained by Pendry *et al.* (1999) whose resonator (see Fig. 2.12) bore close resemblance to resonators already shown in Figs. 4.3(a), 4.4(a), 4.5(b) and 4.6(a). There was, however, a difference. The design of Pendry *et al.* (1999) was specifically aimed at producing an effective negative permeability. The authors also introduced a new embodiment of the swiss roll (Fig. 4.12) that was used earlier for generating homogeneous magnetic fields (Sasaki *et al.*, 1995).

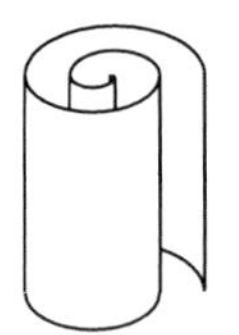

Fig. 4.12 Swiss roll

The subject had not been called metamaterials as yet but there was already a central focus: how to achieve negative material parameters and how to demonstrate negative refraction? It turned out that practically all conceivable magnetic resonators were capable of producing negative permeability, provided the magnetic field could have access to the element.

Next, we shall show three magnetic resonators investigated by Baena *et al.* (2005*a*). The first one (Fig. 4.13(a)) is the celebrated split-ring resonator,[4] the other two are of fairly similar design. Figure 4.13(b) shows a quadruply split double ring, and that in Fig. 4.13(c) is a somewhat distorted spiral (see also Baena *et al.* 2004). For the SRR of Pendry *et al.* (1999), as mentioned already in Section 2.5, the inductance can be regarded as L_{av}, the average of that of the inner and outer rings. Also, the two inter-ring capacitors (those between the splits) are in series resulting in the equivalent circuit of Fig. 4.13(a), giving a resultant capacitance of $C_0/4$, where C_0 is the total inter-ring capacitance. In the quadruply split double ring of Fig. 4.13(b) there are four inter-ring sections that can be regarded as being responsible for four capacitances in series, each of them equal to $C_0/4$. In the spiral of Fig. 4.13(c) there is no gap, hence the total capacitance is equal to C_0. Clearly, for the same physical size the lowest resonant frequency can be achieved with the spiral resonator. We shall quote here two of their results for the SRR and the quadruply split double ring for resonators having the dimensions, average radius = 3.55 mm, metal width = inter-ring separation = 0.3 mm. Resonant frequencies calculated from this theory gave 3.33 GHz for the SRR and, as expected, 6.66 GHz for the quadruply split double ring. The corresponding measured resonant frequencies were 3.40 and 6.77 GHz, showing excellent agreement.

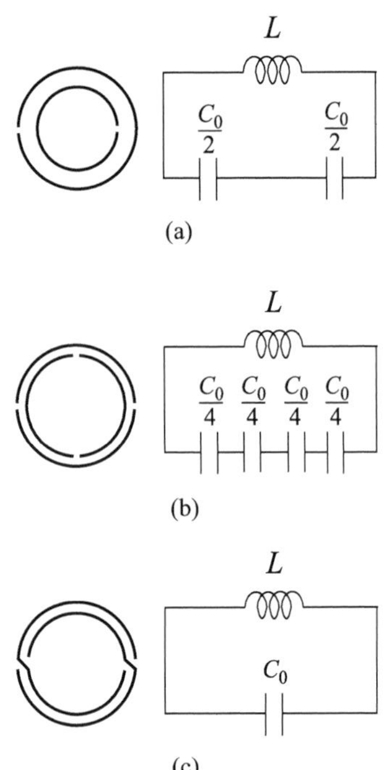

Fig. 4.13 Magnetic resonators and their equivalent circuits: SRR (a), quadruply split double ring (b) and spiral resonator (c). From Baena *et al.* (2005*a*). Copyright © 2005 IEEE

[4]From now on the term split-ring resonator, abbreviated as SRR, will be reserved for the resonator shown in Fig. 4.13(a)

A resonator that may be regarded as the dual of the SRR was also

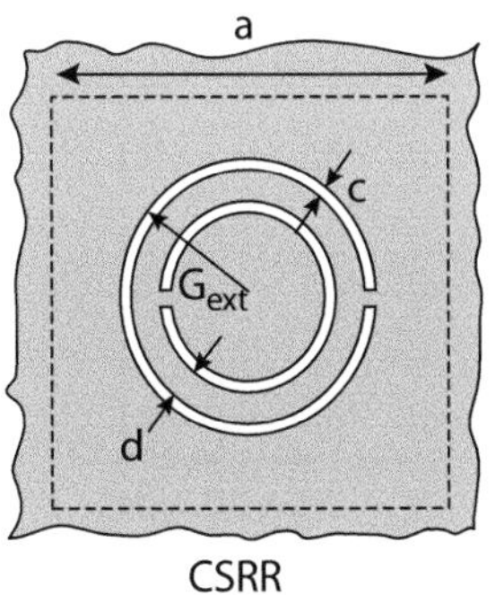

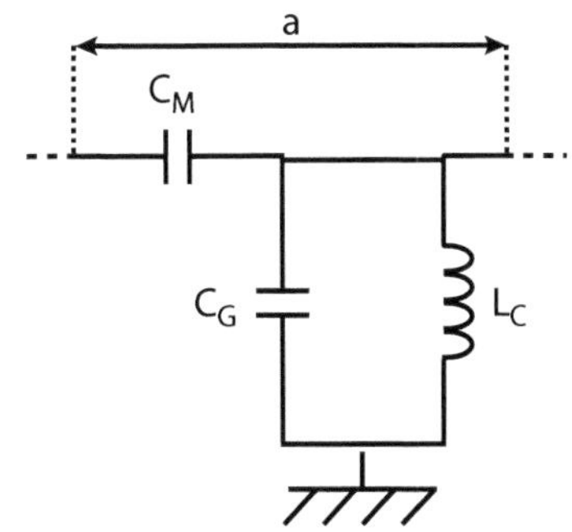

Fig. 4.14 Complementary split-ring resonator (a) and its equivalent circuit (b). From Beruete *et al.* (2006). Copyright © 2006 American Institute of Physics

proposed by Baena *et al.* (2005*b*) and used as an element in an array by Beruete *et al.* (2006). It is shown in Fig. 4.14 with the equivalent circuit. The additional capacitor C_{M} is for coupling to the next element. It may be seen that metal and air have been interchanged, hence the name of complementary split-ring resonator (already introduced in Section 2.5 as an example of a resonant element). For another study of rings and their equivalent circuits see Rogla *et al.* (2007).

Another set of thin-film split-ring resonators were investigated by Aydin *et al.* (2005). The first three (Figs. 4.15(a)–(c)) are single rings split one, two and four times. The fourth resonator is a quadruply split double ring, the same type as shown in Fig. 4.13(b) and the fifth one is another double ring split eight times with the gaps lined up. The basic structure of the SRR has dimensions external radius = 3.6 mm, gap width = inter-ring separation = 0.2 mm, metal width = 0.9 mm. With these parameters the resonant frequency of the SRR is 3.63 and 3.60 GHz obtained by measurement and simulation, respectively. The resonance curves for gap widths equal to 0.2, 0.3, 0.4 and 0.5 mm are shown in Fig. 4.16, the measured results in Fig. 4.16(a) and the simulated ones in Fig. 4.16(b). The agreement may be seen to be good for the resonant frequencies, differs somewhat for the depth of the resonances, and differs considerably for the widths of the resonance curves. Apparently, the simulated results show wider resonance curves, indicating higher losses. For the resonant frequencies the conclusion is clear: it increases as the gap increases, which follows of course from the fact that larger gap means smaller capacitance.

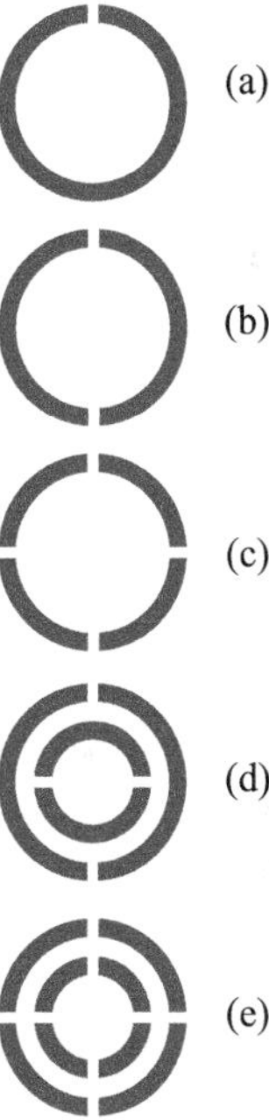

Fig. 4.15 Magnetic resonators: single rings split one, two and four times (a)–(c), quadruply split double ring (d), double ring split eight times (e). From Aydin *et al.* (2005)

Dependence of the resonance curves on ring separation is shown in Figs. 4.17(a) and (b) for measured and simulated results, respectively. Again, the resonant frequency increases as the separation increases, and the reason is again that increased separation reduces the inter-ring capacitance and leads therefore to higher resonant frequency. The agreement between measured and simulated results is again good for the resonant frequencies, reasonable for the resonance dip and less good for the width of the resonant curves.

Next, we shall look at a comprehensive numerical study carried out by Kafesaki *et al.* (2005) in which the authors explore a wide frequency range. The resonator investigated is shown in Fig. 4.18 (dimensions

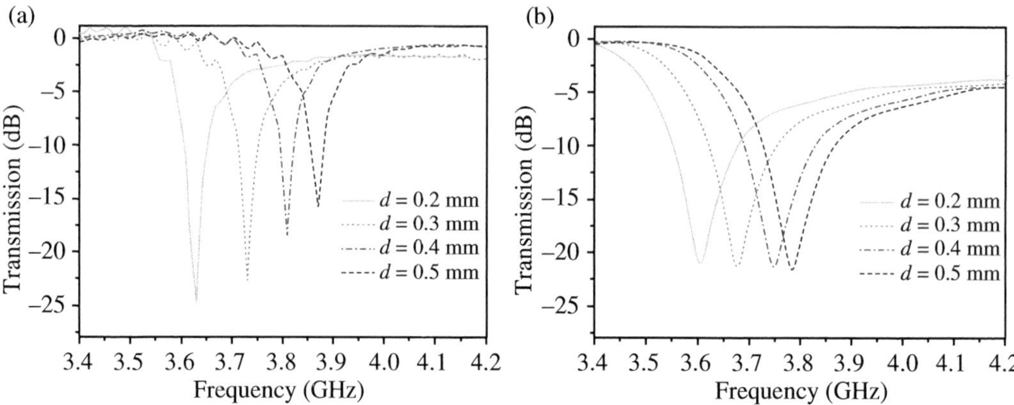

Fig. 4.16 Transmission spectra of individual SRRs with d, the width of the gap, as a parameter, obtained by (a) experiment and (b) simulation. From Aydin *et al.* (2005)

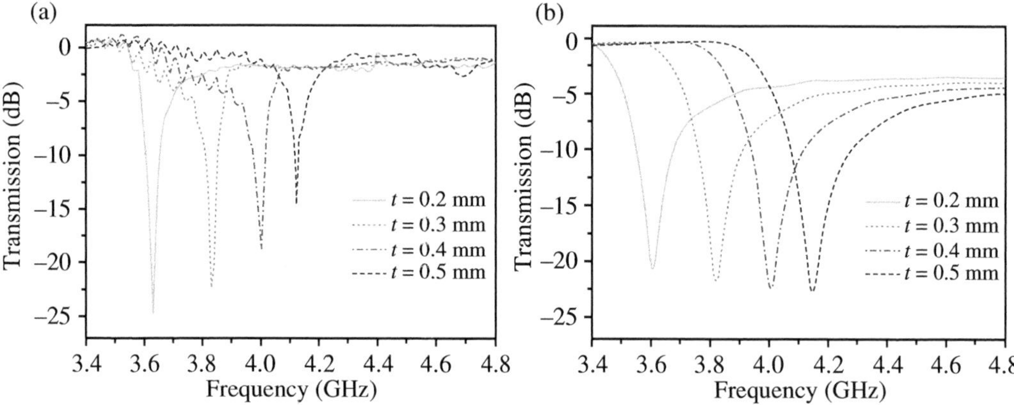

Fig. 4.17 Transmission spectra of individual SRRs with t, the inter-ring separation, as a parameter, obtained by (a) experiment and (b) simulation. From Aydin *et al.* (2005)

$l = 1.8$ mm, $g = t = w = 0.2$ mm). The transmission of a plane wave through this resonator is then determined by a numerical package. The relevant resonance curves are shown in Fig. 4.19(a) for the case when the electric field is parallel with the gaps. There are three resonances at around 22, 50 and 70 GHz. These results may then be compared with those exhibited in Fig. 4.19(b) that are calculated for the inner and outer rings separately. If we reduce the inter-ring separation both resonances tend to move towards lower frequencies: the upper resonance becomes weaker and the lower resonance stronger. At a value of $t = 0.05$ mm the upper resonance practically disappears and the lower resonance shifts to 15 GHz (Tatartschuk, 2007). The single resonance at 15 GHz or the two resonances at 22 and 50 GHz are magnetic resonances in the sense that they are caused by circulating currents that then produce magnetic dipoles. Alternatively, they could be regarded as LC resonances in the sense that they are caused by the inductances and capacitances of the

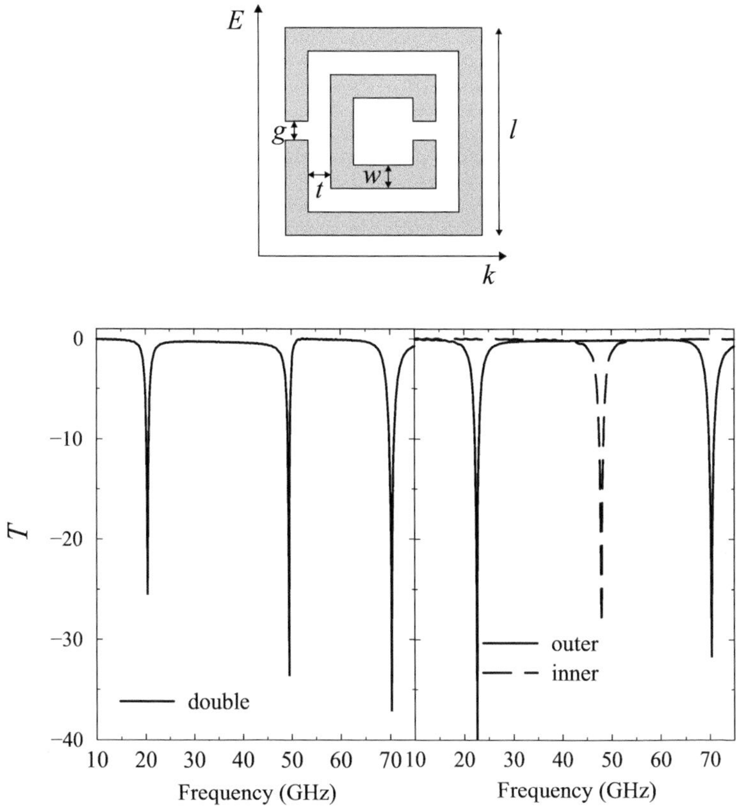

Fig. 4.18 Double-ring SRR. $g = t = w = 0.2$ mm; metal depth $= 0.2$ mm. The SRR side $l = 1.8$ mm. From Kafesaki *et al.* (2005). Copyright © 2005 IOP Publishing Ltd

Fig. 4.19 Transmission (T, in dB) versus frequency (in GHz) for the double-ring SRR shown in Fig. 4.18(a), and for its isolated outer and inner ring resonators (b). The background material is air. From Kafesaki *et al.* (2005). Copyright © 2005 IOP Publishing Ltd

SRR (cf. the physical arguments advanced in Section 2.5 for the resonant frequency of a SRR).

We may then ask the question what the third resonance at 70 GHz (Fig. 4.19) is due to? It seems very likely to be an electrical resonance akin to that of an electric dipole antenna. It needs to be recalled that a centre-fed (or short-circuit) electric dipole antenna has resonances when its length is approximately equal to an integral multiple of half-wavelengths. More accurately, the resonant length of an infinitely thin electric dipole is a little below half-wavelength and the resonant length is further reduced as the diameter to length ratio is increasing. 70 GHz corresponds to a free-space wavelength of 4.3 mm. Thus, a side of the SRR in Fig. 4.18 being equal to 1.8 mm is about 0.42 wavelength. So the electrical resonance hypothesis is plausible.

Kafesaki *et al.* (2005) also determined the transmission of a plane wave for three further orientations of the SRR, as illustrated in Fig. 4.20(a). In configuration (A) the electric field is perpendicular to the gap. The corresponding resonances are shown in Fig. 4.20(b). Four resonances may now be seen. The two magnetic resonances are at the same frequencies, as may be expected, since the magnetic field is perpendicular to the plane of the SRR. The physical mechanism for the two resonances

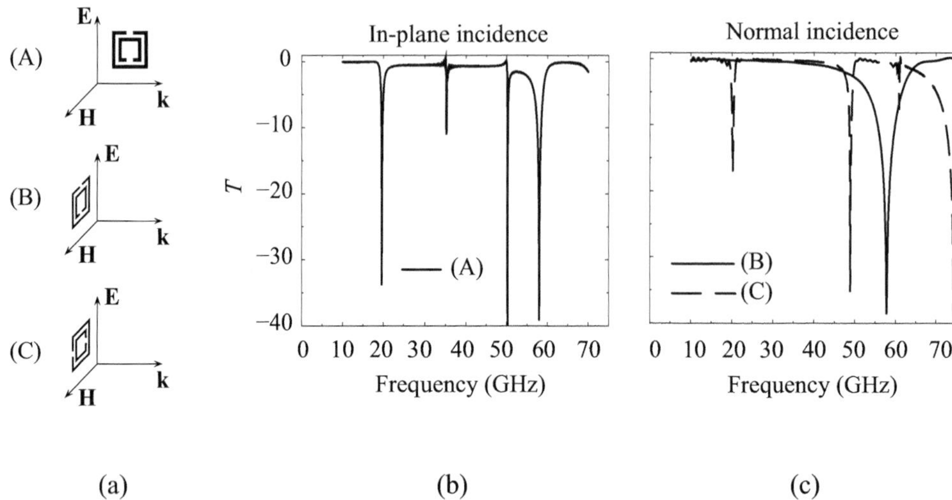

Fig. 4.20 Plane-wave transmission. (a) Configurations (a). Transmission spectra for in-plane incidence (b) and for normal incidence (c). From Kafesaki *et al.* (2005). Copyright © 2005 IOP Publishing Ltd

at 35 and 58 GHz is not clear, it needs further analysis. In configuration (B) the magnetic field is in the plane of the SRR, hence it cannot induce a current. There are no magnetic resonances as shown clearly in Fig. 4.20(c) but there is an electrical resonance at 57 GHz (solid line) that must be the same type as in configuration (A). In configuration (C) there are four resonances, as plotted with dashed lines in Fig. 4.20(c). The magnetic resonances at about 20 and 50 GHz have reappeared in spite of the fact that the magnetic field is not perpendicular to the plane of the SRR. The resonances are caused by the electric field that, in this case, is parallel with the gap. The physical mechanism is that the electric field sets up an electric dipole moment that then causes currents to flow in both the inner and the outer ring, so the conditions are right again for a magnetic resonance. This is certainly one way of looking at the physics but it is probably much better to consider how a magnetic field can set up an electric dipole and then invoke reciprocity. This is done in Appendix H. The other two resonances are slightly shifted relative to those shown in Fig. 4.20(b).

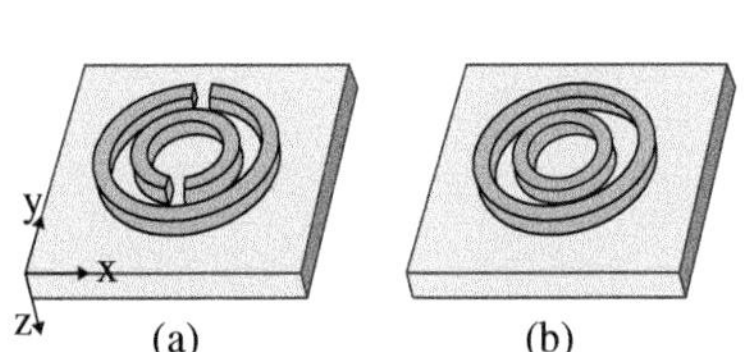

Fig. 4.21 SRR (a) and full ring (b) configurations. From Aydin and Ozbay (2007*b*). Copyright © 2007 Wiley-VCH Verlag GmbH & Co. KGaA

We have talked about magnetic and electric resonances. The lower ones were regarded magnetic and the higher ones electrical. A better criterion is the one used by Koschny *et al.* 2004*a*; Koschny *et al.* 2004*b*, and Aydin and Ozbay 2006; Aydin and Ozbay 2007*b*. The essential physical mechanism in the magnetic resonance of SRRs is the non-uniform distribution of currents along the inner and outer rings. If those currents could be made uniform the magnetic resonances would be absent. We can indeed have uniform distribution of currents by removing the gaps, and we are then left with two full concentric rings. In the experiments of Aydin and Ozbay (2007*b*) the transmission was measured both for a SRR and for a full ring (called by the authors a closed-ring resonator

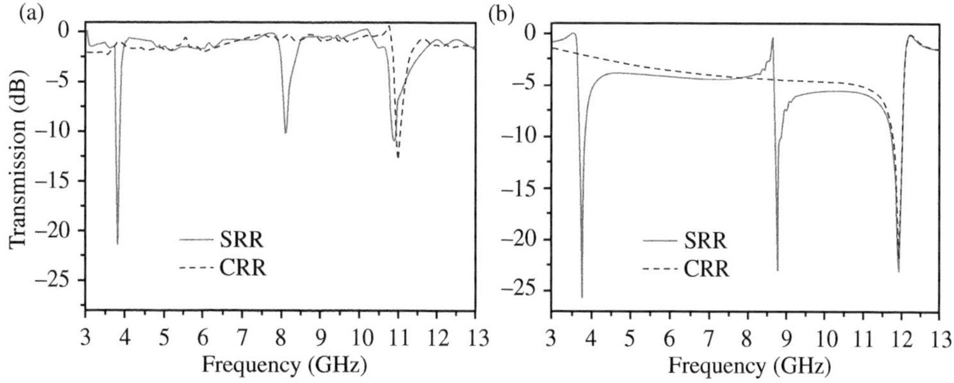

Fig. 4.22 Frequency response of an individual SRR and CRR (full ring) obtained from (a) experiments and (b) simulations. From Aydin and Ozbay (2007*b*). Copyright © 2007 Wiley-VCH Verlag GmbH & Co. KGaA

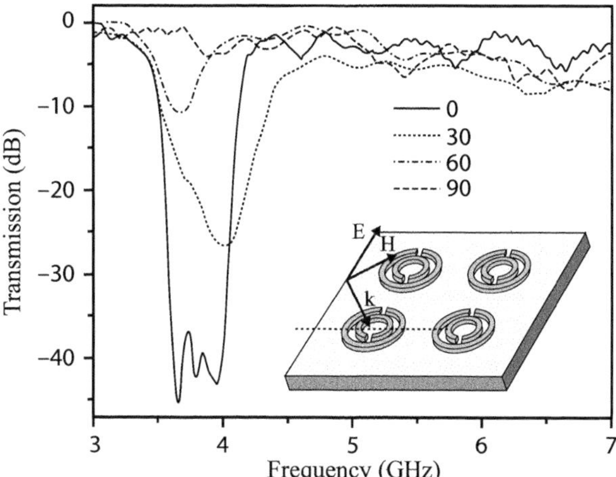

Fig. 4.23 Transmission spectra for oblique incidence. From Aydin and Ozbay (2007*b*). Copyright © 2007 Wiley-VCH Verlag GmbH & Co. KGaA

and abbreviated by CRR) in the configuration when the electric field is in the y direction and the magnetic field in the z direction (see Figs. 4.21(a) and (b)). The dimensions of the ring were: outer radius 3.6 mm, inter-ring separation and the gap 0.2 mm and the metal width 0.9 mm. The experimental results are shown both for the SRR and the full ring in Fig. 4.22(a) and the corresponding simulations in Fig. 4.22(b). There are three resonances for the SRR and only one for the full ring. The agreement between the measured and simulated results may be seen to be very close. Classification is now easy: the resonance that is common for the SRR and full ring is an electrical one, the two others are magnetic.

Aydin and Ozbay (2007*b*) also measured the transmission of microwaves incident at an angle upon a 3D array of SRRs in the configuration shown in the inset of Fig. 4.23. The electric field is in the y direction parallel with the gaps and the direction of propagation varies between 0° and 90°. At 0° the magnetic field is perpendicular to the

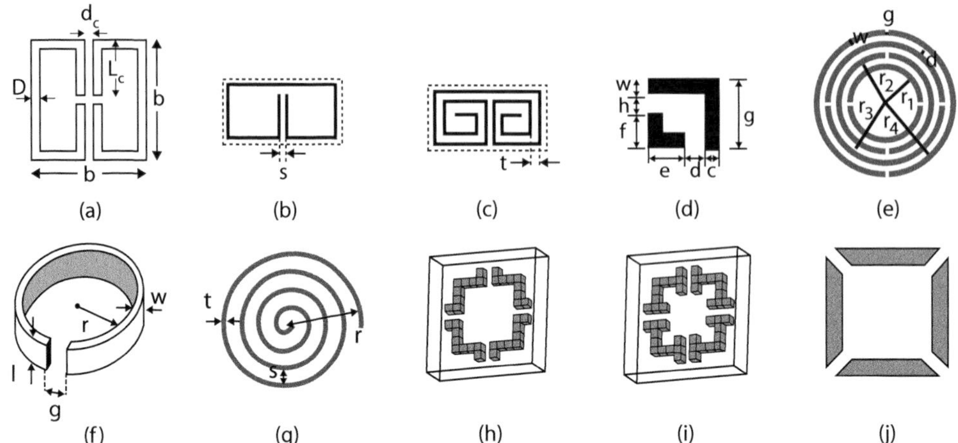

Fig. 4.24 A variety of resonators (a) by O'Brien and Pendry (2002), copyright © 2002 IOP Publishing Ltd, (b) and (c) by Guo *et al.* (2005), copyright © 2005 IEEE, (d) by Hsu *et al.* (2004), copyright © 2004 American Institute of Physics, (e) by Bulu *et al.* (2005*a*), copyright © 2006 Optical Society of America, (f) and (g) by Radkovskaya *et al.* (2007*b*), copyright © 2007 Wiley-VCH Verlag GmbH & Co. KGaA, (h), (i) and (j) by Kafesaki *et al.* (2005), copyright © 2005 IOP Publishing Ltd

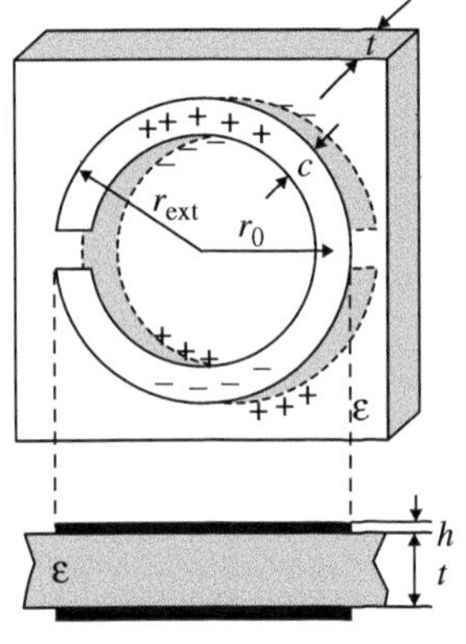

Fig. 4.25 Broadside-coupled split-ring resonator. From Marques *et al.* (2003). Copyright © 2003 IEEE

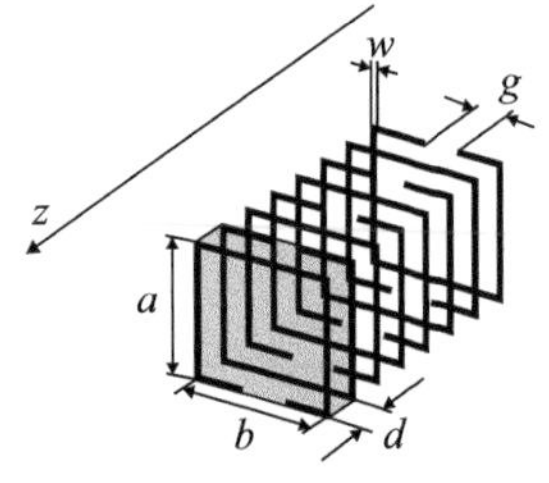

Fig. 4.26 Metasolenoid. From Maslovski *et al.* (2005)

plane of the SRR causing, as mentioned several times before, a magnetic resonance. There is large absorption in the corresponding stop band. As the direction of the wave propagation declines from 90° the component of the magnetic field perpendicular to the SRR plane is becoming smaller and smaller, and correspondingly the resonant dip becomes less and less deep. At 0° incidence the magnetic field no longer threads the SRR; the magnetic resonance is completely absent, the stop band disappears.

A number of other resonators of interest are shown in Fig. 4.24. They look quite different but they all obey the basic rule: loops (mostly broken) to provide the inductance, and surfaces close to each other to provide the capacitance.

The modified split-ring resonator of Marques *et al.* (2002*b*), rechristened later as the broadside-coupled split-ring resonator (Marques *et al.*, 2003), is of particular importance because it has no bianisotropy: there is no magneto-electric coupling. The element is shown schematically in Fig. 4.25. It consists of two split rings in which splits are 180 degrees apart. The rings are then placed parallel to each other upon opposite sides of a dielectric substrate. It has the further advantage that the opposing ring surfaces can be made large, the dielectric substrate can be made thin, so that the resonant frequency can be much higher than for the usual SRRs. A similar design, elements parallel to each other with splits opposite, was suggested by Maslovski *et al.* (2005). It differs from Marques *et al.*'s broadside-coupled split-ring resonator by having a large number of elements lined up, as may be seen in Fig. 4.26. For their resemblance to a solenoid the authors called them metasolenoids. Two interesting designs with symmetry in the horizontal plane and showing some new aspects of SRRs, are due to Gay-Balmaz and Martin (2002), and to Chen *et al.* (2006). They may be seen in Figs. 4.27(a) and (b),

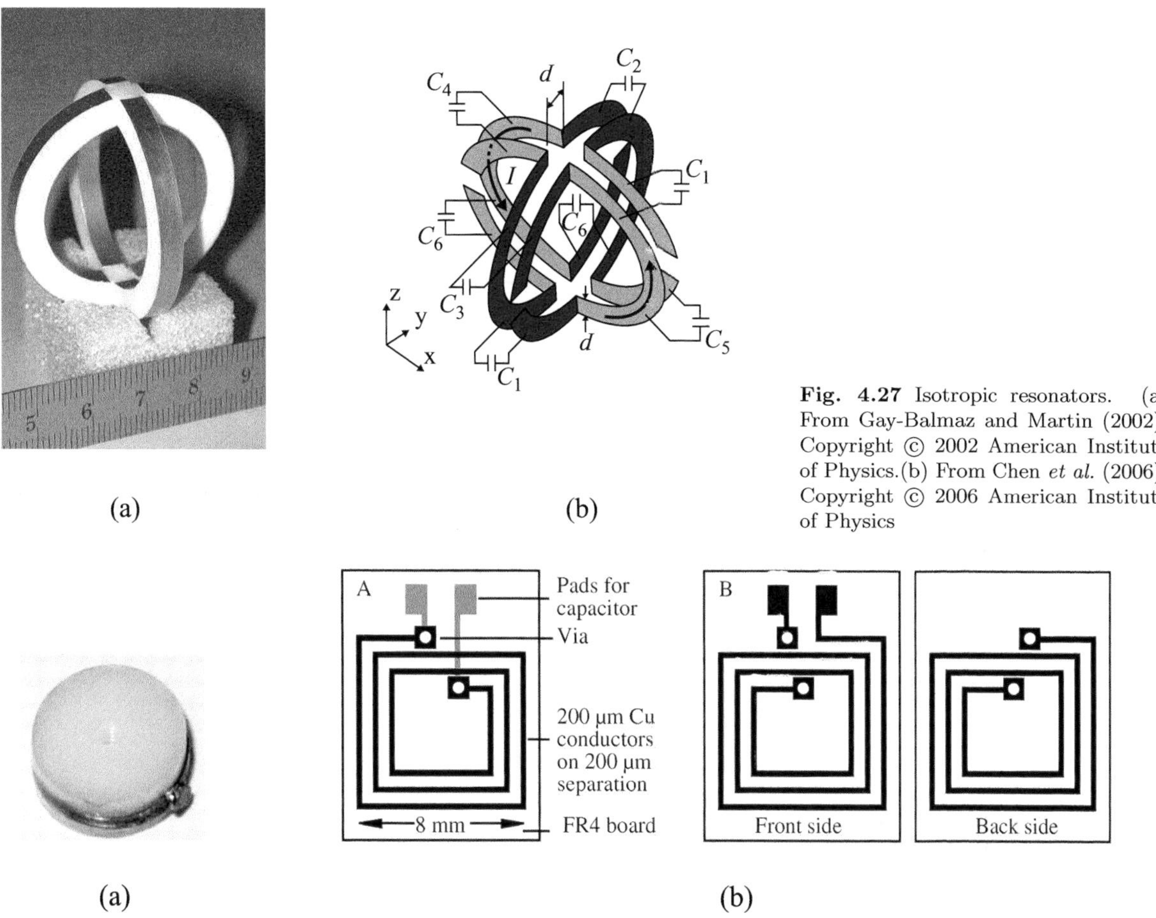

Fig. 4.27 Isotropic resonators. (a) From Gay-Balmaz and Martin (2002). Copyright © 2002 American Institute of Physics. (b) From Chen *et al.* (2006). Copyright © 2006 American Institute of Physics

Fig. 4.28 Capacitively loaded resonators. (a) From Wiltshire *et al.* (2003*b*). (b) From Syms *et al.* (2006*b*)

respectively.

A simple way of building a small resonator to which magnetic fields have access is to take a metallic loop and insert a lumped capacitance. This was done by Wiltshire *et al.* (2003*b*) who wound two turns of 1 mm diameter copper wire on a dielectric rod (9.6 mm diameter) and tuned the frequency to 60 MHz by inserting a capacitor (nominally 100 pF) between the ends of the wire as shown in Fig. 4.28(a). Syms *et al.* (2006*b*) used rectangular coils on one (design A) or both sides (design B) of a PCB as shown in Fig. 4.28(b). A resonator that the authors called an open split-ring resonator was used by Martel *et al.* (2004) as a component in a microstrip line for realizing a filter. It is shown with its equivalent circuit in Figs. 4.29(a) and (b).

It is also possible to augment the inter-ring capacitance in a SRR by adding an external capacitor (Aydin *et al.*, 2005). In one of their examples the resonant frequency of the SRR was 3.63 GHz, which they could change to 2.87 GHz and to 1.63 GHz by inserting capacitors of 0.1

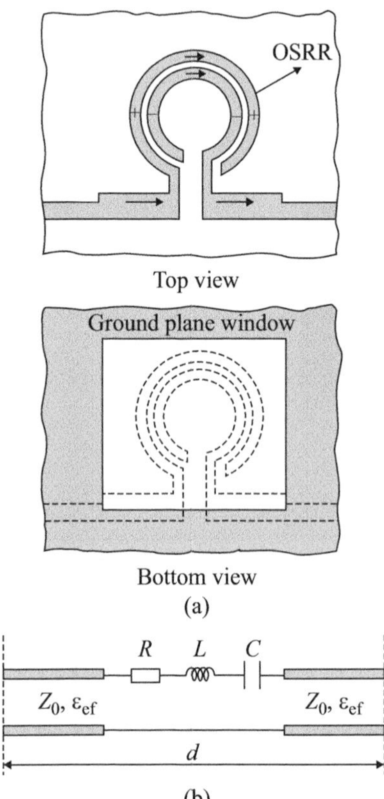

Fig. 4.29 Open split-ring resonator. From Martel *et al.* (2004)

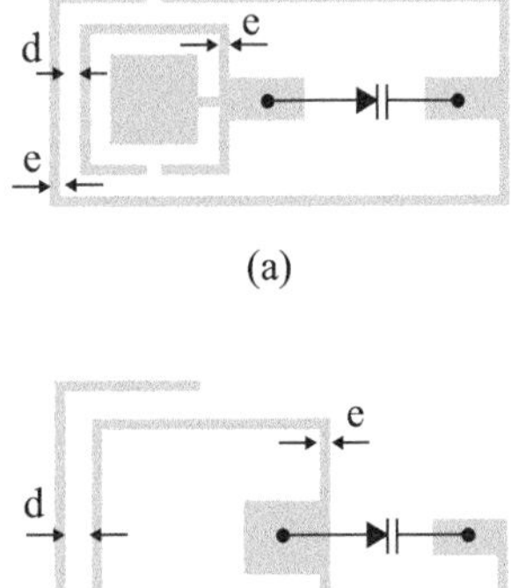

Fig. 4.30 Capacitively loaded split rings. From Gil *et al.* (2006). Copyright © 2006 IEEE

pF and 0.8 pF, respectively, across the outer gap. In another publication Aydin and Ozbay (2007*a*) reported a reduction of the resonant frequency from 3.82 GHz to 0.99 GHz by using a $C = 2.2$ pF capacitor in the gap. The authors also attempted to reduce the resonant frequency by inserting capacitors between the rings. They succeeded, but found that it was a less efficient way of reducing the resonant frequency. Another motivation for inserting a capacitance into a SRR is to tune the resonant frequency. This can be done with the aid of a varactor diode as shown by Gil *et al.* 2004; Gil *et al.* 2006. They were able to tune their element from about 2.5 GHz to 3.1 GHz by applying a voltage to the diode in the range of 0 to 30 V. The element is designed with the need in mind to accommodate the diode and to facilitate its biasing. The two models considered, both elongated, are shown in Figs. 4.30(a) and (b).

All the elements discussed so far had anisotropic properties. If the aim is to construct an isotropic medium then the unit cell should display some 3D symmetry. Examples are the unit cube with six SRRs on its surface (see Fig. 4.31(a)) as proposed by Pendry *et al.* (1999) and further investigated by Baena *et al.* (2006), and the 3D symmetric element of Gay-Balmaz and Martin (2002) (Fig. 4.31(b)). A fully symmetric design by Soukoulis *et al.* (2006) and Padilla (2007) is based on quadruply split

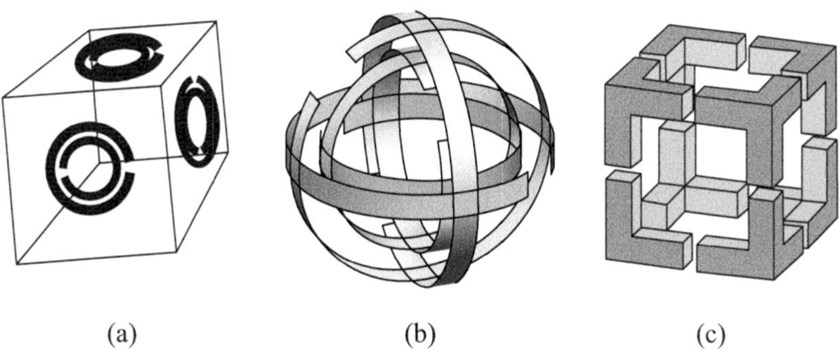

Fig. 4.31 Isotropic resonators. (a) From Pendry *et al.* (1999). Copyright © 1999 IEEE. (b) From Gay-Balmaz and Martin (2002). Copyright © 2002 American Institute of Physics. (c) From Padilla (2007). Copyright © 2007 Optical Society of America

single rings as shown in Fig. 4.31(c).

The metamaterial concept has proved useful for inspiring applications both in the RF and in the microwave region. It is certainly a desirable thing to manipulate fields on a scale much smaller than the wavelength, and there is no difficulty doing that up to the microwave region but should we carry on and explore the possibilities at higher frequencies as well? Is there a need for it? Yes, if we are envisaging optical circuits as proposed, for example, by Engheta *et al.* (2005). How can we realize higher-frequency resonators? Simply by reducing the dimensions of the resonator. It was indeed found by Yen *et al.* (2004), Moser *et al.* (2005), and Xu *et al.* (2006) that they could increase the resonant frequency up to several THz by making the same sort of split-ring resonators smaller, reducing the overall size of the element below 30 μm.

Could SRRs be made even smaller? Should we rely on SRRs anyway? At the time of writing it is not clear which kind of element will win the race towards the highest frequencies. There is, however, a strong argument in favour of the single ring, the magnetron-type open resonator, since for the same size it has a higher resonant frequency. Indeed the use of this type of resonators enabled Gundogdu *et al.* (2006), Zhang *et al.* (2005*b*), Linden *et al.* (2004), Enkrich *et al.* (2005*a*) and Klein *et al.* (2006) to reach 6, 65, 100, 250 and 330 THz, respectively.

We have to add here that Enkrich *et al.* (2005*a*) investigated a range of open resonators as shown in Fig. 4.32, where going from right to left the 'U' is filled in with metal. The metal square on the left is known to give a plasma resonance that hardly changes with the 'filling' but the magnetic resonance keeps on increasing to about 100 THz as the 'filling' becomes smaller and the capacitance increases. A detailed investigation of the resonant frequency of the U-shaped element was performed by Rockstuhl *et al.* (2006). They obtained plasma resonances[5] up to several hundred THz. For an incident plane wave in which the electric field is across the gap and for a particular resonance the same component of the electric field is plotted in Figs. 4.33(a) and (b) and compared with the corresponding distributions in a short, thin metallic rod. The plots

[5] We know from previous studies of the Full SRR that there is no need for any kind of splits to obtain electrical resonances (King, 1969; Tretyakov *et al.*, 1996) and that applies also to the optical range. Aizpurua *et al.* (2003) measured plasma resonances in a gold ring of 60 nm radius in the frequency range of 200–300 THz.

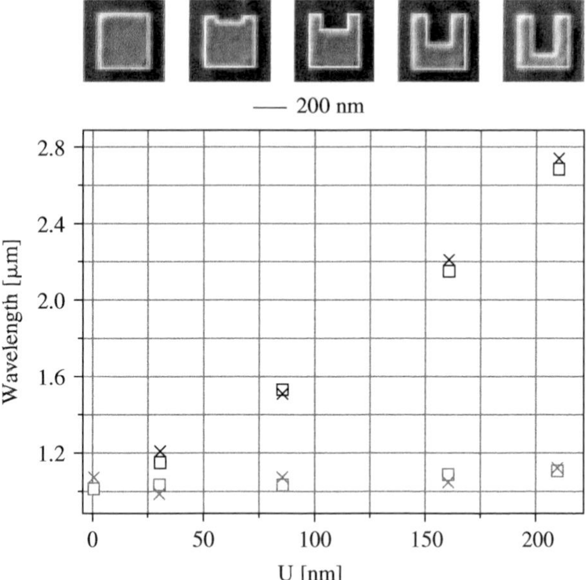

Fig. 4.32 Measured (crosses) and calculated (squares) spectral positions of the magnetic resonance (black) and the electric resonance (gray) as a function of the depth u (from left to right: $u = 0$, 30, 85, 160 and 210 nm). The corresponding five SRR designs are depicted at the top. From Enkrich *et al.* (2005*a*). Copyright © 2005 Wiley-VCH Verlag GmbH & Co. KGaA

offer good physical insight into the emergence of the resonances.

Can even higher frequencies be achieved? Enkrich *et al.* (2005*b*) reached 370 THz. Is there a limit? This question was addressed in a number of publications by Ishikawa *et al.* (2005), Zhou *et al.* (2005*a*), Marques and Freire (2005) and Tretyakov (2007). All agree that the limitation comes about due to the inertia of the electrons, and all agree that the limit for magnetic resonance comes at about 350 THz. In order to reach this conclusion the concept needed is that of kinetic inductance discussed in Section 1.20, which has a very simple expression. Marques and Freire argue that the SRR might still be the preferred candidate as we try to reach higher frequencies. Tretyakov also comes to the conclusion that the limit is about 350 THz for magnetic resonance, but he shows

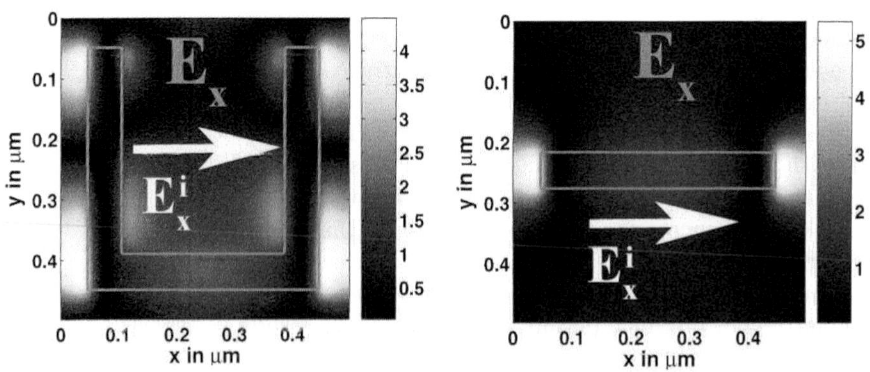

Fig. 4.33 Electric-field distribution at the resonance. From Rockstuhl *et al.* (2006). Copyright © 2006 Springer Science + Business Media. For coloured version see plate section

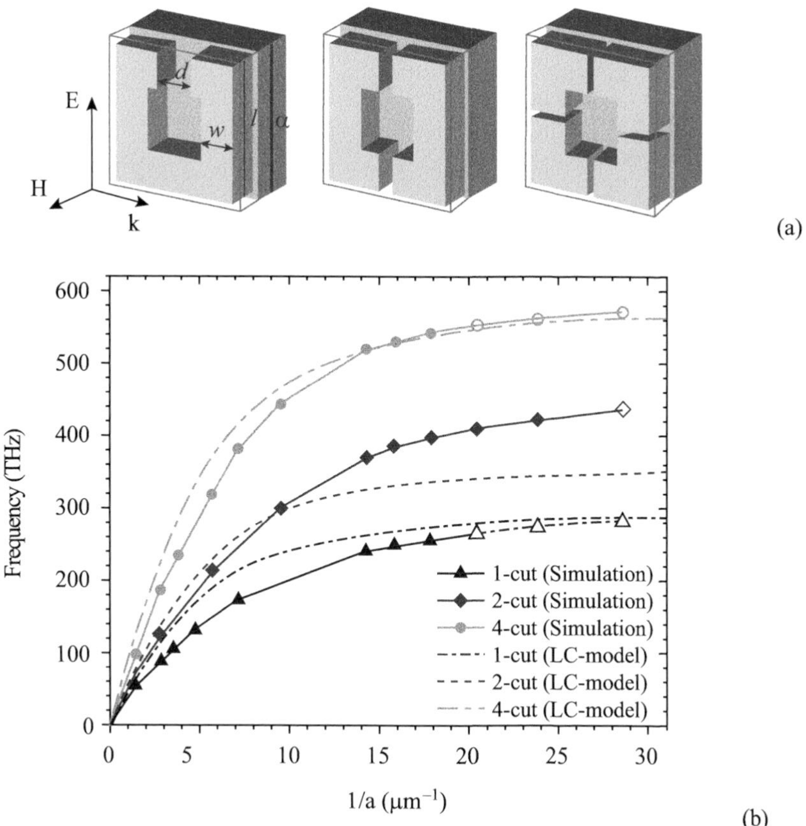

Fig. 4.34 The scaling of the simulated magnetic resonance frequency as a function of the linear size a of the unit cell for the 1-, 2-, and 4-cut SRRs (solid lines with symbols). Up to the lower terahertz region, the scaling is linear. The maximum attainable frequency is strongly enhanced with the number of cuts in the SRR ring. The hollow symbols as well as the vertical line at $1/a=17.9\ \mu\text{m}^{-1}$ indicate that $\mu < 0$ is no longer reached. The non-solid lines show the scaling of the magnetic resonance frequency calculated through the *LC* circuit model. From Zhou *et al.* (2005*a*). Copyright © 2005 by the American Physical Society. For coloured version see plate section

analytically that plasma resonances may occur as long as the operating frequency is below the plasma frequency. Calculations by Sondergaard and Bozhevolnyi (2007) indicated that in a silver strip of 10×20 nm dimensions resonant frequencies as high as 830 THz may be obtained.

We shall show here some curves showing the saturation of the magnetic resonance frequency by Zhou *et al.* (2005*a*). They consider three configurations, as shown in Fig. 4.34(a): the open resonator, the doubly split single ring and the quadruply split single ring. The resonant frequencies coming from their simulations are plotted in Fig. 4.34(b) against inverse size. It may be seen that the resonant frequency in all three cases saturates as the size becomes smaller and smaller. As may be expected, the elements having more splits have higher resonant frequencies.

An interesting and rather unusual set of resonators was realized by Zhang *et al.* (2005*c*). Their design is shown in Figs. 4.35(a) and (b). The

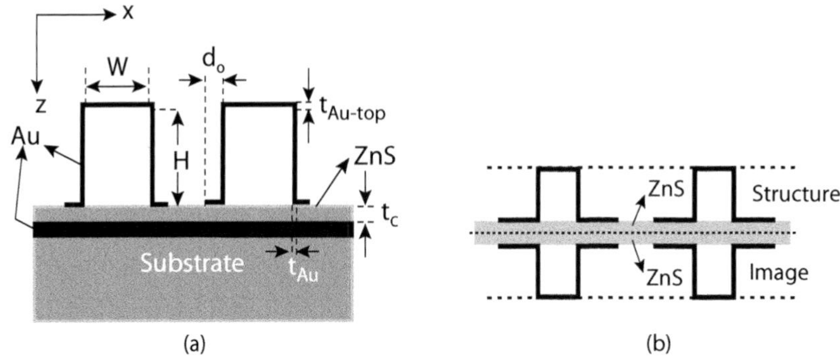

Fig. 4.35 Staple structure. (a) Schematic of the nanostructures. The light is incident from the top: for TM polarization, H is in the y direction, perpendicular to the loop. (b) The equivalent LC circuit formed between the top staple structure and its image in the metal. The dashed lines are the reference planes for the effective permeability calculation. From Zhang *et al.* (2005*c*). Copyright © 2005 by the American Physical Society

Table 4.2 Sample geometrical parameters (all dimensions in nm). From Zhang *et al.* (2005*b*)

Sample	W	H	d_c	t_{Au}	t_{Au-top}	t_c
A	130	280	190	~ 15	~ 30	80
B(D)	130	180	190	~ 15	~ 30	80
C	130	280	90	~ 15	~ 30	80

bold lines show contours by gold. The dielectric between the elements and the reflector is ZnS. The authors call the resonators, quite aptly, 'staples'. Taken together with their mirror image it is easy to see that they possess inductances and capacitances. They were produced in 2D arrays in the x and z directions.

The parameters for three structures realized are given in Table 4.2. The experimental and simulation results for reflection at normal incidence are plotted in Figs. 4.36(a) and (b). A large dip may be clearly seen for configuration A at a frequency of about 45 THz for an incident TM wave (magnetic polarization in the y direction). For configurations B and C the dip (for the same polarization) occurs at higher frequencies. The reason is clearly that for configuration B the height of the 'staple' H is smaller hence the inductance is smaller, and for configuration C the flange is smaller, yielding reduced capacitance. Note that the dip occurs only for the TM polarization when the magnetic field can access the open loops. As may be seen in Fig. 4.36(a), the reflection is frequency-independent for the TE polarization. The effective permeability determined from their model is plotted in Fig. 4.36(c). A negative permeability region for configuration C around 64 THz may be seen. Clearly, the idea of having a reflector leads to interesting physics but it makes it impossible to measure transmission. Absorption and effec-

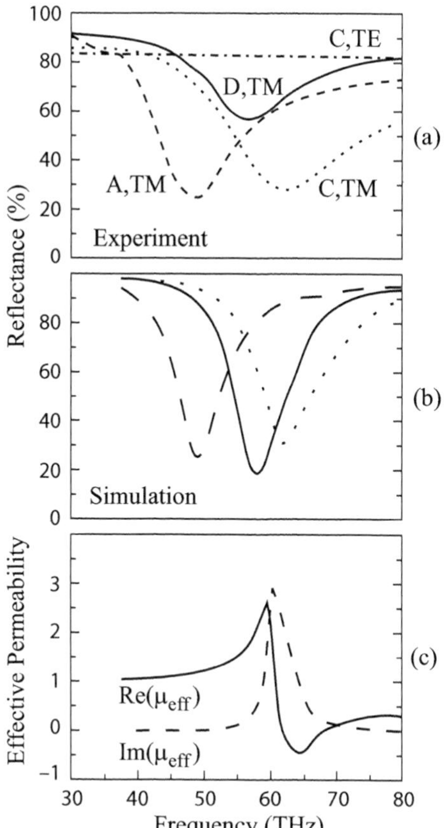

Fig. 4.36 Experimental (a) and simulation (b) results for the reflectivity as a function of frequency for three different structures of Table 4.2. The resonant frequency is a function of the nanostructure parameters and is independent of array period, which is the same for all three structures. The resonance is observed only for TM polarization, i.e. incident magnetic field coupled into the inductive structure. (c) Effective permeability extracted from the model for sample C. From Zhang *et al.* (2005*b*). Copyright © 2005 by the American Physical Society

tive permeability would be of interest for waves propagating in the x direction but those were not available in the paper.

Having considered split-ring resonators, open resonators, various U geometries, the 'partially filled U' and the 'staples' we shall next discuss a very promising candidate for high-frequency resonance: the short-rod pairs (see Figs. 4.37(a) and (b)). It started with sets of metal rods investigated both theoretically Lagarkov and Sarychev (1996) and experimentally Lagarkov *et al.* (1998) in the context of the percolation problem.[6] Another approach to the use of small metallic components was that of Svirko *et al.* (2001) who proposed pairs of metallic stripes at an angle to each other (see Fig. 4.38) for chiral applications in the optical region. Renewed interest came when it was realized (Podolskiy *et al.*, 2002; Podolskiy *et al.*, 2003; Panina *et al.*, 2002) that a pair of short metallic rods will have resonances below the half-wavelength associated with dipoles at lower frequencies. For a set of experimental results on resonances using short metal pillars see Grigorenko *et al.* (2005).

[6] A dielectric suddenly becomes conductive when the density of metallic inclusions reaches a critical value.

One may argue that short-rod pairs do not appreciably differ from the open resonators discussed by Linden *et al.* 2004; Enkrich *et al.* 2005*a*; Rockstuhl *et al.* 2006; Sarychev *et al.* 2006. They have both plasma and LC type of resonances. The plasma resonance follows from the fact that

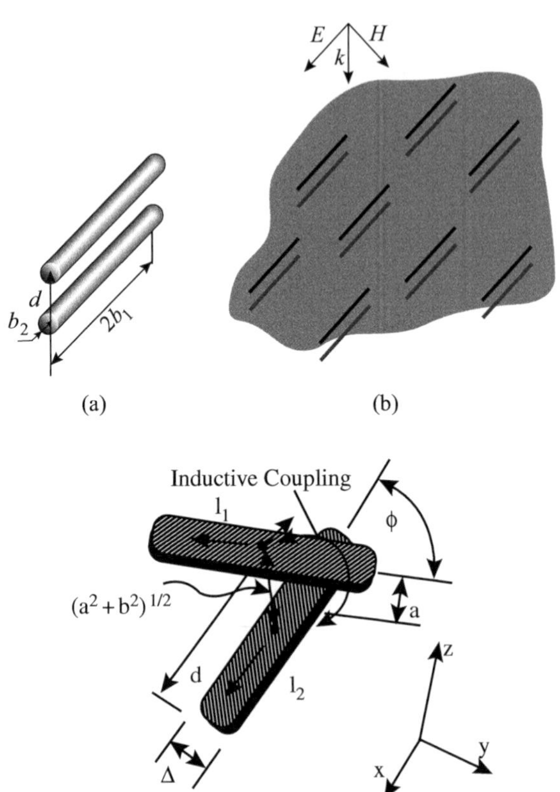

Fig. 4.37 Short-rod pair. (a) Single element. (b) Elements arranged into a medium

Fig. 4.38 An element of the bilayered planar chiral metallic microstructure with inductive cross-layer coupling. Two subwavelength metallic stripes of size $d \times \Delta$ are mutually shifted in the xy plane on distance b and spaced by a dielectric layer of thickness a (not shown). From Svirko *et al.* (2001). Copyright © 2001 American Institute of Physics

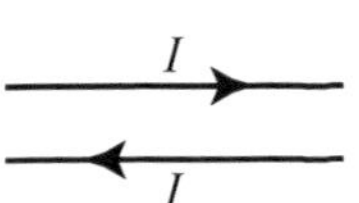

Fig. 4.39 Conduction current in a short-rod pair

[7]The U-shaped resonator, second from the left, is reminiscent of the hairpin resonator of Hong and Lancaster (1998).

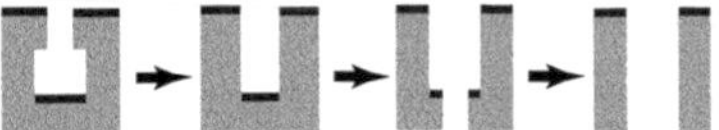

Fig. 4.40 Transition from the open resonator to the short-rod pair. From Dolling *et al.* (2005). Copyright © 2005 Optical Society of America

the frequency is high enough to be not too far from the metal's plasma frequency, and for the LC resonance the following simple argument can be presented. The two rods of Fig. 4.39 have self-inductances, there is a mutual inductance between them and there is a capacitance between the rods, analogously to a split-ring resonator. There are conduction currents (denoted by I flowing in the opposite directions, and displacement currents as well flowing between the rods.

Another illustration of the transition[7] from the open resonator to the short-rod pair due to Dolling *et al.* (2005) is shown in Fig. 4.40.

A set of short-rod pairs turn out to be suitable not only to provide negative permeability but negative permittivity as well. This was shown theoretically by Podolskiy *et al.* (2002) and is illustrated in Figs. 4.41(a) and (b). The corresponding refractive index is plotted in Fig. 4.41(c). Experimental results by Shalaev *et al.* (2005) and Kildishev *et al.* (2006) exhibited negative indices of refraction at a frequency of 200 THz but other experimenters (Dolling *et al.*, 2005; Garwe *et al.*, 2006) did not succeed in obtaining a negative index in that frequency range.

Negative index due to a structure of short wire pairs was found by Zhou *et al.* (2006) in a narrow frequency range near 14 GHz. A design by Zhang *et al.* (2005*a*), shown in Fig. 4.42 (top) relies, instead of short rods, on short but wider surfaces facing each other. This structure is then responsible for the magnetic resonance. Rods reappear to provide

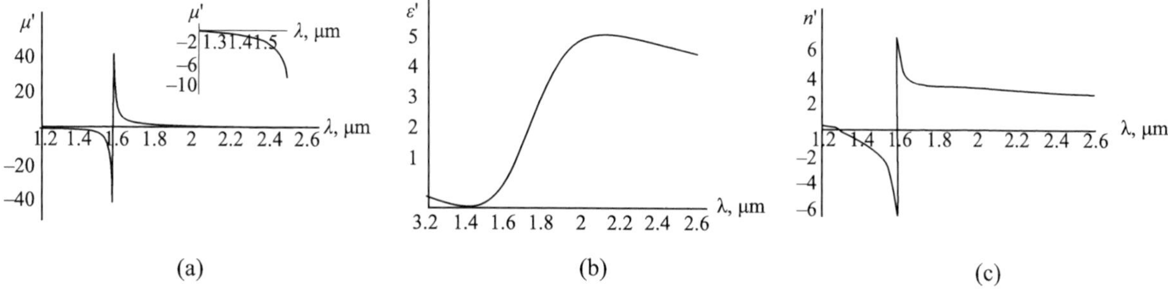

Fig. 4.41 Permeability (a), permittivity (b) and refractive index (c) for the short-rod-pair medium. From Podolskiy *et al.* (2002)

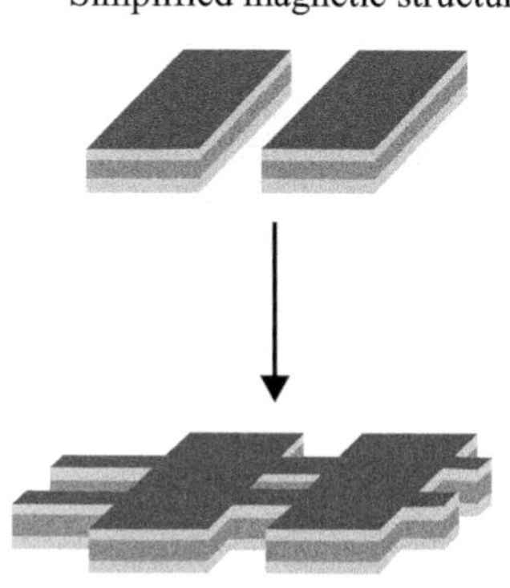

Fig. 4.42 Transition from short-rod pairs to a grid structure. From Zhang *et al.* (2005*a*). Copyright © 2005 Optical Society of America

the electrical resonance but in a somewhat different form, as may be seen in Fig. 4.42 (bottom). Simulations by the authors predict the existence of a frequency region around 150 THz that has both negative permittivity and negative permeability. The structure is becoming increasingly popular and is often called a fishnet.

Zhang *et al.* (2005*a*) also succeeded in demonstrating experimentally a negative index in the same frequency range. They used an entirely different structure as shown in Figs. 4.43(a) and (b), where a dielectric layer between two perforated gold films can be seen. The authors measured the phases and amplitudes of the reflected and transmitted waves in response to a perpendicularly incident beam, and then used the algorithm of Smith *et al.* (2002) to find the negative index.

Further advance was reported recently by Valentine *et al.* (2008) based on the theoretical predictions of Zhang *et al.* (2006). The idea is to use a multilayer structure shown schematically in Fig. 4.44(a). The scanning electron microscope image may be seen in Fig. 4.44(b). The structure consists of alternating layers (21 altogether) of 30 nm silver (Ag) and 50 nm magnesium fluoride (MgF_2). The size of the unit cell is $p = 860$ nm, and the other dimensions are $a = 565$ nm and $b = 265$ nm. The variation of the index of refraction was investigated in a series

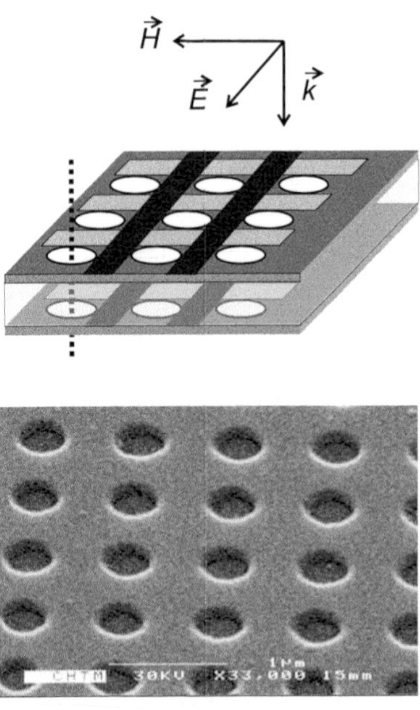

Fig. 4.43 Top: Schematic of the multilayer fishnet structure consisting of an Al_2O_3 dielectric layer between two Au films perforated with a square array of holes (838 nm pitch; 360 nm diameter) atop a glass substrate. For the specific polarization and propagation direction shown, the active regions for the electric (dark regions) and magnetic (hatched regions) responses are indicated. Bottom: SEM picture of the fabricated structure. From Zhang *et al.* (2005*a*). Copyright © 2005 Optical Society of America. For coloured version see plate section

Fig. 4.44 Diagram (a) and SEM image (b) of fabricated fishnet structure. The inset shows a cross-section of the pattern taken at a 45° angle. From Valentine *et al.* (2008). Copyright © 2008 Nature Publishing Group

of experiments, in which an input optical beam was incident upon the material shaped in the form of a prism. In a control experiment the prism was removed letting the beam through a window. The position of the beam, as may be expected, turned out to be independent of the wavelength of the input beam as shown in Fig. 4.45(a). In the presence of the prism the beam was displaced by increasing amounts as shown in Fig. 4.45(b). From the amount of displacement it was possible to determine the index of refraction. It is interesting to note that at a wavelength of 1465 nm the refractive index is close to zero, which means that the optical path through the prism is everywhere the same (i.e. zero), hence the prism loses its ability to displace the beam. The variation of the refractive index as a function of wavelength is plotted in Fig. 4.45(c). The measurement results are denoted by circles and the corresponding error bars. The theoretical curve (continuous line) was obtained by using the rigorous coupled-wave analysis (Moharam *et al.*, 1995).

Plasma resonances of small 2D particles of silver were investigated numerically by Kottmann *et al.* 2000*b*; Kottmann *et al.* 2000*a*; Kottmann

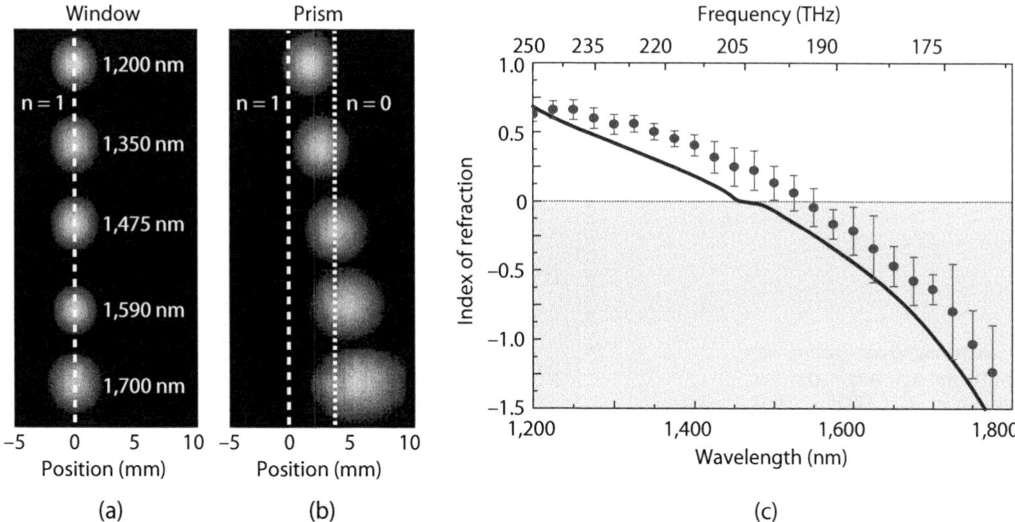

Fig. 4.45 Experimental results and FDTD simulations. Fourier-plane images of the beam for the window (a) and prism sample (b) for various wavelengths. The horizontal axis corresponds to the beam shift d, and positions of $n = 1$ and $n = 0$ are denoted by the white lines. The image intensity for each wavelength has been normalized for clarity. (c) Measurements (circles wuth error bars after four measurements) and simulation (solid line) of the fishnet refractive index. From Valentine *et al.* (2008). Copyright © 2008 Nature Publishing Group. For coloured version see plate section

and Martin 2001; Kottmann *et al.* 2001 both of regular and non-regular shape. They concluded that there are many more resonances for a non-regular shape, as shown in Fig. 4.46 where scattering cross-section is plotted against wavelength for plane-wave incidence. The particle is a right-angled triangle of 10 nm base and 20 nm side. The polarization of the incident electric field is in the plane of the figure perpendicular to the arrow in the inset. As many as five resonances may be seen, whereas an ellipse has only one resonance. The electric-field distribution is shown by a colour code in Fig. 4.47 at the main resonance of 458 nm. Normalization is to the amplitude of the incident field. The amplitude remains ten times that of the incident field at a 10 nm distance from the sharp corner.

The effect of the coupling between two cylinders of 25 nm radii made of silver was also investigated (Kottmann *et al.*, 2000*a*) as a function of the distance between cylinders. A plane wave in the direction of the axis of symmetry was incident with the electric field perpendicular to the axes of the cylinders. The resonant wavelengths were found to be 350 nm, 358 nm, 368 nm, 380 nm and 404 nm for distances between the cylinders of 50 nm, 20 nm, 10 nm, 5 nm and 2 nm. A further radiation-induced resonance was found by Kottmann and Martin (2001).

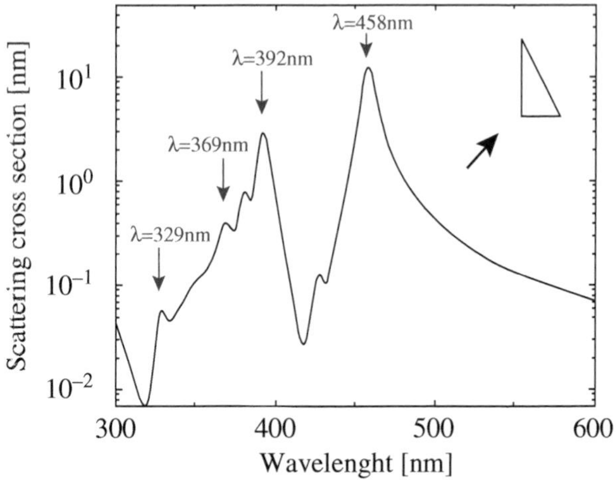

Fig. 4.46 Scattering cross-section versus wavelength for a triangle particle. From Kottmann *et al.* (2000*b*). Copyright © 2000 Optical Society of America

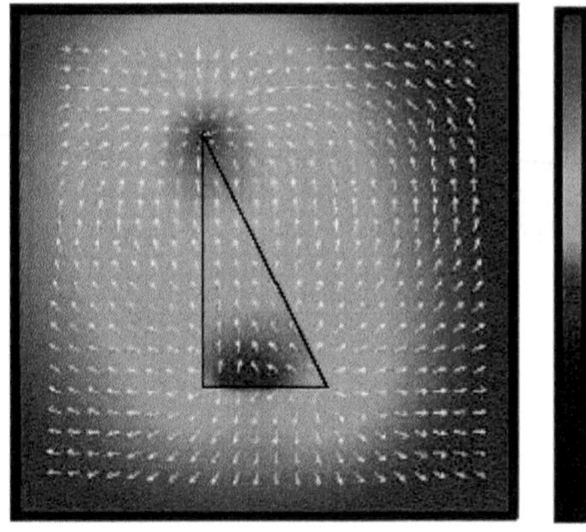

Fig. 4.47 Field map of the electric field at the main resonance for a triangle particle. From Kottmann *et al.* (2000*b*). Copyright © 2000 Optical Society of America. For coloured version see plate section

4.4 A mathematical model and further experimental results

4.4.1 Distributed circuits

Once the size of a resonant element is small relative to the wavelength the chances of a successful representation in terms of circuits becomes high. This was indeed regarded by many authors as an aid to understanding the physics. Most of the representations were in terms of lumped elements (see, for example, Baena *et al.* 2005*b*). Our aim in this section is to introduce a distributed circuit model on the lines of Shamonin *et al.* 2004; Shamonin *et al.* 2005 to describe the properties of the split-ring resonator, shown once more in Fig. 4.48(a). The physical relationships to be embodied in the model are as follows: (i) Conduction currents I_1 and I_2 in the outer and inner rings vary as a function of the azimuthal angle φ, and are complemented by displacement currents in the gaps in order to form closed circuits, (ii) the currents in the outer and inner rings are coupled to each other by a mutual inductance, M, (iii) a dis-

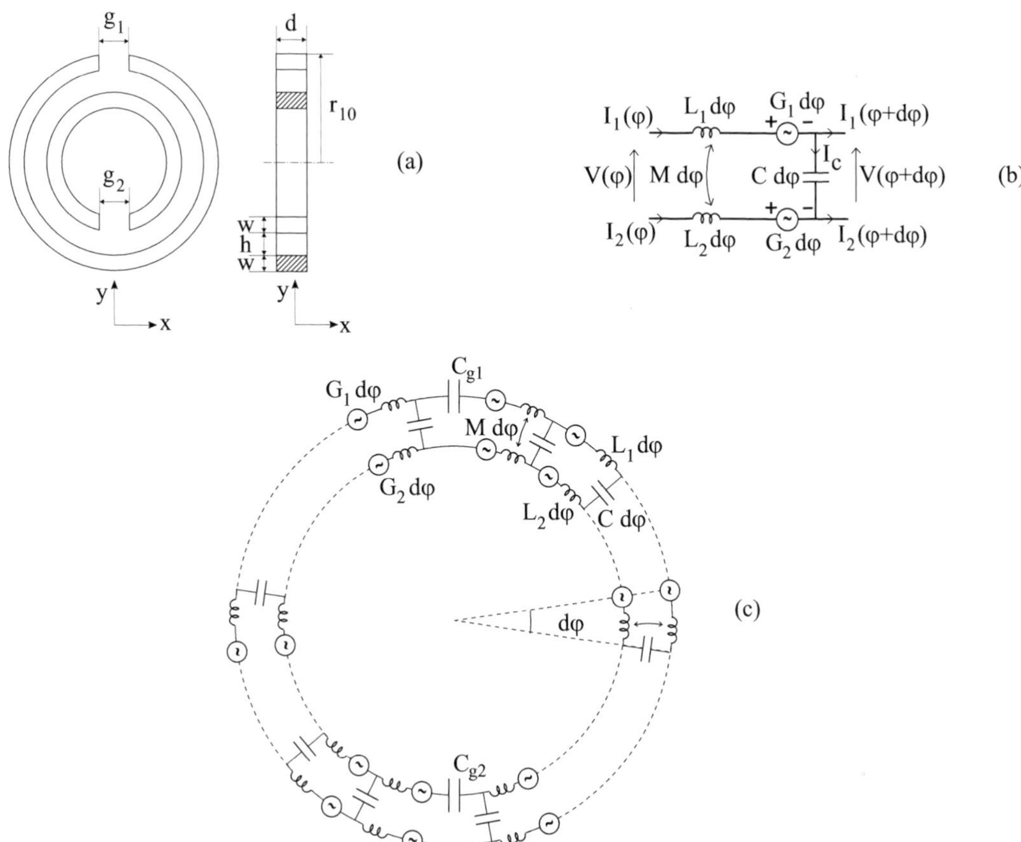

Fig. 4.48 Split-ring resonator. (a) A schematic view. (b) The equivalent circuit of a dφ element. (c) The equivalent circuit of the entire resonator. From Shamonin *et al.* (2004), copyright © 2004 American Institute of Physics, and Shamonin *et al.* (2005), copyright © 2005 Wiley-VCH Verlag GmbH & Co. KGaA

placement current, I_c, flows between the outer and inner rings, (iv) the inner and outer rings have inductances L_1 and L_2 and resistances R_1 and R_2, respectively, (v) the capacitances across the gaps in the outer and inner rings are C_{g1} and C_{g2}, respectively, (vi) a spatially constant and in time harmonically varying magnetic field will induce voltages both in the outer and inner rings.

We shall now write Kirchhoff's equations for a dφ element of the SRR shown in Fig. 4.48(b). The current equation for the upper line is

$$I_1(\varphi + \mathrm{d}\varphi) = I_1(\varphi) - I_{\mathrm{c}}(\varphi)\,, \tag{4.6}$$

where

$$I_{\mathrm{c}}(\varphi) = \mathrm{j}\omega C V(\varphi), \tag{4.7}$$

$V(\varphi)$ is the voltage across the ring and C is the inter-ring capacitance per unit radian. The current equation for the lower line is

$$I_2(\varphi + \mathrm{d}\varphi) = I_2(\varphi) + I_\mathrm{c}(\varphi)\,. \tag{4.8}$$

From eqns (4.6)–(4.8) we obtain the differential equations

$$\frac{\mathrm{d}I_1}{\mathrm{d}\varphi} = -\mathrm{j}\omega CV \qquad \text{and} \qquad \frac{\mathrm{d}I_2}{\mathrm{d}\varphi} = \mathrm{j}\omega CV, \tag{4.9}$$

whence the physically obvious relationship

$$I_1 + I_2 = I_0 \tag{4.10}$$

follows. I_0 is a constant that will have to be determined from the boundary conditions. We can similarly write the voltage equation in the form

$$V(\varphi+\mathrm{d}\varphi) = V(\varphi) + I_1 Z_1 - \mathrm{j}\omega M I_1 - I_2 Z_2 + \mathrm{j}\omega M I_2 + \frac{F_1 - F_2}{2\pi}\,, \tag{4.11}$$

where

$$Z_1 = R_1 + \mathrm{j}\omega L_1, \qquad Z_2 = R_2 + \mathrm{j}\omega L_2\,. \tag{4.12}$$

F_1 and F_2 are the induced voltages in the outer and inner rings, and are equal to

$$F_1 = \mathrm{j}\omega\mu_0 H\pi r_1^2 \qquad \text{and} \qquad F_2 = \mathrm{j}\omega\mu_0 H\pi r_2^2\,, \tag{4.13}$$

where r_1 and r_2 are the inner radii of the outer and inner rings. Note further that the symbols in eqns (4.9)–(4.12), C, R_1, R_2, L_1, L_2 and M all refer to values per radian.

Expressing I_2 in terms of I_1 from eqn (4.10) we can derive a differential equation from eqn (4.11) in the form

$$-\frac{\mathrm{d}V}{\mathrm{d}\varphi} = I_1(Z_1 + Z_2 - 2\mathrm{j}\omega M) - I_0(Z_2 - \mathrm{j}\omega M) - \frac{F_1 - F_2}{2\pi}\,. \tag{4.14}$$

We have now two first-order differential equations in I_1 and V from which we may obtain a second-order differential equation with constant coefficients in either variable that can be solved in terms of constants and trigonometric functions. To continue we need the full equivalent circuit shown in Fig. 4.48(c) which will help to formulate the boundary conditions.

Next, let us look in more detail at what happens at $\varphi = 0$ (note that in order to exploit symmetry our azimuthal co-ordinate system runs from $-\pi$ to 0, and from 0 to π. According to our model the gap capacitor $C_{\mathrm{g}1}$ is exactly at the position $\varphi = 0$ and has zero spatial extent. There is, however, bound to be a voltage drop across it due to the displacement current flowing through it. Consequently, the inter-ring voltage cannot be continuous at $\varphi = 0$. The current must be continuous but not necessarily differentiable. It follows then that we need to solve the differential equations separately for the $-\pi$ to 0, and for the 0 to π regions. The corresponding functions will be called V_-, I_{1-}, V_+, I_{1+}, respectively. The boundary conditions will then take the form: first for the currents

$$\begin{aligned} I_{1+}(0) &= I_{1-}(0)\,, & I_{2+}(0) &= I_{2-}(0)\,, \\ I_{1+}(\pi) &= I_{1-}(-\pi)\,, & I_{2+}(\pi) &= I_{2-}(-\pi)\,, \end{aligned} \tag{4.15}$$

and for the voltages

$$\begin{aligned} \mathrm{j}\,\omega C_{\mathrm{g1}}\,(V_-(0) - V_+(0)) &= I_1(0)\,, \\ \mathrm{j}\,\omega C_{\mathrm{g2}}\,(V_-(-\pi) - V_+(\pi)) &= I_2(\pi)\,. \end{aligned} \tag{4.16}$$

We have taken care of the currents and voltages but not, as yet, of the excitation. For that we shall have to write Kirchhoff's voltage law either along the inner or the outer ring, which will include the excitation. The condition for the total voltage to be zero along the outer ring may then be written as

$$(Z_1+\mathrm{j}\,\omega M)\left(\int_{-\pi}^{0} I_{1-}\mathrm{d}\varphi + \int_{0}^{\pi} I_{1+}\mathrm{d}\varphi\right) - 2\mathrm{j}\,\omega M\pi I_0 + \frac{I_1(0)}{\mathrm{j}\,\omega C_{\mathrm{g1}}} = F_1\,. \tag{4.17}$$

In principle everything is easy. We are concerned only with differential equations of constant coefficients, and the boundary conditions are also constants. Hence, the analytic solution is straightforward although rather laborious, and the resulting expressions are rather long. We shall not show them here but will plot a number of relevant curves later.

Next, we shall look at the eigensolutions that occur in the absence of excitation. The linear equations for the voltages and currents are then homogeneous, and the condition for a solution to exist is that the determinant has to vanish. The resulting characteristic equation then takes the form (Shamonin *et al.*, 2004),

$$\begin{aligned} &\kappa \sin \kappa\pi [4\pi\kappa^2 - \pi\gamma_1\gamma_2 - 2\gamma_1\nu_2 - 2\gamma_2\nu_1] \\ &+ \cos \kappa\pi[-2\pi\kappa^2(\gamma_1+\gamma_2) + \gamma_1\gamma_2(\nu_1+\nu_2)] = -2\gamma_1\gamma_2\nu_{12}\,, \end{aligned} \tag{4.18}$$

where

$$\begin{aligned} \kappa^2 &= \omega^2 L_{\mathrm{eq}} C\,, \quad \gamma_1 = \frac{C}{C_{\mathrm{g1}}}\,, \quad \gamma_2 = \frac{C}{C_{\mathrm{g2}}}\,, \\ \nu_1 &= \frac{(L_1-M)^2}{L_1L_2-M^2}\,, \quad \nu_2 = \frac{(L_2-M)^2}{L_1L_2-M^2}\,, \\ \nu_{12} &= \sqrt{\nu_1\nu_2}\,, \quad L_{\mathrm{eq}} = L_1 + L_2 - 2M\,. \end{aligned} \tag{4.19}$$

We shall look at the solution of this equation for a set of parameters further below. First, let us find an approximate solution for the lowest resonant frequency, i.e. for the lowest value of κ. We can get that approximate solution from eqn (4.19) by assuming that κ is small. Expanding then $\sin\kappa\pi$ and $\cos\kappa\pi$ in a series and neglecting terms in κ^4 and higher, eqn (4.19) reduces to

$$\omega^2 = \frac{1}{\frac{\pi^2}{2}(L_1+L_2)C + 2\pi L_1 C_{\mathrm{g1}} + 2\pi L_2 C_{\mathrm{g2}}}. \tag{4.20}$$

The remarkable thing is that L_{eq} (the only term that contains the mutual inductance M) does not appear in eqn (4.20), i.e. the lowest resonant frequency is independent of the mutual inductance and depends only on the individual inductances. Since in an SRR the radii of the outer and inner rings are close to each other we can make the further approximation that $L_1 = L_2 = L$, in which case eqn (4.20) takes the form

$$\omega^2 = \frac{1}{2\pi L\left(\frac{\pi}{2}C + C_{\mathrm{g1}} + C_{\mathrm{g2}}\right)}. \tag{4.21}$$

Note that L and C are per radian. Thus, if we want to use the total inductance and the total inter-ring capacitance of the SRR then it is $L_{\mathrm{t}} = 2\pi L$ and $C_{\mathrm{t}} = 2\pi C$ with which eqn (4.21) takes the final form

$$\omega^2 = \frac{1}{L_{\mathrm{t}}\left(\frac{C_{\mathrm{t}}}{4} + C_{\mathrm{g1}} + C_{\mathrm{g2}}\right)}. \tag{4.22}$$

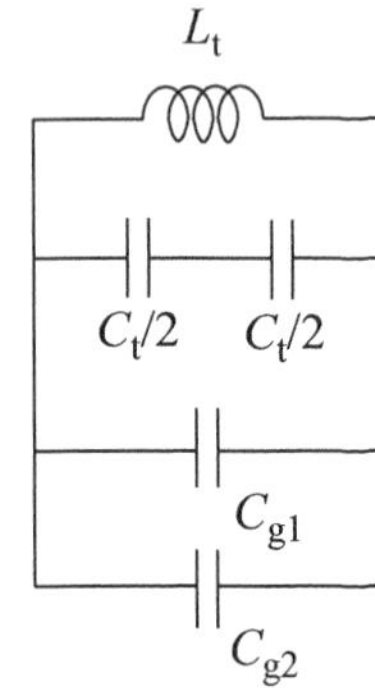

Fig. 4.49 Equivalent circuit

The conclusion is now clear. The lowest resonant frequency of a SRR is given by the equivalent circuit shown in Fig. 4.49, a generalization of that of Fig. 4.13(a). The inter-ring capacitance between two splits is $C_{\mathrm{t}}/2$. These two capacitances in series are then connected in parallel with the two gap capacitances. Note that eqn (4.22) is a generalization of the expression given by eqn (2.26), derived intuitively by Marques *et al.* (2002*b*). It now includes the gap capacitances as well and, interestingly, in a very simple manner.[8]

[8] Having set up a model for a SRR we can easily extend the same principles to the modelling of a singly split double ring that also consists of two concentric rings but differs from the SRR by having only one of the rings split. In fact, the inner ring being split is an old design due to Hardy and Whitehead (1981) that has already been shown in Fig. 4.5. The analysis, quite similar to that presented above, was performed by Shamonin *et al.* (2004). The conclusions were similar too. The main difference relative to the SRR was that having only one split the resonance frequency for the same parameters was higher.

4.4.2 Results

The main interest is in the lowest resonant frequency because the aim is to produce a medium describable by effective material parameters. The smaller the elements the more applicable effective-medium theory is. We shall now look at a few examples. A schematic drawing of the SRR has already been shown in Fig. 4.48(a). We shall now choose the dimensions as follows: Its height is $h = 5$ mm. The external radius of the outer ring is $r_{\mathrm{e1}} = 11$ mm and the wall thickness is 1 mm. The gaps are 1 mm both in the inner and in the outer ring. The wall thickness of the inner ring is 0.8 mm. We shall look at three examples in which the dimensions enumerated above are all identical but the external radius of the inner ring takes three different values, namely $r_{\mathrm{e2}} = 7.5$, 8.5 and 9

ble 4.3 Parameters and resonant frequencies of split-ring resonators with variable inner-ring lius

r_{e2} [mm]	L_2 [nH/rad]	M [nH/rad]	C [pF/rad]	f_0 [GHz]	f_1 [GHz]	f_2 [GHz]
7.5	2.74	1.99	0.239	1.50	3.88	7.97
8.5	3.34	2.70	0.382	1.19	3.80	7.72
9	3.69	3.49	0.550	0.97	4.50	8.96

mm. Having specified the elements we need to determine the parameters that come into the analytical theory presented above: the capacitances and inductances. They can be determined with the aid of formulae and tables available in the literature (see, e.g., Grover 1981; Hammond and Sykulski 1994). The inductance of the outer ring is found as $L_1 = 4.90$ nH/rad, and the gap capacitances as $C_{g1} = 0.106$ pF and $C_{g2} = 0.092$ pF. The other parameters that depend on r_{e2} are given in the second to fourth columns of Table 4.3, all per radian. Having obtained the values of all the parameters we can determine the resonant frequencies by solving numerically eqn (4.18) for ω. The equation, being transcendental, has an infinite number of solutions. The first three resonant frequencies obtained are given in columns 5–7 of Table 4.3. As expected, the lowest resonant frequency occurs for case 3 where the inter-ring separation is minimum and the inter-ring capacitance is maximum.

A qualitative explanation in terms of circuits has already been given for the fundamental resonance. For the first and second higher-order resonance the physical explanation is that an electromagnetic wave travels round in the inter-ring space. The mechanism is somewhat similar to that in a strip-line resonator (Chang, 1996) but in the present case they are capacitively loaded. The result is that the first higher-order resonance occurs at a wavelength considerably less than the perimeter. On the other hand, the second higher-order resonance has a frequency twice as large as the first-order resonance, showing clearly that wave propagation plays a dominant role.

To check these predictions and to have a better understanding of the physics we shall plot the variation of the voltages and currents as a function of angle in Fig. 4.50 as follows from our theory. The parameters are those given above for $r_{e2} = 7.5$ mm. The corresponding resonant frequencies are given in Table 4.3 as $f_0 = 1.5$ GHz, $f_1 = 3.88$ GHz and $f_2 = 7.97$ GHz. The frequencies at which the variables are plotted are 0.05, 1.5, 2.5, 3.02, 3.88 and 7.97 GHz corresponding to Figs. 4.50(a)–(f).

Let us look at the currents first (upper figures). Our normalization is to the sum of the two currents: we shall take $I_0 = 1$. At 0.05 GHz the linear approximation may be seen to be quite good. The surprising thing is that the approximation is still good at the lowest resonant frequency of 1.5 GHz and it is still not too bad at a frequency of 3.02 GHz, although by then the currents are more like sinusoidals. The major change

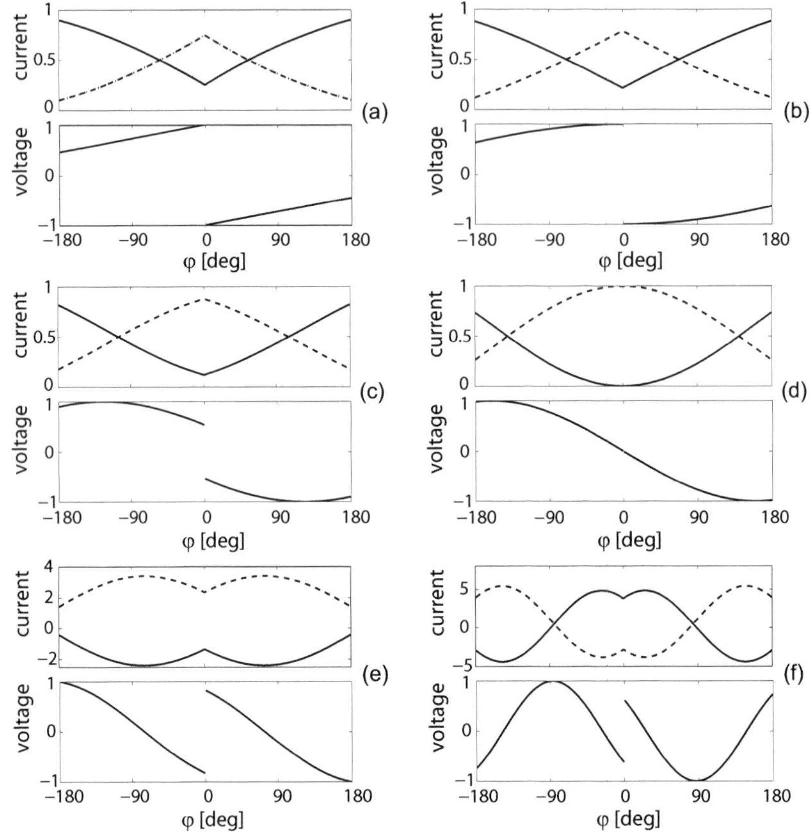

Fig. 4.50 Voltage and current distributions. $f_r = 50$ MHz, 1.5, 2.5, 3.02, 3.88, 7.97 GHz (a)–(f)

in the current distribution occurs at the first higher resonance of 3.88 GHz.[9] Note that the currents are now in the opposite direction, a clear indication that the physics has changed. We have moved away from the LC resonance to a resonance that is more akin to those in transmission lines. At f_2, the second higher resonance, as we may expect, the current has a double period.

The voltages, plotted in the lower figures, tell a similar story. Note that normalization is now to the maximum amplitude. The inter-ring voltage is supposed to be constant according to the qualitative explanation. In fact, it is somewhat different from constant even at the lowest frequency investigated. On the other hand, we may see that at 50 MHz and even at 1.5 GHz, the lowest resonant frequency, the voltage jumps at the two gaps (at 0° and at 180°) from positive maximum to negative maximum and vice versa. At 2.5 GHz the voltage jump at 0° is smaller and it is entirely absent at 3.02 GHz.[10] After 3.02 GHz the voltage jump reverses. At the first higher resonance the voltage period halves and it halves again at the second higher resonance.

Next, we shall discuss a set of experiments by Radkovskaya *et al.*

[9] We are actually slightly off resonance because in the absence of losses all the amplitudes would be infinitely large at the exact resonant frequency.

[10] This particular value of frequency was chosen with the explicit aim of finding a continuous variation of voltage with just one discontinuity at 180°.

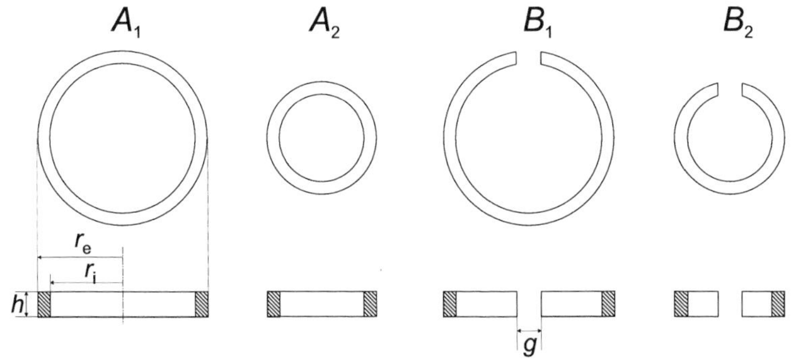

Fig. 4.51 Schematic representation of the elements from which resonators are constructed. From Radkovskaya *et al.* (2005). Copyright © 2005 Wiley-VCH Verlag GmbH & Co. KGaA

Table 4.4 Values of measured and calculated resonant frequencies in GHz for the various resonators. From Radkovskaya *et al.* (2005)

resonator	experimental	analytical	numerical
B_1	1.74		1.77
B_2	2.59		2.67
B_1A_2	1.89	2.17	1.94
A_1B_2	2.78	2.89	2.84
B_1B_2	1.44	1.50	1.438

(2005) in which four copper rings (Fig. 4.51) were used in different combinations. The dimensions of the rings may be recognized as those we used in our earlier theoretical calculations with $r_{e2} = 7.5$ mm. Out of the four rings it is possible to construct five resonators, as shown schematically in Table 4.4. $\mathrm{B_1A_2}$ is the notation for the full ring $\mathrm{A_2}$ placed concentrically inside the split ring $\mathrm{B_1}$. Similarly, $\mathrm{A_1B_2}$ means the split ring $\mathrm{B_2}$ inside the full ring $\mathrm{A_1}$, and $\mathrm{B_1B_2}$ means the smaller split ring $\mathrm{B_2}$ inside the larger split ring $\mathrm{B_1}$ with splits at opposite sides. The last one is of course a SRR.

The purpose of the experiments was to measure the resonant frequencies of all four configurations.[11] They are shown in column 3 of Table 4.4, whereas the resonant frequencies obtained by the MICRO-STRIPES[12] numerical package are in column 5. As expected, $\mathrm{B_2}$, the smaller split

[11] For the same sets of resonators further results (measuring the quality factor, Q) were obtained by Hao *et al.* 2005*b*; Hao *et al.* 2005*a*.

[12] MICRO-STRIPES is a registered trademark of Flomerics Ltd., Surrey, UK.

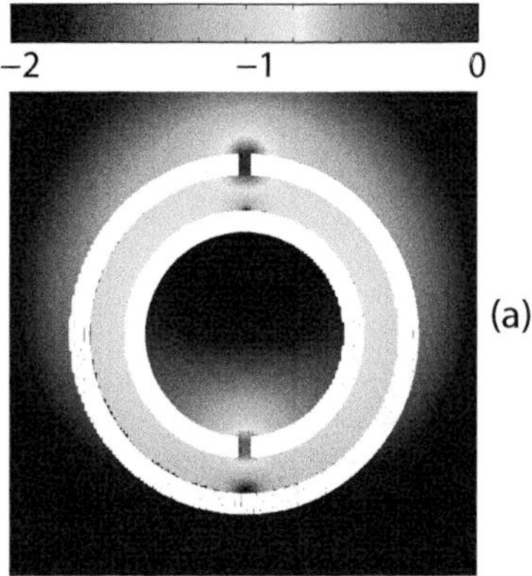

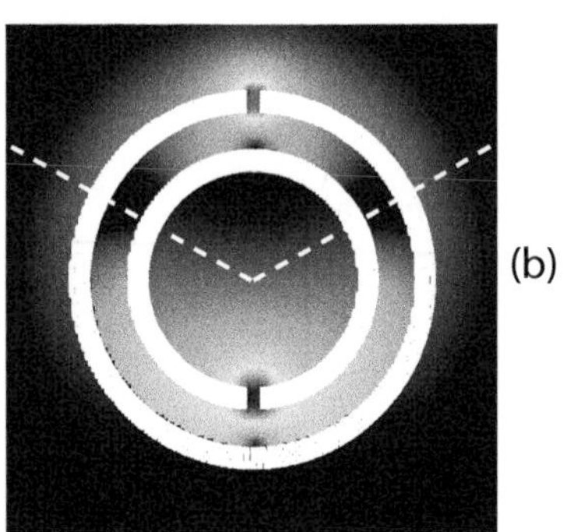

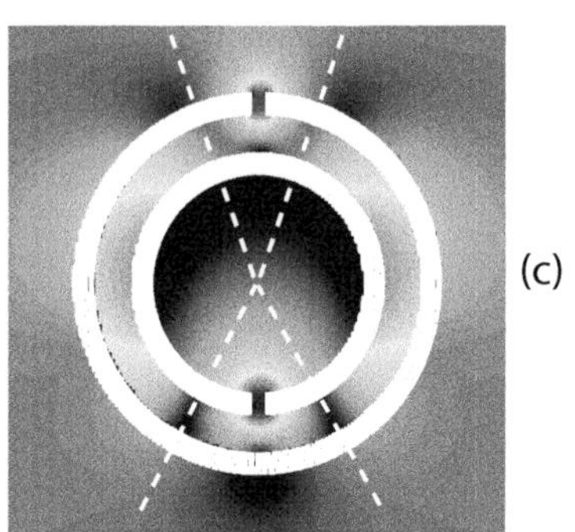

Fig. 4.52 Electric-field distributions at the fundamental (a), the first (b) and the second higher resonance (c). Positions of the voltage zeros are shown by white dashed lines. From Hesmer (2008). For coloured version see plate section

ring has a higher resonant frequency than B_1. The smaller the rings the higher the resonant frequencies. One would indeed expect simple inverse scaling of the resonant frequency with the radii. Note from Table 4.4 that the ratio of the inner radius of ring B_1 to that of B_2 is 1.49, and the inverse ratio of the corresponding resonant frequencies is also 1.49. It may also be seen that in the configurations B_1A_2 and A_1B_2 the resonant frequency is determined by the size of the split ring. The ratio of the corresponding resonant frequencies of 2.78 GHz and 1.89 GHz is 1.47, very close to the ratio of the radii. Thus, the influence of the full inner and outer rings is merely to increase a little the resonant frequency. The lowest resonant frequency is 1.44 GHz, obtained by the SRR. The agreement between the experimental and simulation results is quite close. The maximum deviation is for the B_2 case, which amounts to about 3%.

We have no analytic expressions for the resonance frequencies of the single rings B_1 and B_2. The characteristic equation (eqn (4.18)) makes it possible to determine the resonant frequency, as follows from our distributed circuit model. The resonant frequency obtained from that is 1.50 GHz, in contrast to the measured 1.44 GHz. We have not shown the analytic expressions for the singly split double rings of the A_1B_2 and the A_2B_1 configurations, but they are available in Shamonin *et al.* (2004). The resonant frequencies calculated from there are given in Table 4.4 as 2.17 and 2.89 GHz, which are still reasonable approximations.

For the resonant frequency of the SRR three more approximate expressions are available, that of eqn (4.22), and those of Pendry *et al.* (1999) and Sauviac *et al.* (2004). We shall give here the expression of Pendry *et al.* (1999) which is in the simple form of

$$\omega_{\mathrm{r}} = \frac{c}{r}\sqrt{\frac{3d}{\pi^2 r}}\,, \tag{4.23}$$

where $d = r_{\mathrm{i}1} - r_{\mathrm{e}2}$, $r = (r_{\mathrm{i}2} - r_{\mathrm{e}2})/2$. The above equation yields a resonant frequency of 1.62 GHz, far too high, whereas the expression given by Sauviac *et al.* (2004) yields 1.18 GHz. On the other hand, eqn (4.22) yields quite a good approximation of 1.35 GHz. For the A_1B_2 configuration there is also a simple analytical formula as given by eqn (4.3). It gives a resonant frequency of 6.4 GHz, quite far from the measured 2.78 GHz. The probable reason for the large discrepancy is that they attributed too much importance to the gap capacitance, which was indeed quite large in their design shown in Fig. 4.5.

4.4.3 A note on higher resonances

In Table 4.3 we gave the analytical results both for the fundamental and for the higher-order resonances. We have also determined them by simulations using the numerical package CST MICROWAVE STUDIO[13] (Hesmer, 2008). It is reasonable to expect good agreement for the fundamental resonance, but the distributed circuit model is less likely to be valid for the higher-order resonances. The analytical results from

[13] CST MICROWAVE STUDIO is a registered trademark of CST–Computer Simulation Technology AG, Darmstadt, Germany

Table 4.3 are 1.50, 3.88 and 7.97 GHz. Our numerical simulations gave 1.46, 3.42 and 7.68 GHz, which shows larger discrepancies for the higher orders.

The simulations also provided the field distributions at the resonant frequencies. They are shown in Fig. 4.52. It may be immediately seen that the field distribution has no zeros for the fundamental resonance, two zeros for the first higher-order resonance and four zeros for the second higher-order resonance. Looking at Figs. 4.50(e) and (f) we can see that the number of zero points is the same as in the analytical solution. We can also compare the positions of the zeros. The analytical calculations give (they are symmetric around the zero angle so we shall give only the positive values) for f_1 74.2°, and 26.2° and 147° for f_2. In contrast the simulations yield 62° for the first and 19.5° and 145° for the second higher resonance. Not a close agreement, but all are in the same ballpark.

Subwavelength imaging

5

5.1 Introduction

The subject of metamaterials started with Veselago's (1967) introduction of negative refractive index and his simultaneous proposal for a flat lens. We have already discussed the flat-lens family in Section 2.11.4. Its characteristic feature is the existence of a focus inside the lens. The basic principle of its operation has already been shown in Fig. 2.24; for convenience it is shown again in Fig. 5.1. The essential requirement is that the index of refraction is $n = -1$. The ray trajectories follow then in the geometrical optics approximation. A focal point is brought to a focal point and an object in the object plane is reproduced perfectly in the image plane. This is an obvious advantage of the flat lens. It has no optical axis. A simple illustration in Fig. 5.1 shows an object consisting of three points in the object plane that is perfectly reproduced. At the same time we need to mention a serious disadvantage of the flat lens. It can work only at a single frequency because the $n = -1$ condition is very strongly frequency-dependent. In traditional imaging language this means that the flat lens has very large chromatic aberration.

As follows from the construction in Fig. 5.1, Veselago's lens is perfect when we consider propagating waves only. But, if the object is small relative to the wavelength it will have both propagating and evanescent components in its Fourier spectrum. The image is not perfect because the subwavelength features have not been reproduced. It was shown by Pendry (2000) that a flat lens in the Veselago geometry can reproduce

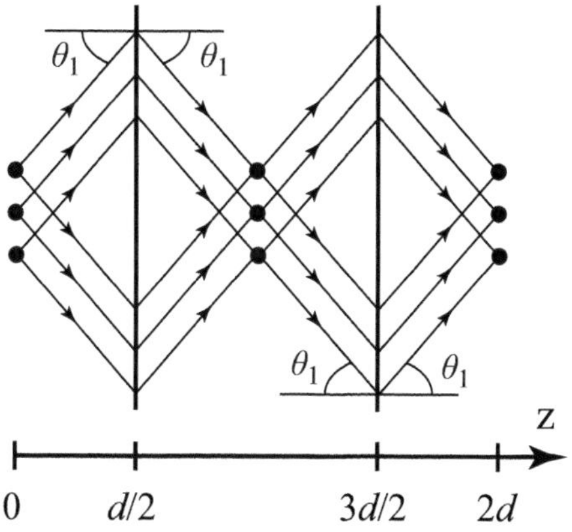

Fig. 5.1 Imaging three point sources with Veselago's flat lens

subwavelength features on account of the evanescent waves growing[1] in the lens. This was already discussed in Section 2.12. Here, we shall assume that the reader has already read that section, and is also familiar with surface plasmons discussed in Chapter 3. Later in this chapter we shall of course look at subwavelength imaging from a number of different angles but here we shall start with a brief summary in the form of questions and answers.

Firstly, what is the difference between Veselago's flat lens and Pendry's perfect lens? The geometry is the same but not the parameters of the lens. Veselago's lens will work whenever the index of refraction is $n = -1$. For Pendry's perfect lens it is required for both material parameters, ε_r and μ_r, to be equal to -1. Pendry's lens is always matched. If in Veselago's lens the $n = -1$ condition is obtained by both material parameters being equal to -1 then for a lossless lens there is no difference between the Veselago and Pendry variety.

Secondly, why is the total phase change from object to image in Pendry's lens equal to zero? Because the wave in the lens is a backward wave that has the same phase velocity as the forward waves from object to lens, and from lens to image. Since the path in the lens is equal to the path in air the total phase change is zero.

Thirdly, why is the total amplitude of a non-propagating space harmonic from object to image equal to unity? This is because the rate of increase in the lens is equal to the rate of decrease in air.

Fourthly, is a flat transfer function possible in practice? It is not possible in practice and actually not in theory either because it implies infinitely high fields.

In the fifth place, what will prevent the transfer function from being flat? Losses, because the growing non-propagating waves will be attenuated and therefore perfect compensation of amplitudes becomes impossible. Also, any deviation from the ideal conditions will cause the appearance of a cutoff in the transfer function.

In the sixth place, why is there a chance to build a subwavelength lens from a material in which only ε_r or only μ_r is equal to -1? It follows from the properties of TM waves that their amplitudes depend strongly only on ε and vice versa for TE waves.

In the seventh place, why does the perfect lens need both material parameters to be -1 if with the right polarization the amplitudes depend strongly only on one of the material parameters? The reason is that the phase velocity of propagating waves and the rate of decay (or growth) of non-propagating waves depend on both material parameters.

In the eighth place, why do we have growing non-propagating waves in the epsilon-negative-only lens? Because surface plasmon–polaritons (SPPs) are excited at the far boundary and they cannot be excited by a propagating wave owing to their shorter wavelength.

In the ninth place, why is it necessary for the epsilon-negative-only lens to be thin? This is because the SPP on the rear surface can only be excited by an input wave if the front and rear surfaces are strongly coupled, and they are strongly coupled only when they are near to each

[1]There is some problem here with terminology. In Section 2.12 we have already objected to the term 'amplification' because there is no source of additional power there. Strictly speaking we should not talk about growing evanescent waves either. To evanesce means to fade away. A growing evanescent wave is a contradiction in terms. So does it grow or does it evanesce? It can't do both. The correct description would be the growth of the amplitude of the non-propagating waves in the $+z$ direction. Unfortunately, that's quite a mouthful and goes against current practice. Thus, with regret, we shall often refer to growing evanescent waves.

other.

In the tenth place, why is a multilayer lens preferable to a single-layer lens? Because the SPP interaction between the front and rear surfaces is still active for consecutive thin lenses, and at the same time the distance between the object and image increases.

The above is a rather rough and very concise summary of what has been said so far in this book about imaging by a flat lens. The rest of the chapter will of course go into a lot of details. Section 5.2 will be concerned with history, with the controversy that arose in the wake of the publication of Pendry's paper (2000). The most important part of this chapter is probably Section 5.3, which will describe in great detail what happens to the transfer function when the parameters deviate from the ideal ones. Also in the same section, we shall establish relationships between phenomena in the perfect lens and the physics of SPPs. In Section 5.4 we shall investigate the quality of images under conditions when only the permittivity is equal to -1. Section 5.5 is concerned with the improvements a multilayer lens can offer. Section 5.6 discusses magnification. Section 5.7 is an attempt to show and briefly discuss some of the widely held beliefs that may not stand closer scrutiny.

5.2 The perfect lens: controversy around the concept

Pendry's paper on the perfect lens has been the main driving force in the rapid development of the field of metamaterials.[2] It literally stirred up the scientific community and caused an avalanche of papers devoted to various aspects of the concept of the perfect lens. At first, the concept was challenged by a large number of authors who questioned the correctness of the approach and looked at flaws in Pendry's arguments. The controversy around the idea of the perfect lens has had many facets.

[2]Published in October 2000, this paper was quoted over 2000 times in the less than eight years that passed since.

5.2.1 Battle of wits

The possibility of negative refraction and that of the perfect lens were attacked afterwards in several publications (see, e.g., Williams 2001; Garcia and Nieto-Vesperinas 2002; Valanju *et al.* 2002; Pokrovsky and Efros 2002*c*). An observer could witness a battle of research papers with quick-witted titles, which showed that scientists are good at puns and at repartees: 'Negative reaction to negative refraction' (Cartlidge, 2002), 'Wave refraction in negative-index media: always positive and very inhomogeneous' (Valanju *et al.*, 2002), 'Perfect lens in a non-perfect world' (Kik *et al.*, 2002), 'Near-sighted lens' (Podolskiy and Narimanov, 2005), 'Entering the Negative Age' (Pendry, 2001), 'Optics: positively negative' (Pendry, 2003*a*). 'Perfect lenses made with left-handed materials: Alice's mirror?' (Maystre and Enoch, 2004). For a while, the controversy raged unabated, the 'perfect lens' was demoted to a 'near-perfect lens' (Ramakrishna and Pendry, 2002). Fairly soon the tone of the papers changed: The goal was no longer solely to prove that the perfect lens concept has flaws (it surely does, as nothing is ever perfect in nature), but rather to understand the underlining physical mechanisms of subwavelength imaging and to explore the possibilities offered by metamaterial near-field lenses. Most researchers agree by now that the concept of subwavelength imaging by a flat lens is both novel and useful.

5.2.2 Non-integrable fields

The concern about non-integrable quantities was raised, among others, by 't Hooft 2001; Pokrovsky and Efros 2002*a*; Pokrovsky and Efros 2002*b*; Garcia and Nieto-Vesperinas 2002 and Haldane 2002. The essence of the problem is as follows. Considering wave propagation from an object plane via air, metamaterial slab, and air again, the field is decomposed into Fourier components with k_x values ranging from zero to infinite values. The total electrical field at each point in space is calculated by performing integration (or summation in the discretized formulation of the problem) over all possible wave vectors. The transfer function from the object plane to the image plane being flat means that all evanescent (non-propagating) components, no matter how large their k_x vectors, are all growing inside the lens. As already described in Section 2.12.2, for the case $\varepsilon_\mathrm{r} = \mu_\mathrm{r} = -1$, the k_z components of the wave vectors in the air and in the metamaterial are purely imaginary for sufficiently large values of k_x. This turns a square-integrable incoming wave into a non-integrable one. The mathematics is quite simple. We consider a square-integrable field with Fourier components $E_x(k_x)$ at the source plane $z = 0$. Each wave component that declined by a factor $\exp(-|k_z|d/2)$ before reaching the entrance surface of the metamaterial is then growing in the slab and reaches at the rear surface $z = 3d/2$ a value of $\exp(|k_z|d)$ times higher than in the input signal. A summation over all Fourier components yields an integral that diverges for large wave vectors. This problem of non-integrable quantities occurs, however, solely for the case with ε_r and μ_r being precisely equal to -1. For any other combination of ε_r and μ_r the remedy is provided by the high-frequency cutoff.

5.2.3 High spatial frequency cutoff

Common sense suggests that the divergences mentioned above can be cured by losses. Due to losses the amplitudes of the various space harmonics are not faithfully reproduced causing a high spatial frequency cutoff. An immediate consequence of the existence of this high-frequency cutoff is that the image is no longer perfect, as the spatial components above the cutoff die out before having a chance to reach the image plane. Are there any other reasons for a cutoff frequency? We shall mention some of the factors below.

Periodic structure. A constraint that comes immediately to mind is that the metamaterial is not homogeneous as discussed already in

Section 2.12. It is a heterogeneous structure composed of unit cells with finite dimensions. There must be a length scale for any practical realization of a metamaterial below which the homogeneous description is no longer applicable. Thus, components with too large k_x values above the characteristic value in the reciprocal space related to the periodicity of the material will be subject to a cutoff. If the period is a, there can be no chance whatsoever to reproduce a Fourier component at $k_x = 2\pi/a$.

Dispersion. The material constants ε and μ are bound to be dispersive in the region where they are negative. We know further (Kramers–Kronig relations, see, e.g., Landau and Lifschitz 1984) that any variation of the real parts of ε and μ with frequency is accompanied by variation in the imaginary part.

Deviations from the ideal conditions. The three reasons for the existence of a high spatial frequency cutoff enumerated above are perhaps not trivial but with a little thought devoted to the problems arising they are quite predictable. The effect of deviations from ideal conditions of $\varepsilon_r = -1$ and $\mu_r = -1$ is less obvious but, as will be shown later, it also leads to the appearance of a cutoff frequency.

Build-up time. The starting point of Gomez-Santos' (2003) analysis was the belief that the concept of the perfect lens can be rescued. He follows Haldane (2002) in attributing a crucial role to the surface plasmons and particularly to their near degeneracy as $k_x \to \infty$. Working in the time domain and using a model of coupled oscillators he proves that high resolution is obtained at the price of delay. Smaller details take longer and longer time to develop. Infinitely high resolution needs infinitely long time for its realization.

Only one material parameter being equal to -1*, say* $\varepsilon_r = -1$, $\mu_r = 1$. If only ε is negative and μ is positive then, as mentioned earlier, the rate of decay in air is not the same as the rate of growth in the lens, leading to a cutoff in the transfer function. The claim (Pendry, 2000) that for a lossless lens with $\varepsilon_r = -1$, $\mu_r = 1$ perfect imaging is still possible for an incident TM wave is not valid in general.

5.3 Near-perfect lens

5.3.1 Introduction

Let us sum up the conclusions about the perfect lens we have reached so far. The new feature we have learned is that when the ideal lens ($\varepsilon_r = \mu_r = -1$) is excited by an incident field from a source with sub-wavelength features, the 'perfect lens' structure would not provide at once a perfect image in the image plane. Instead, as time goes by, a field distribution corresponding to a solution with continuously improving resolution would build up. The higher the desired resolution, the more patience (and more energy from the source field) is required.

So there is one more reason why the perfect lens is an idealization that can never be reached. The next question that can be legitimately asked is how well a near-perfect lens would operate whose material parameters differ slightly from the ideal constellation. In other words to what extent imaging beyond the diffraction limit is possible using a fabricated material and to what extent the process can be simulated by numerical methods that inevitably approximate the ideal situation.

To familiarize ourselves with the operation of a homogeneous metamaterial slab acting as a near-perfect lens we will from now on consider only the stationary state. It follows from our previous discussion in Section 5.2 that we can do it with a clear conscience: a steady state would exhibit a high-frequency cutoff ensuring that all fields are square-integrable.

We shall now investigate the functioning of the lens by making use of various tools of analysis. Firstly, the transfer function provides the information on how well (or how badly) the input field in the plane $z = 0$ is reproduced in the image plane $z = 2d$. Secondly, the reflection coefficient (identically zero in the perfect-lens idealization) when it is different from zero indicates the existence of a surface mode at the front surface. Thirdly, variation of individual Fourier components across the slab helps to visualize the mechanism of what is going on inside the slab.

This section will be divided into a number of subsections. In Section 5.3.2 we shall provide the equations needed. The effect of losses on the transfer function, and in particular on the cutoff frequency, is treated in Section 5.3.3, then in Section 5.3.4 we shall return to the lossless case and investigate the effect of deviations from the ideal values of $\varepsilon_{\mathrm{r}} = \mu_{\mathrm{r}} = -1$ followed by an investigation of the deviations from the ideal refractive index of $n = -1$, and finally the case is examined when ε and μ vary in such a manner that their product remains the same, i.e. the refractive index does not change. In Section 5.3.5 the effect of lens thickness is investigated in the presence of losses. The influence of the various kinds of deviations is summarized in Section 5.3.6. An interesting and important point is raised in Section 5.3.7 that may cheer up those sceptical about the usefulness of numerical simulations. The problem seems to be an inherent mismatch in the constitutive parameters of the slab and the surrounding medium that may lead to incorrect results.

5.3.2 Field quantities in the three regions

We need the expressions for all components of the field amplitudes. We consider for simplicity TM polarization keeping in mind that similar expressions can be easily derived for TE polarization. We have already derived the solution to Maxwell's equations in a slab in Chapter 1 and discussed eigenmodes of the solution, SPPs, in Chapter 3. We quote here expressions needed for our analysis.

The fields from a source in the plane $z = 0$ are expanded in a Fourier series over propagating and evanescent waves, i.e. those with $k_x \leq k_0$ and with $k_x > k_0$, all of which are proportional to $\exp(\mathrm{j}\,\omega t)$ in the steady state,

$$H_y(x, z=0) = \sum_{k_x} A(k_x)\,\mathrm{e}^{-\mathrm{j}\,k_x x}\,. \tag{5.1}$$

The field expansion of the full solution everywhere in the xz plane takes the form

$$\begin{aligned}
H_{y1}(x,z) &= \sum_{k_x}\left(A\,\mathrm{e}^{-\mathrm{j}\,k_{z1}z} + B\,\mathrm{e}^{\mathrm{j}\,k_{z1}z}\right)\mathrm{e}^{-\mathrm{j}\,k_x x} \\
&\quad (\text{medium } 1)\,, \\
H_{y2}(x,z) &= \sum_{k_x}\left(C\,\mathrm{e}^{-\mathrm{j}\,k_{z2}z} + D\,\mathrm{e}^{\mathrm{j}\,k_{z2}z}\right)\mathrm{e}^{-\mathrm{j}\,k_x x} \\
&\quad (\text{medium } 2)\,, \\
H_{y3}(x,z) &= \sum_{k_x} F\,\mathrm{e}^{-\mathrm{j}\,k_{z3}z\,-\,\mathrm{j}\,k_x x} \\
&\quad (\text{medium } 3)\,.
\end{aligned} \tag{5.2}$$

$A = A(k_x)$ and $B = B(k_x)$ are the amplitudes of the incident and reflected waves of the Fourier component k_x in medium 1, $C = C(k_x)$ and $D = D(k_x)$ are the corresponding amplitudes of the transmitted and reflected waves in medium 2 (slab) and $F = F(k_x)$ is the amplitude of the transmitted wave in medium 3. The corresponding components of the electric field are

$$\begin{aligned}
E_{x1} &= \sum_{k_x}\frac{k_{z1}}{\omega\varepsilon_1}\left(A\,\mathrm{e}^{-\mathrm{j}\,k_{z1}z} - B\,\mathrm{e}^{\mathrm{j}\,k_{z1}z}\right)\mathrm{e}^{-\mathrm{j}\,k_x x} \\
&\quad (\text{medium } 1)\,, \\
E_{x2} &= \sum_{k_x}\frac{k_{z2}}{\omega\varepsilon_2}\left(C\,\mathrm{e}^{-\mathrm{j}\,k_{z1}z} - D\,\mathrm{e}^{\mathrm{j}\,k_{z1}z}\right)\mathrm{e}^{-\mathrm{j}\,k_x} \\
&\quad (\text{medium } 2)\,, \\
E_{x3} &= \sum_{k_x}\frac{k_{z3}}{\omega\varepsilon_3}F\,\mathrm{e}^{-\mathrm{j}\,k_{z3}z\,-\,\mathrm{j}\,k_x x} \\
&\quad (\text{medium } 3)
\end{aligned} \tag{5.3}$$

and

$$\begin{aligned}
E_{z1} &= -\sum_{k_x}\frac{k_x}{\omega\varepsilon_{\mathrm{r}1}}\left(A\,\mathrm{e}^{-\mathrm{j}\,k_{z1}z} + B\,\mathrm{e}^{\mathrm{j}\,k_{z1}z}\right)\mathrm{e}^{-\mathrm{j}\,k_x x} \\
&\quad (\text{medium } 1)\,, \\
E_{z2} &= -\sum_{k_x}\frac{k_x}{\omega\varepsilon_{\mathrm{r}2}}\left(C\,\mathrm{e}^{-\mathrm{j}\,k_{z1}z} + D\,\mathrm{e}^{\mathrm{j}\,k_{z1}z}\right)\mathrm{e}^{-\mathrm{j}\,k_x x} \\
&\quad (\text{medium } 2)\,, \\
E_{z3} &= -\sum_{k_x}\frac{k_x}{\omega\varepsilon_{\mathrm{r}3}}F\,\mathrm{e}^{-\mathrm{j}\,k_{z3}z\,-\,\mathrm{j}\,k_x x} \\
&\quad (\text{medium } 3)\,.
\end{aligned} \tag{5.4}$$

We wish to see what impact minor deviations from the ideal perfect lens constellation would have, therefore medium 1 and 3 are assumed to be identical (in the simplest case with $\varepsilon_{\rm r1} = \varepsilon_{\rm r3} = 1$, $\mu_{\rm r1} = \mu_{\rm r3} = 1$). The amplitudes B, C, D, F are given by eqns (1.75)–(1.78). In particular, the transfer function for the distance $2d$ is

$$\begin{aligned} T &= \frac{F}{A}\,\mathrm{e}^{-\mathrm{j}\,k_{z1}d} \\ &= \frac{4\zeta_{\rm e}\,\mathrm{e}^{-\mathrm{j}\,k_{z1}d}}{(1+\zeta_{\rm e})^2\,\mathrm{e}^{\mathrm{j}\,k_{z2}d} - (1-\zeta_{\rm e})^2\,\mathrm{e}^{-\mathrm{j}\,k_{z2}d}}\,, \end{aligned} \tag{5.5}$$

and the reflection from the front surface of the slab is

$$R = \frac{B}{A} = \frac{2\mathrm{j}\,(1-\zeta_{\rm e}^2)\sin(k_{z2}d)}{(1+\zeta_{\rm e})^2\,\mathrm{e}^{\mathrm{j}\,k_{z2}d} - (1-\zeta_{\rm e})^2\,\mathrm{e}^{-\mathrm{j}\,k_{z2}d}}\,. \tag{5.6}$$

Note that the presence of a pole in the expressions for transmission and reflection indicates excitation of a SPP mode at the frequency ω considered.

Next, we shall go through a number of examples of minor perturbations in the perfect-lens parameters.

5.3.3 Effect of losses: Transfer function, cutoff, electrostatic limit

Losses will be characterized by assigning imaginary parts to the values of permittivity and permeability

$$\varepsilon_{\rm r2} = -1 - \mathrm{j}\,\varepsilon''\,, \qquad \mu_{\rm r2} = -1 - \mathrm{j}\,\mu''\,, \tag{5.7}$$

with $\varepsilon'', \mu'' \ll 1$. As it happens, the transfer function behaves quite differently in the cases when *(i)* k_x is large but not too large and *(ii)* $k_x \to \infty$.

The electrostatic limit has been briefly discussed in Section 2.12.4. The essential part of the argument was that with good approximation

$$k_{z1} = k_{z2} = k_{z3} = -\mathrm{j}\,k_x\,, \tag{5.8}$$

and expression (5.5) for the transfer function can roughly be approximated as

$$T = \frac{4\zeta_{\rm e}\,\mathrm{e}^{-k_x d}}{(1+\zeta_{\rm e})^2\,\mathrm{e}^{k_x d} - (1-\zeta_{\rm e})^2\,\mathrm{e}^{-k_x d}}\,. \tag{5.9}$$

Consequently, the first term in the denominator of eqn (5.5) is regarded as being close to zero. If k_x is not too large but still $k_x \gg k_0$ then the second term of the denominator in eqn (5.9) dominates, leading to a transfer function

$$T \simeq \frac{4\,\mathrm{e}^{-k_x d}}{4\,\mathrm{e}^{-k_x d}} = 1 \tag{5.10}$$

that is entirely flat. But this is only true for this particular range of not too large values of k_x. The situation is quite different in the limit $k_x \to \infty$. The second term in the denominator of the transfer function, which declines exponentially with k_x, becomes smaller than the first term and eventually vanishes in the limit $k_x \to \infty$. The resulting transfer function in the limit of high k_x values is thus given by

$$T \simeq \frac{4\,\mathrm{e}^{-k_x d}}{(1+\zeta_\mathrm{e})^2\,\mathrm{e}^{k_x d}} \simeq -\frac{4\,\mathrm{e}^{-2k_x d}}{(\varepsilon'')^2}\,. \tag{5.11}$$

Thus, in the presence of loss the transfer function declines exponentially as k_x approaches infinity.

The value of k_x at which the transfer function (eqn (5.5)) stops being flat and starts to decline with k_x is called the cutoff frequency and can be estimated from the condition that the first term in the denominator (which is the dominant one at not too large k_x) and the second term in the denominator (which dominates as $k_x \to \infty$) are equal (so that the magnitude of the total transfer function reduces to 1/2),

$$(1+\zeta_\mathrm{e})^2 \simeq (1-\zeta_\mathrm{e})^2\,\mathrm{e}^{-2k_x d}\,, \tag{5.12}$$

yielding the cutoff value of the spatial frequency at

$$k_x^{\mathrm{cutoff}} \simeq \frac{1}{d}\ln\frac{2}{\varepsilon''}\,. \tag{5.13}$$

This approximate expression gives already an idea about how much the deviation of the metamaterial parameters from the ideal situation $\varepsilon_\mathrm{r} = \mu_\mathrm{r} = -1$ and the thickness of the metamaterial slab as well affect the cutoff frequency. Note that loss in permeability does not enter the expression for the cutoff frequency in the electrostatic limit. It would of course appear for TE polarization. We shall discuss this a little later.

A more general formulation can also be easily derived (see Podolskiy and Narimanov 2005). Introducing

$$\chi = \sqrt{\frac{k_x^2}{k_0^2} - 1}\,, \tag{5.14}$$

we rewrite κ_1 and κ_2 as

$$\begin{aligned} \kappa_1 &= k_0\chi\,, & (5.15)\\ \kappa_2 &\simeq k_0\chi\left(1 - \mathrm{j}\,\frac{\varepsilon'' + \mu''}{2\chi^2}\right), & (5.16) \end{aligned}$$

so that

$$\begin{aligned}\zeta_{\rm e} &\simeq -1+{\rm j}\,\frac{\varepsilon''(1+2\chi^2)+\mu''}{2\chi^2}\,, &(5.17)\\ (1+\zeta_{\rm e})^2 &\simeq -\left[\frac{\varepsilon''(1+2\chi^2)+\mu''}{2\chi^2}\right]^2, &(5.18)\\ (1-\zeta_{\rm e})^2 &\simeq 4\,. &(5.19)\end{aligned}$$

The transfer function (eqn (5.5)) takes the form

$$T \simeq \frac{1}{{\rm e}^{2k_0\chi d}\left[\dfrac{\varepsilon''(1+2\chi^2)+\mu''}{4\chi^2}\right]^2+1}\,. \tag{5.20}$$

A direct comparison with the exact solution eqn (5.5) proves that this approximate expression for the transfer function fails for small k_x in the vicinity of k_0, but it does indeed provide excellent results for the cutoff frequency as was proven by Podolskiy and Narimanov (2005). The cutoff frequency can be found again from the condition that the transfer function halves there,

$${\rm e}^{2k_0\chi d}\left[\frac{\varepsilon''(1+2\chi^2)+\mu''}{4\chi^2}\right]^2 = 1\,. \tag{5.21}$$

Contrary to the electrostatic limit, now both losses in permeability and permittivity appear in the expression that determines the cutoff frequency. In particular, if only permittivity is lossy, $\mu''=0$, we find

$${\rm e}^{2k_0\chi d}\left[\frac{\varepsilon''(1+2\chi^2)}{4\chi^2}\right]^2 = 1\,, \tag{5.22}$$

and in addition, assuming $k_x/k_0 \gg 1$, so that $\chi \simeq k_x/k_0$, the cutoff frequency agrees with the value predicted in the electrostatic limit,

$$k_x^{\rm cutoff} = \frac{1}{d}\ln\frac{2}{\varepsilon''}\,, \tag{5.23}$$

with logarithmic sensitivity to losses and an inverse dependence on the thickness. If only permeability is lossy, $\varepsilon''=0$, the cutoff is given by

$$k_x^{\rm cutoff} = \frac{1}{d}\ln\left[\frac{4}{\mu''}\left(\frac{k_x^{\rm cutoff}}{k_0}\right)^2\right]\,, \tag{5.24}$$

showing a weaker sensitivity to losses.

Case A. Lossy permittivity

Figures 5.2 to 5.5 illustrate in a number of examples the deterioration of the transfer function with loss. The parameters are d/λ, the thickness of the slab in terms of wavelength, and the imaginary part of the dielectric constant that we denote here by δ. The exact solution is plotted by a solid line and the electrostatic approximation by a dotted line. For

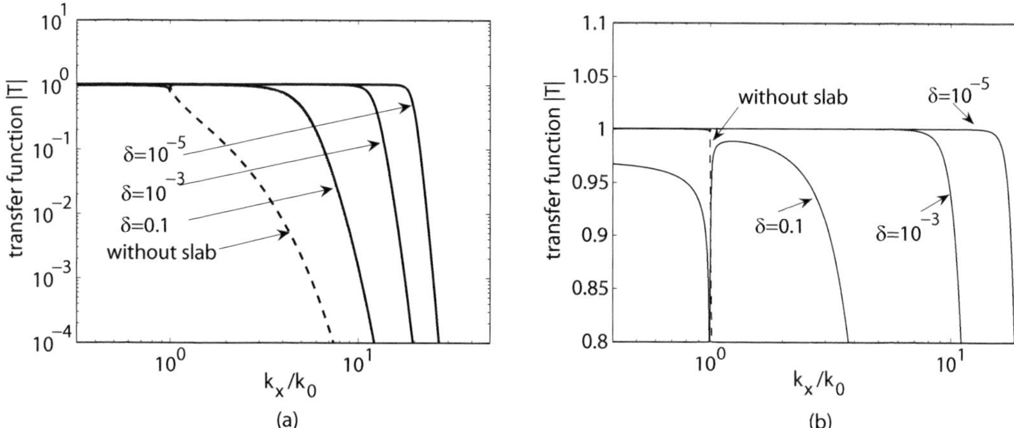

Fig. 5.2 Transfer function of a near-perfect lens. $d/\lambda = 0.1$, losses only in ε, $\delta = 10^{-5}$, 10^{-3} and 0.1. (a) Exact solution (solid curves) and electrostatic approximation (dotted curves) are shown together with the transfer function for the case when the slab is removed and the total distance from object to image plane is still the same, $2d$ (dashed curve). Note that curves obtained within the electrostatic limit coincide with the exact solution. (b) Detailed view of a part of the graph; only exact solutions shown

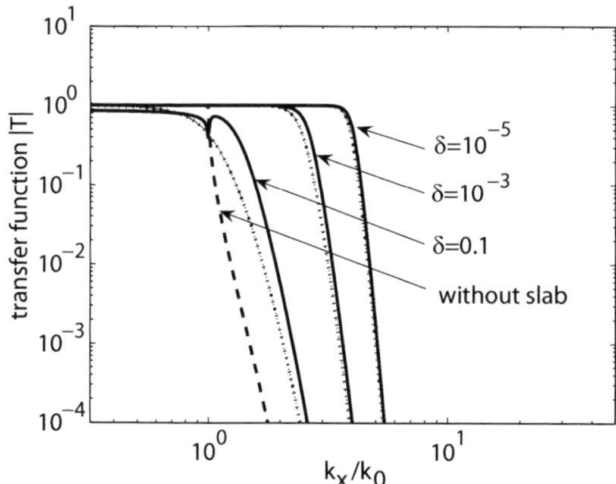

Fig. 5.3 Same as in Fig. 5.2(a), losses only in ε, but for $d/\lambda = 0.5$. Exact solution (solid curves) and electrostatic approximation (dotted curves). Electrostatic limit is good for large k_x and for small loss. It is less accurate for a loss of $\delta = 0.1$

comparison, the transfer function for the case when the slab is removed is plotted as well (dashed line). In Fig. 5.2(a) a thin slab is taken, $d/\lambda = 0.1$. For a small amount of loss, $\delta = 10^{-5}$, the transfer function exhibits a cutoff at $k_x^{\text{cutoff}}/k_0 = 100$. This means a slab of 0.1λ thickness with $\varepsilon_\text{r} = -1-\text{j}\,0.00001$ and $\mu_\text{r} = -1$ would reproduce accurately objects with a resolution of $\lambda/100$. That is pretty good. Increasing the loss reduces the resolution. For $\delta = 10^{-3}$ the cutoff frequency reduces to $k_x^{\text{cutoff}} = 10k_0$, and for $\delta = 0.1$ to $k_x^{\text{cutoff}} = 3k_0$. The resolution of the slab with $\delta = 10^{-3}$ is therefore $\lambda/10$ and for $\delta = 0.1$, which is not an unrealistic value, only $\lambda/3$.

Is such a device better than empty space? Certainly yes; the transfer function is flatter and more evanescent components can make it to the image plane even for such a high loss of $\delta = 0.1$. The situation is quite different for propagating components with $k_x < k_0$. After propagation

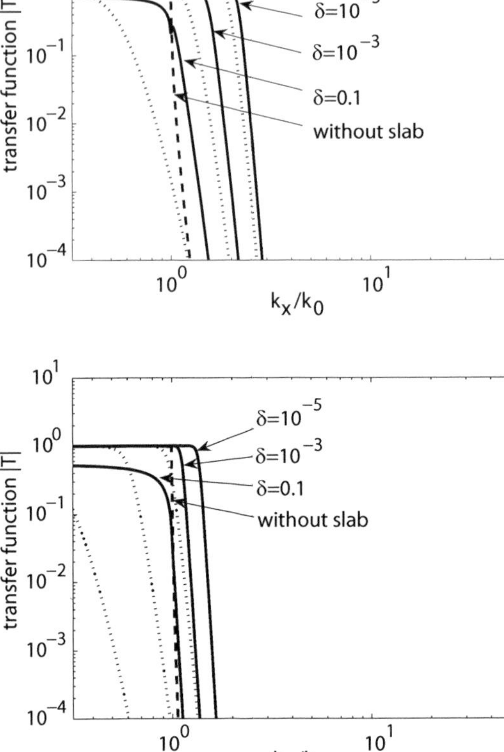

Fig. 5.4 Same as in Figs. 5.2(a) and 5.3, losses only in ε, but for $d/\lambda = 1$. Exact solution (solid curves) and electrostatic approximation (dotted curves). Electrostatic limit is only reliable for large k_x

Fig. 5.5 Same as in Figs. 5.2(a), 5.3 and 5.4, losses only in ε, but for $d/\lambda = 2$. Electrostatic limit is no longer a reliable approximation

through a lossy slab their amplitude decays, naturally, so the propagating part of the spectrum suffers, while the evanescent part of the spectrum actually improves in comparison to the empty-space case. To make this clear in Fig. 5.2(b) only the relevant part of the transfer function from $k_x/k_0 = 0.3$ to 20 is plotted so that small details can be seen. Does the electrostatic approximation work in this case? Yes, it does, the exact transfer function and the one obtained in the electrostatic approximation practically coincide (see Fig. 5.2(a)), and the cutoff frequency can be estimated from the approximate expression (eqn (5.13)).

The transfer function in Figs. 5.3 to 5.5 plotted for thicker slabs, $d/\lambda = 0.5$, 1 and 2, can be seen to deteriorate increasingly faster with loss. For $d/\lambda = 0.5$ the cutoff frequency k_x^{cutoff} is $5k_0$ for $\delta = 10^{-5}$, it is $3.5k_0$ for $\delta = 10^{-3}$ and just over k_0 for $\delta = 0.1$. For a thicker slab the cutoff frequency is quite close to k_0 so that only marginally subwavelength resolution can be expected, and, in addition, the slab is getting more and more opaque for propagating components that lose more and more energy as the slab thickness is increasing. For $d/\lambda \geq 0.5$ the slab with $\delta = 0.1$ does not seem to give much advantage over empty space. For $d/\lambda = 2$, even with small losses, $\delta = 10^{-5}$ the cutoff of

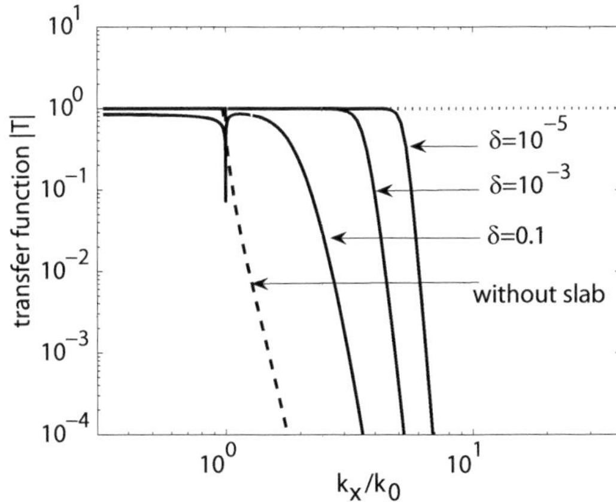

Fig. 5.6 Same as in Fig. 5.3, but with losses in μ instead of ε. Electrostatic limit (dotted horizontal line) fails completely

$k_x^{\text{cutoff}}/k_0 = 1.5$ does not make the imaging device look feasible.

As far as the electrostatic limit is concerned, it becomes, as thickness increases, less and less accurate (Figs. 5.3–5.5). For $d/\lambda = 0.5$ the electrostatic limit can be seen to differ considerably from the exact solution for $\delta = 0.1$. For $d/\lambda = 1$ it occurs at $\delta = 10^{-3}$ and for $d/\lambda = 2$ the electrostatic limit $\delta = 10^{-5}$ is no longer capable of reproducing the real situation.

Case B. Lossy permeability. TM polarization

In the previous example we assumed that only the permittivity is lossy. What happens if we introduce loss to the permeability instead, taking $\mu_{\text{r}} = -1 - \text{j}\,\delta$ and $\varepsilon_{\text{r}} = -1$? Figure 5.6 shows the transfer function for the 0.5λ thick slab with $\delta = 10^{-5}$, 10^{-3} and 0.1, with the electrostatic limit and with the transfer function of empty space. The tendency is the same, but the cutoff frequency is less sensitive to the loss in permeability as can be clearly seen by comparing Fig. 5.6 to Fig. 5.3. The cutoff frequency k_x^{cutoff} is $6k_0$ for $\delta = 10^{-5}$, it is $4k_0$ for $\delta = 10^{-3}$ and $2k_0$ for $\delta = 0.1$. Note that the electrostatic limit still predicts an ideal flat transfer function. It is not particularly surprising. We should remember that we consider TM polarization. The only place where μ occurs is in κ_1 and κ_2. By neglecting any deviation in κ_1 and κ_2 from k_x we disregard any deviations of μ_{r} from minus unity. The conclusion follows: the transfer function for TM polarization in the electrostatic approximation is unaware of any variation in μ and is bound to fail providing unphysical and unreliable results. This conclusion would apply to any value of μ. By employing the electrostatic limit we disregard the effect of μ. Are we allowed to do it? Obviously, we are not: as clearly follows from Fig. 5.6 even minor losses result in a cutoff and this cutoff goes missing in the electrostatic approximation.

The electrostatic approximation gives simpler formulae and simpler physics. Unfortunately, it may lead to wrong conclusions. We firmly believe that there is still a need to clarify the situation. We will address this important issue later in this chapter when discussing properties of the silver superlens.

The same would apply in the case of TE polarization to losses introduced to permittivity. The transfer function in that magnetostatic limit would not account for any change in ε. Clearly, a lens with losses in permittivity would not be correctly described in the magnetostatic limit for incident TE polarized light.

5.3.4 Lossless near-perfect lens with $\varepsilon_{\mathrm{r}} \simeq -1$, $\mu_{\mathrm{r}} \simeq -1$

Let us now look at other possible scenarios for a near-perfect lens, this time disregarding losses[3] and introducing small variations to the real parts of ε_{r} and μ_{r} from being minus unity.

[3] As soon as we go off the perfect lens frequency that provides $\varepsilon_{\mathrm{r}} = -1$, $\mu_{\mathrm{r}} = -1$, surface plasmon–polaritons are excited corresponding to the poles in the transfer function resulting in infinitely large amplitudes. Some authors, in order to facilitate numerical calculations, introduce some loss, with the argument that if it is small enough it would not have any influence on the position of the cutoff frequency (it is true when the imaginary parts in ε and μ are much smaller than the perturbations in their respective real parts).

Case C. $\varepsilon_{\mathrm{r}} = -(1 \pm \delta)$, $\mu_{\mathrm{r}} = -1$

This case is analogous to that considered first by Smith *et al.* (2003) who for TE polarization assumed the real part of the permeability slightly different from the ideal value, keeping the real part of the permittivity at -1. Starting again from the expression (5.5) the transfer function can easily be shown to simplify to

$$T \simeq \frac{1}{\dfrac{\delta^2}{4}\,\mathrm{e}^{2k_x d} - 1}\,. \tag{5.25}$$

The cutoff frequency can be obtained as

$$k_x^{\mathrm{cutoff}} \simeq \frac{1}{d}\ln\frac{2}{|\delta|}\,. \tag{5.26}$$

The transfer function (eqn (5.5)) plotted in Figs. 5.7(a) and (b) for a number of values of δ (both positive and negative) has some common features with previously shown results for the lossy near-perfect lens. The cutoff is inversely proportional to the thickness and depends logarithmically on the deviation from $\varepsilon_{\mathrm{r}} = \mu_{\mathrm{r}} = -1$.

The new feature is that the transfer function exhibits resonances, one resonance in Fig. 5.7(a) and two in Fig. 5.7(b). The origin of these resonances is the excitation of eigenmodes of the slab, the SPP modes. The direct excitation of the SPP resonances is, especially if losses are low, undesirable for imaging. The corresponding k_x components will be disproportionately represented in the image. Yet, the existence of these surface modes for imaging is essential, as the recovery of the evanescent modes can be seen as the result of driving the surface modes off resonance.

Shouldn't we expect to see always two resonances, one close to the light line, and another one at high k_x? Not necessarily. Looking at Figs. 3.35 and 3.36 helps to understand this. For those frequencies for which both $\varepsilon_{\mathrm{r}2}$ and $\mu_{\mathrm{r}2}$ are negative and their product is larger than

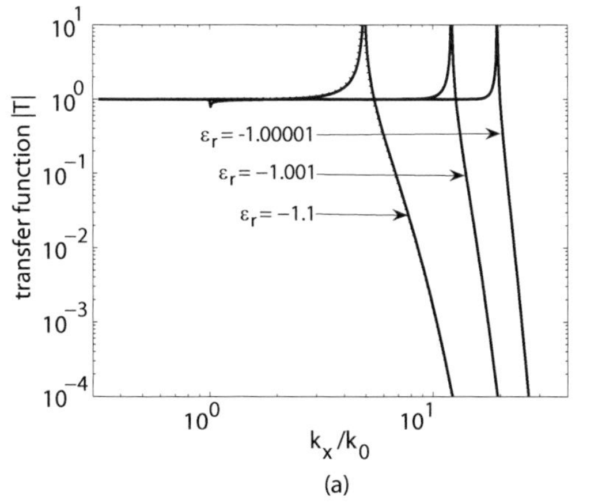

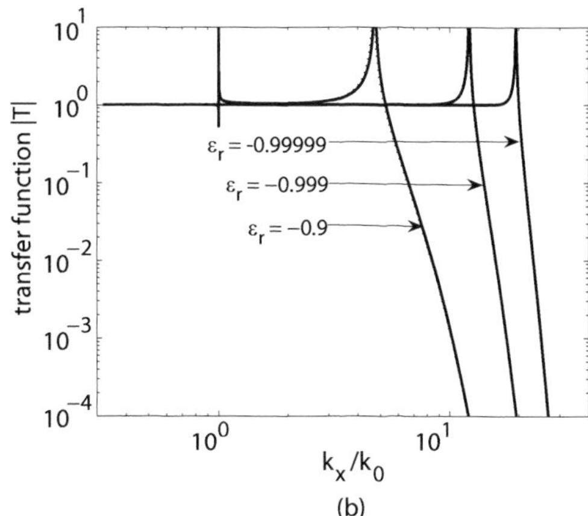

Fig. 5.7 Effect of deviations in the real part of permittivity. (a) $\varepsilon_{\mathrm{r}} = -1 - \delta$, $\mu_{\mathrm{r}} = -1$. Only one resonance can be seen. (b) $\varepsilon_{\mathrm{r}} = -1 + \delta$, $\mu_{\mathrm{r}} = -1$. Two resonances are excited. Cut-off is sensitive to variations in ε. The electrostatic limit is working well

unity, there is only one surface plasmon–polariton mode excited by the TM polarized light, namely the $\omega^{(-)}$ mode. Mathematically, the solution for the $\omega^{(+)}$ branch happens to lie to the left of the 'left-handed light line', $k_x^2 = \varepsilon_{\mathrm{r}2}\mu_{\mathrm{r}2}k_0^2$. This means that for this part of the spectrum evanescent components in vacuum turn into propagating ones inside the slab without fulfilling the condition for a resonant excitation of surface modes. Mathematically, although there is a singularity in the denominator at a value k_x close to k_0 from the $\omega^{(+)}$ branch close to the light line, this value of k_x corresponds to $\kappa_2 = 0$ (transition frequency from the propagating to evanescent components inside the slab). Taking the limit $\kappa_2 \to 0$ in eqn (5.5), it can be shown that the transfer function is finite and continuous there. Note that this is quite a special situation. Figure 5.7(b) shows what happens if ε_{r} is slightly larger than -1. In this case the refractive index is larger than -1 and there are two resonances in the transfer function. Note that the electrostatic limit, that in this example works well for large values of k_x, is not capable of describing the first resonance close to k_0.

There is a direct connection between the achievable resolution and the SPP resonances. The flat bit of the transfer function stretches up to the value of large k_x corresponding to the excitation of a SPP mode. It is the position of this resonance that ultimately determines the resolution of the slab. It is the excitation of these short waves that enables the recovery of the subwavelength part of the spectrum of an object. The cutoff can therefore be, alternatively, attributed to the value of k_x for the second resonance, instead of the condition $|T| = 1/2$.

Similar results can be obtained for deviations in μ_{r}, in both directions away from the value -1, see Fig. 5.8. In this case the transfer function can be shown to take the form

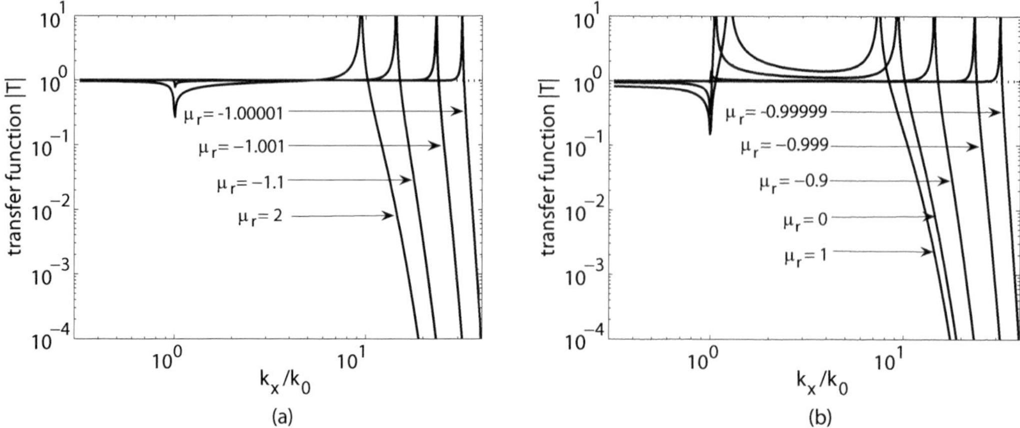

Fig. 5.8 Effect of deviations in the real part of permeability. (a) $\varepsilon_r = -1$, $\mu_r = -1 - \delta$. Only one resonance can be seen. (b) $\varepsilon_r = -1$, $\mu_r = -1 + \delta$. Two resonances are excited. Thickness $d/\lambda = 0.1$. Note that the electrostatic limit (dotted line) fails completely

$$T \simeq \frac{1}{-\left(\frac{\delta}{4}\right)^2 \left(\frac{k_0}{k_x}\right)^4 \mathrm{e}^{2k_x d} + 1}. \tag{5.27}$$

The cutoff frequency can be obtained as

$$k_x^{\text{cutoff}} = \frac{1}{d} \ln \left[\frac{4}{|\delta|} \left(\frac{k_x^{\text{cutoff}}}{k_0} \right)^2 \right]. \tag{5.28}$$

Again, if $\mu_r < -1$ there will be only one surface mode that can exist, and if $\mu_r > -1$, there can be two resonances. Note that the effect of variations in μ is weaker than the effect of variations in ε, in agreement with the cutoff expressions (5.26) and (5.28).

An interesting feature of the transfer function is that its magnitude can be larger than unity, not only at the two resonances but also everywhere in-between, the plateau of the transfer function means the evanescent part of the spectrum gained in amplitude as compared to its values coming from the object. The question is whether or not there is anything wrong with energy conservation (see also Section 5.7). The electrostatic limit fails as expected. It does not know about variations in μ.

Figures 5.9(a) and (b) show that a thicker slab fails to be a good lens; the transfer function deteriorates too quickly. The cutoff is too short and no reasonable resolution can be obtained, even with variations as small as $\delta = 10^{-5}$! In addition, in the case of variations in ε only, the electrostatic limit begins to fail as the cutoff moves to smaller values of k_x.

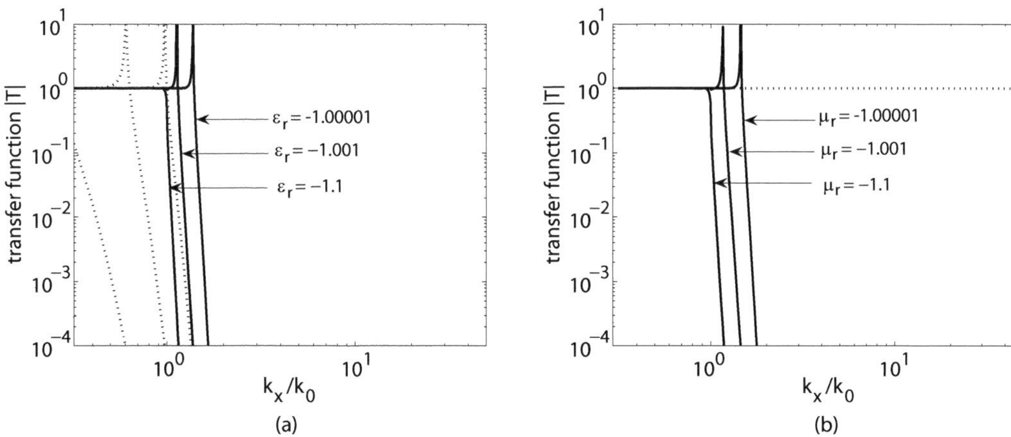

Fig. 5.9 (a) Same as in Fig. 5.7(b) but for thickness $d/\lambda = 2$. The only good of the electrostatic limit is that it also predicts a cutoff, but it fails to give any reasonable quantitative agreement with the full solution. The cutoff is less than $2k_0$ even for a deviation from $\varepsilon_r = -1$ as small as 10^{-5}. (b) Same as in Fig. 5.8(b) but for thickness $d/\lambda = 2$. The electrostatic limit fails. The cutoff is less than $2k_0$ even for a deviation as small as 10^{-5} from $\mu_r = -1$

Case D. Small deviation in refractive index

It is also possible to look only at the deviation in the refractive index. This was analyzed by Merlin (2004) who assumed the refractive index in the form

$$n = -\sqrt{1-\sigma} \tag{5.29}$$

and obtained analytical expressions for $\sigma \ll 1$. In agreement with other studies (Haldane, 2002; Gomez-Santos, 2003) he stressed the importance of the excitation of surface plasmon–polaritons as $k_x \to \infty$. He found that depending on the sign of σ the near-field distribution is either symmetric or antisymmetric.

Case E. The refractive index remains at -1 but both ε and μ undergo small variations

Smith *et al.* (2003) and Merlin (2004) approached the perfect lens by perturbing the refractive index $n = \sqrt{\varepsilon_r \mu_r}$ and therefore the perfect lensing effect is lost for propagating components. Lu *et al.* (2005), Cui *et al.* (2005) and French *et al.* (2006) proposed a different route of addressing deviations from the perfect-lens scenario by introducing deviations in ε and μ correlated in such a way that their product remains unity. This can be achieved by taking

$$\varepsilon_r = -\frac{1}{1 \pm \delta}, \qquad \mu_r = -(1 \pm \delta), \tag{5.30}$$

with δ being real. Any modifications in resolution are therefore not due to loss effects within the lens.

Figures 5.10(a) and (b) show the transfer function for $d/\lambda = 0.1$ and 2, respectively. But does this configuration bring advantages relative

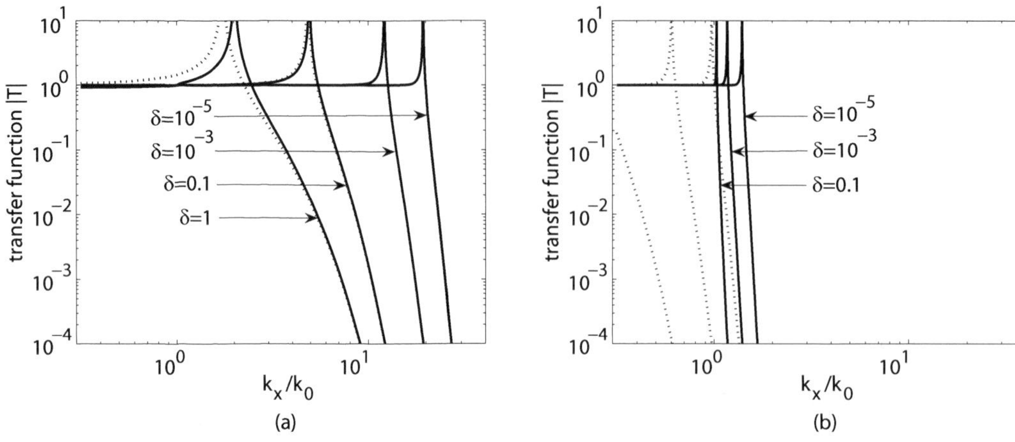

Fig. 5.10 Keeping refractive index $n = -1$. Deviations in ε and μ correlated. Thickness $d/\lambda = 0.1$ (a) and 2 (b). Electrostatic limit works only for a thin slab (a); it is too inaccurate for a thick slab (b)

to the cases when only one material parameter is different from that of the perfect lens? The transfer function for the propagating components is, apparently, better than in previous examples. The reason is quite obvious. The refractive index n being minus unity means merely that the propagating components exhibit negative refraction at an angle equal to the angle of incidence. So refraction is at the desired angle but the reflection coefficient is no longer zero, as for the perfect transfer function, because the impedance of the negative-index medium,

$$\eta = \sqrt{\frac{\mu}{\varepsilon}} = \sqrt{\frac{\mu_0}{\varepsilon_0}}(1 \pm \delta) \neq \sqrt{\frac{\mu_0}{\varepsilon_0}} = \eta_0\,, \tag{5.31}$$

differs from that of free space. The good news is that the electrostatic limit seems to work. At least as long as the cutoff frequency is large relative to k_0. Remember, the electrostatic limit implies that both κ_2 and κ_1 are equal and can be approximated by k_x. In the case considered κ_1 and κ_2 are indeed equal to each other. But the other assumption, that they can be approximated by k_x, is only valid for $k_x \gg k_0$. So when the cutoff is too low—and this is, for instance, the case for the curve with $\delta = 1$, e.g. $\varepsilon_r = -2$, $\mu_r = -0.5$—the electrostatic limit fails yet again. This failure of the electrostatic limit is even more obvious for a thicker slab (see Fig. 5.10(b)). As far as resolution is concerned, just as in previous examples, it deteriorates logarithmically with $|\delta|$ and is inversely proportional to the slab thickness.

5.3.5 Near-perfect? Near-sighted!

Suppression of the high-frequency wave components limits the resolution of the lossy slab acting as an imaging system since the spatial size of the image (say, its half-width Δ) and the spectral width of the corresponding spectrum, δ, are inversely proportional to each other.[4] The

[4] Many authors like to refer to this fact as an 'uncertainty principle' using the quantum-mechanical language to describe the fact that the narrower the object the broader is its Fourier spectrum.

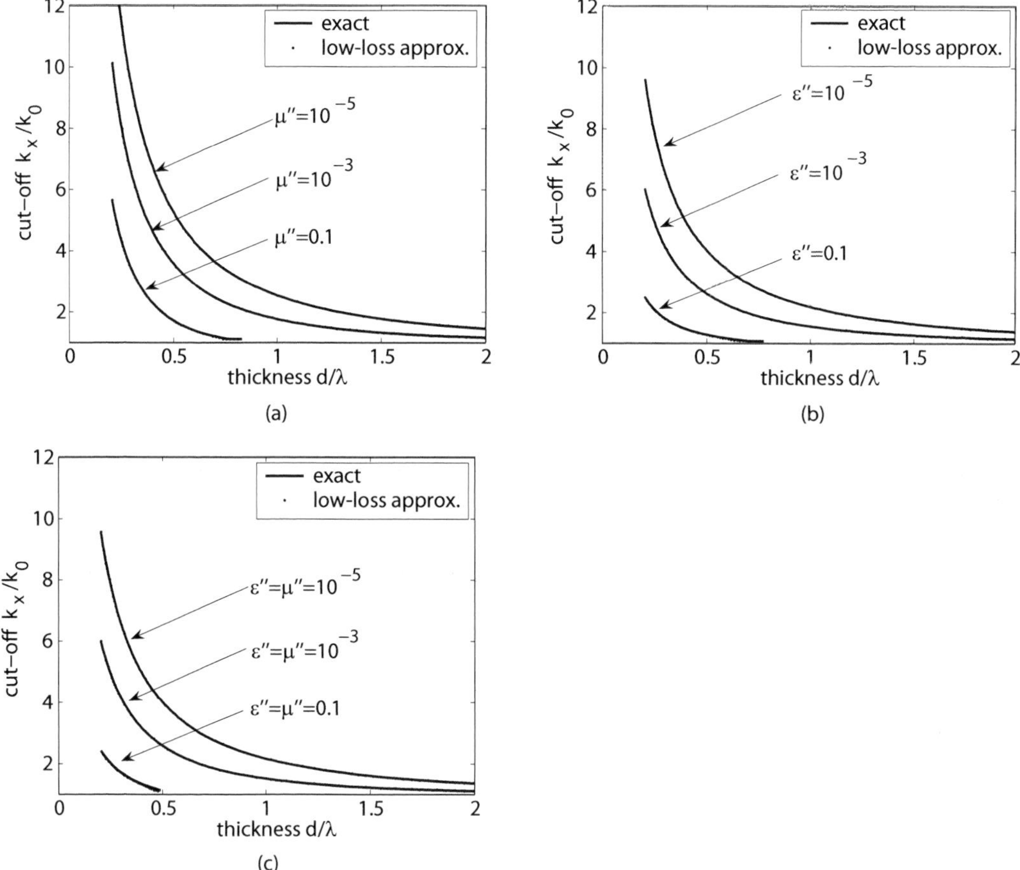

Fig. 5.11 Near-sighted lens. Cut-off frequency versus slab thickness. Losses in μ only (a), in ε only (b) and in both ε and μ (c). Solid line: exact calculation, circles: low-loss approximation from Podolskiy and Narimanov (2005)

strong dependence of the cutoff and of the resulting resolution on the slab thickness means that there is a certain maximum thickness allowed to achieve a specific resolution.

Estimating roughly the resolution as being inversely proportional to the cutoff frequency, $\Delta = 2\pi/k_x^{\mathrm{cutoff}}$, it follows after some algebra from eqn (5.21) that

$$d = \frac{\Delta}{2\pi\sqrt{1-\left(\dfrac{\Delta}{\lambda}\right)^2}} \ln \frac{4\left[\left(\dfrac{\lambda}{\Delta}\right)^2 - 1\right]}{\varepsilon''\left[2\left(\dfrac{\lambda}{\Delta}\right)^2 - 1\right] + \mu''}. \tag{5.32}$$

The dependence of the cutoff frequency on the slab thickness is plotted in Figs. 5.11(a)–(c) for various amounts of loss. By comparing Fig. 5.11(a) and Fig. 5.11(b) it can be seen that, in agreement with eqns (5.23) and (5.24), the cutoff frequency for thinner slabs is higher for lossy μ than for lossy ε. As thickness increases, even small losses either in ε (Fig. 5.11(a)) or in μ (Fig. 5.11(b)) or in both ε and μ (Fig. 5.11(c)) eventually bring the cutoff down to k_0. Note that the definition chosen for the cutoff frequency assumes that the transfer function reaches the magnitude of 1/2. For any loss, there is a critical thickness for which the definition of a cutoff becomes meaningless as $|T| < 1/2$ for any k_x (noticeable in each of Figs. 5.11(a)–(c)[5] as an abrupt end for curves with the largest value of loss). An important conclusion follows: while a thin lens of $d \ll \lambda$ can provide subwavelength resolution, the resolution of a lossy far-field lens with $d \geq \lambda$ does not perform better than usual optical devices (Podolskiy and Narimanov, 2005). Thus, the expression for the resolution limit of a LHM-based lens proves that the area of its subwavelength performance is usually limited to the near-field zone.[6]

[5] Note that the plots have been calculated from the exact expressions but the results calculated from the approximation of Podolskiy and Narimanov (2005) are practically identical.

[6] Podolskiy and Narimanov (2005) introduced as 'near-sighted' the flat near-perfect lens, a catchy name that has often been used. Interestingly, it was Narimanov's group that later and in parallel to Engheta's group proposed the concept of a 'far-sighted' cylindrical lens described below in Section 5.6 (see also Jacob *et al.* 2006; Salandrino and Engheta 2006).

We can put it in a different way. The consequence of the extreme dependence of the resolution on the deviation from the perfect-lens condition is that in order to achieve a cutoff frequency not lower than 10 k_0, a slab of 0.1 λ thickness would require a loss level of about 0.002, but a slab of 0.67 λ would tolerate a loss of no greater than 6×10^{-19} (see also Smith *et al.* 2003; French *et al.* 2006).

5.3.6 General cutoff frequency relationships

We have seen that any deviation from the ideal conditions causes the appearance of a cutoff in the spatial frequency spectrum. This was shown in a number of figures. A general feature found was that the cutoff was higher when μ deviated from its ideal value than when ε did so. However, for deviations in the same material parameter it is not obvious which deviation is more harmful. Is it the real part or the imaginary part? Or possibly what matters is the deviation from the $n = -1$ condition? The answer may be obtained from Fig. 5.12 where the normalized cutoff frequency is plotted against the logarithm of the deviation δ for the case when $d/\lambda = 0.1$, i.e. when the lens is thin enough for good resolution. It turns out that the main difference is whether it is ε that changes or not. The two cases differ quite substantially. But apart from that it hardly matters what kind of deviation occurs. All deviations seem to have identical effects. Comparing expressions (5.23) and (5.26) it becomes obvious that deviations in either the real or the imaginary part of the permittivity would result in the same resolution. The same would apply to TE polarization and variations in the permeability, respectively.

The relationship between the higher SPP resonance and the cutoff in spatial frequency has been noted before. We have done the calculations for all the deviations investigated before, in order to find out how close the relationship is. It turns out to be so close that it is not worth plotting the results. It is though worth pointing out that the resonance is not necessarily provided by the $\omega^{(-)}$ branch. It can be the $\omega^{(+)}$ branch.[7]

[7] For more details see Chapter 3 and also discussions on SPPs for the silver lens later in this chapter. $\omega^{(-)}$ is shown there to be the higher k_x resonance that determines the cutoff only if $\varepsilon_r < -1$, $\mu_r = 1$, whereas it is $\omega^{(+)}$ for $-1 < \varepsilon_r < 0$, $\mu_r = 1$.

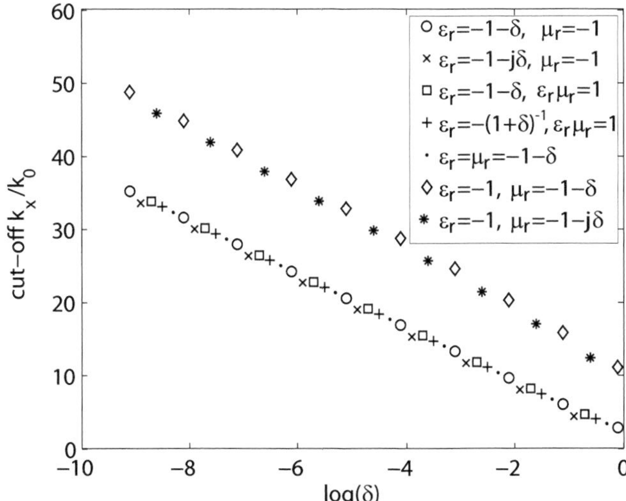

Fig. 5.12 Relative cutoff frequency against the logarithm of the deviation δ from the ideal conditions. $\varepsilon_r = -1 - \delta$, $\mu_r = -1$ (circles); $\varepsilon_r = -1 - j\,\delta$, $\mu_r = -1$ (crosses); $\varepsilon_r = -1 - \delta$, $\varepsilon_r\mu_r = 1$ (squares); $\varepsilon_r = -1/(1+\delta)$, $\varepsilon_r\mu_r = 1$ (pluses); $\varepsilon_r = \mu = -1 - \delta$ (dots); $\varepsilon_r = -1$, $\mu_r = -1 - \delta$ (diamonds); $\varepsilon_r = -1$, $\mu_r = -1 - j\,\delta$ (stars)

5.3.7 Effect of discretization in numerical simulations

In the wake of the availability of various numerical packages simulations have become very fashionable. It is much less effort to do the simulation than to perform the corresponding experiment. A sign of this way of thinking is the introduction of the term ‘numerical experiment’. The term is justified in the sense that by solving Maxwell’s equations we can get the truth, the whole truth and nothing but the truth. That is the principle anyway. The reality is often different. Just as with real experiments one needs green fingers to perform numerical experiments. The expertise needed is usually acquired by devoting a long time to the resolution of mysteries inherent in all numerical packages. Paul *et al.* (2001) found that the resolution of the perfect lens is not better than the diffraction limit. Ziolkowski and Heyman (2001) maintained that no superlens effect occurs when the lens medium is dispersive, which of course it is bound to be. In a later publication Chen *et al.* (2005) investigate the ability of the finite-difference time-domain method to model a perfect lens. They show that because of the frequency-dispersive nature of the medium and the time discretization, an inherent mismatch in the constitutive parameters exists between the slab and its surrounding medium. This mismatch in the real part of the permittivity and permeability is shown to have the same order of magnitude as the losses typically used in numerical simulation. When the LHM slab is taken as lossless this mismatch is shown to be the main factor contributing to the image resolution loss of the slab.

Koschny *et al.* (2006) investigate the transfer function of the discretized perfect lens in finite-difference time-domain (FDTD) and transfer matrix method (TMM) simulations; the latter allow us to eliminate the problems associated with the explicit time dependence in FDTD simulations. They find that the finite discretization mesh acts like imaginary deviations from $\mu_r = \varepsilon_r = -1$ and leads to a cross-over in the transfer function from constant values to exponential decay around k_x^{cutoff} limiting the attainable super-resolution. A qualitative model is proposed that is capable of describing the impact of the discretization. k_x^{cutoff} is found to depend logarithmically on the mesh constant, in qualitative agreement with the transfer matrix method simulations.

5.4 Negative-permittivity lens

5.4.1 Introduction

It should have become clear from previous discussions that it is not necessary at all for a material to have negative refractive index to provide subwavelength resolution. The most prominent example is a slab of silver that supports SPP modes with high k_x values due to its negative permittivity. In Engheta's (2002) notation this is called an epsilon-negative material and abbreviated as ENG. Such a slab can be used to replicate field distribution from the objects with subwavelength features for TM polarization. Of course the images produced cannot be expected to be perfect: propagating modes emanating from a source do not contribute to the image because inside the slab they turn into evanescent waves (the k vector is purely imaginary as $k^2 = \omega^2 \varepsilon\mu < 0$) so they can barely make their way to the image plane.[8]

[8] Losses will of course contribute to the decay but the major factor is reflection. If the slab is not too thick, the wave can tunnel through the slab, emerging at the output surface but with much reduced amplitude.

As far as the evanescent part of the spectrum is concerned, perfect imaging requires a flat transfer function for all values of k_x. For silver, the obstacles to achieving that aim, SPP resonances causing the high spatial frequency cutoff, cannot be disregarded and would never disappear as would be in the case of the perfect lens $\varepsilon_r = \mu_r = -1$ (Shamonina *et al.*, 2001). Retardation effects (i.e. not assuming the electrostatic limit) were also investigated by Shen and Platzman 2002; Ramakrishna *et al.* 2003; Jiang and Pike 2005.

5.4.2 Dependence on thickness

The lens, being near-sighted, applies to the 'epsilon-negative-only' case as well. Thinner lenses are better than thicker ones. An example is shown in Fig. 5.13 where the material is assumed slightly lossy: $\varepsilon_r = -1 - \mathrm{j}\,0.01$. The object positioned at $z = 0$ is of the shape of a Gaussian. The y component of the magnetic field varies as

$$H_y = \exp\left[-\left(\frac{x}{\tau}\right)^2\right]. \tag{5.33}$$

In our example τ is chosen as $\lambda/15$ so that the half-width of the Gaussian is equal to 30 nm. The aim is to look at an object that is small relative to the wavelength but not too small.

The object, the image and the transfer function are shown in Figs. 5.13(a)–(d) for thicknesses of d = 60, 40, 20 and 10 nm, respectively.

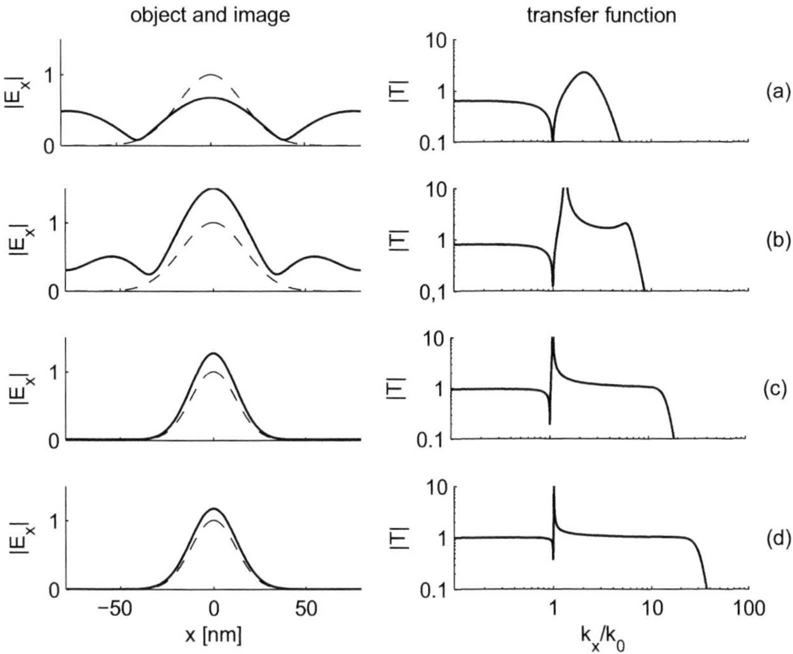

Fig. 5.13 Imaging for a low-loss ($\varepsilon'' = 10^{-2}$) silver slab of various thicknesses. The object (dashed lines) and image (solid lines) are shown in the left-hand column. The amplitude of the transfer function is shown in the right-hand column. $d = 60$, 40, 20, 10 nm (a)–(d)

It may be seen that at 60 nm thickness (twice the half-width of the Gaussian) the transfer function looks entirely unsuitable for imaging and the image is so poor that the original Gaussian cannot be recognized. For $d = 40$ nm the situation is better. The transfer function has not got any flat parts but at least the cutoff frequency is further away. Now, with a little imagination one can see some distorted Gaussian. Proceeding to a thickness of $d = 20$ nm the transfer function is now flat for a range of spatial frequencies and we have a good Gaussian. Reducing the thickness further to a value of 10 nm the flat part of the transfer function becomes wider and the image practically coincides with the object.

The conclusion may now be drawn that for good imaging the thickness of the lens should be less than the width of the object. Does this conclusion apply to the case when losses are much higher? In our next example we shall take the permittivity as $\varepsilon_r = -1 - j0.4$ and look at the imaging of the same Gaussian variation of the magnetic field for the same four thicknesses. A brief look at Fig. 5.14 reveals that the situation has considerably deteriorated. Thicknesses of $d = 60$ and 40 nm are obviously not suitable but even for $d = 20$ nm (which is less than the width of the object) the resulting image is much wider than the object. For $d = 10$ nm there is some flat region in the transfer function and the image is quite good. The conclusion is clearly that losses are bad. High

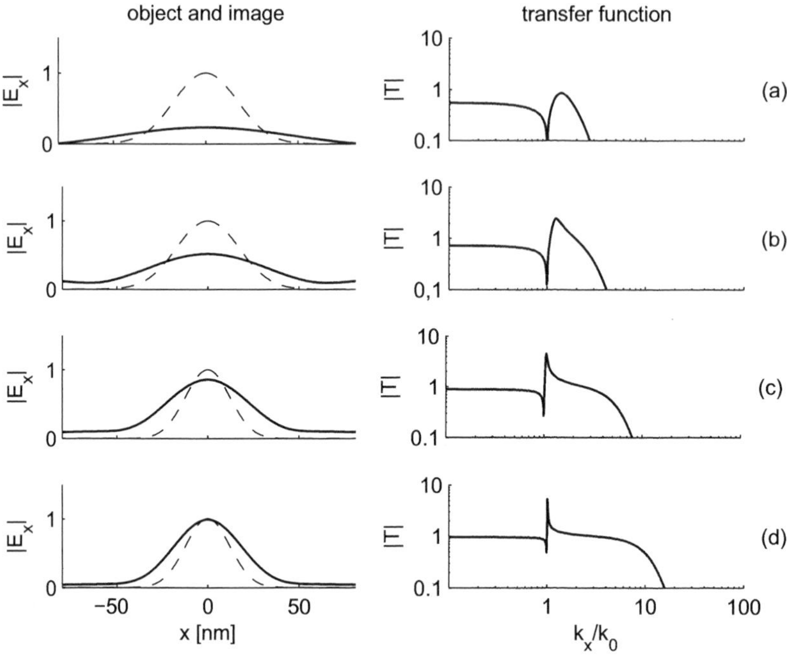

Fig. 5.14 Same as Fig. 5.13 but for $\varepsilon'' = 0.4$

losses lead to further deterioration in image quality.

5.4.3 Field variation in the lens

As discussed in Chapter 2 the field grows exponentially in the ideal lens. For the 'epsilon-negative-only' case this is not true in general. Let us investigate it in an example. We know that we have good imaging for low losses when the lens thickness is $d = 20$ nm. The transfer function is that in the third row of Fig. 5.13, which is replotted in Fig. 5.15. The upper SPP resonance is now at $k_x = 12k_0$. We shall look at three subwavelength components, $k_x/k_0 = 5$, 10 and 15 denoted by the points A, B and C. The corresponding field variations (maximum normalized to unity) are shown in Figs. 5.15(b)–(d). At A, in the flat section of the transfer function, the field variation is the same as in the ideal lens consisting of decay, growth, decay. At B, close to the SPP resonance, there is no pure growth inside the lens: the field first slightly decays. At C, beyond the SPP resonance, the field has two clear maxima inside the lens, the larger one at the input surface. It is clear now that the image reproduction does not in general proceed as predicted by the ideal case but it still applies to lower values of k_x.

The explanation may again be provided in terms of SPPs. The transfer function is flat as long as the rear surface is excited strongly enough. Beyond the cutoff, $k_x d$ becomes too large, so that the two surfaces are

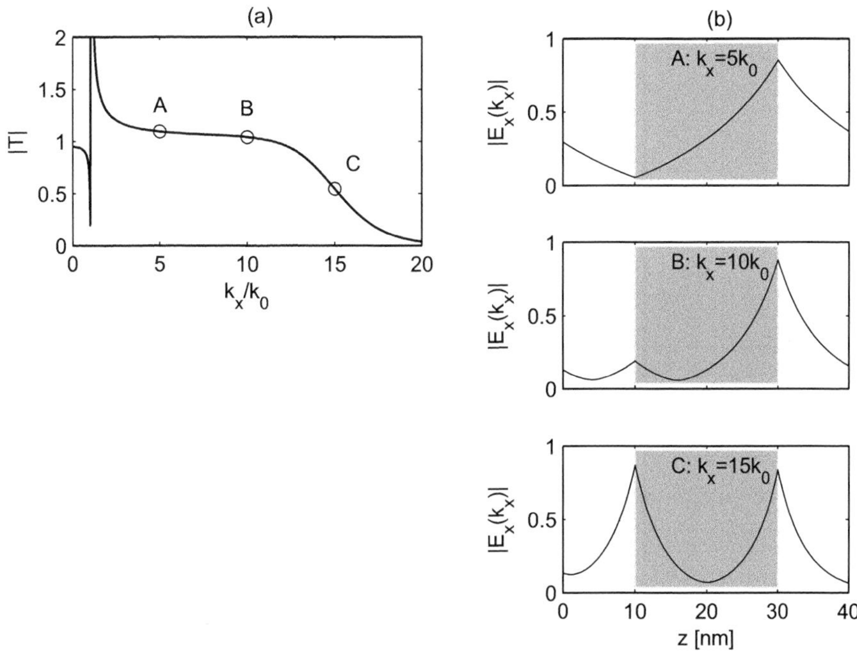

Fig. 5.15 Transfer function (a) and the evolution of individual Fourier components for a number of spatial frequencies indicated by the points A, B and C (b)

effectively decoupled and the imaging mechanism collapses as discussed before.

5.4.4 Other configurations: compression

Is the $d/2$–d–$d/2$ configuration the only one suitable for imaging with a negative-index material? Obviously not. The geometrical optics focus is still there if the lens in the Veselago geometry is shifted sideways keeping the positions of the object and image fixed as shown in Fig. 5.16. The distances are now d_1, d, and $d-d_1$ but the total distance from object to image is still equal to $2d$. The paths are the same in air and in the lens, hence the total phase shift is zero for propagating waves and the total amplitude is unity for non-propagating waves satisfying the conditions for the perfect lens. We are not aware of any detailed investigation of what happens in this lens when conditions differ from the ideal ones. On the whole it seems unlikely that this unsymmetric geometry would have any advantages over the symmetric geometry.

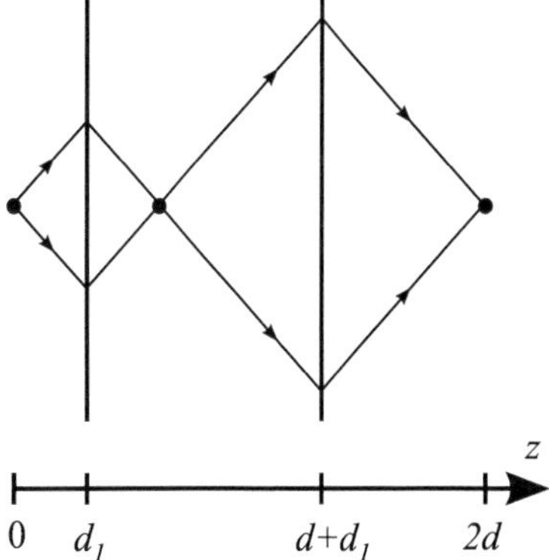

Fig. 5.16 Veselago's flat lens: lens shifted relative to the object. Image at the same position

If our aim is not to faithfully reconstruct but only to process some spatial information in the spatial frequency region then the field is open, any configuration is possible. We shall show here only one example that we reported in our early paper (Shamonina *et al.*, 2001) considering the compression of the signal. It is possible to do that by investigating (in

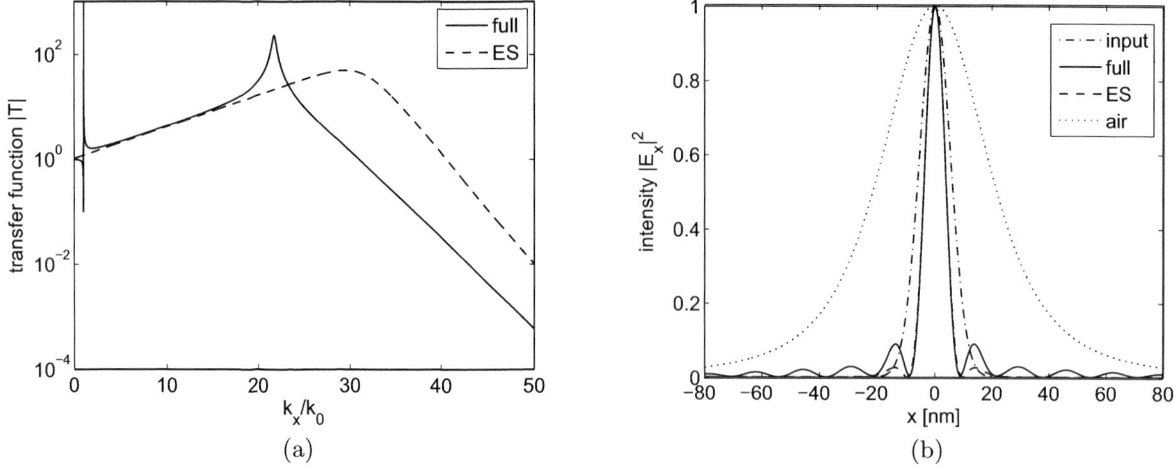

Fig. 5.17 (a) The transfer function and (b) the image for the 5-18-5 nm configuration

our case it was merely by trial and error) how altering the geometry affects the transfer function. We shall show here the object, the image and the transfer function in Fig. 5.17 for the 5–18–5 configuration, all dimensions in nm. Interestingly, the transfer function between the two SPP resonances increases. If higher values of k_x dominate that means that the image is made up by higher harmonics and, consequently, it may be narrower than the object. Compression has been achieved.

5.4.5 The electrostatic approximation revisited

Earlier in this chapter we have seen a number of examples where the electrostatic approximation is compared with the exact result. Sometimes it is good, sometimes reasonably good and sometimes outright bad. In the present section we shall re-examine the approximation related to the first publication of the perfect lens (Pendry, 2000). In Fig. 5.18 we reproduce the original figure from Phys. Rev. Lett. The object consists of a pair of step functions 14 nm wide centred at $x = \pm 33$ nm. The material constants are $\mu_\mathrm{r} = 1$ and $\varepsilon_\mathrm{r} = -1 - 0.4j$. The thickness of the lens is $d = 40$ nm and the object–lens and lens–image distances are 20 nm, i.e. it follows the Veselago geometry. The image (intensity, not amplitude) is then plotted by Pendry as shown in Fig. 5.18(c). The two peaks are nicely resolved. The image is also shown in the absence of the lens that has only a central maximum. This figure is quite well known. Unfortunately, at some stage a numerical mistake must have been committed. The correct figure for the same parameters is shown in Fig. 5.19(a). The image in the absence of the lens looks similar but not in the presence of the lens. The two peaks are no longer resolved either in the electrostatic approximation or in the exact calculation. We can, however, resolve the two peaks in the object if we reduce all the dimensions by a factor of two, i.e. taking a 20-nm thick lens, as may be seen

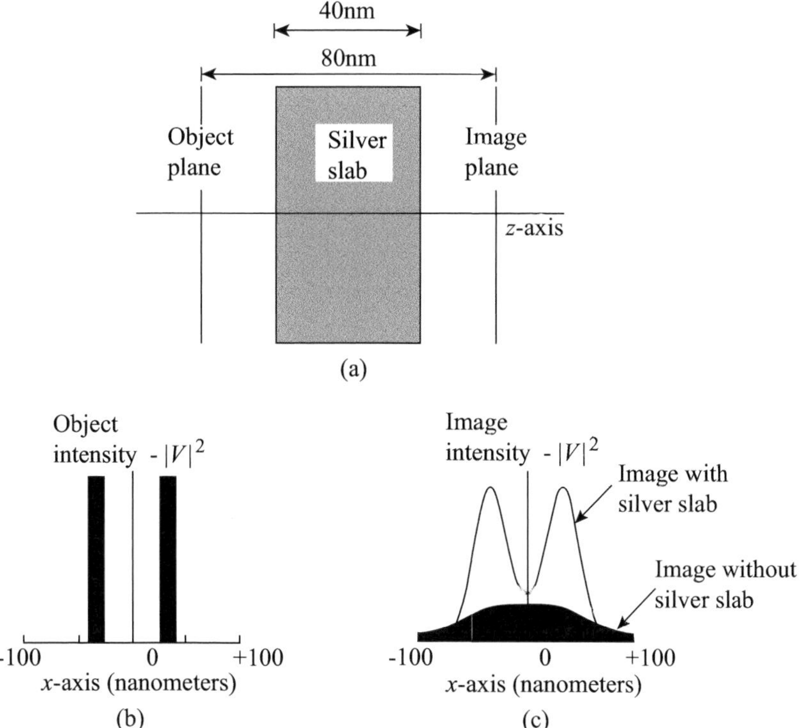

Fig. 5.18 Reproduction of Fig. 2 of Pendry's paper on the perfect lens (2000). (a) Plan view of the new lens in operation. A quasi-electrostatic potential in the object plane is imaged by the action of a silver lens. (b) The electrostatic field in the object plane. (c) The electrostatic field in the image plane with and without the silver slab in place. The reconstruction would be perfect were it not for finite absorption in the silver. Copyright © 2000 by the American Physical Society

in Fig. 5.19(b). The same conclusion can be drawn from the transfer functions plotted in Figs. 5.20(a) and (b). The one for the 40-nm lens is much narrower. There is roughly a factor of two between the cutoff frequencies.

The numerical mistake is unfortunate but that is not the point we wish to make. It is the electrostatic approximation that we wish to examine more closely. Pendry based the electrostatic approximation on the fact that in the examples investigated all dimensions were much smaller than the electromagnetic wavelength. The question is whether small dimensions in themselves are sufficient for the accuracy of the electrostatic approximation to the exact result. Pendry certainly came to the conclusion that for a thin lens the electrostatic approximation, which in the lossless case predicts a flat transfer function and perfect reconstruction of the image, is correct. This is suggested both in the figure caption of his Fig. 2 (our Fig. 5.18) and in the text where he writes: 'Evidently, only the finite imaginary part of the dielectric function prevents ideal reconstruction'. This statement can be checked. Comparing the electrostatic approximation with the exact result in Fig. 5.19(a) we can see that the plots are similar. The same is true if we look at the transfer functions:

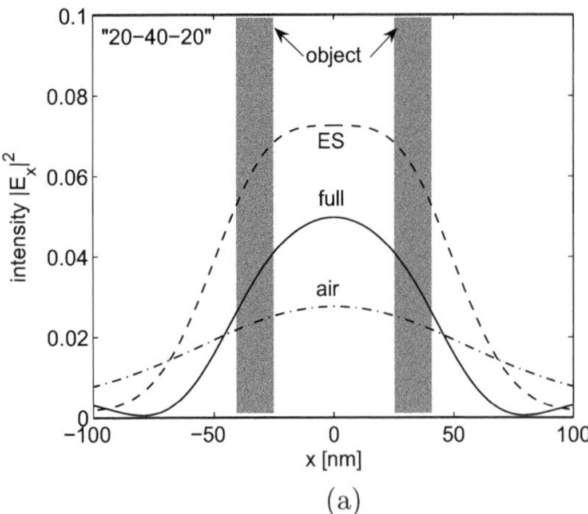

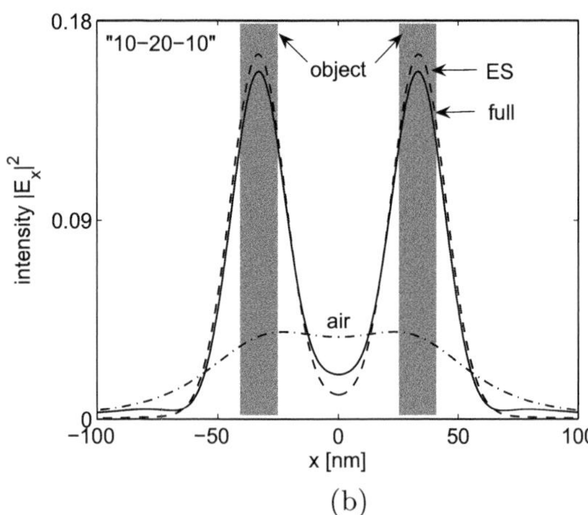

Fig. 5.19 The object (Pendry's double step function shown as gray bars), the image from the full solution (solid line) and from the electrostatic approximation (dashed line) and the image in the absence of the slab (dashed-dotted line). Silver slab with $\varepsilon_{\mathrm{r}} = -1 - \mathrm{j}\,0.4$ of the thickness $d = 40$ nm (a) and $d = 20$ nm (b)

the electrostatic approximation is quite close to the exact result, see Fig. 5.20(a). Thus, the claim that the electrostatic approximation is applicable might be acceptable. However, if we think of the physical reason for the growth of the non-propagating waves, namely the excitation of SPP resonances, then this argument cannot be correct. It is incorrect on two counts. In the absence of losses the transfer function will tend to infinity at the value of k_x that satisfies the resonance condition. Secondly, the resonance is always followed by a cutoff. The cutoff will limit the resolution but the resonance will do something worse: it will lead to an image in which the spatial harmonic causing the resonance dominates.

In the next example we shall show what happens when the object is the same and the thickness of the lens is the same but we shall drastically reduce losses by taking $\varepsilon'' = 10^{-4}$. The field distribution in the image, and the transfer function are shown in Fig. 5.21. The electrostatic approximation yields a very good reproduction of the object, much better than the one achieved for $\varepsilon'' = 0.4$. Alas, the results calculated from the exact solution show a much wider image caused not so much by the cutoff but by the SPP resonance. There is also considerable discrepancy now between the cutoff points of the approximate and exact results.

We may thus say categorically that the electrostatic approximation fails when the losses are zero or very small. For high enough losses the electrostatic approximation works reasonably well. High losses are good because they blunt the resonances.

The mathematical reason for the failure of the electrostatic approximation for 'epsilon-negative-only' materials (first pointed out by Shamonina *et al.* (2001)) was already given in Section 2.12.4. It comes down to the case of which one of two small quantities is negligible.

Whether the electrostatic approximation works or not is not an important matter. One can always rely on the exact expression. The

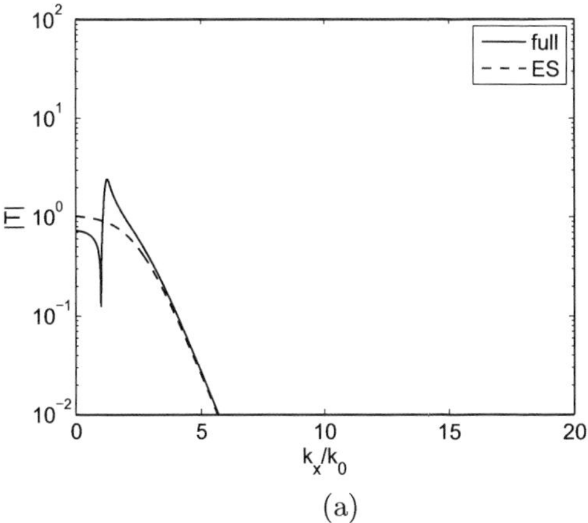

(a)

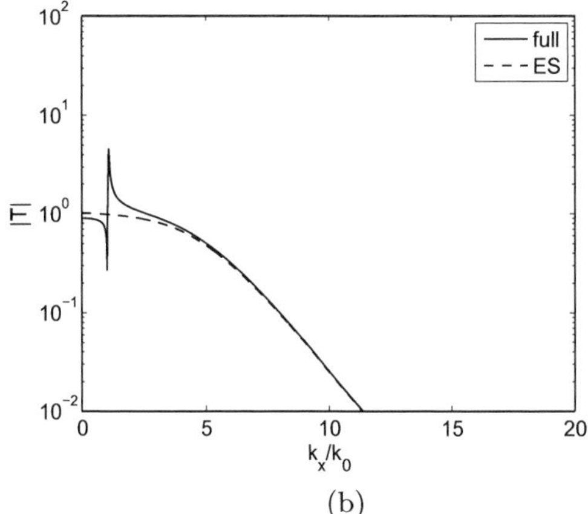

(b)

Fig. 5.20 (a) Transfer function for the same object for a slab of silver, $\varepsilon_{\rm r} = -1 - {\rm j}\,0.4$. Thickness of lens (a) $d = 40$ nm and (b) $d = 20$ nm

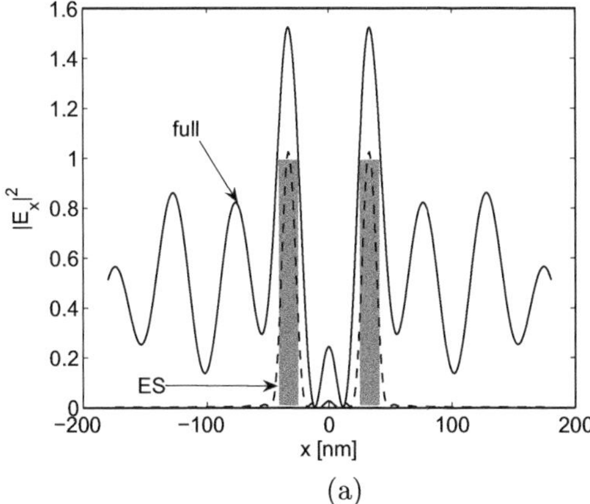

(a)

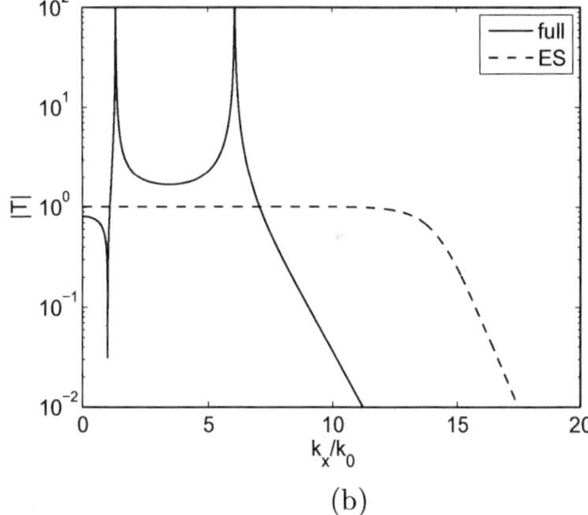

(b)

Fig. 5.21 (a) Image and (b) transfer function for a slab of silver, $\varepsilon_{\rm r} = -1 - {\rm j}\,10^{-4}$. Lens thickness $d = 40$ nm

effect of the SPP resonance on the image is not necessarily significant either. See, for example, Fig. 5.13(c) where the sidelobes are kept low in spite of the SPP resonances. The reason we devoted a whole section to the accuracy of the electrostatic approximation is that those inaccurate statements were published in a paper that, at the time of writing, had over 2000 citations. It is an influential paper, the statements made there have gained wide currency: it needed a correction. We shall also mention it in Section 5.7, the last section of this chapter concerned with widely held beliefs that do not stand closer scrutiny.

5.4.6 Experimental results

Theoreticians pride themselves on being able to deliver a new theory in the course of a single afternoon, provided it is rainy and they have

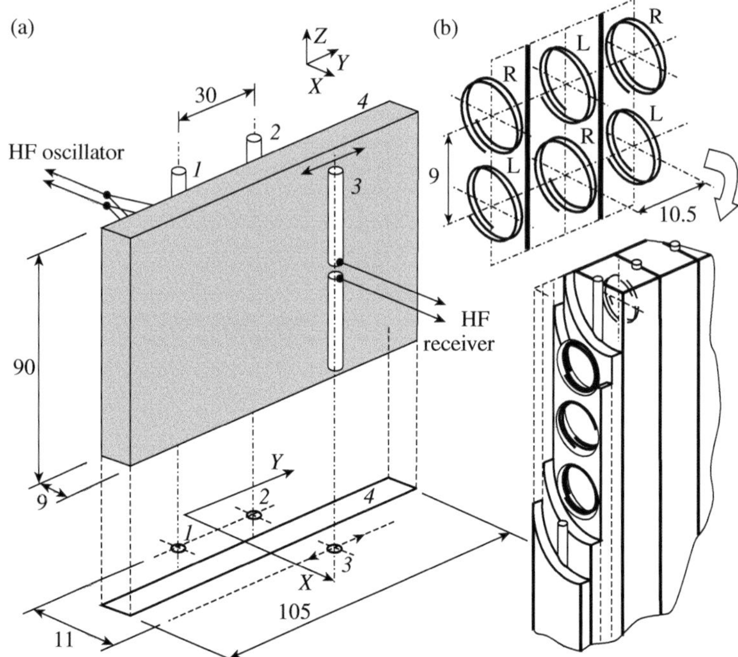

Fig. 5.22 Schematic representation of the experimental setup. From Lagarkov and Kissel (2005). Copyright © 2005 by the American Physical Society

nothing better to do. To prepare a new experiment takes longer. This is the principal reason why it took nearly four years before experimental papers on the superlens came out. The first paper concerned with imaging in a negative-index material was written by Lagarkov and Kissel (2005). The experiments were performed with microwaves at around a frequency of 1.7 GHz using the experimental setup shown in Fig. 5.22. The negative-index material was similar to that of Shelby *et al.* (2001*b*) but instead of SRRs the authors used spirals. The object was made up by two vertical half-wave dipoles 30 mm apart. The received power was measured by a similar dipole in the image plane. Measurements conducted in the absence and presence of the lens showed that the two transmitting dipoles $\lambda/6$ apart could be resolved with the aid of the lens, whereas no feature was seen in the absence of the lens. Similar experiments at frequencies around 3.5 GHz were reported by Aydin *et al.* (2007) using the rod+SRR structure of Shelby *et al.* (2001*a*). They were able to resolve two dipole sources separated by a distance of $\lambda/8$.

All other reports we know of were concerned with imaging at much shorter wavelengths from the infra-red to the ultraviolet region relying on epsilon-negative-only materials. Most of the experiments were done in three groups, those of Blaikie, Zhang and Hillebrand and Shvetz. The first attempt to use silver as the lens material was made by Melville *et al.* (2004). The lens chosen was 120-nm thick, unsuitable for subwavelength imaging but served as a practice run. The authors were able to show that the image was better in the presence than in the absence of the lens. In the next paper (Melville and Blaikie, 2005) the lens configuration was

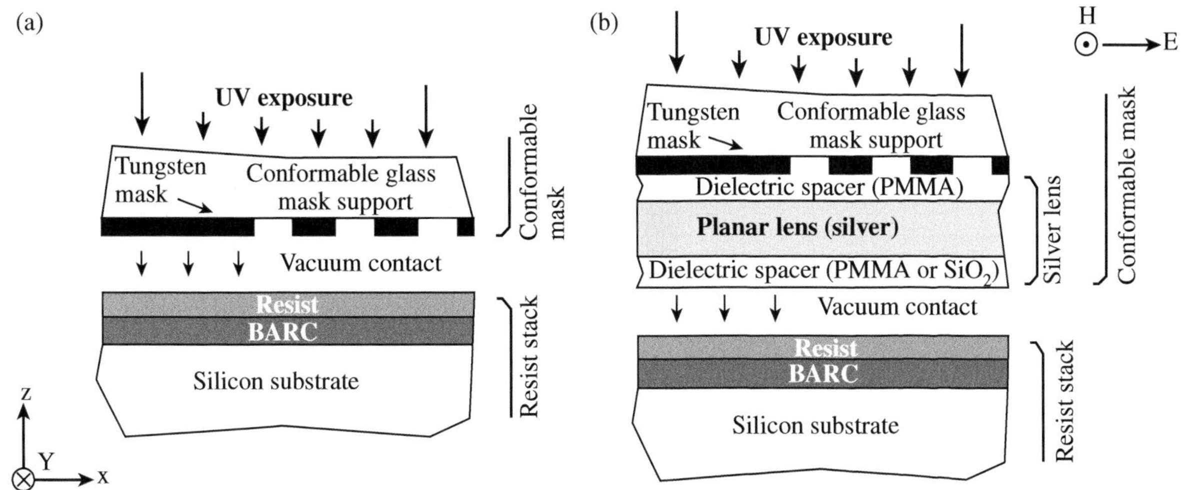

Fig. 5.23 Schematic representation of two experimental setups comparing hard-contact and flat-lens lithography. From Melville *et al.* (2006). Copyright © 2006 Elsevier Ltd

PMMA[9]–Ag–SiO_2 with thicknesses 25–50–10 nm. The wavelength of operation was 365 nm. The patterns that were imaged included isolated lines, line pairs, gratings and arrays of dots with some feature sizes smaller than 100 nm. Gratings with periods down to 145 nm were resolved.

[9]poly-methyl metachrylate

In a later paper by Melville *et al.* (2006) the relative merits of hard-contact lithography (often referred to as evanescent near-field optical lithography) and flat-lens lithography were explored. The aim was to project an image into the photoresist. Schematic drawings of the two somewhat different experimental setups are shown in Figs. 5.23(a) and (b). They found that above the diffraction limit both hard-contact lithography and the silver lens perform in a similar manner. Below the diffraction limit the hard/contact lithography produces clearer images in the resist. The proposed explanation is that it is due to losses in the silver and to (insufficient) quality of the silver deposition. Both single- and double-layer lenses were examined by Melville and Blaikie (2006). They found that for the same total thickness of silver, the resolution limit is qualitatively better for the double-layer stack. However, pattern fidelity is reduced in the double-layer experiments because of increased surface roughness. The same group was also concerned with modelling the silver superlens (Blaikie and McNab, 2002; Melville and Blaikie, 2007) in order to assist the experimental work.

In another set of experiments with a silver lens by Fang *et al.* (2005) the object was an array of 60-nm wide slots of 120-nm pitch next to the word NANO inscribed onto a chromium[10] screen (see Fig. 5.24). The illumination coming from below is at a wavelength of 365 nm. The silver lens is separated from the object by a layer of PMMA and the image is projected into a layer of photoresist. Detection is achieved by

[10]Note that the plasma frequency of chromium is sufficiently far away from that of silver so as not to interfere with the superlensing effect.

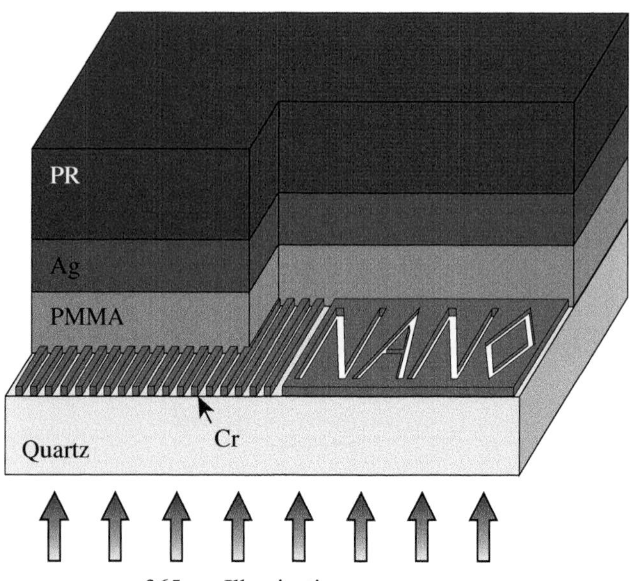

Fig. 5.24 Schematic representation of an experiment imaging the word 'NANO'. From Fang *et al.* (2005). Copyright © 2005 AAAS

converting the image into topographic modulations and is then mapped using atomic force microscopy. The same group (Lee *et al.*, 2005) reported further improvement in their experimental procedure achieving the image of a 50-nm half-pitch object at $\lambda/7$ resolution.

Webb-Wood *et al.* (2006) performed frequency-dependent near-field scanning optical microscopy measurements of plasmon-mediated near-field focusing using a 50-nm thick Au film. In these studies the tip aperture of a probe acts as a localized light source, while the near-field image formed by the metal lens is detected *in situ* using nanoscale scatterers placed in the image plane. By scanning the relative position of object and probe, the near-field image generated by the lens is resolved. Scans performed at different illumination frequencies reveal an optimum near-field image quality at frequencies close to the SPP resonant frequency.

A different material, SiC, in a different wavelength spectrum, mid-infra-red, was utilized by Korobkin *et al.* 2006*a*; Korobkin *et al.* 2006*b* for producing subwavelength imaging based on earlier work of their group (Shvets, 2003; Shvets and Urzhumov, 2004). SiC is a polar material. The analogy between plasmon–polaritons and phonon–polaritons was briefly discussed in Section 3.2 and the dependence of its dielectric constant on frequency was given in eqn (3.8). It may be seen there that its dielectric constant is negative in the *Reststrahlen* band between the frequencies of the transverse and longitudinal optical phonons. If the superlens is made up by a SiO_2–SiC–SiO_2 sandwich then the dielectric constant of the SiC slab should match that of SiO_2, which is $\varepsilon_d = 4$. Very conveniently the dielectric constant of SiC is equal to -4 at $\lambda = 11\mu$m and that wavelength, very conveniently again, can be produced by a tunable CO_2 laser. The lens itself is of conventional design with thicknesses of 200–400–200 nm but the detection mechanism is not

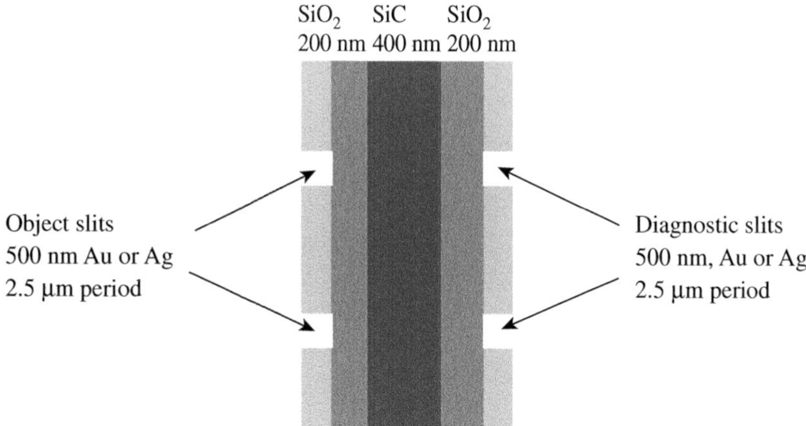

Fig. 5.25 Schematic representation of an imaging experiment with slits for detection. From Korobkin *et al.* (2006*a*). Copyright © 2006 Optical Society of America

conventional (see Fig. 5.25).

There are now additional metallic sheets on both sides of the lens. A set of periodic slits in the front metal sheet constitutes the object and another set of slits of the same period serve for detecting the image. Two kinds of detecting slits are used. The first one (seen in Fig. 5.25) is spatially in-phase with the object slits and a second one (not shown) is produced so that the detecting slits are half-way in between the object slits. The measurement is done with an electromagnetic wave incident perpendicularly upon the lens. Away from the wavelength at which the superlens conditions are satisfied, it makes no difference where the detecting slits are. The output is the same whichever configuration is used. On the other hand, under superlens conditions each of the object slits are imaged exactly at the place where the image slits are, yielding a high output. The output for both slit configurations was measured as a function of wavelength. It was found that under superlens conditions the output was higher by several orders of magnitude than at wavelengths further away, an excellent proof for the existence of the superlensing effect in SiC.

Further experimental evidence was provided by Taubner *et al.* (2006). Their experimental setup is shown in Fig. 5.26. There, the SiO_2–SiC–SiO_2 lens has slightly different thicknesses, 220–440–220 nm, the objects are holes of about 1 μm diameter in a 60-nm thick gold film, and the detection is by a probe. The authors were able to resolve these holes at a wavelength of 10.85 μm.

An entirely new idea for a superlens was reported by Liu *et al.* (2007*a*). Instead of using a flat silver lens the outer boundary was formed into a subwavelength grating. It served to convert the incident evanescent waves into propagating waves that could then be magnified by a conventional microscope. The authors were able to resolve two nanowires separated by 70 nm at a wavelength of 377 nm.

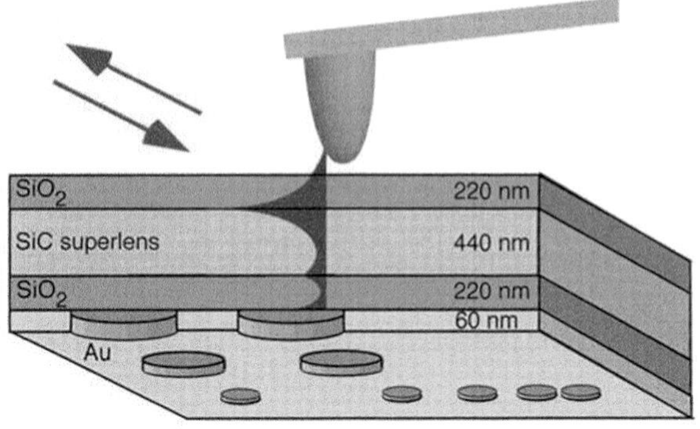

Fig. 5.26 Schematic representation of imaging holes by a SiC experiment. From Taubner *et al.* (2006). Copyright © 2006 AAAS

5.5 Multilayer superlens

We have already mentioned the multilayer lens in Section 2.12.3 and illustrated it in Fig. 2.37. When only ε_r is equal to minus unity then the image is no longer perfect but a multilayer configuration can still significantly improve the imaging abilities of the flat lens. Although the idea was proposed as early as 2001 (Shamonina *et al.*, 2001), the next lot of papers came only in 2003 (Pendry and Ramakrishna, 2003; Ramakrishna *et al.*, 2003) and 2004 (Gao and Tang, 2004; Tang and Gao, 2004). However, by now (writing in early 2008) there is a consensus that the most exciting application, to have pre-magnification by a flat lens followed by a conventional optical microscope, would be impossible without a multilayer configuration. The multilayer lens is now part of conventional wisdom. We shall return to the magnifying variety in Section 5.6. Here we shall discuss the original idea.

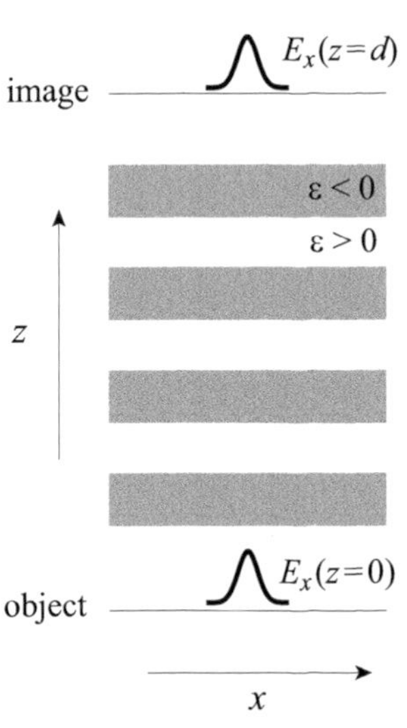

Fig. 5.27 Multilayer superlens

The rationale for the operation of the multilayer lens can be deduced, as in many previous cases, from the properties of SPPs. We know that the excitation of coupled SPPs at both interfaces of the silver slab is a necessary condition for the near-field superlens. The excitation of coupled SPPs ceases, however, if the slab is not thin enough. A silver slab of 60 nm would, e.g., be too thick to act as a superlens at the wavelength of 360 nm. It is, of course, desirable for the image to be transferred over a larger distance. The solution is offered in the form of a multilayer metamaterial made from a series of slabs with negative permittivity separated by layers with positive permittivity (see Fig. 5.27).

In our first example we shall look at the advantage of having more than one layer in the superlens while the total width of the active material remains constant. The object in all four cases shown in Fig. 5.28 is a Gaussian of 14 nm half-width, and the wavelength of operation is 365 nm. For the imaginary part of the dielectric constant we shall choose $\varepsilon'' = 0.1$. This value comes as a compromise. If losses are too low, as explained before, the image quality may be poor on account of the SPP resonance. If losses are too high they will cause a cutoff at a low spatial frequency that would again adversely influence the image quality. The

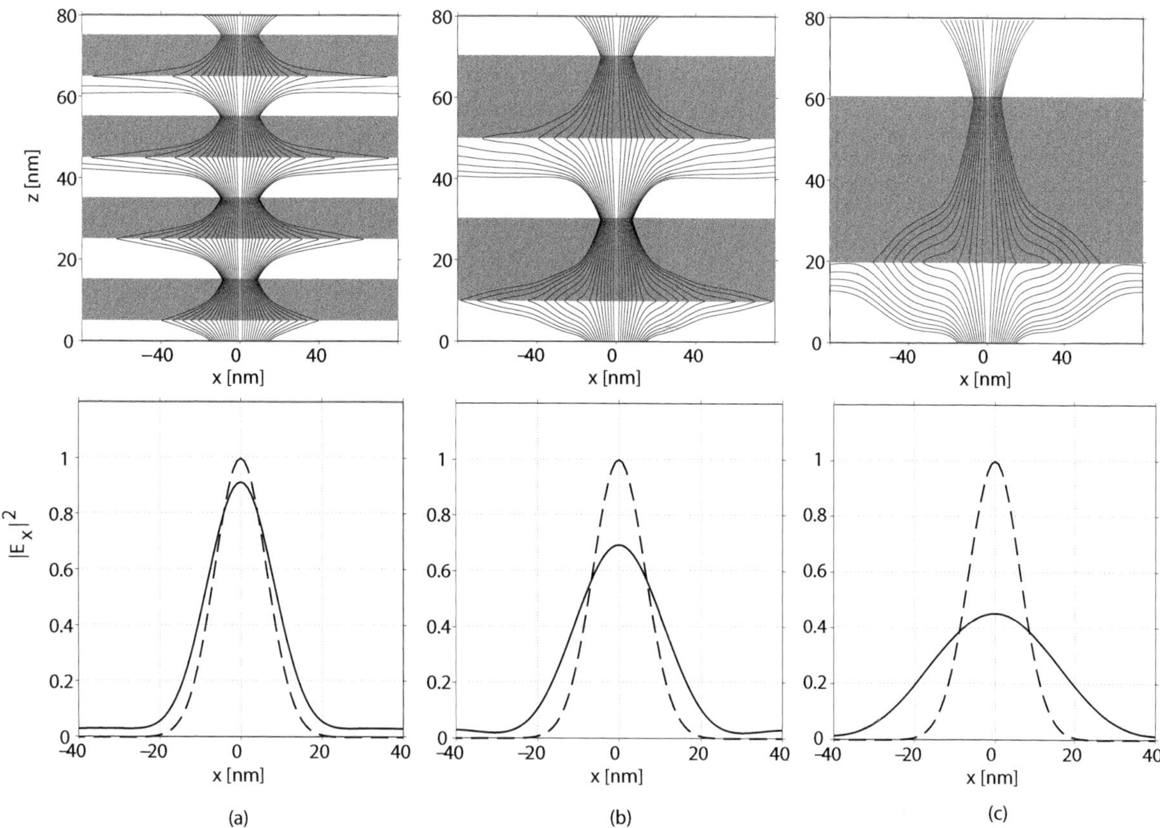

Fig. 5.28 Superlens of the overall thickness of 80 nm. Top: Poynting vector streamlines. Bottom: object (dashed line) and image (solid line). (a) Multilayer lens of four 10-nm silver layers separated by 10-nm layers of air. (b) Two 20-nm silver layers separated by 20-nm air layer. (c) Single lens of 40-nm thick silver slab. $\varepsilon_r = -1 - j\,0.1$

upper figures show the streamlines of the Poynting vector. As shown already in Fig. 2.38(c), the focus moves to the outer boundary of the slab as soon as we take into account evanescent waves. This is confirmed by Fig. 5.28(c) where a single silver layer of 40 nm is considered. For two silver layers (Fig. 5.28(b)) the Poynting vector diverges-converges-diverges-converges-diverges, and the same phenomenon occurs for the four-layer case shown in Fig. 5.28(a). As stressed at several places in this book we should not underestimate the significance of the Poynting vector streamlines. Veselago's construction of the flat lens relied on the concept of rays, clearly unsuited for near-field phenomena. When evanescent waves are included then the ray picture is bound to fail. Its role is taken over by the streamlines of the Poynting vector.[11] The streamlines converging and diverging offer an excellent physical picture of how the image is carried from layer to layer.[12]

The lower figures in Fig. 5.28 show the fidelity of reproduction in the three cases. In the configuration of four 10-nm thick silver layers (Fig. 5.28(a)) the imaging is good, although there is a slight deterioration in

[11]The concept of Poynting vector optics was introduced in a somewhat different context by Russell (1984).

[12]This happens because at the frequency close to the plasmon resonance, dispersion of SPPs approaches high k_x values with the group velocity approaching zero. The net power flow along the boundary is therefore close to zero. This comes about as the power flow outside the slab diverging away from the small object is compensated by the power flow inside the slab.

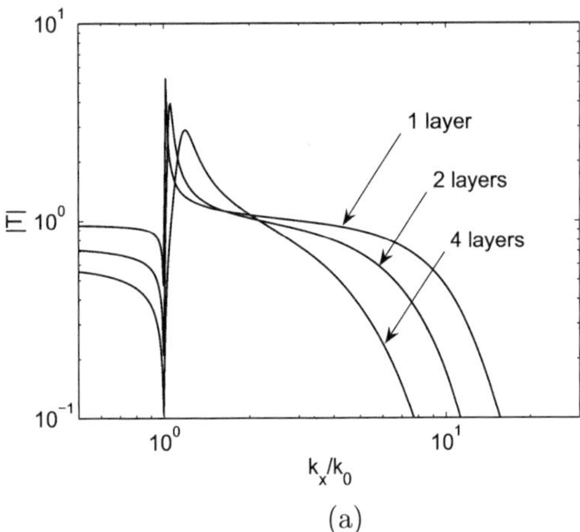

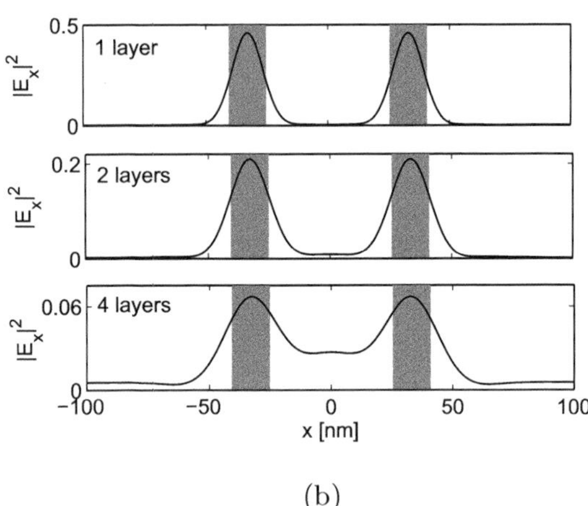

Fig. 5.29 Superlens with one, two and four 10-nm thick layers of silver. Transfer function and image distribution

quality. For two silver layers of 20 nm each (Fig. 5.28(b)) the image becomes wider, and even wider when there is only one single silver layer of 40 nm thickness (Fig. 5.28(c)). The above example offers a clear proof for the efficacy of the multilayer lens both by the Poynting vector picture and by the improved fidelity of reproduction.

There can be no doubt, however, that the multilayer lens is not a magic cure for all the ills of the silver lens. The image quality is bound to deteriorate with the number of layers. We shall show here a further example that considers not only the fidelity of reproduction but resolution as well, and look at the image after 1, 2 and 4 layers. The object consists of the two step functions 50 nm apart shown already in Fig. 5.18(b). For the same amount of loss as in the previous example we plot the transfer function and the image intensity in Figs. 5.29 and 5.30 for d = 10 nm and 20 nm, respectively. For d = 10 nm there is a slight deterioration of the image as the number of layers increases from 1 to 4. For d = 20 nm the two peaks can be clearly resolved for 1 layer, there is some deterioration for 2 layers and the object is unrecognizable when there are 4 layers. The conclusions are the same as before. For good imaging, thin layers are needed.

We have now covered the major principles upon which multilayer lenses work. For further discussions see Feng *et al.* 2005; Feng and Elson 2006; Dorofeenko *et al.* 2006; Webb and Yang 2006; Melville and Blaikie 2007.

Another strand of thought that could lead to multilayer imaging comes from traditional treatments of finding the effective dielectric constant (for a review see, e.g., Bergman 1978). In its electrical engineering context see the detailed analysis by Wait (1962). In optics they appear in the classical work of Born and Wolf (1975). They are called stratified media described by transfer matrices and used mainly as filters and matching elements. For small modulation of the dielectric constant they

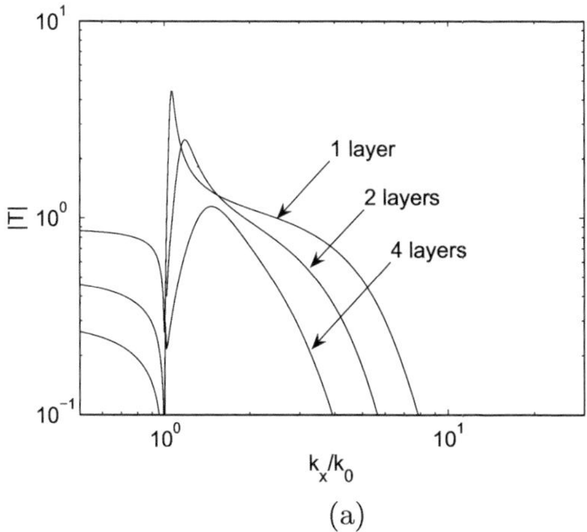

(a)

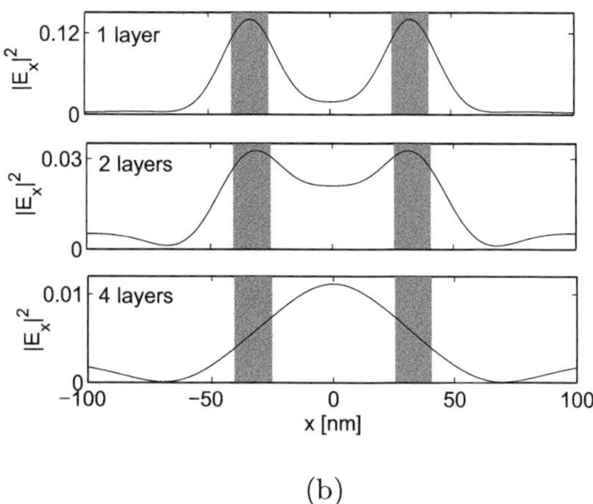

(b)

Fig. 5.30 Superlens with one, two and four 20-nm thick layers of silver. Transfer function and image distribution

provide holographic filters (see, e.g., Solymar and Cook 1981; Solymar *et al.* 1996), and they also appear in photonic bandgap materials in their 1D version[13] (Scalora *et al.*, 1998).

[13] Scalora *et al.* (1998) were the first to realize that the transparency of a stack of metal–dielectric layers could be much higher than that of a single one. They also introduced the term 'resonance-enhanced tunnelling'.

In the simple case of a multilayer made up by isotropic materials of equal thickness the derivation of the effective dielectric constants is given in Appendix I. It results in an anisotropic material with longitudinal and transverse dielectric constants given as

$$\varepsilon_z^{-1} = \frac{1}{2}\left(\varepsilon_1^{-1} + \varepsilon_2^{-1}\right) \qquad \text{and} \qquad \varepsilon_{\mathrm{t}} = \frac{1}{2}(\varepsilon_1 + \varepsilon_2)\,. \tag{5.34}$$

The idea is that for sufficiently thin materials simple averaging works. For $\varepsilon_{\mathrm{r}1} = 1$ and $\varepsilon_{\mathrm{r}2} = -(1+\delta)$ the effective dielectric constants,

$$\varepsilon_z \simeq \frac{2}{\delta} \qquad \text{and} \qquad \varepsilon_{\mathrm{t}} \simeq -\delta\,, \tag{5.35}$$

turn out to be of opposite sign. Taking $\varepsilon_{\mathrm{r}1} = 1$ and $\varepsilon_{\mathrm{r}2} = -1$ we find the interesting outcome that

$$\varepsilon_z = \infty \qquad \text{and} \qquad \varepsilon_{\mathrm{t}} = 0\,. \tag{5.36}$$

The consequence is that an image pasted on the front surface will appear at the back surface. It is an interesting concept worth mentioning but it is a poor approximation. A layer thickness of 20 nm (for which imaging is shown in Fig. 5.30) is already too thick for the approximation to work. For further discussions see Wood *et al.* (2006).

The difference between the two methods may be best seen by looking at the Poynting vector. The above effective-medium approximation gives a straight line, whereas the one based on the transfer matrices leads to the diverging-converging picture of Fig. 5.28. The result of a numerical simulation[14] by CST MICROWAVE STUDIO (Tatartschuk, 2008) is shown in Fig. 5.31.

[14] Although we have a healthy scepticism concerning the results of many simulations we are happy with this one because it comes from a very reliable source (our group) and provides further proof for the correctness of the physical picture presented in Fig. 5.28.

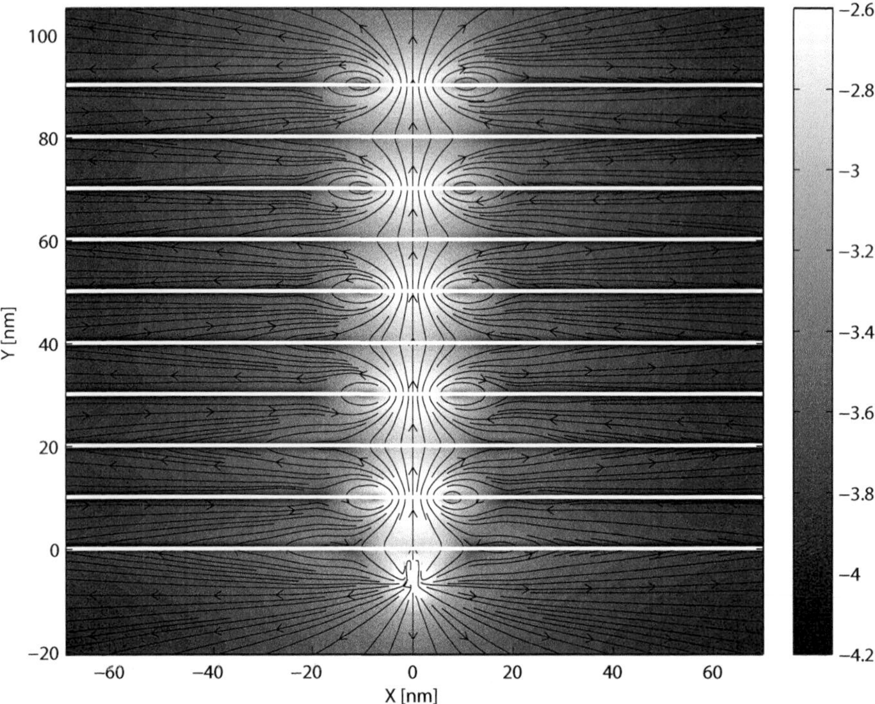

Fig. 5.31 Numerical simulation of the streamlines and amplitude of the electric field for a multilayer silver lens. From Tatartschuk (2008). For coloured version see plate section

5.6 Magnifying multilayer superlens

In Fig. 5.31 we have seen that in an epsilon-negative multilayer structure the spatial information is preserved by the Poynting vector converging in the negative-epsilon material and diverging in the dielectric. With proper design it can be ensured that the reproduction of the subwavelength image is still good. What could we expect if instead of the multiple planar slabs we have multiple cylindrical annuli? Common sense suggests that the image will simply expand according to the rules of geometrical optics. Numerical simulations using CST MICROWAVE STUDIO confirm this expectation, as may be seen in Fig. 5.32 (Tatartschuk, 2008). Taking two objects some distance apart very close to the surface of the first annulus the variation of the Poynting vector streamlines may be seen to follow a regular pattern. The image is magnified. Again, common sense suggests that the magnification is equal to the ratio of the radii of the outer ring to the inner ring.

The idea that a cylindrical lens may give a magnified image was first proposed by Pendry and Ramakrishna (2002). They used conformal transformation from rectangular to cylindrical co-ordinates to find the equivalent of the perfect flat lens in cylindrical form. A somewhat different derivation of the magnification effect is due to Pendry (2003*b*). In

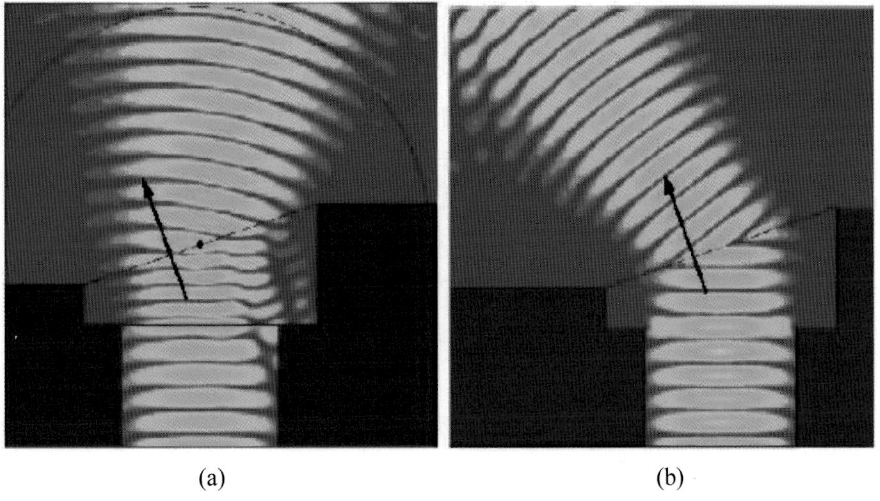

Fig. 2.35 Numerical simulations of the refraction of a finite beam by a wedge, (a) $\varepsilon_\mathrm{r} = 2.2$, $\mu_\mathrm{r} = 1$, (b) $\varepsilon_\mathrm{r} = -1$, $\mu_\mathrm{r} = -1$. From Kolinko and Smith (2003). Copyright © 2003 Optical Society of America

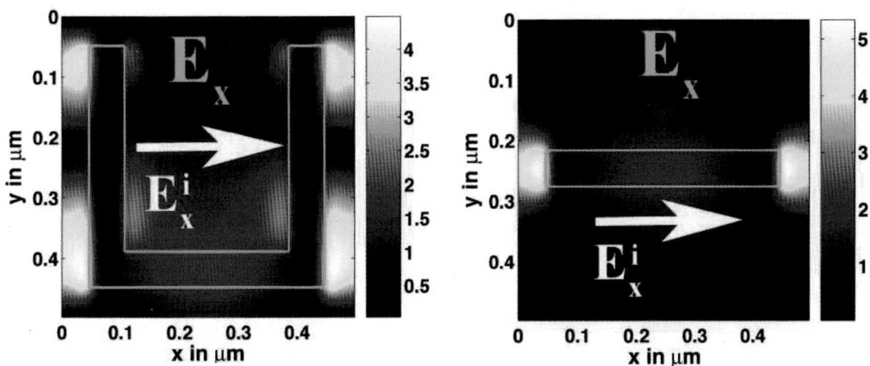

Fig. 4.33 Electric-field distribution at the resonance. From Rockstuhl *et al.* (2006). Copyright © 2006 Springer Science + Business Media

Fig. 4.34 The scaling of the simulated magnetic resonance frequency as a function of the linear size a of the unit cell for the 1-, 2-, and 4-cut SRRs (solid lines with symbols). Up to the lower terahertz region, the scaling is linear. The maximum attainable frequency is strongly enhanced with the number of cuts in the SRR ring. The hollow symbols indicate that $\mu < 0$ is no longer reached. The non-solid lines show the scaling of the magnetic resonance frequency calculated through the LC circuit model. From Zhou *et al.* (2005*a*). Copyright © 2005 by the American Physical Society

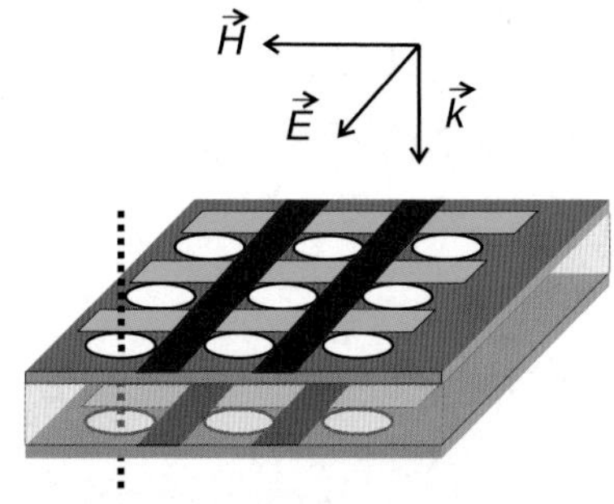

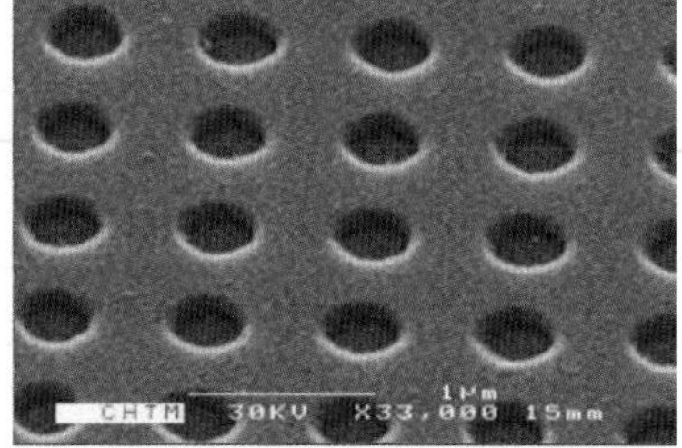

Fig. 4.43 Top: Schematic of the multilayer fishnet structure consisting of an Al_2O_3 dielectric layer between two Au films perforated with a square array of holes (838 nm pitch; 360 nm diameter) atop a glass substrate. For the specific polarization and propagation direction shown, the active regions for the electric (dark regions) and magnetic (hatched regions) responses are indicated. Bottom: SEM picture of the fabricated structure. From Zhang *et al.* (2005*a*). Copyright © 2005 Optical Society of America

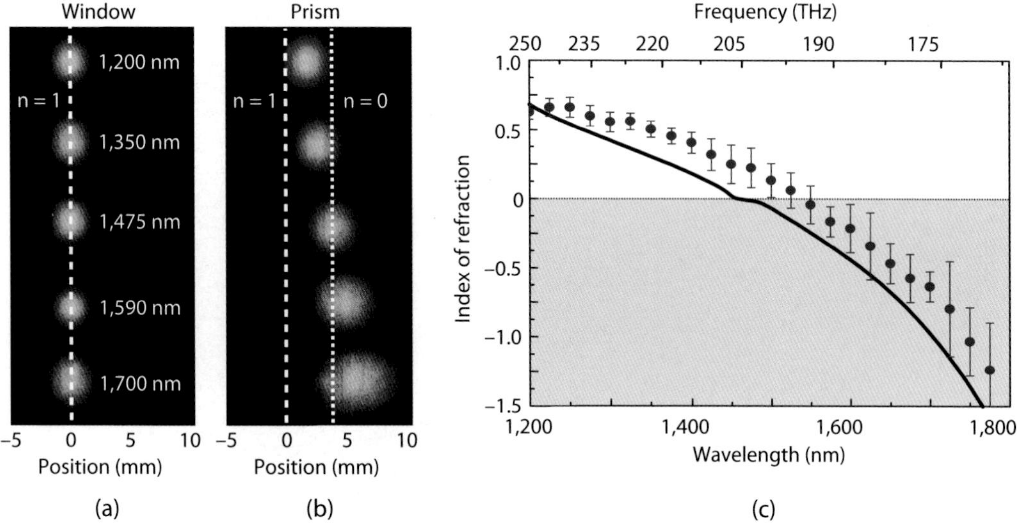

Fig. 4.45 Experimental results and FDTD simulations. Fourier-plane images of the beam for the window (a) and prism sample (b) for various wavelengths. The horizontal axis corresponds to the beam shift d, and positions of $n = 1$ and $n = 0$ are denoted by the white lines. The image intensity for each wavelength has been normalized for clarity. (c) Measurements (circles with error bars after four measurements) and simulation (solid line) of the fishnet refractive index. From Valentine *et al.* (2008). Copyright © 2008 Nature Publishing Group

Fig. 4.47 Field map of the electric field at the main resonance for a triangle particle. From Kottmann *et al.* (2000*b*). Copyright © 2000 Optical Society of America

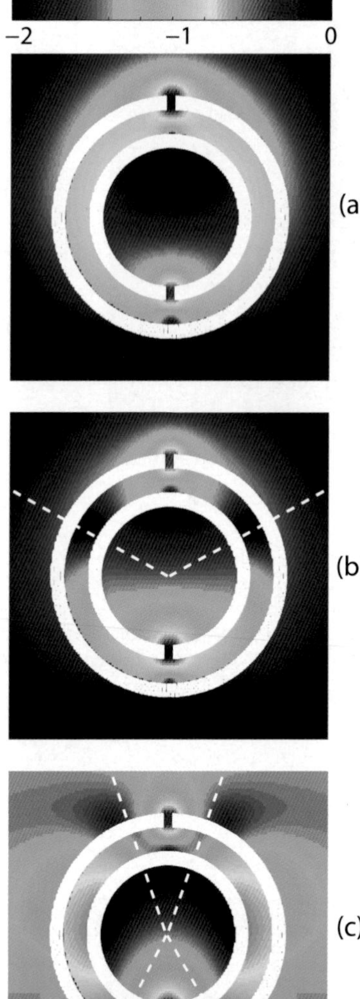

Fig. 4.52 Electric-field distributions at the fundamental (a), the first (b) and the second higher resonance (c). Positions of the voltage zeros are shown by white dashed lines. From Hesmer (2008)

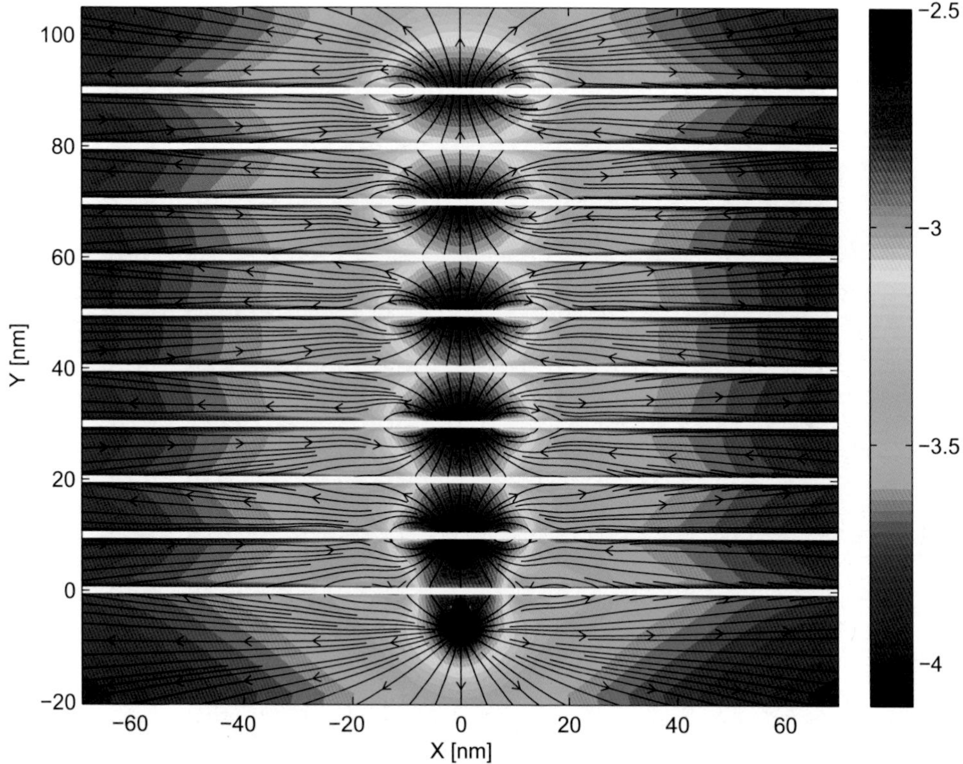

Fig. 5.31 Numerical simulation of Poynting vector streamlines and amplitude for a multilayer silver lens. From Tatartschuk (2008)

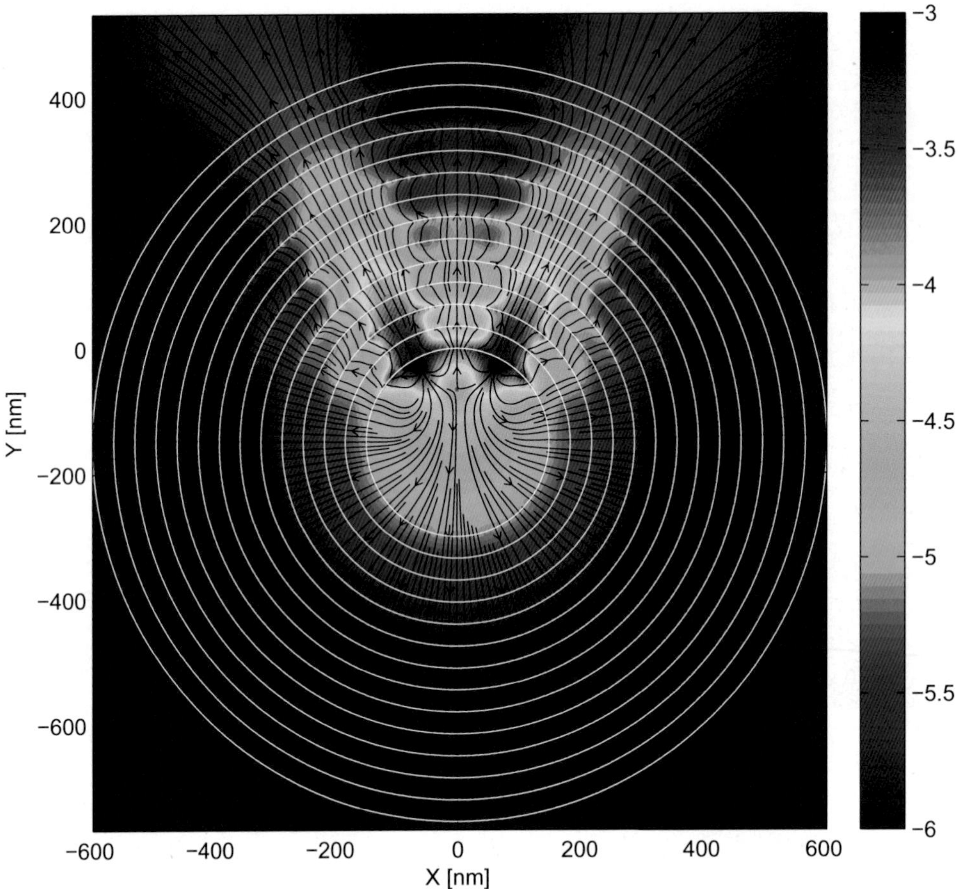

5.32 Magnifying superlens. Numerical simulation of Poynting vector streamlines emanating from two subwavelength objects. $\varepsilon_{\mathrm{r}} = -1 - \mathrm{j}\,0.4$. From Tatartschuk (2008)

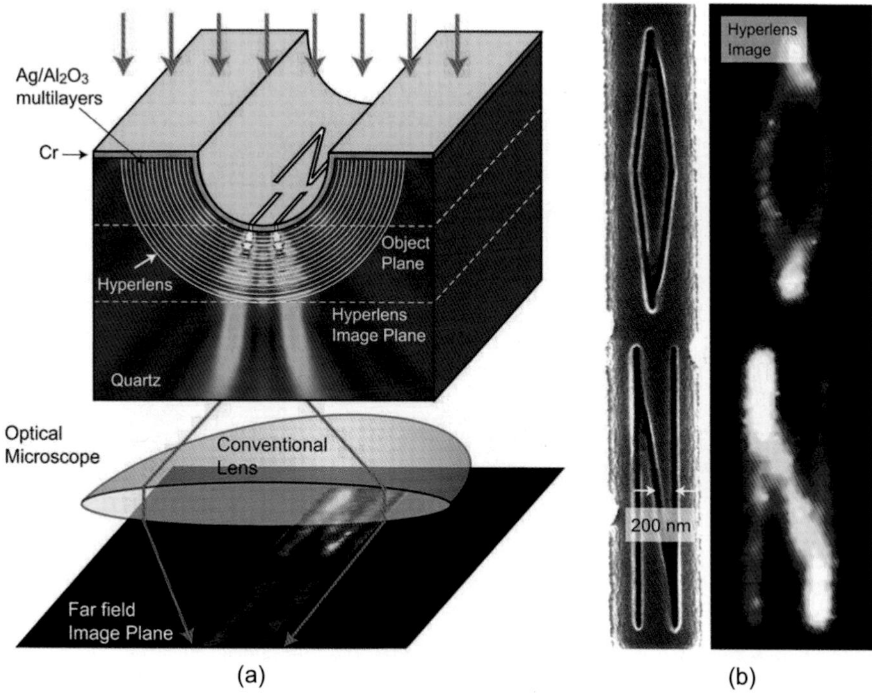

Fig. 5.33 (a) Schematic of magnifying optical hyperlens and numerical simulation of imaging of subdiffraction-limited objects. (b) An arbitrary object 'ON' imaged with subdiffraction resolution. Line width of the object is about 40 nm. The hyperlens is made of 16 layers of Ag-Al_2O_3. From Liu *et al.* (2007*b*). Courtesy of Prof. Xiang Zhang of University of California at Berkeley. Copyright © 2007 AAAS

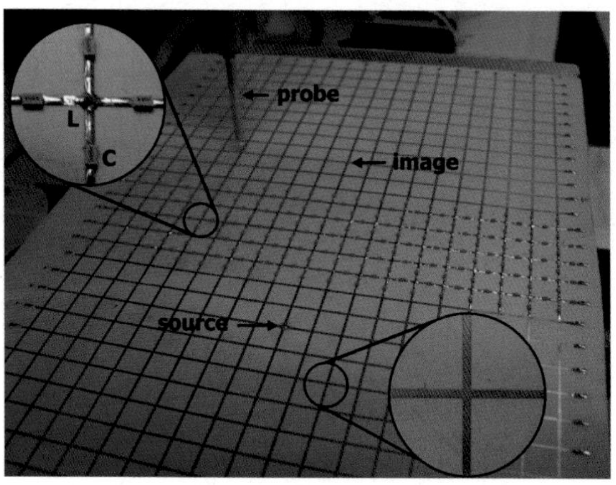

Fig. 6.14 The left-handed planar transmission-line lens. The unit cell of the left-handed (loaded) grid is shown in the top inset, that of the right-handed (unloaded) grid is shown in the bottom inset. From Grbic and Eleftheriades (2004). Copyright © 2004 by the American Physical Society

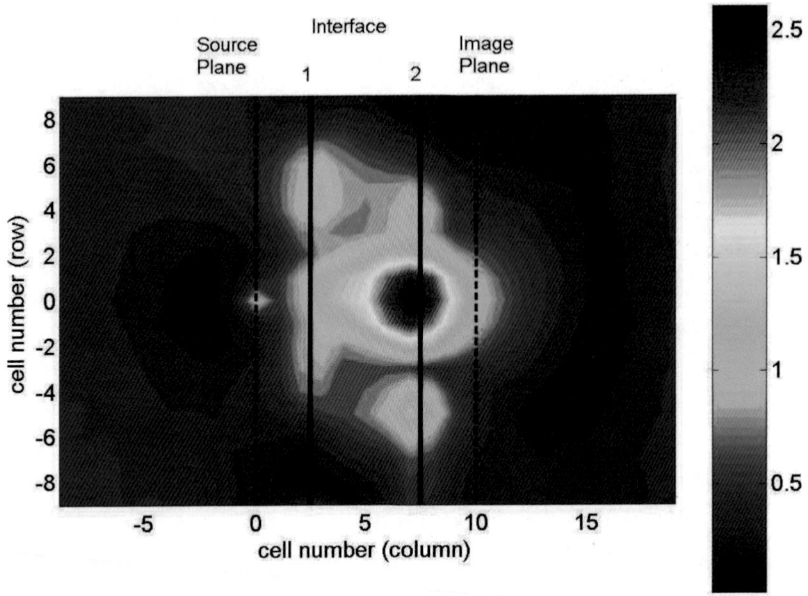

Fig. 6.16 The measured vertical electric field detected 0.8 mm above the surface of the entire structure at 1.057 GHz. The plot has been normalized with respect to the source amplitude (linear scale). From Grbic and Eleftheriades (2004). Copyright © 2004 by the American Physical Society

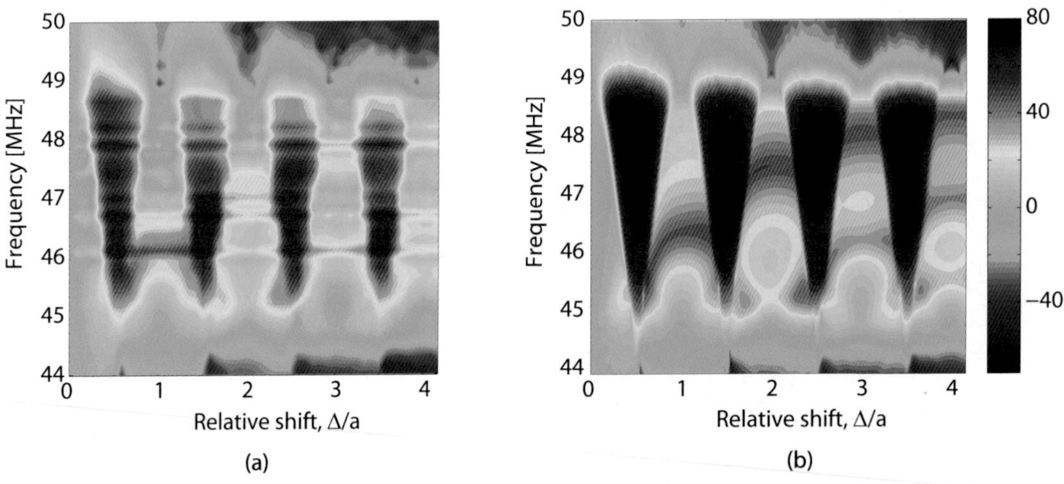

Fig. 7.39 Contour plots of transmission between the split-pipe arrays as a function of frequency and shift, (a) experiment, (b) theory. From Radkovskaya *et al.* (2007*b*). Copyright © 2007 Wiley-VCH Verlag GmbH & Co. KGaA

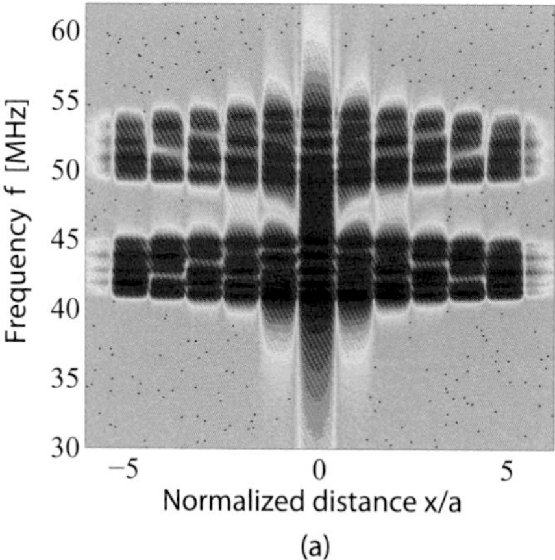

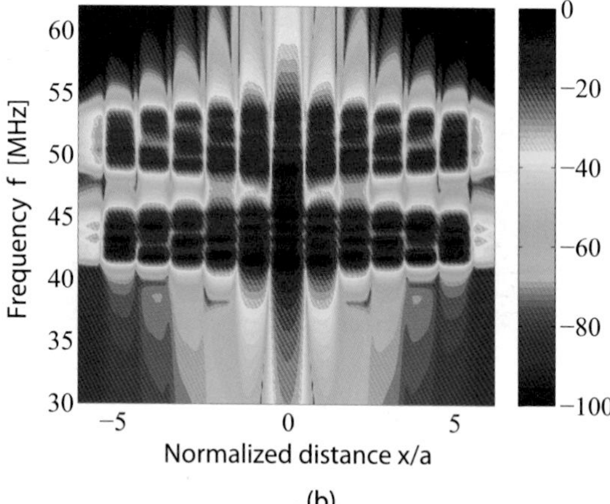

Fig. 7.46 Near-field imaging for the double lens with $h = 10$ mm. Magnetic-field distribution in the image plane versus frequency (contour plot). Experiment (a) and theory (b). From Sydoruk *et al.* (2007*b*). Copyright © 2007 American Institute of Physics

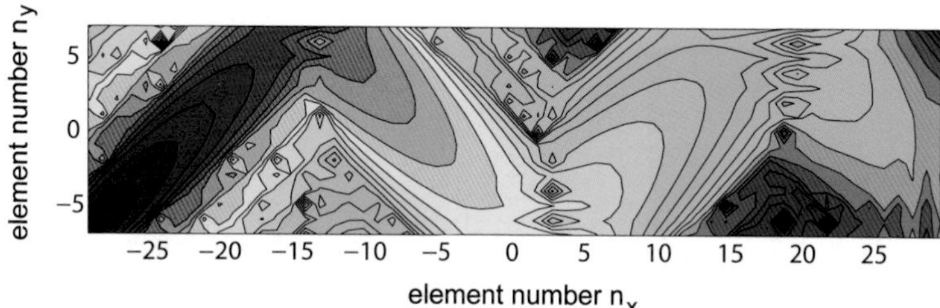

Fig. 8.8 Contour plot of currents in a 2D array of resonators showing reflection of a magnetoinductive wave

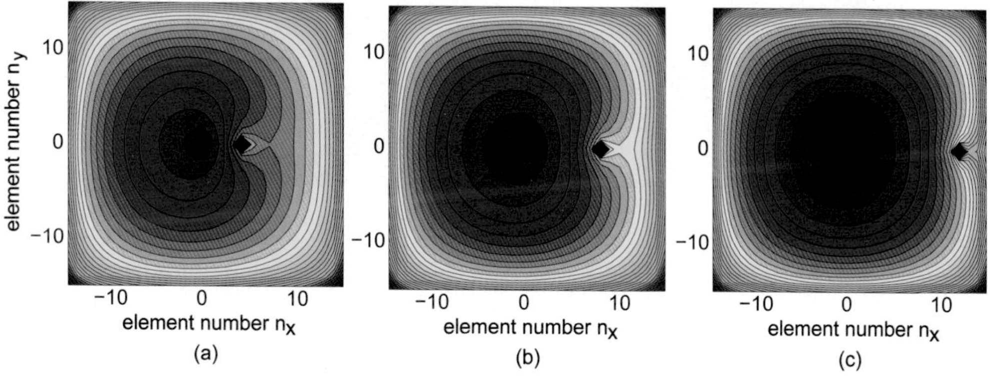

Fig. 8.9 Contour plot of currents in a 2D array of resonators showing diffraction on a defect

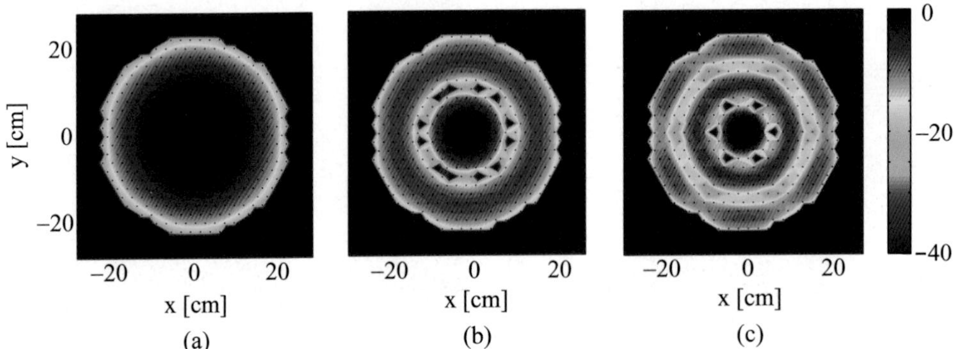

Fig. 8.12 Numerically obtained current distributions for circular boundary conditions at $\omega/\omega_0 =$ (a) 1.207, (b) 1.192, (c) 1.167. Black dots show positions of the elements. From Zhuromskyy *et al.* (2005*a*). Copyright © 2005 Optical Society of America

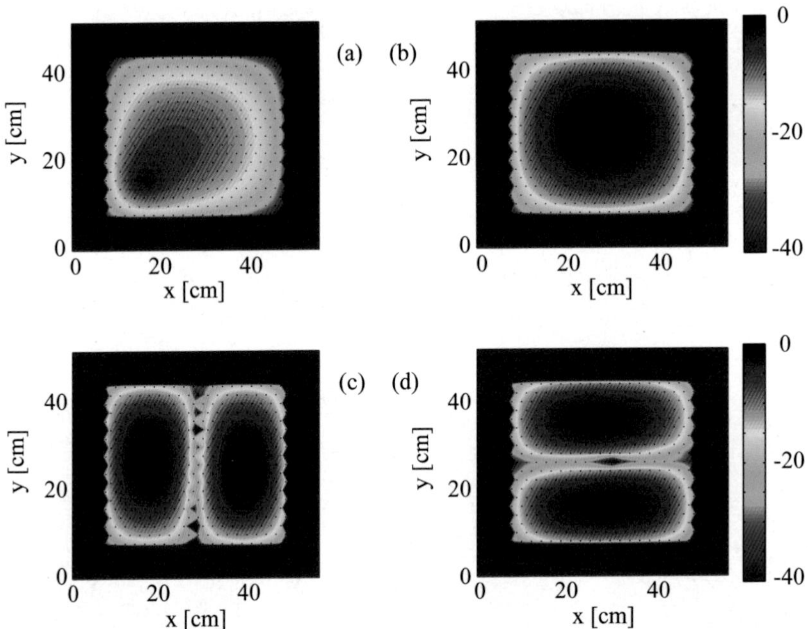

Fig. 8.13 Numerically obtained current distributions for rectangular boundary conditions, (a) non-resonant distribution for asymmetric excitation, (b)–(d) resonant excitation at $\omega/\omega_0 = 1.207$, 1.202 and 1.201, respectively. Black dots show positions of the elements. From Zhuromskyy *et al.* (2005*a*). Copyright © 2005 Optical Society of America

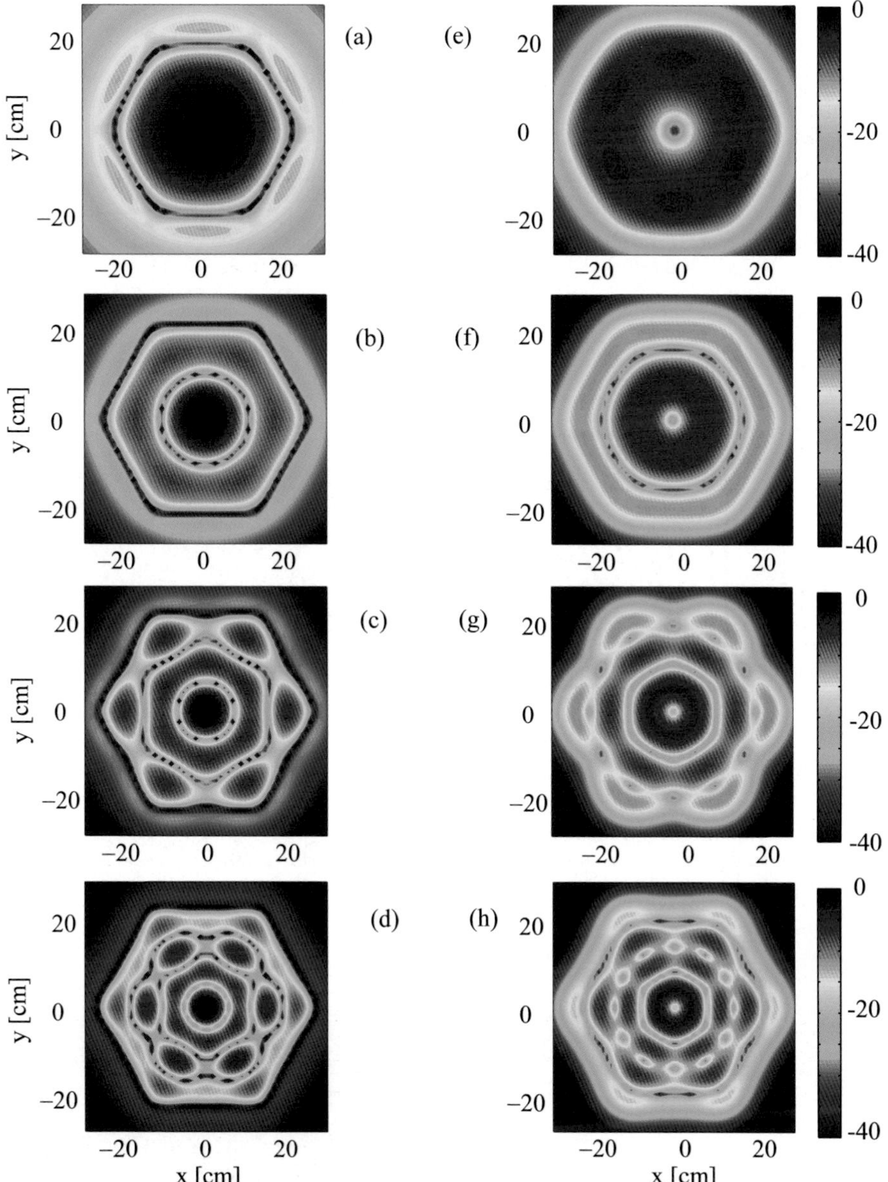

Fig. 8.14 Magnetic-field distribution at four resonant frequencies: (a) and (e) $\omega/\omega_0 = 1.206$, (b) and (f) $\omega/\omega_0 = 1.186$, (c) and (g) $\omega/\omega_0 = 1.154$, (d) and (h) $\omega/\omega_0 = 1.114$; normal component (a)–(d) and tangential component (e)–(h). From Zhuromskyy *et al.* (2005*a*). Copyright © 2005 Optical Society of America

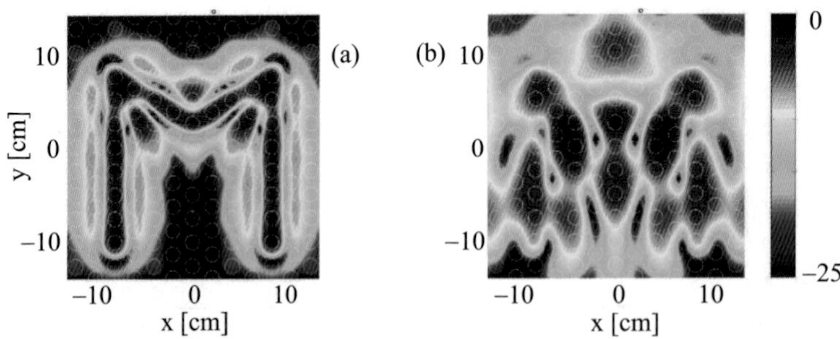

Fig. 8.15 The normal component of magnetic field at $\omega/\omega_0 = 0.98$ (a) and 1.01 (b). From Zhuromskyy *et al.* (2005*a*). Copyright © 2005 Optical Society of America

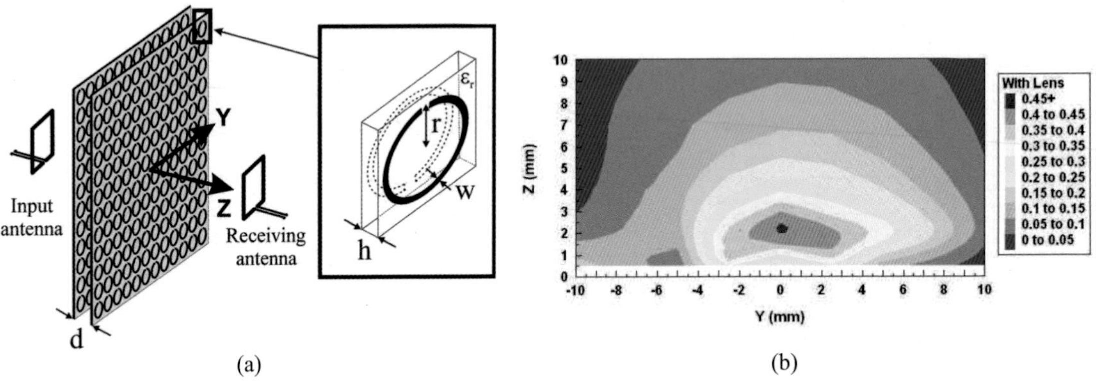

Fig. 8.16 (a) Experimental setup for imaging with magnetoinductive waves. (b) The magnitude of the transmission coefficient between the input and output antennas at a frequency of 3.23 GHz. From Freire and Marques (2005). Copyright © 2005 American Institute of Physics

Fig. 8.37 Experimental setup of a magnetoinductive ring resonator. From Syms *et al.* (2008)

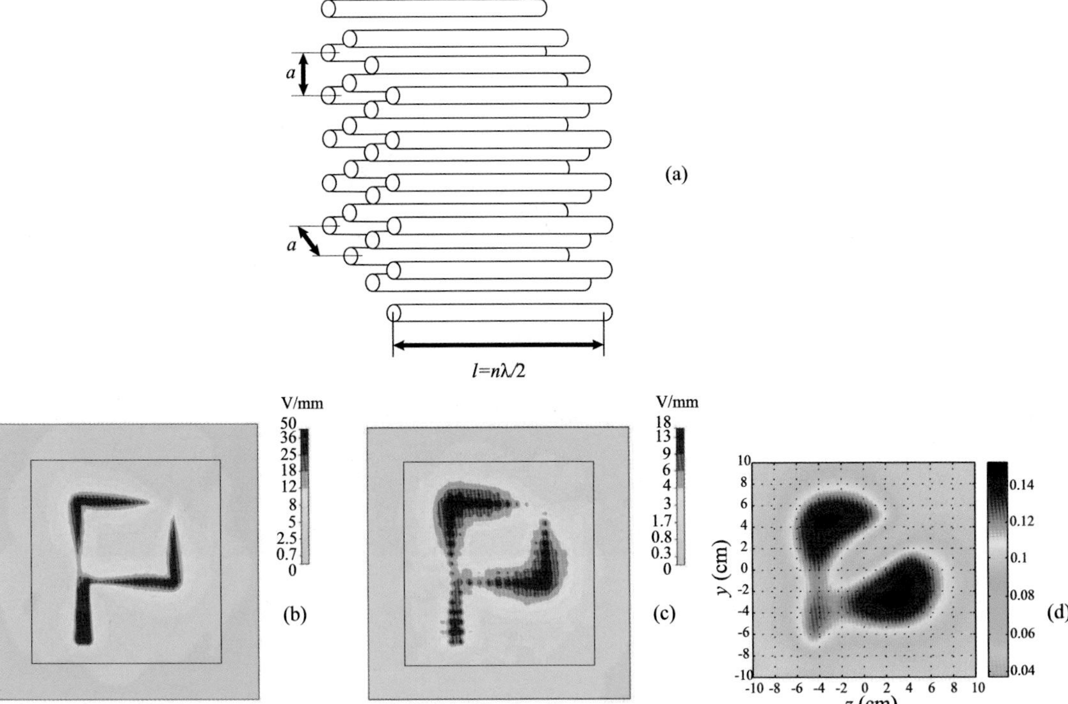

Fig. 9.5 (a) Horizontal stack of a square lattice of thin wires. Length chosen so as to satisfy the Fabry–Perot condition of resonance. (b) Image of letter P 2.5 mm in front of the object obtained by numerical simulation. (c) Image 2.5 mm behind the lens obtained by numerical simulation. (d) Electric-field distribution in the image plane measured by near-field scanning. From Belov *et al.* (2006*a*). Copyright © 2006 by the American Physical Society

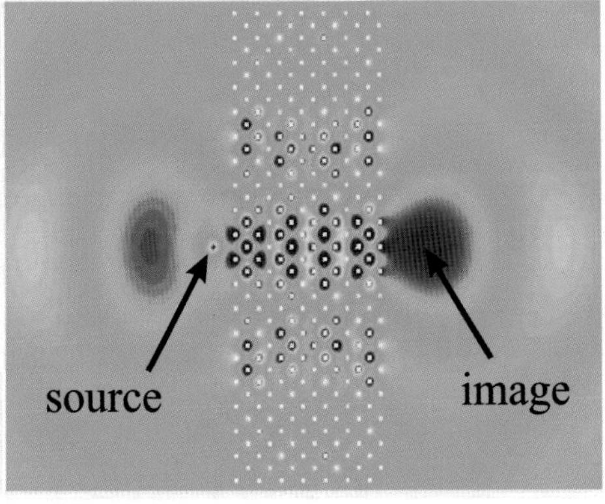

Fig. 9.8 Distribution of the electric field showing the source and the image. From Belov *et al.* (2005). Copyright © 2005 by the American Physical Society

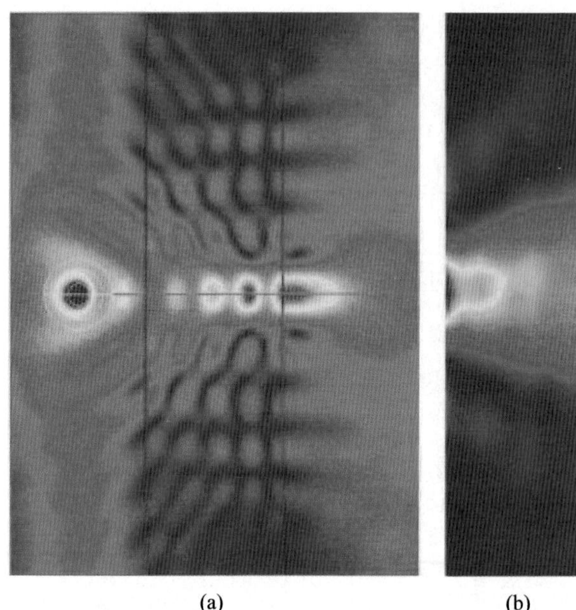

Fig. 9.11 (a) Numerical simulation of the distribution of the electric field. The slab indicated by the solid lines has $\varepsilon_r = -1$ and a diagonal permeability tensor for which the longitudinal component $\mu_z = -1$ and $\mu_x = \mu_y = 1$. The slab is 16 cm long, with a line source placed 2 cm from the slab. The slab thickness is 4 cm. (b) Experimentally obtained field distribution in the plane starting 4 cm away from the output of the slab. From Smith *et al.* (2004*b*). Copyright © 2004 American Institute of Physics

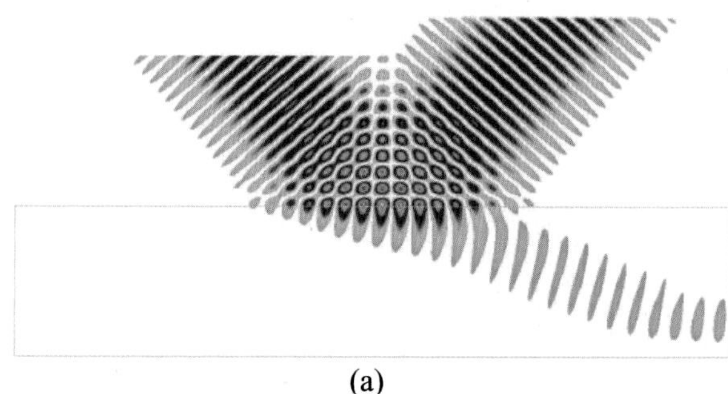

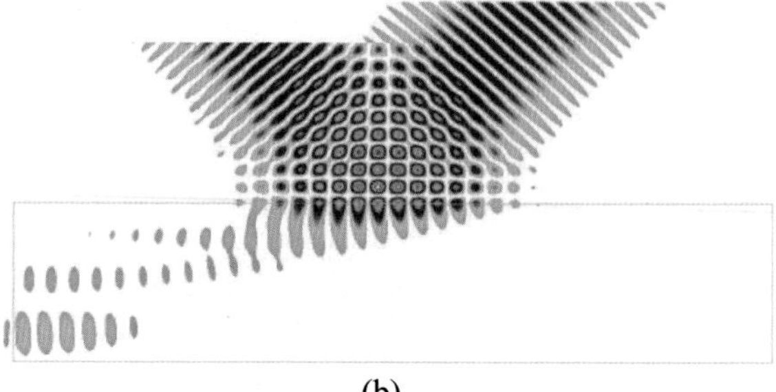

Fig. 9.13 Intensity distribution for a Gaussian beam incident at 40° from a medium with $\varepsilon_r = 9$, $\mu_r = 1$ upon a slab of (a) $\varepsilon_r = 3$, $\mu_r = 1$ and (b) $\varepsilon_r = -3$, $\mu_r = -1$. From Ziolkowski (2003*a*). Copyright © 2003 Optical Society of America

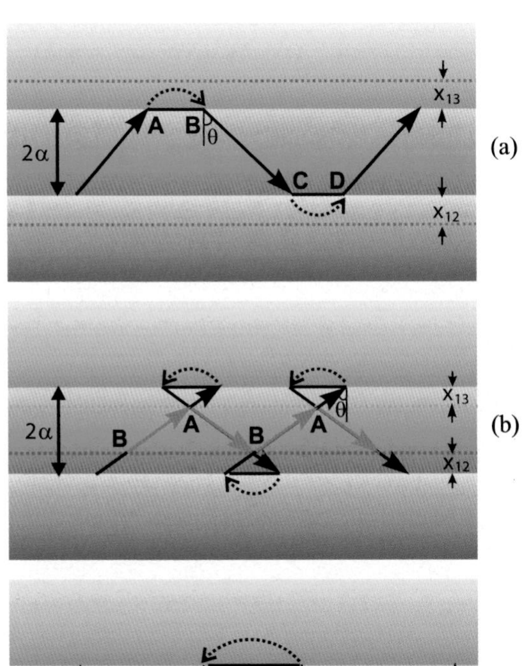

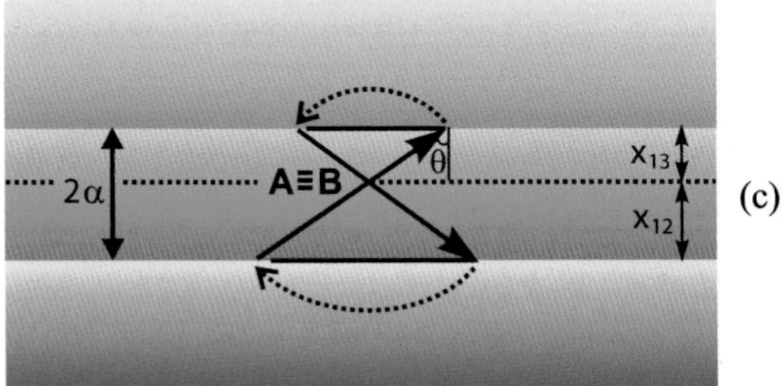

Fig. 9.14 A waveguide consisting of a negative-index material bounded by two positive-index materials. Parameters are chosen so that (a) the wave is forward, (b) the wave is backward, and (c) the wave is stationary. From Tsakmakidis *et al.* (2007). Copyright © 2008 Nature Publishing Group

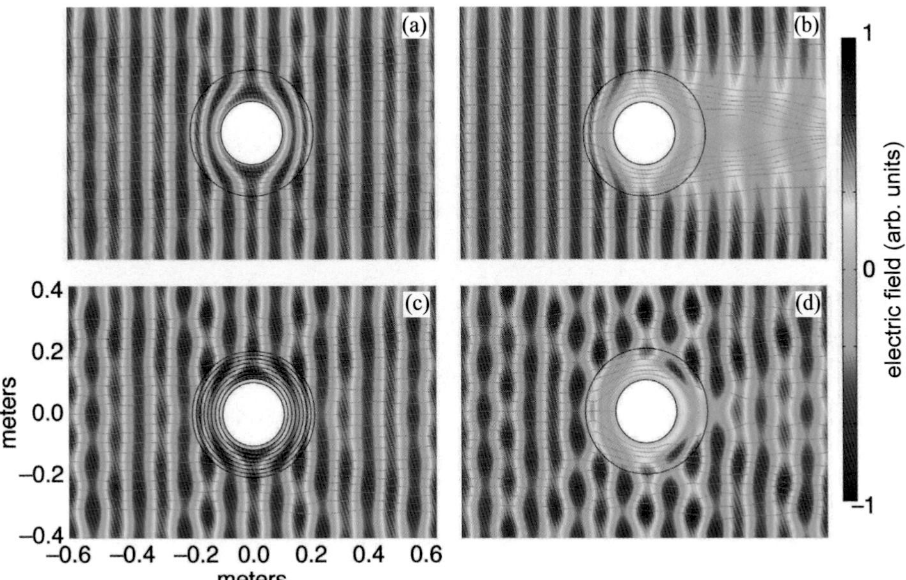

Fig. 9.23 Electric-field distribution for an incident plane wave in the vicinity of a perfectly conducting shell and a cloak. (a) ideal parameters, (b) with a loss tangent of 0.1, (c) with an 8-layer approximation to the desired distribution of the material parameters, (d) with a simplified cloak in which only μ_r is varying spatially. From Cummer *et al.* (2006). Copyright © 2006 by the American Physical Society

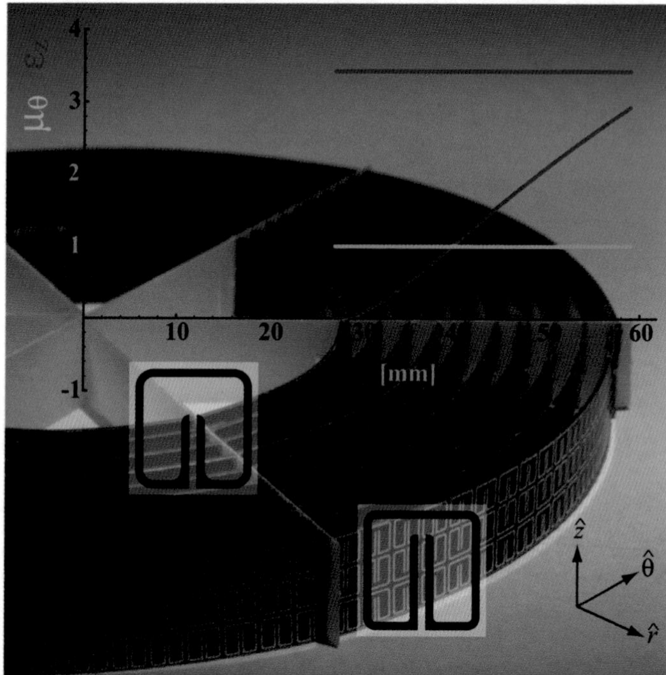

Fig. 9.24 2D microwave cloaking structure with a plot of the material parameters implemented. μ_r (red line) is multiplied by a factor of 10 for clarity. μ_θ (green line) $= 1$, $\varepsilon_z = 3.423$. The SRRs of cylinder 1 (inner) and cylinder 10 (outer) are shown in expanded schematic form. From Schurig *et al.* (2006). Copyright © 2006 AAAS

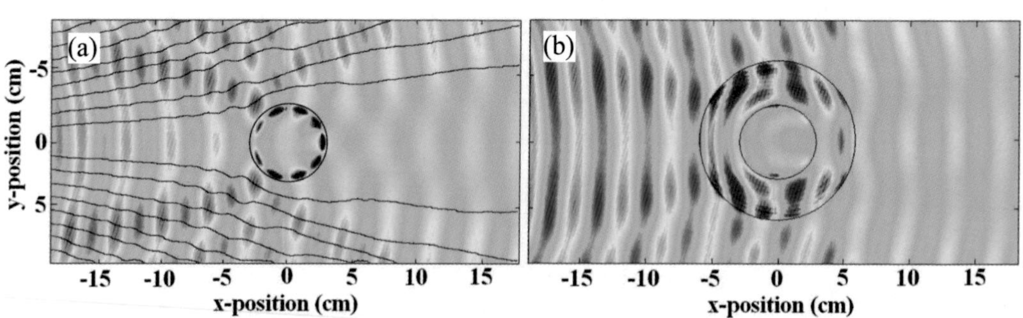

Fig. 9.25 Experimental field distribution for a copper cylinder (to be hidden) and a ten-layer cloak made up by resonating elements. (a) In the absence and (b) in the presence of the cloak. From Schurig *et al.* (2006). Copyright © 2006 AAAS

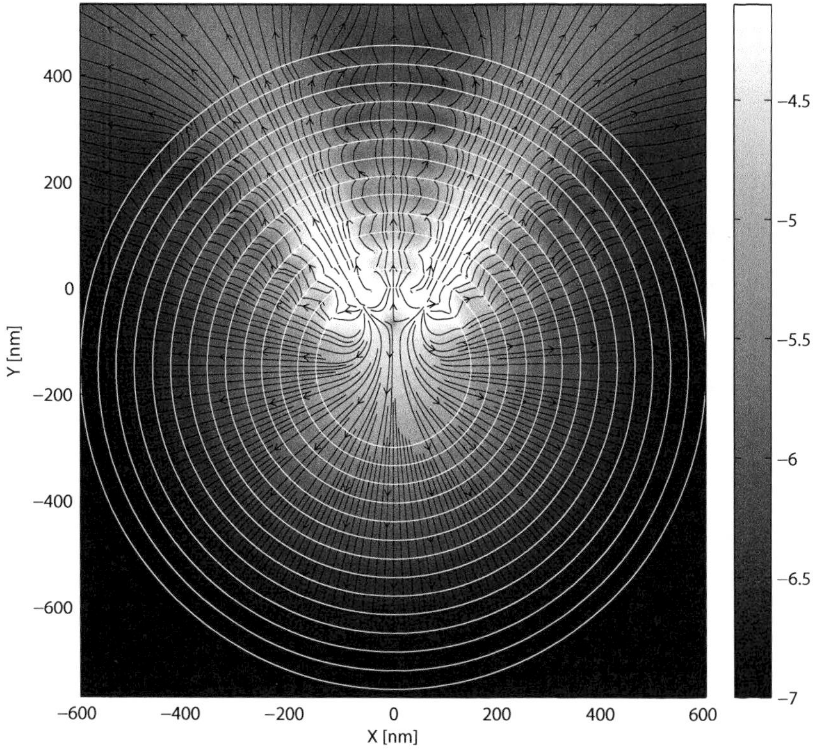

Fig. 5.32 Magnifying superlens. Numerical simulation of Poynting vector streamlines emanating from two subwavelength objects. $\varepsilon_r = -1 - j\,0.4$. From Tatartschuk (2008). For coloured version see plate section

that solution, however, the permittivity and permeability had to change as a function of space in order to provide a perfect image.

The possibility of using a cylindrical multilayer structure for magnifying the image was proposed by Salandrino and Engheta (2006). A theory of the magnification process for a cylindrical multilayer lens in terms of higher-order scattering of cylindrical waves was formulated by Jacob *et al.* (2006). They also found the 2D dispersion equation for the cylindrical geometry. The isofrequency curves, plotted in the k_r (wave vector in the radial direction), k_θ (wave vector in the azimuthal direction) plane, turn out to be hyperbolas. We may remember from Section 5.5 that a planar multilayer material in the effective-medium approximation has diagonal elements of the opposite sign in the permittivity tensor, and that leads to a hyperbolic dispersion equation. It is not unreasonable to expect that the same kind of relationship is valid in a multilayer cylindrical structure as well. See the analysis by Jacob *et al.* (2006) for the cylindrical case, and also Sections 8.1.2 and 9.4 for similar dispersion curves.

A practical realization of the magnifying lens with subwavelength resolution has been achieved by Liu *et al.* (2007*b*). The lens consisted

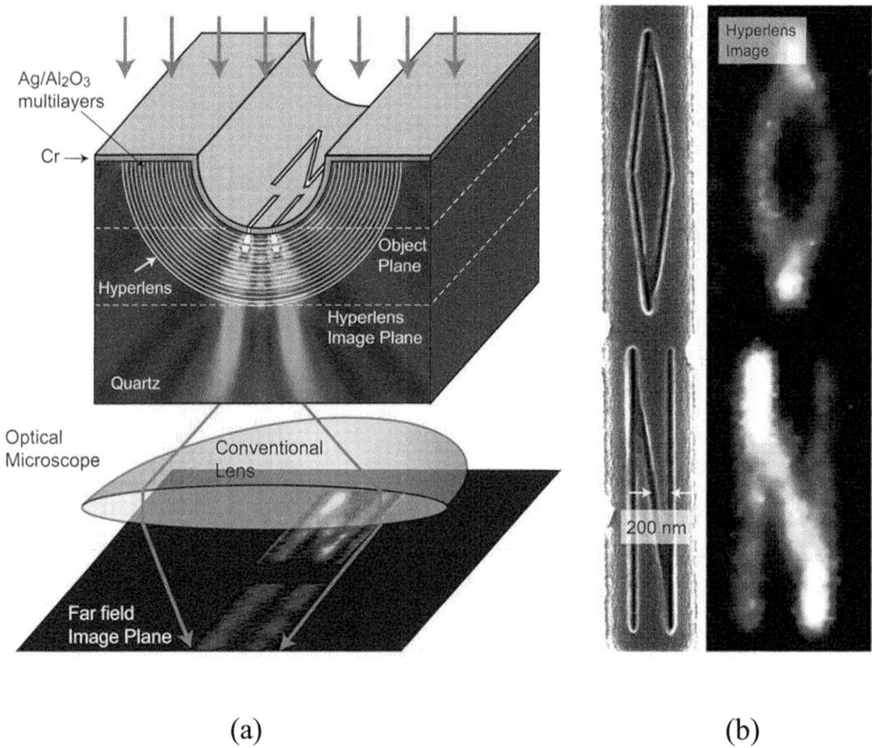

Fig. 5.33 (a) Schematic of magnifying optical hyperlens and numerical simulation of imaging of subdiffraction-limited objects. (b) An arbitrary object 'ON' imaged with subdiffraction resolution. Line width of the object is about 40 nm. The hyperlens is made of 16 layers of Ag-Al_2O_3. From Liu *et al.* (2007*b*). Courtesy of Prof. Xiang Zhang of University of California at Berkeley. Copyright © 2007 AAAS. For coloured version see plate section

of alternate cylindrical layers of Ag and Al_2O_3 deposited on a half-cylindrical cavity (see Fig. 5.33). There were 16 layers of both materials with thicknesses of 35 nm each. Finally, a 50-nm thick chromium layer is deposited upon the last layer of the lens. The object is the two letters ON inscribed in the chromium layer. The smallest feature, the linewidth, is 40 nm and the lines are 150 nm apart. In the magnified image that spacing becomes 350 nm. The wave illuminating the object has a wavelength of 356 nm. It is the wavelength where the dielectric constant of silver[15] is equal to -1. The spacing of 350 nm is close to the wavelength of the incident wave, hence the output image can be magnified by a conventional microscope. This was indeed the aim of the authors. The hope is that lenses of this type will be available in the future with higher initial magnification making it possible to view subwavelength objects by a conventional microscope. The main limitation is that the object has to be very close to the first layer of the lens. A generalization may also be possible. One may argue that if such a magnifying system can be realized by a cylindrical lens then a spherical lens should also be within the realm of possibilities, although in that case the polarization of the

[15] See discussion in Section 3.3.

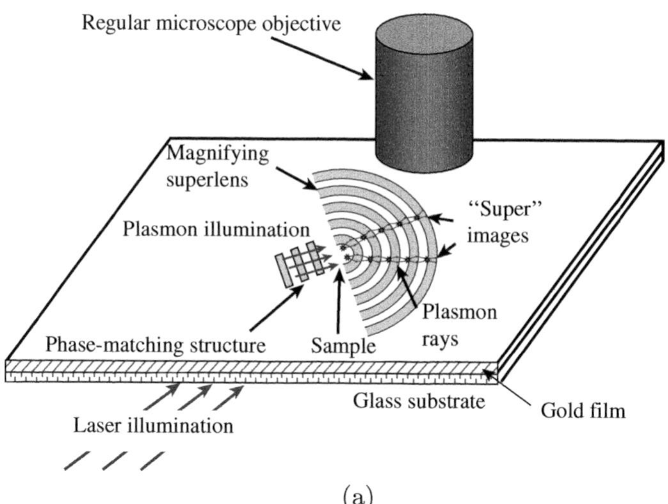

(a)

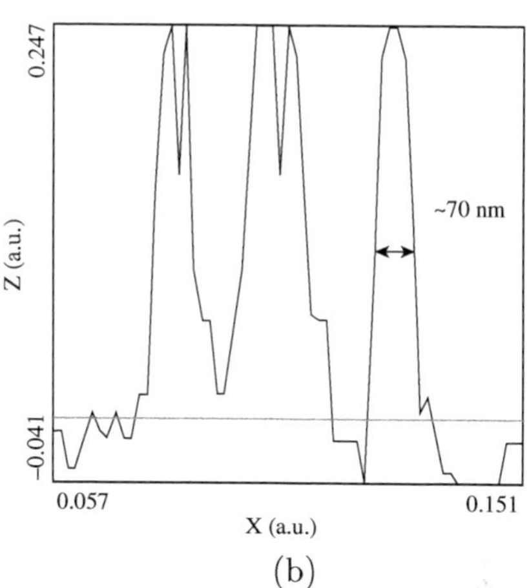

(b)

Fig. 5.34 (a) Schematic of the magnifying superlens integrated into a conventional microscope. The plasmons generated by the phase-matching structure illuminate the sample positioned near the centre of the superlens. The lateral distance between the images produced by the alternating layers of materials with positive and negative refractive index grows with distance along the radius. The magnified images are viewed by a regular microscope. (b) The cross-section of the optical image indicates resolution of at least 70 nm or $\sim \lambda/7$. Z, optical signal; X, distance. From Smolyaninov *et al.* (2007). Copyright © 2007 AAAS

wave scattered by the object might adversely influence the fidelity of reproduction.

Another realization of image magnification by a multilayer cylindrical lens is due to Smolyaninov *et al.* (2007). Their experimental setup is shown in Fig. 5.34(a). The lens consists of cylindrical annuli of PMMA deposited on a gold film. The wave propagating in the lens is a SPP excited by a laser at a wavelength of 495 nm. In what sense is this a lens? The principle is still the same as in the previous example in which layers of positive and negative ε alternated. Considering the properties of SPPs it is possible to assign to them an index of refraction on the basis that by how much is the velocity of the SPP reduced. We have two regions: (i) the gold film on its own, and (ii) the gold film covered by PMMA. For the gold film on its own the SPP resonance is above the frequency of excitation. The SPP is a forward wave with a positive index of refraction, n_1. For the PMMA-covered gold film the SPP resonance[16] is below the frequency of excitation hence the wave propagating in this region is a backward wave to which it is possible to assign a negative index of refraction, n_2. Now the condition for imaging action is that

$$n_1 d_1 = -n_2 d_2 \,, \tag{5.37}$$

where d_1 and d_2 are the widths of regions 1 and 2. The parameters were so chosen by the authors that the above relationship is satisfied at the wavelength of excitation.

The objects were 2 or 3 rows of PMMA dots in the radial direction positioned near the centre of the lens. For the three-row case the distance

[16] Remember the SPP resonant frequency is $\omega_{\rm s} = \omega_{\rm p}/\sqrt{1+\varepsilon_{\rm d}}$, where $\varepsilon_{\rm d}$ is the relative dielectric constant of the adjoining dielectric.

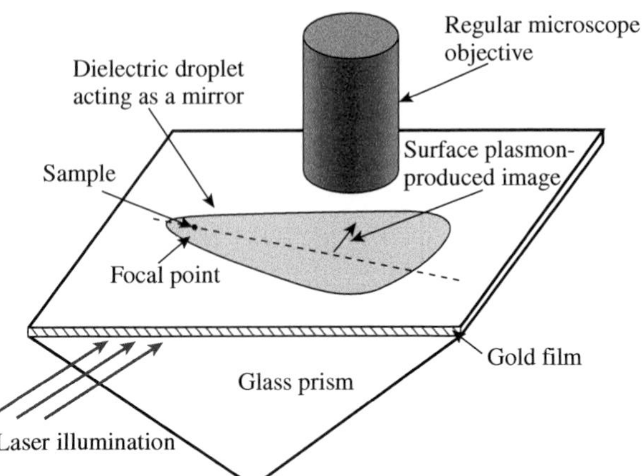

Fig. 5.35 Magnifying superlens. From Smolyaninov *et al.* (2005). Copyright © 2005 by the American Physical Society

between them at the input of the lens was 70 nm. Upon illumination they gave rise to three divergent SPP 'rays'. The field distribution at the output of the lens was then large enough (10 times as large as at the object) to be further magnified by a conventional microscope. The microscope output showing three distinct peaks is plotted in Fig. 5.34(b).

An entirely different approach to image magnification was proposed by Smolyaninov *et al.* (2005) based on earlier work (Smolyaninov, 2003; Smolyaninov and Davis, 2004). The essential ingredient is a parabolic-shaped glycerine droplet on a metal film as shown in Fig. 5.35. The waves involved are SPPs excited by laser light. In the experiments, light of 502 nm wavelength was used at which the SPP wavelength is about 70 nm giving an effective index of refraction of about 7. Imaging is due to reflection of the SPPs by the parabolic boundary. An object placed in the vicinity of the focal point will have an image as shown in Fig. 5.35. The authors successfully imaged arrays of nanometric pinholes in the gold film. The field distribution is then viewed from above by a conventional microscope.

5.7 Misconceptions

Whatever our chosen field is we like to understand it in simple qualitative terms. What can we say about the superlens? It gives a perfect image but a perfect image cannot be obtained in practice due to losses, imperfections, time delay and the granular nature of the metamaterial. It is easy to understand the consequence of granular structure. Surely, we cannot obtain a resolution that is better than the period in the metamaterial. The influence of other effects is less obvious. Some of the relationships are quite complicated and superficial references to them may not be adequate. The misconceptions, quoted below, may only be entertained by a small minority but we think it is worth including them in this section.

Let us first look at the superlens with $\varepsilon_r = \mu_r = -1$.

Believed by many: The physical picture of imaging shown, e.g., in Fig. 2.38(a) with the aid of geometrical optics is still approximately true in the presence of evanescent waves.

Not true, not even approximately. There is no internal focus when the object is subwavelength. The Poynting vector streamlines converge upon the outer boundary of the lens.

Believed by many: The superlens not only images but focuses as well.

A focus means that the intensity is high in a small region around it, and in whichever direction we move away from the focus the intensity declines. This does not occur in the superlens. If, for example, we measure the intensity along the axis beyond the outer boundary of the lens it turns out to be a monotonically declining function. It has no maximum. It has no focal point.

Believed by many: Evanescent waves do not carry power.

Indeed, they do not carry power in the lossless case but they do in the presence of losses because there is then a phase difference between the coefficients of the various evanescent waves.

Believed by many: The presence of evanescent waves does not influence the flow of power.

Not true. The streamlines of the Poynting vector are quite different when the evanescent waves are taken into account. There are then vortices that may be closed inside the negative-index material but they may also extend into the neighbouring positive-index medium.

Beliefs that would not stand closer scrutiny extend also to the silver lens, i.e. when only the relative dielectric constant is equal to minus unity and the relative permeability remains at plus unity.

Believed by many: Resolution of the silver superlens is merely restricted by losses. The image is perfect without loss.

Not true. There is a cutoff spatial frequency even in the absence of losses.

Believed by many: The smaller the loss in a silver lens the more faithfully is the image reproduced.

Not true. If losses are absent or too small the transfer function exhibits resonances due to the excitation of the SPPs. If the amplitude of the resonance peak is large, then the image is not good at all: it shows only the oscillatory field distribution of the corresponding SPP mode.

Believed by many: The electrostatic approximation is good if losses are small.

Not true. The SPP resonances need to be damped for the electrostatic approximation to apply. Hence, losses have to be sufficiently large.

6 Phenomena in waveguides

6.1 Introduction

The term waveguide is often used in a general sense meaning a structure that guides a wave from point A to point B. Some waveguides are open; the fields may extend to a considerable distance away or they may be closed, confining the fields to the interior of the waveguide. Both varieties will be discussed in the present chapter. For experiments on metamaterials it is an advantage to confine the waves. This was the reason, for example, why in the first experiments showing negative refraction (Shelby *et al.*, 2001*a*), the waves propagated between metal plates (see Fig. 2.34).

The waveguides discussed fall into two categories: they have either some special features or have device applications. In Section 6.2 the special feature is that under certain well-defined circumstances waves may propagate in a hollow waveguide (see Section 1.5) below their cutoff frequency. In Section 6.3 we shall present filters using metamaterial elements in coplanar and strip waveguides. Section 6.4 is concerned with phase shifters, Section 6.5 with couplers. In Section 6.6 we look in some more detail at the 2D experiments of Grbic and Eleftheriades (2004) where microstrip waveguides loaded by lumped elements are used for demonstrating subwavelength imaging.

6.2 Propagation in cutoff waveguides

We have learned in Section 1.5 that electromagnetic waves can propagate in a hollow metallic waveguide, provided the dimensions of the waveguide are large enough. The question may then be asked: how will the waves be affected if we insert in the waveguide an array of small resonators? In the metamaterial context the first experiments were conducted by Marques *et al.* (2002*a*). The experimental setup is shown in Figs. 6.1(a) and (b). The waveguide has a square cross-section of $a \times a$ where $a = 6$ mm. The corresponding cutoff frequency (see eqn (1.38)) is 25 GHz. The SRRs inserted have diameters of 5.6 mm and a resonant frequency of about 6 GHz. With a distance of a between them they nearly touch each other. The input and output are via standard coaxial to rectangular waveguide junctions.

The experimental results for a waveguide of 36 mm length housing nine elements are shown in Fig. 6.2 where transmission between input and output is plotted against frequency. It may be clearly seen that

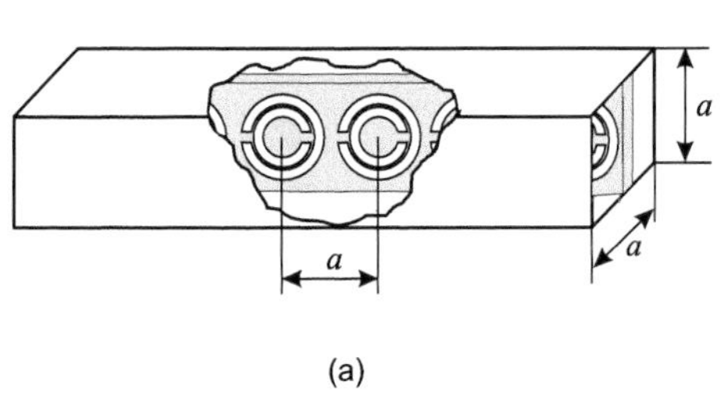

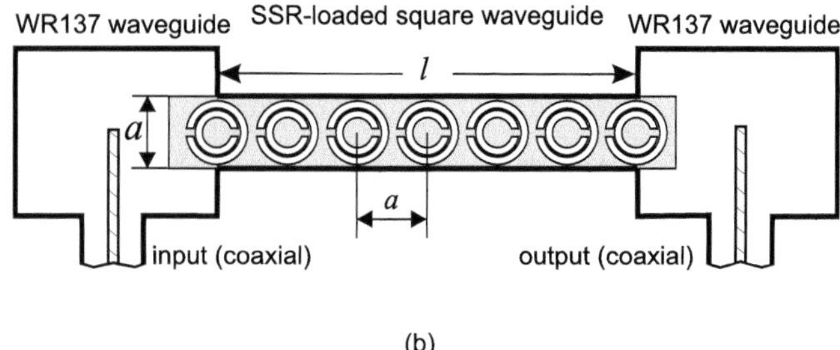

Fig. 6.1 (a) The SRR-loaded square waveguide. (b) Sketch of the experimental setup. From Marques *et al.* (2002*a*). Copyright © 2002 by the American Physical Society

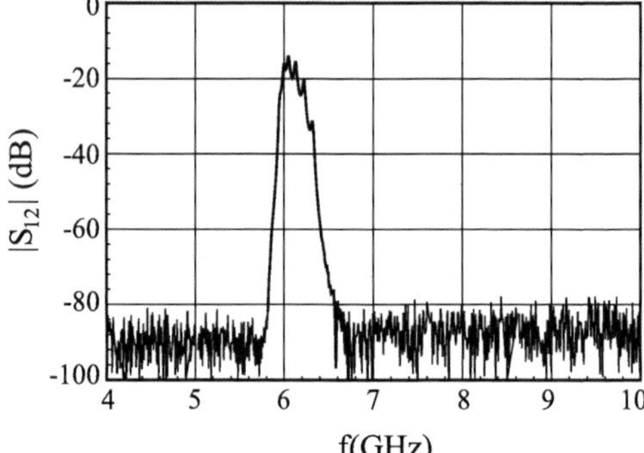

Fig. 6.2 Measured transmission coefficient for l = 36 mm. From Marques *et al.* (2002*a*). Copyright © 2002 by the American Physical Society

there is a transmission peak in the frequency region between about 5.8 and 6.5 GHz. This is a remarkable result. In this region the waveguide is well below cutoff and yet there is transmission through it owing to the effect of the SRRs. The pass band is more than 60 dB above the noise level that characterizes the cutoff waveguide.

The experiments were repeated by Hrabar *et al.* (2005), also using square waveguides and SRRs with a resonant frequency of 7.8 GHz. Their experimental results are shown in Figs. 6.3(a) and (b), where the scattering coefficient S_{21} (as discussed in Section 1.24, $|S_{21}|^2$ gives the output power relative to the input power) is plotted against frequency. The results for waveguide A (dimensions, a = 35 mm and b = 15 mm, cutoff frequency 4.3 GHz) are shown in Fig. 6.3(a). In that case, the cutoff frequency is below the element's resonant frequency and the presence of the SRRs appears in a stop band. Note that the SRRs also influence the cutoff frequency, which moved towards higher frequencies. Waveguide B has a cutoff frequency of about 12.5 GHz, well above the resonant frequency of the element. Now, the presence of the SRRs appears again in an upward shift of the cutoff frequency, and, more significantly, in a pass band, as shown in Fig. 6.3(b); the same result as obtained by Marques *et al.* (2002*a*).

The theoretical description of these phenomena is quite straightforward, provided we realize that the effective permeability may be a tensor (see Section 1.15). For a SRR this is discussed in Section 2.8 where the

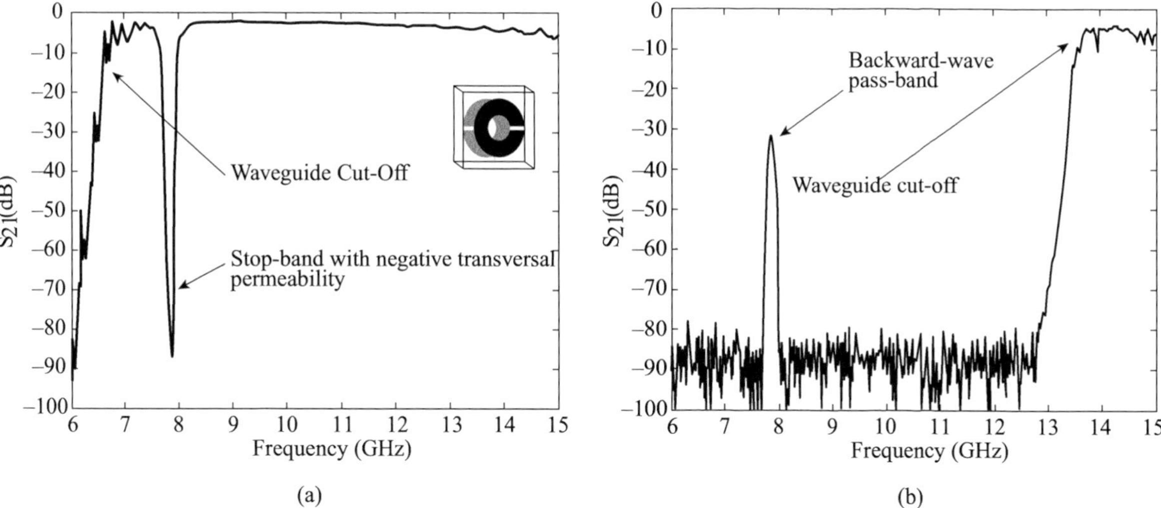

Fig. 6.3 Measured transmission coefficient. There is (a) a stop band when the cutoff frequency of the waveguide is below the elements' resonant frequency; (b) a pass band when the cutoff frequency of the waveguide is above the elements' resonant frequency. From Hrabar *et al.* (2005). Copyright © 2005 IEEE

effective permeability is determined in the direction perpendicular to the plane[1] of the SRR.

The corresponding theory (Kondrat'ev and Smirnov, 2003; Hrabar *et al.*, 2005) follows from that introduced in Section 1.5 considering further that the relative magnetic permeability is a tensor (see eqn (1.97)), which may now be written as

$$\mu = \mu_0 \begin{pmatrix} 1 & 0 & 0 \\ 0 & \mu_{yy} & 0 \\ 0 & 0 & 1 \end{pmatrix}, \tag{6.1}$$

where μ_{yy} is the relative permeability in the y direction. Equations (1.32)–(1.34) may now be rewritten as follows

$$\frac{\partial H_z}{\partial y} - \frac{\partial H_y}{\partial z} = \mathrm{j}\,\omega\varepsilon_0 E_x\,, \tag{6.2}$$

$$\frac{\partial E_x}{\partial z} = -\mathrm{j}\,\omega\mu_0\mu_{yy} H_y\,, \tag{6.3}$$

$$\frac{\partial E_x}{\partial y} = \mathrm{j}\,\omega\mu_0 H_z\,. \tag{6.4}$$

The above set of differential equations are still linear with constant coefficients, hence they can be solved without difficulty. Using again the assumption of a wave solution and satisfying the boundary conditions on the waveguide wall we end up with the relationship

$$k_z^2 = \mu_{yy}\left[k_0^2 - \left(\frac{\pi}{a}\right)^2\right]. \tag{6.5}$$

[1] Unfortunately, it happens sometimes that customary notations clash and therefore at least one of them has to be modified. The direction of anisotropy for an SRR is usually taken in the z direction, which is the notation used in Section 2.8. We have also chosen the z axis as the direction of wave propagation. In order to resolve the conflict we take in this chapter the y direction as perpendicular to the plane of SRRs. Hence, the anisotropic term in eqn (6.1) is referred to as μ_{yy}.

This equation differs from that of eqn (1.37) by having μ_{yy} on the right-hand side. But that makes all the difference. We can now explain the experimental results shown in Figs. 6.2 and 6.3. Above the cutoff frequency $k^2 - (\pi/a)^2 > 0$, hence μ_{yy} being negative will result in k_z^2 being negative and, consequently, in a stop band. Below the cutoff frequency $k^2 - (\pi/a)^2 < 0$, hence μ_{yy} being negative results in k_z^2 being positive, and, consequently, in a pass band.

The original explanation provided by Marques *et al.* (2002*a*) is somewhat different. They quoted Rotman (1962) who simulated plasmas not only by metallic rods (discussed in Section 2.4) but also by cutoff waveguides. The claim was that a cutoff waveguide behaves analogously to a plasma below its plasma frequency. Hence, Marques *et al.* argued, if a set of elements, capable of producing negative permeability, is inserted into a cutoff waveguide that should lead to propagation, the situation being equivalent to a medium having both negative ε and negative μ. This argument can be made quantitative by assuming, with Rotman, that a cutoff waveguide could be regarded as having an effective relative dielectric constant of

$$\varepsilon_{\mathrm{r\,eff}} = 1 - \left(\frac{\lambda}{2a}\right)^2 = 1 - \left(\frac{\omega_{\mathrm{p\,eff}}}{\omega}\right)^2, \tag{6.6}$$

where

$$\omega_{\mathrm{p\,eff}} = \frac{c\pi}{a} \tag{6.7}$$

is the effective plasma frequency.

In general, the propagation coefficient of an electromagnetic wave in a homogeneous medium can be written as

$$k_z^2 = \mu_{\mathrm{r}}\varepsilon_{\mathrm{r}}\frac{\omega^2}{c^2}, \tag{6.8}$$

where the wave happens to propagate in the z direction.

Using the definition of the effective relative dielectric constant in the form of eqn (6.6) we find that

$$k_z^2 = k_0^2\mu_{\mathrm{r}}\left[1 - \left(\frac{\lambda}{2a}\right)^2\right] = \mu_{\mathrm{r}}\left[k_0^2 - \left(\frac{\pi}{a}\right)^2\right], \tag{6.9}$$

which is of the same form as eqn (6.5). Thus, we may conclude that the heuristic derivation in terms of the effective dielectric constant leads to the same result as the one based on the field equations. The advantage of the heuristic derivation is that it links the phenomena observed in waveguides to those in left-handed materials allowing the possibility that both ε and μ may be negative.

Note, however, that eqn (6.9) does not represent the whole truth. It has μ_{r} instead of μ_{yy}. Looking at eqn (6.9) one might come to the conclusion that propagation is possible in a cutoff waveguide filled with an isotropic negative-permeability material. And that would be the wrong conclusion. Equation (6.5) suggests that in order to have propagation

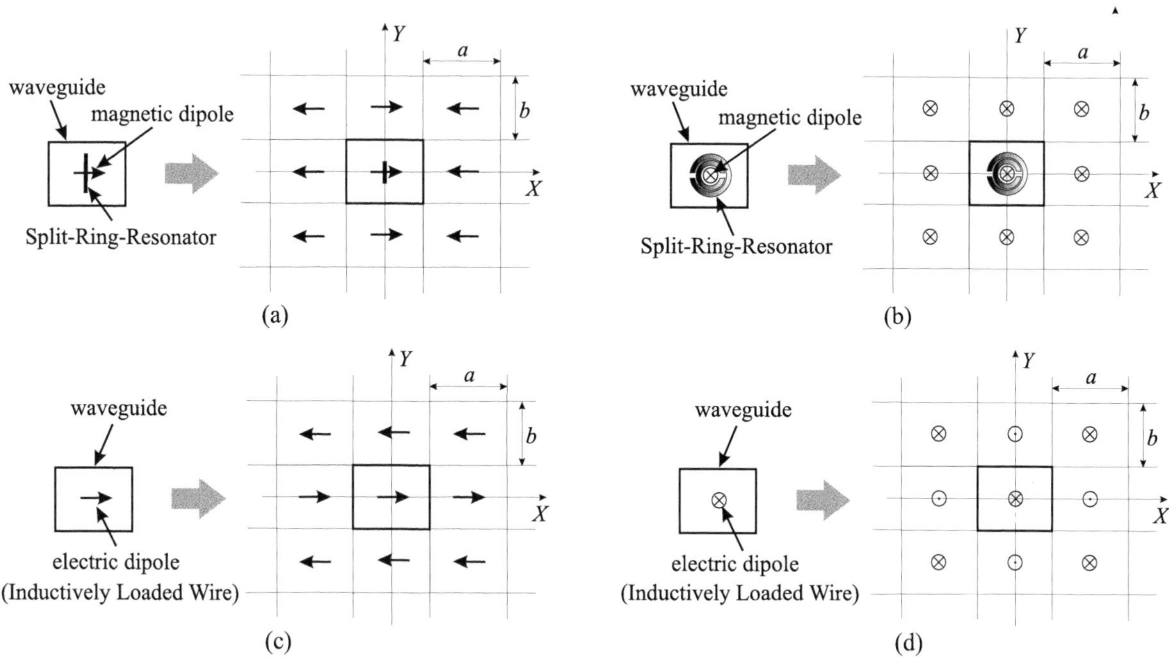

Fig. 6.4 Transformation of the waveguide problem using the image principle. From Belov and Simovski (2005*b*). Copyright © 2005 by the American Physical Society

we need specifically a negative *transverse* permeability. For that reason it is better to rely on the derivation offered at the beginning of this section.

It has been argued by Belov and Simovski (2005*b*) that neither of the previously offered explanations represent the whole truth. Their model is quite different. First, they resort to a simplification. Instead of talking about actual resonant elements, e.g. SRRs or some other varieties, they assume that the waveguide is filled with resonant arrays of either magnetic dipoles or short electric dipoles. The latter are made resonant by loading them with inductances (their properties were investigated in detail by Tretyakov *et al.* (2003)). Both types of dipoles can have polarizations either in the axial or in the transverse directions. The model is based on mirror images in the waveguide walls as may be seen in Figs. 6.4(a)–(d). Thus, instead of a waveguide problem they solve the problem of propagation in an infinite three-dimensional lattice. They derive the relevant dispersion relationships, i.e. the variation of wave number against frequency. It is essentially the same kind of approach as that of Shamonina *et al.* (2002*b*) for MI waves in 3D that was based on nearest-neighbour interaction. However, the treatment of Belov and Simovski is much more rigorous. They include interaction between all elements, and also include radiation effects, i.e. take into account the dipole fields varying with the inverse of the distance (see Section 1.12). Their conclusion is that all four configurations may lead to pass bands in cutoff waveguides.

It should be mentioned that cutoff waveguides that can propagate electromagnetic waves have potential for applications because their dimensions may be much smaller than waveguides working above cutoff. However, devices in cutoff waveguides are in no sense new. In the 1960s and 1970s it was a hot subject. The research resulted in a number of microwave devices: not only broadband filters for which it is most suitable but also circulators, frequency mixers, multipliers, phase shifters, etc. The main aim was to reduce weight and, often, trade weight for loss. For a review see Craven (1972).

Finally, we need to face the undisputable truth that the explanation of propagation in a cutoff waveguide in terms of negative transverse permeability is not new either. The same explanation was suggested by Thompson (1955) whose paper in Nature covered both theory and experiment. The experiments were conducted in a circular metallic waveguide loaded by a longitudinally biased ferrite rod. The phenomenon was further investigated by Thompson (1963).

Hollow circular waveguides with perfectly conducting walls and loaded along their axes by a periodic array of thin dielectric ($\varepsilon > 0$) and metallic ($\varepsilon < 0$) layers, were investigated by Govyadinov and Podolskiy (2006) in the infra-red region. Note that the anisotropy is this time in the axial direction. The authors showed that it was possible to obtain both positive and negative effective indices of refraction. When $n < 0$ then propagation is only possible when the waveguide radius is sufficiently small. Hence, if the radius of the waveguide (with the load inside) is gradually reduced waves can still propagate and the result is high power compression.

6.3 Filters in coplanar and microstrip waveguides

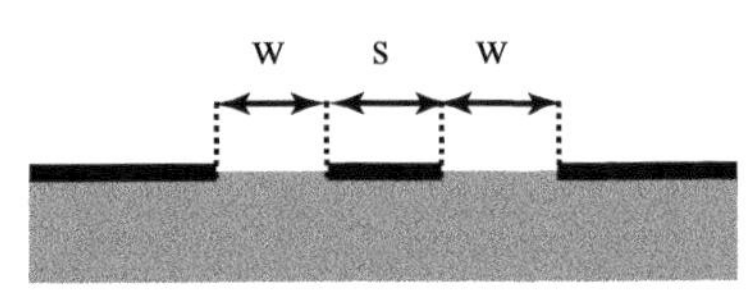

Fig. 6.5 Coplanar waveguide geometry

We have discussed a number of very interesting phenomena in hollow metallic waveguides in the previous section. When it comes to cutoff waveguides the size is certainly smaller but unlikely to be small enough for practical applications under present conditions when one of the requirements is compatibility with planar technology. A waveguide particularly suitable for this purpose is the coplanar waveguide shown schematically in Fig. 6.5. It consists of a centre conductor (signal strip) and ground planes on both sides. Loading this waveguide by series capacitors and shunt inductors Grbic and Eleftheriades (2002) realized a backward-wave line that, they showed, could radiate in the reverse direction.

A series of studies concerned with microwave filters were carried out by a team from three Spanish Universities, those of Navarra, Barcelona and Sevilla (Martin *et al.*, 2003*a*; Martin *et al.*, 2003*b*; Falcone *et al.*, 2004*a*; Falcone *et al.*, 2004*b*; Falcone *et al.*, 2004*c*; Falcone *et al.*, 2004*d*; Martel *et al.*, 2004; Garcia-Garcia *et al.*, 2004; Baena *et al.*, 2005*a*). They

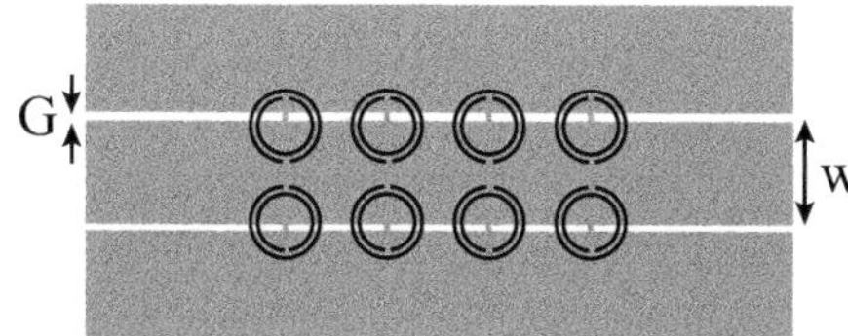

Fig. 6.6 Layout of the SRR loaded coplanar waveguide. From Martin *et al.* (2003*a*). Copyright © 2003 American Institute of Physics

(a)

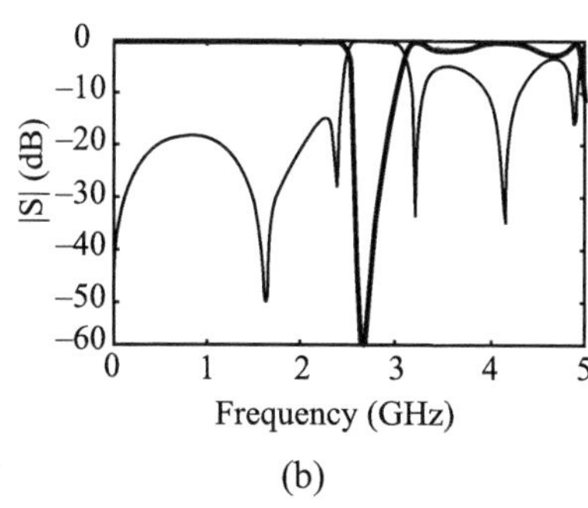

(b)

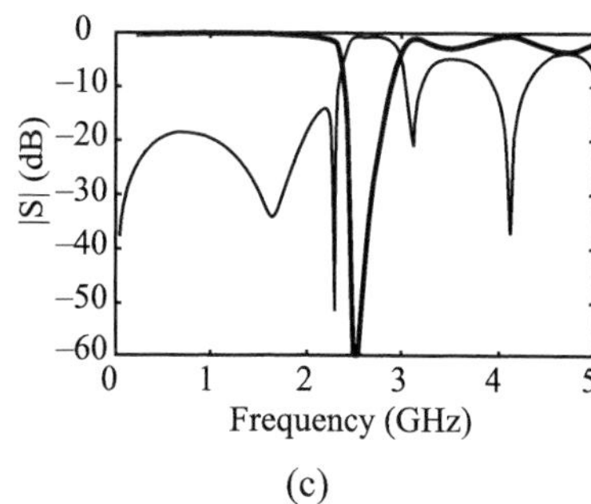

(c)

Fig. 6.7 Photograph of a four-element filter based on complementary SRRs. (b) Simulated and (c) measured S_{11} (thin line) and S_{21} (thick line) coefficients. From Falcone *et al.* (2004*a*). Copyright © 2004 IEEE

loaded transmission lines (mainly in the form of coplanar waveguides) by SRRs and rods and relied on the previously established experimental results that SRRs on their own provided a stop band and the stop band turned into a pass band when short metallic rods were added to the structure. In one of the realizations (Martin *et al.*, 2003*a*) SRRs are placed symmetrically upon the reverse side of the substrate of a coplanar waveguide, as shown in Fig. 6.6. Note further that at the centre of the SRRs a thin metal wire connects the signal strip to ground. In this configuration the magnetic field is perpendicular to the plane of the SRRs and the electric field is between the signal and ground conductors. In the experiments both transmission and reflection were measured. Similarly to the experiments of Shelby *et al.* (2001*a*) they found that in the range around 8 GHz, which corresponds to the resonant frequency of the SRRs, there is a pass band when the wires are present, and a stop band when the wires are removed.

In the study by Falcone *et al.* (2004*a*) complementary SRRs (see Section 4.3) were used in producing a four-element filter in a microstrip waveguide (Fig. 6.7(a)). Their simulation and experimental results for reflection (S_{11}) and transmission (S_{21}) are remarkably close to each other, as may be seen in Figs. 6.7(b) and (c). In another similar development Falcone *et al.* (2004*b*) used spiral resonators (see Section 4.3) instead of SRRs. Their advantage is that for the same resonant frequency they have a smaller size. We should also note that in order to achieve a wider bandwidth Martin *et al.* (2003*a*) used SRRs tuned to slightly different frequencies.

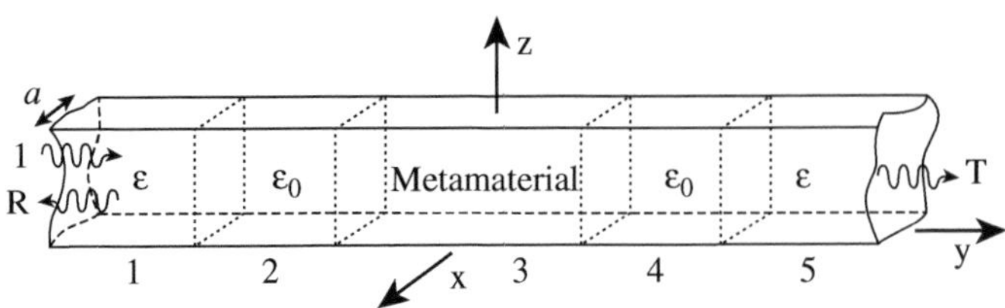

Fig. 6.8 A metamaterial-filled square waveguide sandwiched between two empty waveguides operating below cut-off and two input/output waveguides above cutoff. The length of each waveguide section is l_i. One side of the waveguide is $a = 24$ mm. From Baena *et al.* (2005*c*). Copyright © 2005 by the American Physical Society

6.4 Tunnelling

Tunnelling is a well-known phenomenon in a cutoff waveguide. It means that in spite of the wave being evanescent a certain proportion of the input power can get through to the output. As we know, there can be a growing wave in a negative-index material. Would the same thing apply in a cutoff waveguide? The analysis was done by Baena *et al.* (2005*c*). A schematic representation of the waveguide is shown in Fig. 6.8. As may be seen, there are 5 sections. The waveguide dimensions are such that in the empty sections, 2 and 4, the wave cannot propagate. In sections 1 and 5, both taken as infinitely long, the waveguide is above cutoff by choosing ε to be high enough. In section 3 there is a negative-index material. The length of each section is l_i. The electromagnetic wave is incident from section 1 in the TE_{10} mode. Since the size of the waveguide is the same everywhere the TE_{10} mode remains unchanged in all the other sections.

It is shown that in the lossless case for isotropic materials the conditions for perfect tunnelling (100% transmission) are as follows:

$$l_2\alpha_2 + l_3\alpha_3 + l_4\alpha_4 = 0 \tag{6.10}$$

and

$$Z_2 = Z_3 = Z_4\,, \tag{6.11}$$

where Z_i is the impedance of the ith section and α_i is the attenuation coefficient. This is of course attenuation under cutoff conditions. It does not imply any ohmic loss. In fact, in order to satisfy eqn (6.10), α_3 must be negative, which means that the wave is growing in section 3, and

$$l_3 = l_1 + l_2\,. \tag{6.12}$$

To satisfy eqn (6.11) we have to choose the materials constants in section 3 as

$$\varepsilon_3 = -\varepsilon_0 \quad \text{and} \quad \mu_3 = -\mu_0\,. \tag{6.13}$$

It may now be noticed that the conditions for perfect tunnelling are the same as for perfect imaging.

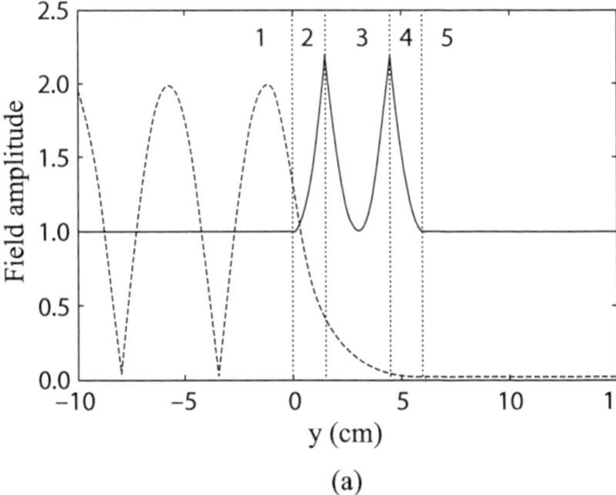

(a)

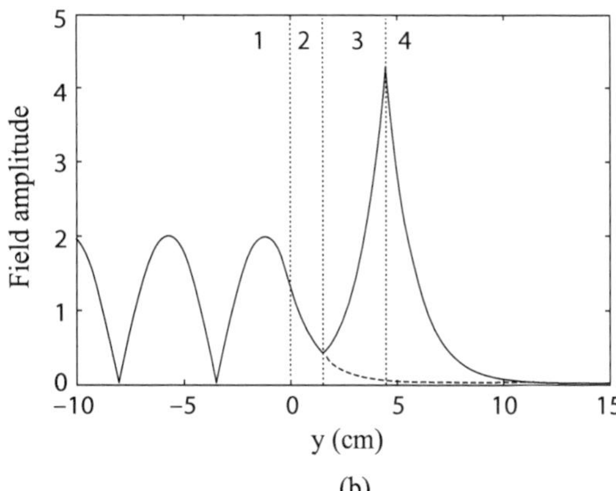

(b)

Fig. 6.9 Field distribution inside the structure of Fig. 6.8 when section 3 is filled by an isotropic metamaterial with $\varepsilon_r = \mu_r = -1$. (a) Perfect tunnelling when $l_3 = 2l_2 = 2l_4$. (b) Field distribution when $l_4 \to \infty$. Dashed lines show the field amplitude when section 3 is empty. From Baena *et al.* (2005*c*). Copyright © 2005 by the American Physical Society

The distribution of the electric field in the waveguide under these conditions was calculated by Baena *et al.* (2005*c*). The geometrical dimensions are $a = 24$ mm, $l_3 = 30$ mm and the frequency is 5 GHz. We have the conventional tunnelling situation when $\varepsilon_2 = \varepsilon_3 = \varepsilon_4$. Then, there is a standing-wave pattern in section 1, and the wave declines beyond that, as shown in Fig. 6.9(a) (dashed line). Under optimum conditions (continuous line) there is now no standing-wave pattern and the input amplitude may be seen to be the same as the output amplitude, the sign of perfect tunnelling. Any deviation from the optimum conditions leads to rapid degradation of the transmitted power. In Fig. 6.9(b) the field distribution is shown for two different conditions: when section 3 is empty (dashed line) and when section 4 extends to infinity (continuous line). As may be expected, there is an exponential decay in both cases, but in the latter one the growing wave in section 3 is still there.

A somewhat analogous situation was investigated by Alu and Engheta (2003) in free space for a periodic array of ε-negative materials paired with μ-negative materials. Among others they found conditions for perfect tunnelling.

Perfect tunnelling in free space through a negative-positive-negative permittivity medium was investigated by Zhou *et al.* (2005*b*) both experimentally and by simulation. They found that perfect tunnelling occurs at two frequencies at which there is an increase in the magnetic field. They showed also that the perfect tunnelling effect is insensitive to the input angle of the incident wave.

A related problem, perfect tunnelling under conditions of frustrated total internal reflection for pairs of metamaterials was investigated by Zhou and Hu (2007).

6.5 Phase shifters

A transmission line loaded by a shunt inductance and a series capacitor was discussed in Section 2.13 and is shown schematically in Fig. 2.39(a). The circuits were considered for applications as phase shifters by Antoniades and Eleftheriades (2003) in the symmetric form shown in Fig. 6.10. The corresponding dispersion curve was already plotted in Fig. 2.39(b), exhibiting a stop band between ω_1 and ω_2. In order to address phase shifts around zero the stop band needs to be eliminated. This occurs when the shunt circuit and the parallel circuit have the same resonant frequencies. The phase shift per unit cell is then given by

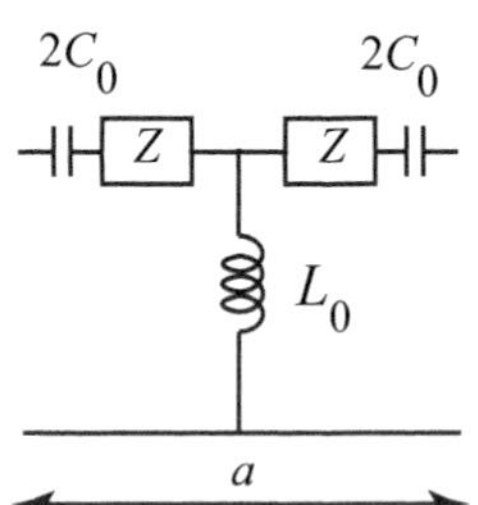

Fig. 6.10 1D phase shifter unit cell

$$ka = \omega\sqrt{L_\mathrm{t} a C_\mathrm{s}} - \frac{1}{\omega}\sqrt{L_\mathrm{sh} C_\mathrm{t} a}\,, \tag{6.14}$$

i.e. a backward phase shift is subtracted from a forward phase shift.

Such phase shifters consisting of two and four stages were realized by Antoniades and Eleftheriades (2003) using coplanar waveguides. The advantage relative to transmission-line phase shifters is that phase shifts can be not only positive but also negative or zero. This is a major advantage if small phase shifts around zero are required. Moreover, the phase incurred is independent of the length of the structure. For further work on phase shifters see Islam and Eleftheriades 2004; Antoniades and Eleftheriades 2005; Abdalla *et al.* 2005; Eleftheriades and Islam 2007; Eleftheriades and Balmain 2005; Caloz and Itoh 2006.

There is further discussion of the properties of phase shifters using backward waves by Nefedov and Tretyakov (2005). They point out that a metamaterial phase shifter of $-10°$ is bound to have a broader band than a transmission-line phase shifter of 350° on account of its shorter length. However, they prove that any combination of forward- and backward-wave sections that produce a positive phase shift is less broadband than a transmission-line phase shifter producing the same phase shift. They further show that dispersion can be reduced by lines exhibiting positive anomalous dispersion (see Section 1.18).

This is probably the best place to introduce the proposal by Engheta (2002) for a thin subwavelength cavity resonator. It consists of a cavity into which two materials of thicknesses d_1 and d_2 are placed side-by-side. One of the materials has a conventional positive index and the other one a negative index. It is shown there that the negative-index material in the cavity may serve as a phase compensator, leading to a resonant frequency that depends not on the sum of the thicknesses but on their ratio. Thus, in principle, the cavity can be arbitrarily small. It is further shown that under certain approximations the ratio of the two thicknesses is independent of the dielectric constants and depends only on the permeabilities as

$$\frac{d_1}{d_2} = \left|\frac{\mu_2}{\mu_1}\right|\,. \tag{6.15}$$

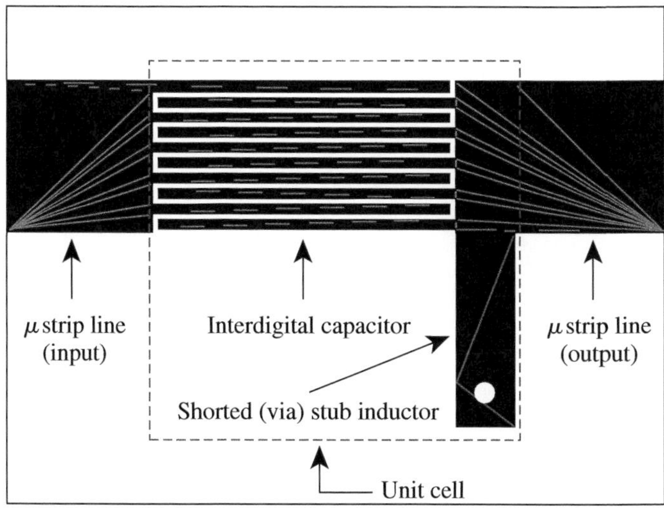

Fig. 6.11 Layout of the unit cell of the microstrip left-handed transmission line. From Caloz *et al.* (2004). Copyright © 2004 IEEE

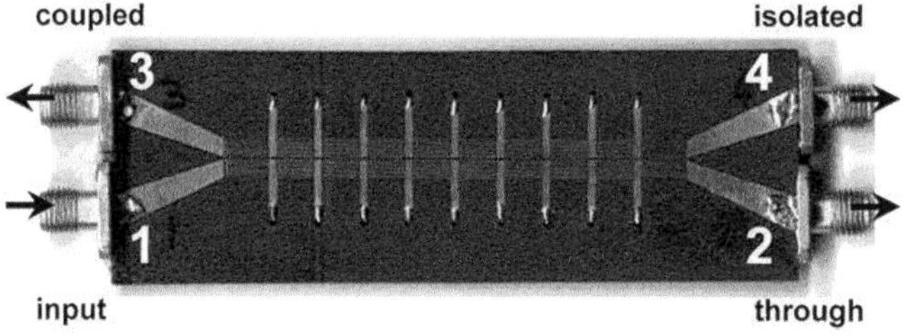

Fig. 6.12 A 0-dB (9-cell) edge-coupled directional coupler comprising of two interdigital/stub composite right-left-handed transmission line. From Caloz *et al.* (2004). Copyright © 2004 IEEE

6.6 Waveguide couplers

A left-handed transmission line in terms of a series capacitor and a shunt inductance, incorporated in a microstrip line, was realized in the form shown in Fig. 6.11 (Liu *et al.*, 2002; Caloz *et al.*, 2004; Caloz and Itoh, 2006). The series capacitor is an interdigital one in order to increase the capacitance. The shunt inductance is implemented by a stub connected to the ground plane. However, the authors argue that neither the capacitance nor the inductance can be regarded as pure. Both will be accompanied by parasitic reactances that will constitute elements of a right-handed transmission line. The parasitic inductance L_{R} is caused by the current flowing along the digits of the capacitor and the parasitic capacitance C_{R} is due to the electric field between the metal structure forming the elements and the ground plane. This is the justification for the use of the equivalent circuit shown already in Section 2.13 in which there is a series resonant circuit consisting of the series capacitance C_{L} and the parasitic inductance L_{R}, and a shunt resonant circuit consisting of an inductance L_{L} in parallel with the parasitic capacitance C_{R}. When two such lines are laid sufficiently close to each other (see Fig. 6.12) they are coupled via a mutual inductance and a coupling edge capacitance. The result is a waveguide coupler that connects the various ports to each

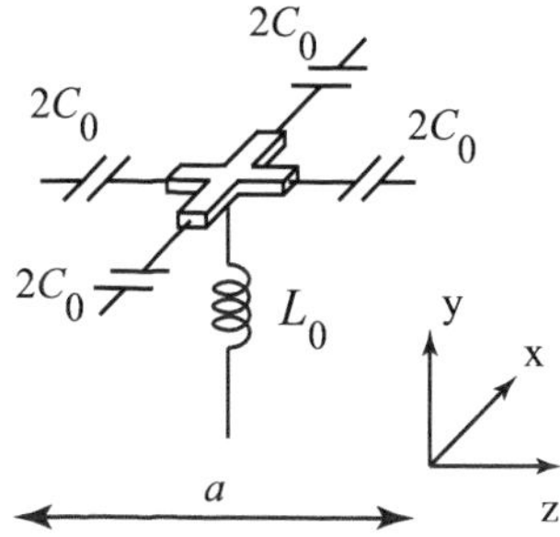

Fig. 6.13 2D realization of the 1D circuit of Fig. 6.10

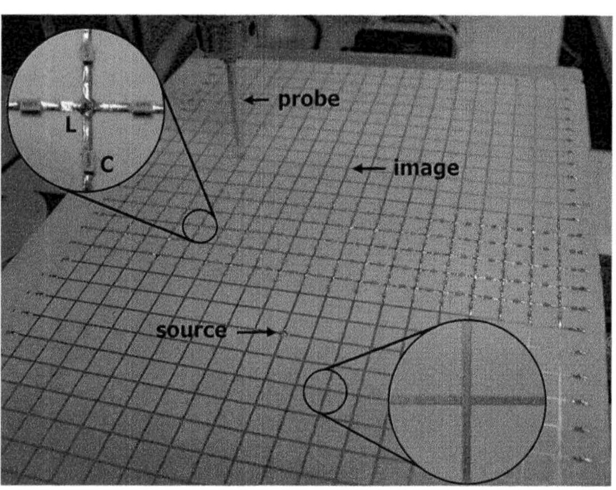

Fig. 6.14 The left-handed planar transmission-line lens. The unit cell of the left-handed (loaded) grid is shown in the top inset, that of the right-handed (unloaded) grid is shown in the bottom inset. From Grbic and Eleftheriades (2004). Copyright © 2004 by the American Physical Society. For coloured version see plate section

other. For a detailed description of this coupler, for its design, theory of operations and measurement results see Caloz *et al.* 2004; Caloz and Itoh 2006.

6.7 Imaging in two dimensions: transmission-line approach

One-dimensional representations of negative-index materials in terms of loaded transmission lines were discussed in Sections 2.6 and 2.13. They were suitable for finding dispersion characteristics and demonstrating the backward-wave properties of metamaterials. However, if the aim is to demonstrate negative refraction or subwavelength focusing, it is necessary to resort to no less than two dimensions. The generalization from one to two dimensions is quite straightforward. The two transmission lines are then rectangular to each other and loading is introduced in both directions as shown in Fig. 6.13, which is a direct generalization of the 1D circuit of Fig. 6.10 to two dimensions. A number of aspects of these 2D loaded transmission lines were investigated by Eleftheriades and coworkers in a series of papers (Eleftheriades *et al.*, 2002; Grbic and Eleftheriades, 2003*b*; Grbic and Eleftheriades, 2003*c*; Grbic and Eleftheriades, 2003*d*; Grbic and Eleftheriades, 2004; Iyer *et al.*, 2003).

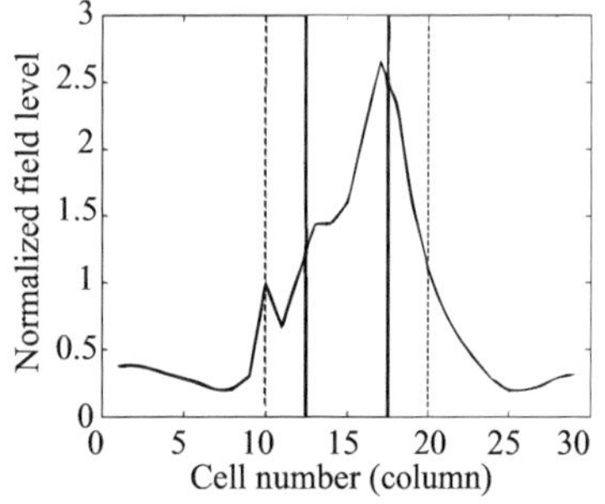

Fig. 6.15 The measured vertical electric field above row 0 at 1.057 GHz. The vertical dashed lines identify the source (column 0) and image (column 10) planes, while the vertical solid lines identify the interfaces of the left-handed planar slab. The growth of the evanescent waves in the left-handed lens is clear. From Grbic and Eleftheriades (2004). Copyright © 2004 by the American Physical Society

We shall describe here in some detail one of their imaging experiments (Grbic and Eleftheriades, 2004) in which microstrip lines (dielectric constant, 3.0, substrate thickness 1.52 mm) are loaded by 2-pF capacitors in parallel, and by 18-nH inductors in series. The size of the unit cell is equal to 8.4×8.4 mm. There are 19×5 unit cells for the loaded transmission line (see Fig. 6.14 for the experimental setup), and 19×12 unit cells on each side for the bare (unloaded) transmission lines that are analogous to right-handed media. The effective propagation coefficients of the loaded and unloaded media were designed to be equal in magnitude but opposite in sign at 1.00 GHz. In the experiment the first unloaded medium is excited by a vertical monopole fed by a coax-

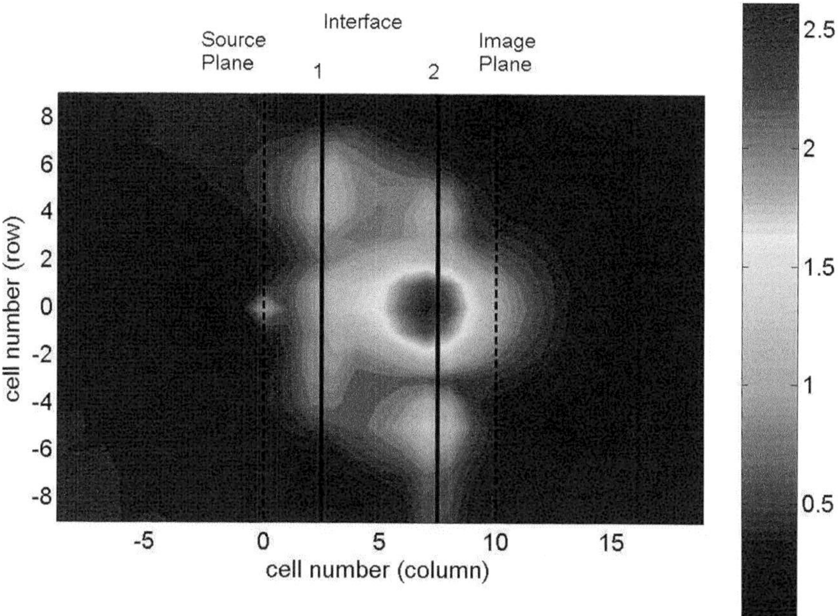

Fig. 6.16 The measured vertical electric field detected 0.8 mm above the surface of the entire structure at 1.057 GHz. The plot has been normalized with respect to the source amplitude (linear scale). From Grbic and Eleftheriades (2004). Copyright © 2004 by the American Physical Society. For coloured version see plate section

ial cable through the ground plane. The monopole attaches the inner conductor of the coaxial cable to the unloaded medium, while the outer conductor of the coaxial cable is attached to the ground plane. The excitation is in the central row 2.5 unit cells away from the interface. The image is then supposed to be 2.5 unit cells away on the other side of the loaded medium. The vertical electric field is detected 0.8 mm above the surface using a short vertical probe connected to a network analyzer. Its value above row 0 is plotted in Fig. 6.15 at the optimum frequency of 1.057 GHz. The source plane at column 0 and the image plane at column 10 are denoted by dotted lines, whereas the boundaries of the loaded line are denoted by thick continuous lines. The value of the field may be seen to be the same at the source and at the image. A two-dimensional plot of the electric field is shown in Fig. 6.16. As may be expected, the field is highest at the output interface. The lateral size of the image was found to be equal to 0.21 effective wavelengths, in contrast to the 0.36 wavelengths that is the diffraction-limited value. This is the first time that the classical limit was beaten by a flat negative-index lens. It was made possible by the small size of the loading inductors and capacitors, by the fact that the non-resonant structure was less lossy than its resonant counterparts and finally because it was relatively easy to produce an isotropic backward-wave medium.

Magnetoinductive waves I

7

7.1 Introduction

Magnetoinductive (MI) waves have come about as a by-product of the research on metamaterials. The magnetic elements used in the first realization of negative refraction by Shelby *et al.* (2001*a*) were split-ring resonators that could be modelled (Marques *et al.*, 2002*c*) as LC circuits, provided the dimensions are small relative to the free-space wavelength. We shall also assume, as we have done so far, that the separation of the elements is also much smaller than the wavelength. This is often referred to as the quasi-static approximation.

The simplest realization of an LC circuit as a metamaterial element is a capacitively loaded loop, shown schematically in Fig. 7.1(a). Two such loops close to each other are coupled to each other due to the magnetic field of one loop threading the other loop and inducing a current in it (Fig. 7.1(b)). The presence of such coupling leads to waves that were called MI waves by Shamonina *et al.* (2002*a*). They belong to the category of slow waves that propagate at a velocity less than that of light.

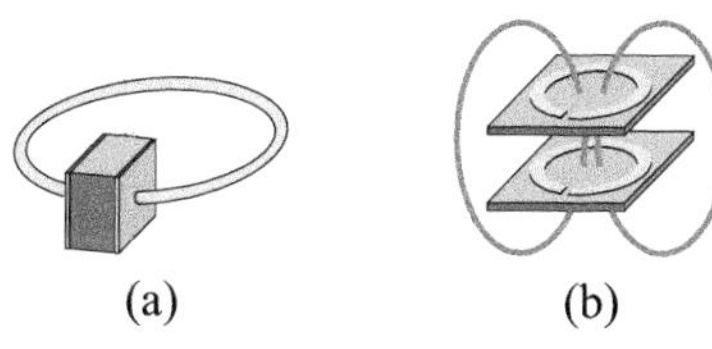

Fig. 7.1 (a) Capacitively loaded loop. (b) Two magnetically coupled elements

The properties of magnetoinductive waves have received considerable attention since they were first proposed (Shamonina *et al.*, 2002*a*). A detailed study of the dispersion characteristics was carried out in Shamonina *et al.* 2002*b*, some other aspects of the theory were treated in Syms *et al.* 2005*a*; Syms *et al.* 2005*b*; Sydoruk *et al.* 2005, experimental results and comparison between experiments and theory were given in Wiltshire *et al.* 2003*b*; Wiltshire *et al.* 2004*b*; Sydoruk *et al.* 2006; Syms *et al.* 2006*b*; Sydoruk *et al.* 2007*b*; Syms *et al.* 2007*a*, devices in MI waveguides were reported in Shamonina and Solymar 2004; Syms *et al.* 2005*c*; Syms *et al.* 2006*a* and review papers were published in Shamonina and Solymar 2006; Shamonina 2008.

Further properties of MI waves concerned with 2D effects and retardation will be presented in Chapter 8. Note that we have already discussed some simple aspects of MI waves earlier in this book in Section 1.23, among wave solutions on four-poles (Brillouin, 1953): an example was given concerned with magnetic coupling within one resonant four-pole. The result was a dispersion equation of the form

$$\omega = \frac{\omega_0}{1 + \dfrac{2M}{L}\cos(ka)}, \tag{7.1}$$

where ω_0 is the resonant frequency, L is the inductance and M is the

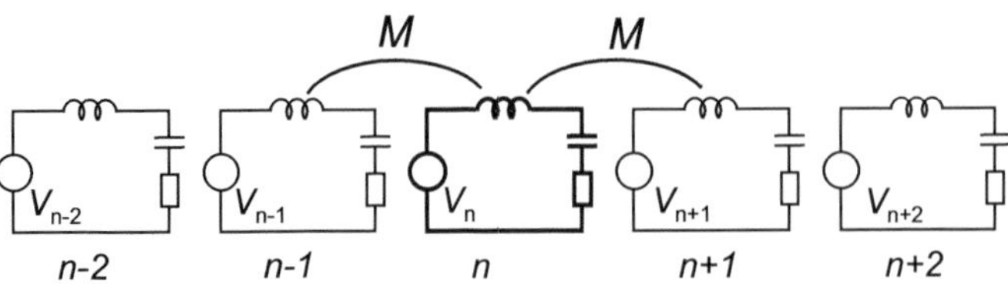

Fig. 7.2 Equivalent circuit of an array of capacitively loaded loops

mutual inductance. The same dispersion equation was also derived in Section 2.2 by applying Kirchhoff's voltage law to magnetically coupled resonant loops assuming that there is coupling only between nearest neighbours. This constitutive (recurrent) relation was given in the form

$$Z_0 I_n + \mathrm{j}\,\omega M(I_{n-1} + I_{n+1}) = 0\,, \tag{7.2}$$

where Z_0 is the self-impedance of the loop and I_n is the current in the nth loop as shown in Fig. 7.2.

Our aim in the present chapter is further to investigate the properties of a one-dimensional array of capacitively loaded loops. In Sections 7.2–7.10 we shall assume nearest-neighbour coupling and give some illustrations how such an array may work as a transmission line. We shall discuss the dispersion relations in Section 7.2, matching by a terminal impedance in Section 7.3, the problem of excitation in Section 7.4, eigenvalues and eigenvectors in Section 7.5, the distribution of current along the line in Section 7.6, the Poynting vector in Section 7.7, the definition of power in Section 7.8, reflection and transmission at a boundary in Section 7.9, and the tailoring of the dispersion curve in Section 7.10. Experimental results and comparisons between theory and experiments are presented in Section 7.11. In Section 7.12 we shall abandon the nearest-neighbour approximation and assume that, in general, each element of the line is coupled to all the other elements, and investigate its effect on the dispersion curve and on the current distribution. Pseudo-one-dimensional cases (when the theoretical treatment requires only some minor modification of one-dimensional theory) will be presented in Sections 7.13 and 7.14, and applications in Section 7.15.

7.2 Dispersion relations

As mentioned before the dispersion relations were derived in Chapters 1 and 2 in two different manners for the simple case when losses are absent. We shall now add losses, i.e. assume the self-impedance in the form

$$Z_0 = \mathrm{j}\,\omega L + \frac{1}{\mathrm{j}\,\omega C} + R\,, \tag{7.3}$$

where R is the resistance. The wave assumption is still taken in the form (see eqn (2.3))

$$I_n = I_0\,\mathrm{e}^{-\mathrm{j}\,kna}\,, \tag{7.4}$$

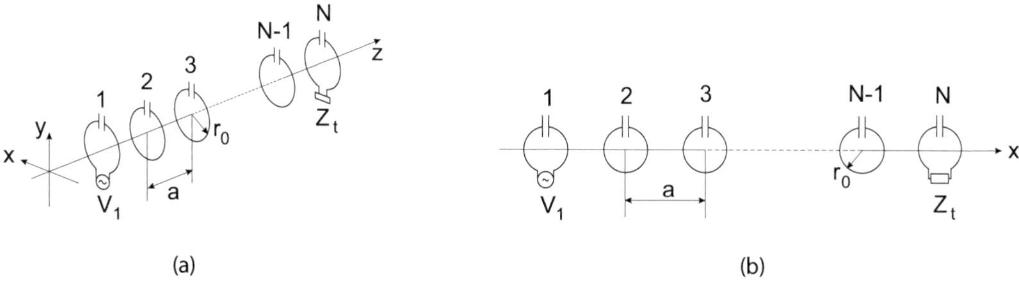

Fig. 7.3 (a) Axial and (b) planar array

but now k is complex

$$k = \beta - \mathrm{j}\alpha , \tag{7.5}$$

where β is the propagation constant and α is the attenuation coefficient. The dispersion equation (7.1) may then be separated into real and imaginary parts yielding

$$1 - \frac{\omega_0^2}{\omega^2} + \kappa \cos(\beta a) \cosh(\alpha a) = 0 , \tag{7.6}$$

$$\frac{1}{Q} - \kappa \sin(\beta a) \sinh(\alpha a) = 0 , \tag{7.7}$$

where κ is the coupling coefficient equal to $2M/L$ and Q is the quality factor of the resonant circuit equal to $\omega L/R$. Note that M may be positive or negative, as discussed in Section 1.23. It is positive when the array is axial (see Fig. 7.3(a)) and negative in the planar configuration[1] (Fig. 7.3(b)).

If losses are small enough (Q is high enough) then

$$\cosh(\alpha a) = 1 , \quad \sinh(\alpha a) = \alpha a , \tag{7.8}$$

which means that the dispersion equation for the phase change per element remains the same, and the losses per element are given as

$$\alpha a = \frac{1}{\kappa Q \sin(\beta a)} . \tag{7.9}$$

It may indeed be expected that losses decline as the coupling coefficient and the Q of the circuit increase. Dependence on βa may also be anticipated, that attenuation should be minimum at the resonant frequency and should increase towards the band edge where $\beta a = 0$ or π. The approximation, however, breaks down right at the band edge where the attenuation should not be infinitely large. We can get the correct result by solving numerically eqns (7.6) and (7.7).

We have already plotted in Fig. 1.18 the dispersion equations for $\kappa = \pm 0.1$ for the lossless case. We shall replot them in Figs. 7.4 and 7.5 for $Q = 40$, 100 and 1000 including both the phase change and attenuation.

[1] The simplicity of derivation and the ease with which such arrays can be constructed (see Section 7.4) makes it likely that MI waves will be included in the near future in the undergraduate syllabus of both physics and electrical engineering courses. It would also help to demystify the concept of backward waves. As things stand the only place where backward waves make an appearance in the physics syllabus is in the derivation and interpretation of the optical branch of acoustic waves. But even then the emphasis is on the diatomic lattice and on the interaction with incident electromagnetic waves. The backward-wave character, if at all, is only casually mentioned. In circuit theory backward waves may appear (as in Section 1.23) but they are usually associated with lumped inductors and capacitors and not as waves in which the magnetic and electric fields vary along the direction of propagation. A look at the MI transmission line and the corresponding dispersion curves would convince students that there is nothing exotic about backward waves.

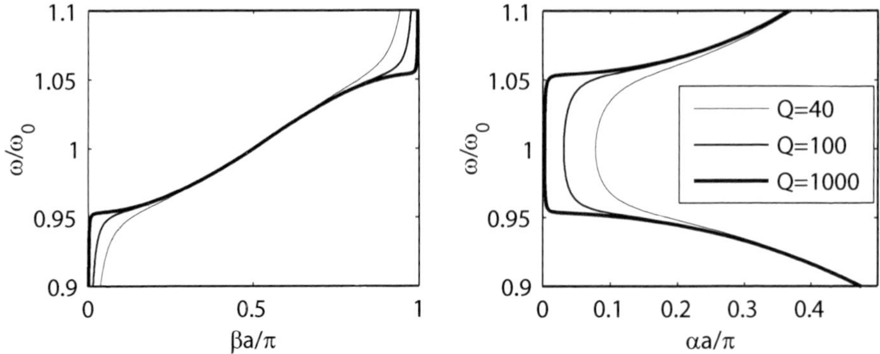

Fig. 7.4 Dispersion for an axial array. $\kappa = 0.1$

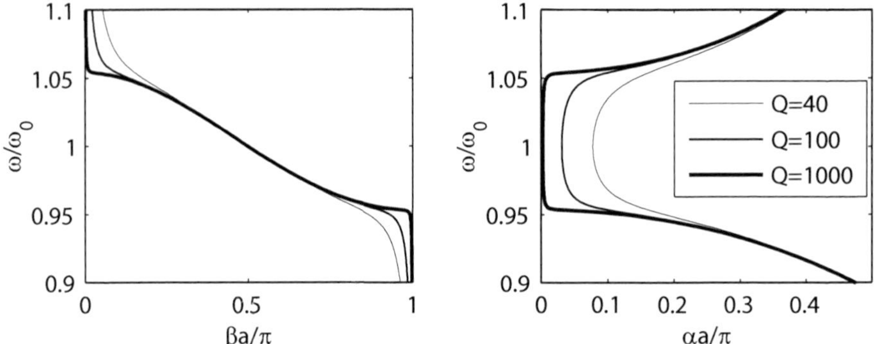

Fig. 7.5 Dispersion for a planar array. $\kappa = -0.1$

It may be seen that (i) the attenuation sharply increases as the quality factor declines, and (ii) both phase change and attenuation vary rapidly near the band edge.

It needs to be noted that the coupling coefficient is not necessarily small. In principle it may be as high as 2. In practice, the highest value measured so far (Syms *et al.*, 2006*b*) is 1.5 in the axial configuration in which the elements can be quite closely packed and -0.7 (Syms, 2006) in the planar configuration. For $\kappa = 1.5$ and $Q = 100$ the dispersion curves look quite different, as may be seen in Fig. 7.6. There is still a lower cutoff frequency but the bandwidth now extends to arbitrarily high frequencies with an asymptote at

$$\beta a = \arccos\left(\frac{1}{\kappa}\right), \tag{7.10}$$

which in the present case comes to $\beta a = 0.73\,\pi$.

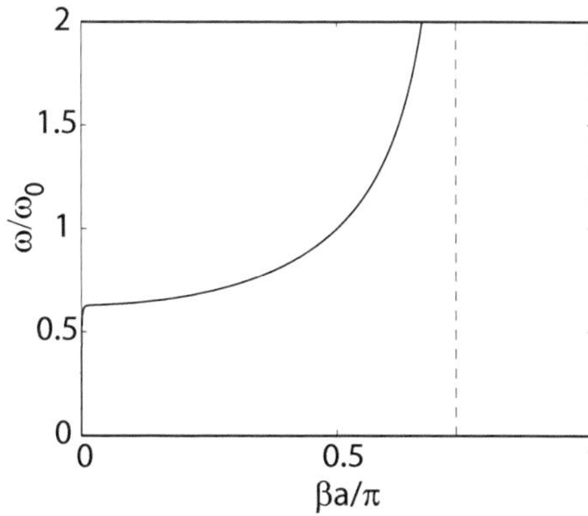

Fig. 7.6 Dispersion for an axial array in the case of a strong coupling. $\kappa = 1.5$

7.3 Matching the transmission line

It is quite clear from what we have done so far that MI waves can propagate along an array whether it is axial or planar. We may therefore regard such arrays as transmission lines having somewhat unusual dispersion relations. We know that for all transmission lines a terminal impedance exists that can absorb all the incident power. When this happens we talk about matching. In other words, in a matched transmission line there is a wave travelling from source to load but there are no reflected waves. How can we find this terminal impedance for a MI transmission line? Let's go back to eqn (7.2), which gives the relationship between currents I_{n-1}, I_n and I_{n+1}. This is true everywhere in the line except at the first and the last element. If the last element is the Nth then it can see a neighbour at the site $N-1$ but the element at $N+1$ is missing. We may then pose the question: can we substitute an impedance for the missing $(N+1)$th element? Let's call this impedance Z_T that is inserted in the Nth element. Hence, Kirchhoff's law for the Nth element takes the form (Shamonina *et al.*, 2002*b*)

$$(Z_0 + Z_T)I_N + \mathrm{j}\,\omega M I_{N-1} = 0\,. \tag{7.11}$$

We want a travelling wave so that I_N and I_{N-1} are related to each other by the factor $\exp(-\mathrm{j}\,ka)$, whence Z_T may be obtained from eqn (7.11), with the aid of the dispersion equation, as

$$Z_T = \mathrm{j}\,\omega M\,\mathrm{e}^{-\mathrm{j}\,ka}\,. \tag{7.12}$$

It turns out that the terminal impedance is not a real constant, as for a coaxial line for example, but it is complex and frequency-dependent.

7.4 Excitation

The dispersion equation is derived on the assumption that there is no external excitation. As it happens, it is quite easy to include possible excitation by ideal voltage sources in any of the elements. The excitation may come from an incident plane wave or from separate voltage sources. The general relationship between the applied voltages and the resulting currents is still described by Kirchhoff's law but now in the form (Shamonina *et al.*, 2002*b*)

$$\mathbf{V} = Z\mathbf{I}\,. \tag{7.13}$$

The above equation may also be regarded as a generalized Ohm's law. $\mathbf{V}$ and $\mathbf{I}$ are N-dimensional vectors

$$\mathbf{V} = (V_1, V_2, \dots V_n, \dots V_N)\,, \qquad \mathbf{I} = (I_1, I_2, \dots I_n, \dots I_N)\,, \tag{7.14}$$

and Z is the $N \times N$ impedance matrix

$$Z = \begin{pmatrix} Z_0 & \mathrm{j}\omega M & 0 & 0 & \dots & 0 & 0 \\ \mathrm{j}\omega M & Z_0 & \mathrm{j}\omega M & 0 & \dots & 0 & 0 \\ \dots & \dots & \dots & \dots & \dots & \dots & \dots \\ 0 & 0 & 0 & \dots & \mathrm{j}\omega M & Z_0 & \mathrm{j}\omega M \\ 0 & 0 & 0 & \dots & \dots & \mathrm{j}\omega M & Z_0 \end{pmatrix}. \tag{7.15}$$

This is clearly a tri-diagonal matrix. The main diagonal elements are all Z_0 and the elements next to them (left, right, up or down) are all $\mathrm{j}\omega M$. If the line is terminated by an impedance Z_T then the last element in the matrix (Nth row, Nth column) should be replaced by $Z_0 + Z_T$.

If the voltage excitation vector is known the current may be obtained by inverting the Z matrix,

$$\mathbf{I} = Z^{-1}\mathbf{V}\,. \tag{7.16}$$

Very often, only the first element is excited. In which case only the first component of the voltage vector is finite and all the others are zero.

7.5 Eigenvectors and eigenvalues

To determine the natural modes of the system we need to work out the eigenvectors and eigenvalues, which we shall do here for the lossless case. The eigenvectors for a line consisting of N elements may be easily guessed from the physics. We may postulate that there are elements at sites zero and $N+1$ but the current is zero there. We may also assume that the currents vary sinusoidally between the two ends. Hence, the elements of the lth eigenvector may be written in the form (Shamonina *et al.*, 2002*b*)

$$I_n^{(l)} = I(N)\sin\left(\frac{nl\pi}{N+1}\right) \qquad (n = 0, 1, 2, \dots N, N+1). \tag{7.17}$$

The value of $I(N)$ can be found from the orthonormality condition

$$\mathbf{I}^{(l)} \cdot \mathbf{I}^{(m)} = \begin{cases} 0 & \text{if} \quad l \neq m \\ 1 & \text{if} \quad l = m \end{cases}, \tag{7.18}$$

yielding

$$I(N) = \frac{2}{N+1} I_0 . \tag{7.19}$$

The corresponding eigenvalues may be obtained from their definition

$$Z\mathbf{I}^{(l)} = \lambda_l \mathbf{I}^{(l)} , \tag{7.20}$$

as

$$\lambda_l = Z_0 + 2\mathrm{j}\,\omega M \cos\left(\frac{l\pi}{N+1}\right) . \tag{7.21}$$

Having got the eigenvectors and eigenvalues we can find, in closed form, the response to a general excitation. A voltage vector $\mathbf{V}$ of dimension N can be expanded in terms of the eigenvectors as

$$\mathbf{V} = \sum_{l=1}^{N} \mu_l \mathbf{I}^{(l)} . \tag{7.22}$$

The unknown coefficients μ_l can be determined from the orthonormality condition as

$$\mu_l = \mathbf{V} \cdot \mathbf{I}^{(l)} . \tag{7.23}$$

Considering further that a matrix can be expanded in terms of its eigenvectors and eigenvalues we obtain for the current vector

$$\mathbf{I} = \sum_{l=1}^{N} \frac{\mu_l}{\lambda_l} \mathbf{I}^{(l)} . \tag{7.24}$$

It may be seen from the above equation that in the general case all the eigenvectors are excited. In order to excite a single mode the corresponding eigenvalue must be close to zero.[2] From eqn (7.21) the lth eigenvalue is zero when

$$Z_0 + 2\mathrm{j}\,\omega M \cos\left(\frac{l\pi}{N+1}\right) = 0 . \tag{7.25}$$

But this is nothing else than our dispersion equation. A single mode, i.e. a single value of ka, can be excited for a discrete set of frequencies.

For $N = 5$ the normalized set of eigenvectors are plotted in Fig. 7.7(a). For the axial configuration and for $\kappa = 0.1$ the corresponding values of ka are shown on the dispersion curve of Fig. 7.7(b).

[2] When the eigenvalue is exactly zero the current amplitude will be infinitely large. This is a consequence of neglecting losses.

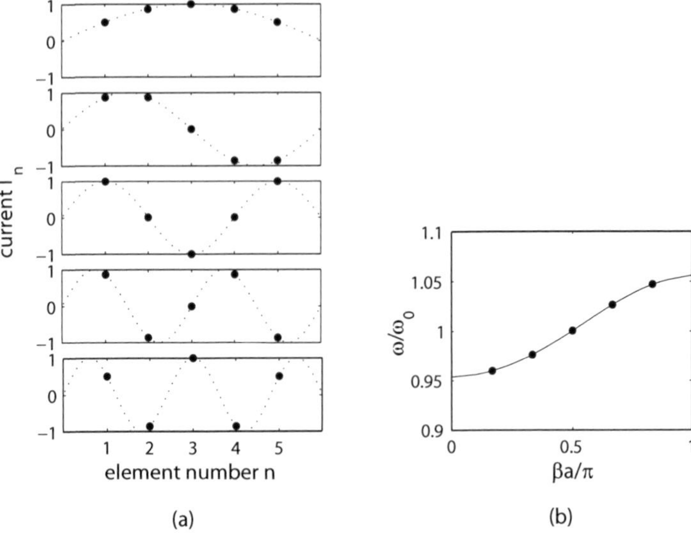

Fig. 7.7 (a) Eigenvectors and (b) the corresponding values of ka for a 5-element array

7.6 Current distributions

We have in the last two sections derived the current distribution in response to a voltage excitation in two different ways: by inverting the impedance matrix (eqn (7.16)) and by relying on eigenvectors and eigenvalues (eqn (7.24)). In the present section we shall use the former method that we have found numerically more convenient, particularly for the lossy case. As examples, we shall show current distributions on arrays of resonant loops for three typical cases: (i) travelling waves, (ii) resonances occurring for standing waves and (iii) evanescent current distributions with implications for near-field pixel-to-pixel imaging (Section 7.15.3). In each of the following examples we will assume that only one or two of the loops are excited by an external voltage source and that the currents in all other loops are induced via the magnetoinductive interactions.

(i) Travelling waves

If the line with a finite number of resonant loops is terminated by its matching terminal impedance (eqn (7.12)) then a single travelling MI wave propagates along the line. As an example, we choose an axial array with the parameters $N = 31$, $\kappa = 0.1$, $Q = 100$, with the first element being driven by a voltage V_1 and with the last element being terminated with the matching terminal impedance. The current distributions in the complex plane for three different frequencies, $\omega/\omega_0 = 0.9757$, 1, 1.0262, are shown in Fig. 7.8. The corresponding values of the propagation constant, $\beta a = \pi/3$, $\pi/2$, $2\pi/3$ and of the attenuation constant, $\alpha a =$ 0.12, 0.1, 0.12, are provided by the dispersion equations (7.6) and (7.7) (see also Fig. 7.4). It may be seen from Fig. 7.8 that the phase angle of the current varies from element to element by the corresponding value of βa and that the amplitude of the current changes exponentially as

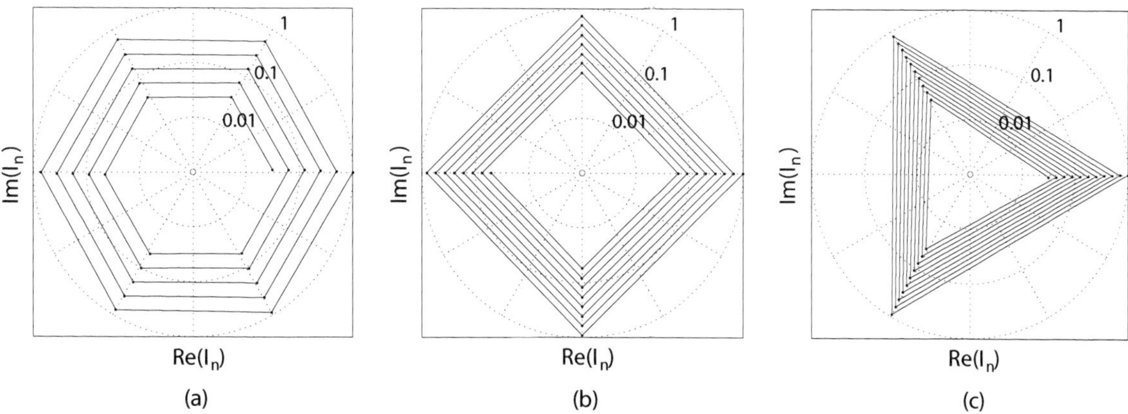

Fig. 7.8 Travelling wave. Complex amplitude of currents along the axial array, normalized to that at element 1, at $\omega/\omega_0 =$ 0.9757, 1 and 1.0262 (a)–(c)

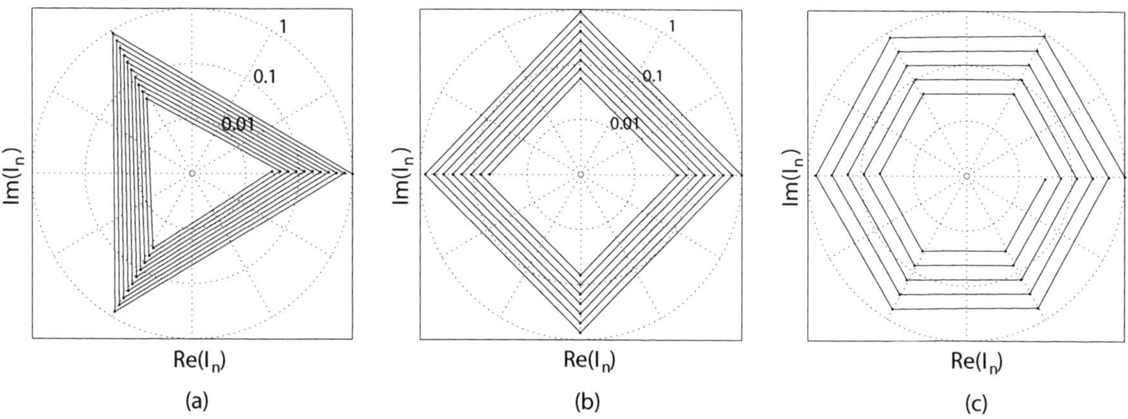

Fig. 7.9 Travelling wave. Complex amplitude of currents along the planar array, normalized to that at element 1, at $\omega/\omega_0 =$ 0.9757, 1 and 1.0262 (a)–(c)

$\exp(-\alpha a)$. Note that the radial co-ordinate is on a logarithmic scale covering three orders of magnitude. All the currents can be seen to lie on an angular spiral (in the lossless case the spiral would turn into a polygon). The higher the frequency, the larger the value of βa, the more tightly wound is the spiral. This is a consequence of the forward nature of the axial array.

Figure 7.9 shows another example, current distributions for a planar line with $\kappa = -0.1$ and for the same three values of ω/ω_0. The corresponding phase change is now $\beta a = 2\pi/3$, $\pi/2$, $\pi/3$ and the attenuation constant $\alpha a = 0.12$, 0.1, 0.12. The current distributions look similar to those in Fig. 7.8, they follow a spiral. There are though two major differences in comparison with the axial case. (i) the sense of rotation of the spiral has changed and (ii) the higher the frequency the smaller the phase change from element to element, the less tightly wound is the

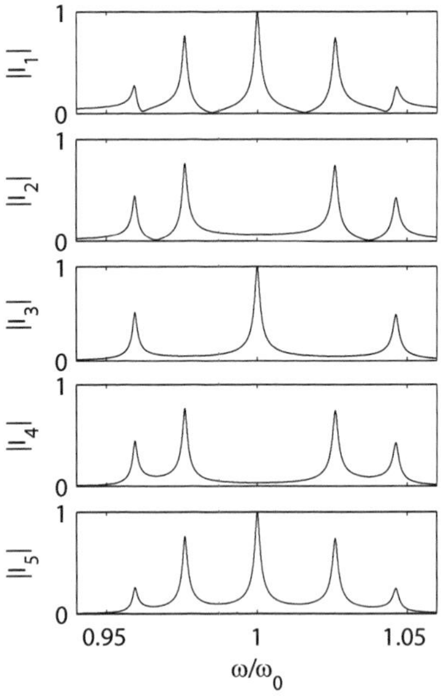

Fig. 7.10 Resonances. Normalized current amplitudes as a function of frequency for a five-element axial structure with element 1 being excited. $Q = 100$. $\kappa = 0.1$

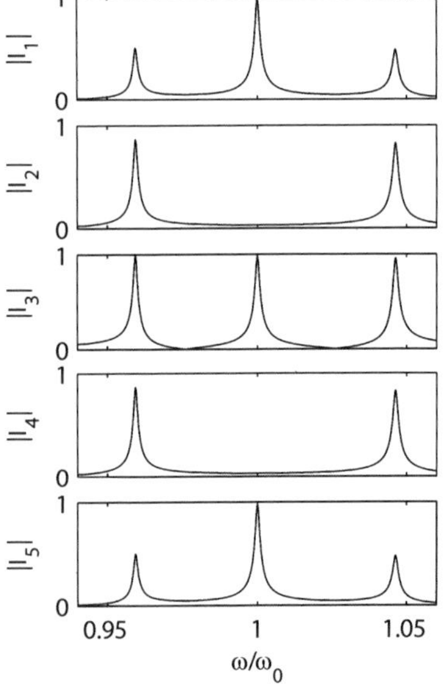

Fig. 7.11 Resonances: symmetric excitation. Normalized current amplitudes as a function of frequency for a five-element axial structure with element 3 being excited. $Q = 100$. $\kappa = 0.1$

spiral. This is a consequence of the backward nature of the planar array.

(ii) Resonances

If the line is not terminated by its matching impedance, reflections of MI waves from the line's ends take place, resulting in a standing-wave pattern for the currents. Standing waves that have an integral number of half-periods correspond to the eigenvectors of the system (see eqn (7.17) and Fig. 7.7) which manifest themselves as resonances. In general, a line consisting of N elements can have up to N resonances at N discrete frequencies given by eqn (7.25). It is clear that all the resonances are within the pass band of the MI waves since the argument of the cosine function can never exceed π.

Whether or not any particular resonant mode can be excited depends on the symmetry of the problem. For example, an antisymmetric mode can never be excited under a symmetric excitation and vice versa. To illustrate this we consider again the five-element axial line (for which the eigenvectors and the corresponding resonances are shown in Fig. 7.7). Figure 7.10 shows the relative amplitudes of the currents in all five elements as a function of frequency for the case when the first current is being driven by an external voltage. There are five resonant frequencies. Note that the third resonance is missing in elements 2 and 4 and the second and fourth resonances are missing in element 3.

We can selectively exclude some of the resonances from being excited by choosing either a symmetric or an antisymmetric excitation. Figure 7.11 shows the currents for the case of a symmetric excitation, when the central loop is excited by an external voltage. Due to the symmetry argument it is quite obvious why the second and the fourth resonances that correspond to the antisymmetric eigenmodes are missing here.

Finally, Fig. 7.12 shows the currents for the case of an antisymmetric

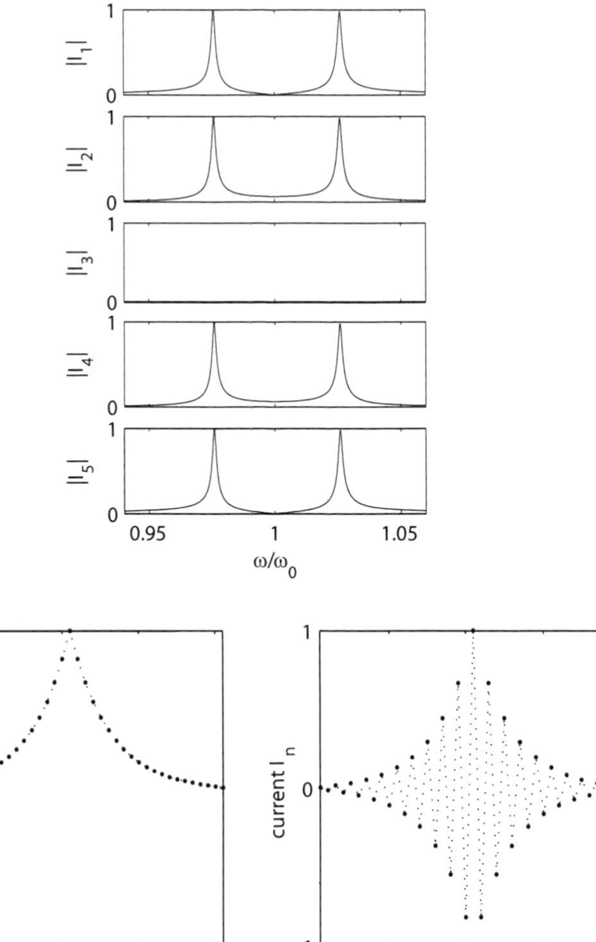

Fig. 7.12 Resonances: antisymmetric excitation. Normalized current amplitudes as a function of frequency for a five-element axial structure with elements 1 and 5 being excited in antiphase. $Q = 100$. $\kappa = 0.1$

Fig. 7.13 Evanescent magnetoinductive waves for a lossless 41-element axial line. $\kappa = 0.1$. $\omega/\omega_0 = 0.95$ (a) and 1.06 (b)

excitation, when the first loop and the last loop are excited by external voltages in antiphase. In this case the central loop is not excited at all, and only the second and the fourth resonances are present.

(iii) Excitation outside the pass band

Excitation of an array outside the pass band results in evanescent magnetoinductive waves. As follows from the dispersion equation (7.1) there are two different branches of evanescent waves. For the branch with $\beta = 0$ currents in all elements are in phase, while for the branch with $\beta a = \pi$ currents in neighbouring elements are always in antiphase. We take as an example a lossless 41-element axial line (which is long enough in order to disregard reflections from the unmatched ends) with the central element being excited. Figure 7.13 shows the current distributions for two values of the frequency, one from the lower and another one from the upper stop band, which correspond to $\alpha a = 0.2$. Obviously

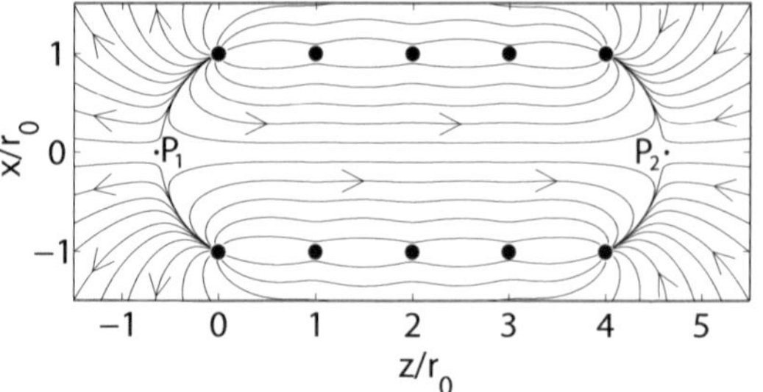

Fig. 7.14 Poynting vector streamlines for a 5-element array. From Shamonina *et al.* (2002*b*). Copyright © 2002 American Institute of Physics

the current amplitudes decay strongly as we move away from the element that is driven by an external source. This result has an important consequence. When operating in the stop band, a magnetoinductive line is able to replicate any pattern of excitation by the 'pixel-to-pixel' mechanism. This significant property is crucial for the design of the magnetoinductive near-field lens (see Section 7.15.3).

7.7 Poynting vector

One of the advantages of MI waves is that their properties can be easily understood both in their circuit and field representations. Currents flowing in loops create magnetic fields. They can be calculated by first determining the vector potential from the current (assumed to be constant along the periphery) with the aid of eqn (1.14) and the magnetic field from the vector potential (eqn (1.10)).[3] Having got the magnetic field the electric field can be obtained from Maxwell's equations. Having both the electric and magnetic fields it is then possible to determine the Poynting vector.

[3]The derivation is available in a number of textbooks (see, e.g., Landau and Lifschitz 1984). The vector potential is obtained in the form of elliptic functions.

The starting point is the excitation that is assumed to be done by a voltage source applied to the first element. From that, and from the inverted impedance matrix, the currents can be found. Next, the magnetic and electric fields need to be calculated from each of the loop currents, and then they have to be added vectorially to find the total electric and magnetic fields in a given point. There is then enough information to determine the direction of the Poynting vector at each point. The calculation was performed both for a 25-element (Shamonina *et al.*, 2002*a*) and for a 5-element line (Shamonina *et al.*, 2002*b*). The Poynting vector streamlines for the latter case are shown in Fig. 7.14. It may be clearly seen that the streamlines originate on the first loop and reach, by following a variety of paths, the last loop, which is terminated by a matched impedance. Note that the streamlines approaching the load from the inside and outside are separated from each other by a so-called P-point, where the Poynting vector is zero.

7.8 Power in a MI wave

One way of determining the power in a MI wave is to take a cross-section somewhere between two elements and integrate the Poynting vector. This can be done only numerically, which would in no way help in building a physical picture. We shall instead resort to an alternative analytical method in which the power is calculated by multiplying the stored energy per unit distance by the group velocity.

The stored energy may be calculated from those stored in the inductance of the loop and in the capacitance, to which the mutual energy due to coupling between the elements should be added. The full expression is of the form (Syms *et al.*, 2005*b*)

$$W = \frac{1}{2}L|I_n|^2 + \frac{1}{2}C|V_n|^2 + \frac{1}{2}M(I_n I_{n-1}^* + I_n I_{n+1}^*)\,. \tag{7.26}$$

Note that V_n here is not the voltage applied to the nth element but the voltage across the capacitor of the nth element, so it is related to I_n as

$$I_n = \mathrm{j}\,\omega C V_n\,. \tag{7.27}$$

Assuming again a travelling wave and substituting the values of I_{n-1} and I_{n+1} in terms of I_n and making use of the dispersion equation the expression for the stored energy simplifies to

$$W = \frac{\omega_0^2}{\omega^2}L|I_n|^2\,. \tag{7.28}$$

The group velocity may be obtained from the dispersion equation (7.1) as

$$v_\mathrm{g} = \frac{\mathrm{d}\omega}{\mathrm{d}k} = \frac{\mathrm{d}}{\mathrm{d}k}\left(\frac{\omega_0}{\sqrt{1+\kappa\cos(ka)}}\right) = \frac{\omega_0 a}{2}k\left(\frac{\omega_0}{\omega}\right)^3\sin(ka)\,, \tag{7.29}$$

whence the power can be found as

$$P = W v_\mathrm{g} = \frac{1}{2}\omega M|I_0|^2\sin(ka)\,. \tag{7.30}$$

Note that no power can be transferred at the band edges and optimum transfer is at the resonant frequency when $ka = \pi/2$.

7.9 Boundary reflection and transmission

Let us assume that two MI transmission lines, say both of them in the axial configuration, are joined together as shown in Fig. 7.15. The distance between the elements is a_1 in line 1, a_2 in line 2 and a_b across the boundary. The mutual inductances are M_1 in line 1, M_2 in line 2 and M_b between the last element of line 1 and the first element of line 2. The elements are numbered in such a way that element 0 is the last element of line 1 and element 1 is the first element of line 2.

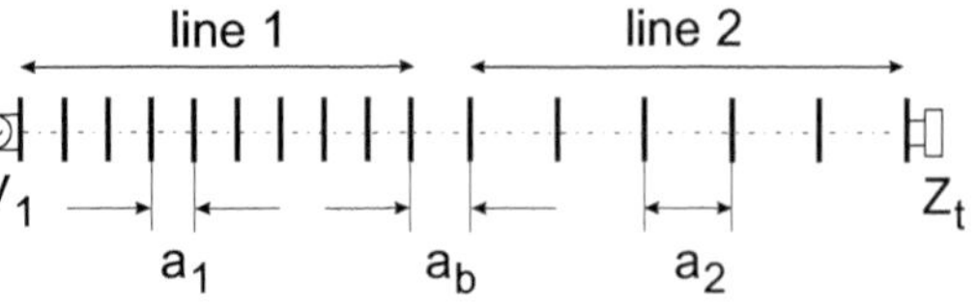

Fig. 7.15 Two magnetoinductive lines joined together

Assume further that a MI wave is incident from the left. In general, when a wave is incident from one medium upon another medium it may be expected that part of the wave will be reflected back into medium 1 and part of it will propagate in medium 2. Hence, the current distributions in lines 1 and 2 will be taken in the form (Syms *et al.*, 2005*b*)

$$I_n = I_{00}\left[\mathrm{e}^{-\mathrm{j}n(ka)_1} + R\,\mathrm{e}^{\mathrm{j}n(ka)_1}\right], \quad n \le 0 \tag{7.31}$$

and

$$I_n = I_{00}\,T\,\mathrm{e}^{-\mathrm{j}n(ka)_2}, \quad n > 0\,, \tag{7.32}$$

where I_{00} is a constant, R and T are the reflection and transmission coefficients, and $(ka)_1$ and $(ka)_2$ are the phase change per element in media 1 and 2, respectively.

We may now write Kirchhoff's law for elements 0 and 1, the elements on the opposite sides of the boundary, in the form

$$Z_{01}I_0 + \mathrm{j}\omega M_1 I_{-1} + \mathrm{j}\omega M_\mathrm{b} I_1 = 0 \tag{7.33}$$

and

$$Z_{02}I_1 + \mathrm{j}\omega M_2 I_2 + \mathrm{j}\omega M_\mathrm{b} I_0 = 0\,. \tag{7.34}$$

Substituting eqns (7.31) and (7.32) into eqns (7.33) and (7.34) the two unknowns R and T may be determined as follows

$$R = \frac{M_\mathrm{b}^2\,\mathrm{e}^{-\mathrm{j}(ka)_2} - M_1 M_2\,\mathrm{e}^{-\mathrm{j}(ka)_1}}{M_1 M_2\,\mathrm{e}^{\mathrm{j}(ka)_1} - M_\mathrm{b}^2\,\mathrm{e}^{-\mathrm{j}(ka)_2}}\,, \tag{7.35}$$

$$T = \frac{2\mathrm{j}\,M_1 M_\mathrm{b}\sin\left[(ka)_1\right]}{M_1 M_2\,\mathrm{e}^{\mathrm{j}(ka)_1} - M_\mathrm{b}^2\,\mathrm{e}^{-\mathrm{j}(ka)_2}}\,. \tag{7.36}$$

In the special case when $M_1 = M_\mathrm{b} = M_2$ eqns (7.35) and (7.36) reduce to

$$R = \frac{\mathrm{e}^{-\mathrm{j}(ka)_2} - \mathrm{e}^{-\mathrm{j}(ka)_1}}{\mathrm{e}^{\mathrm{j}(ka)_1} - \mathrm{e}^{-\mathrm{j}(ka)_2}}\,, \tag{7.37}$$

$$T = \frac{2\mathrm{j}\,\sin\left[(ka)_1\right]}{\mathrm{e}^{\mathrm{j}(ka)_1} - \mathrm{e}^{-\mathrm{j}(ka)_2}}\,. \tag{7.38}$$

This situation can arise only when the sole difference between the two media is that the self-impedances (and hence the dispersion curves) are different.

Note that eqn (7.37) appears in Tretyakov's book (2003) but the underlying physics is quite different in the two cases. Here, we are concerned with the reflection and transmission of waves propagating in different periodic media, whereas Tretyakov's expression is valid when a plane electromagnetic wave is incident upon a periodic medium and higher-order modes at the boundary can be disregarded. It is a coincidence that the expressions are identical.

A further simplification can be obtained by considering the continuous limit when $(ka)_1$, $(ka)_2 \ll 1$. Then, the reflection and transmission coefficients reduce to

$$R = \frac{k_1 - k_2}{k_1 + k_2} \quad \text{and} \quad T = \frac{2k_1}{k_1 + k_2} . \tag{7.39}$$

The above expressions look quite familiar, occurring for example when Schrodinger's equation is solved for an electron wave incident upon a potential barrier (see, e.g., Solymar and Walsh 2004). There are no periodic media in that case, simply a wave incident from a medium with propagation constant k_1 upon a medium with propagation constant k_2.

For the reflection and transmission coefficients to make sense it is a necessary condition that the power flow should be the same in lines 1 and 2. The power in lines 1 and 2 may be written as

$$P_1 = \frac{1}{2}\left(1 - |R|^2\right)\omega M_1 \sin(ka)_1 , \tag{7.40}$$

$$P_2 = \frac{1}{2}|T|^2 \omega M_2 \sin(ka)_2 . \tag{7.41}$$

Substituting eqns (7.34) and (7.35) into the above equations for power it may be shown, using a fair number of algebraic operations, that $P_1 = P_2$, i.e. the power across the boundary is conserved.

7.10 Tailoring the dispersion characteristics: biperiodic lines

For small coupling coefficients, which is usually the case, the pass band of MI waves is narrow. However, in many cases this may not be desirable. Ideally, we would like to tailor the dispersion characteristics to any requirement. We might, for example, wish to realize two pass bands instead of one. Another example, to which we shall return in Chapter 8 where non-linear relations are discussed, is parametric amplification (see, e.g., Sydoruk *et al.* 2007*a*). In that case we have a signal wave that has a propagation coefficient β at a frequency ω. The aim is to amplify this signal wave with the aid of a pump wave propagating on the same structure at twice the frequency and with a propagation

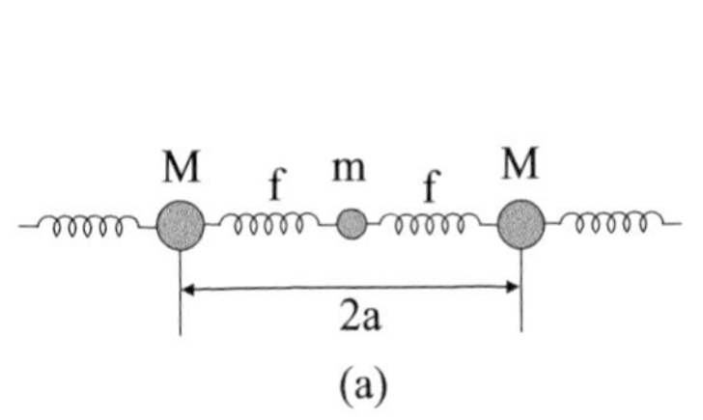

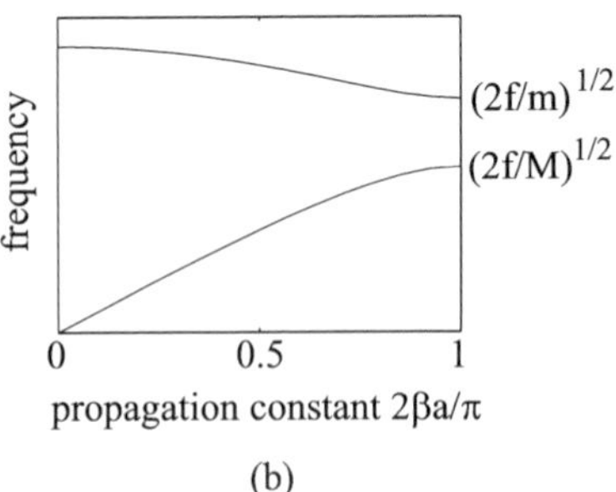

Fig. 7.16 (a) Diatomic chain of atoms and (b) the phonon dispersion curve with two branches

coefficient 2β. The two waves are then in synchronism because they propagate with the same phase velocity

$$v_{\mathrm{p}} = \frac{\omega}{\beta} = \frac{2\omega}{2\beta}\,, \tag{7.42}$$

and the signal wave can be amplified by transferring power from the pump wave.

In this section our aim is to show that we have some freedom over the dispersion characteristics and, in particular, we can realize the synchronism condition. The way to do it is suggested by the close analogy between MI waves and acoustic waves. It is well known that the dispersion characteristics of diatomic solids differ greatly from those having identical elements (see, e.g., Brillouin 1953). The consequence of having a material in which two different atoms of different masses alternate (e.g. NaCl) is the appearance of a new band, known as the optical branch, in the dispersion characteristics. A sketch of the dispersion characteristics of such a diatomic solid (Brillouin, 1953) is shown in Fig. 7.16.

We have now two pass bands. When it comes to acoustic waves in a solid we have very little freedom. The two different masses are provided by Nature and we have very little control over the way such masses arrange themselves. However, when metamaterial elements are put next to each other we can build every single element as we wish.

There are two obvious ways of achieving double periodicity: (i) To change some parameter of the element (L or C resulting in a change of resonant frequency) and (ii) to vary the distance between the elements, which means that there will be two different mutual inductances. These possibilities are shown schematically in Figs. 7.17(a) and (b) for the planar configuration and in Figs. 7.17(c) and (d) for the axial configuration.

Let us look at such a biperiodic line and apply Kirchhoff's equations to elements $2n$ and $2n+1$ (see Fig. 7.18). We find

$$Z_{01} I_{2n} + \mathrm{j}\,\omega M_1 I_{2n-1} + \mathrm{j}\,\omega M_2 I_{2n+1} = 0 \tag{7.43}$$

and

$$Z_{02} I_{2n+1} + \mathrm{j}\,\omega M_2 I_{2n} + \mathrm{j}\,\omega M_1 I_{2n+2} = 0\,. \tag{7.44}$$

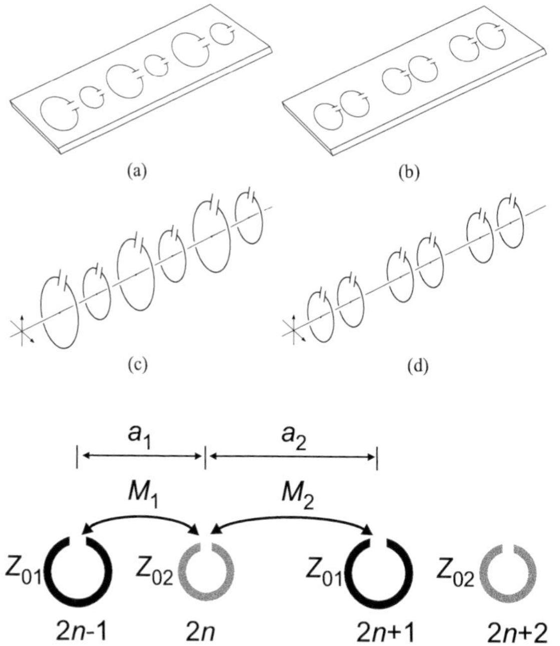

Fig. 7.17 Schematic representation of planar (a) and (b) and axial (c) and (d) biperiodic configurations. (a), (c) Resonant frequency varies from element to element. (b), (d) Distance varies between neighbouring elements. From Sydoruk *et al.* (2005). Copyright © 2005 American Institute of Physics

Fig. 7.18 Schematic view of a biperiodic chain of metamaterial elements

We shall assume now propagating solutions both for the even- and for the odd-numbered elements

$$I_{2n} = A_2\,\mathrm{e}^{-\mathrm{j}\,k2n(a_1+a_2)} \tag{7.45}$$

and

$$I_{2n+1} = A_1\,\mathrm{e}^{-\mathrm{j}\,k(2n+1)(a_1+a_2)}\,. \tag{7.46}$$

Substituting eqns (7.45) and (7.46) into eqns (7.43) and (7.44) we find after some algebraic operations the dispersion equation

$$\cos\left(\frac{k(a_1+a_2)}{2}\right) = \frac{1}{2}\frac{\sqrt{-\dfrac{Z_{01}Z_{02}}{\omega^2} - (M_1-M_2)^2}}{\sqrt{M_1M_2}}\,. \tag{7.47}$$

Note that eqn (7.47) reduces to eqn (7.1) when $Z_{01} = Z_{02} = Z_0$ and $M_1 = M_2 = M$, as it should.

Let us now look at two practical examples, one for the planar, and one for the axial configuration. The parameters chosen are as follows: $r_0 = 10$ mm, wire thickness $d_\mathrm{w} = 2$ mm, $L = 33$ nH, $C_1 = 208$ pF, $C_2 = 177$ pF, $\omega_{01} = (LC_1)^{-1/2} = 0.95\,\omega_0$, $\omega_{02} = (LC_2)^{-1/2} = 1.05\,\omega_0$, $\omega_0/2\pi = 63.87$ MHz.[4] For the axial configuration we choose $a = 10$ mm resulting in $M/L = 0.149$ and for the planar case $a = 20.5$ mm, corresponding to $M/L = -0.104$. Losses are taken into account by the quality factor, $Q = 150$. The dispersion curves with both propagation and attenuation constants are shown in Figs. 7.19 and 7.20 for the axial and planar cases, respectively.

[4]The reason for this choice is that 63.87 MHz corresponds to the magnetic resonance frequency of a proton for a magnetic field of 1.5 T, one of the possible choices for magnetic resonance imaging.

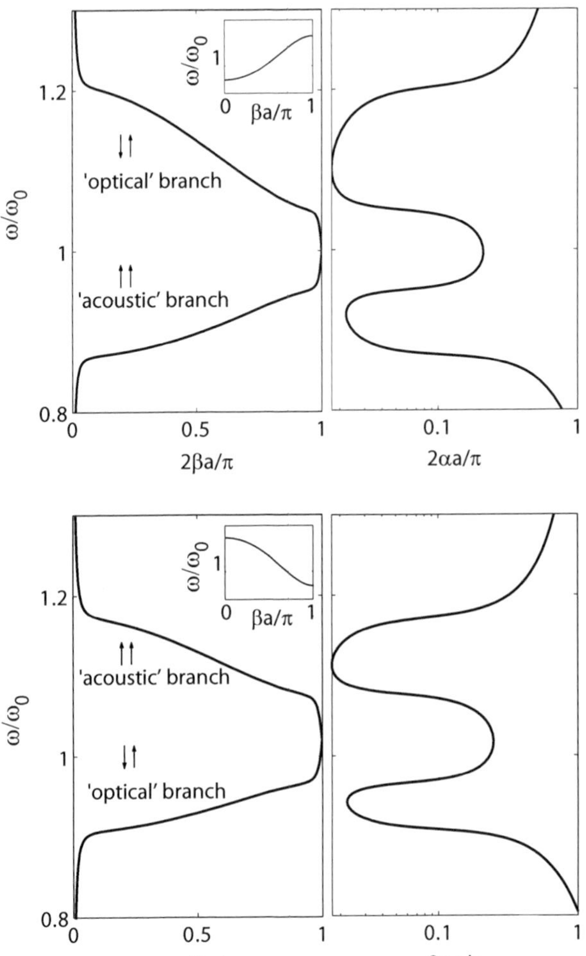

Fig. 7.19 Dispersion curve for the biperiodic axial configuration. Currents within a period are in antiphase in the upper ('optical') and in phase in the lower ('acoustic') branch. Inset shows the dispersion curve for the singly periodic axial configuration with the resonant frequency $\omega_0/2\pi = 63.87$ MHz and coupling between nearest neighbours $2M/L = 0.149$. From Sydoruk *et al.* (2005). Copyright © 2005 American Institute of Physics

Fig. 7.20 Dispersion curve for the biperiodic planar configuration. Currents within a period are in phase in the upper ('acoustic') and in antiphase in the lower ('optical') branch. Inset shows the dispersion curve for the singly periodic planar configuration with the resonant frequency $\omega_0/2\pi = 63.87$ MHz and coupling between nearest neighbours $2M/L = -0.104$. From Sydoruk *et al.* (2005). Copyright © 2005 American Institute of Physics

It may be immediately seen that the major distinction between the axial and planar cases, that one gives a forward wave and the other one a backward wave, is no longer there. In both cases the lower branches are forward waves and the upper branches are backward waves. There is, however, a difference if we consider the phases of the currents within a pair constituting the unit cell. For the axial line the currents of the neighbouring elements in the upper branch are in antiphase and the currents in the lower branch are in phase (Fig. 7.19). Using the analogy with the diatomic model we can refer to the upper branch as 'optical' and to the lower branch as 'acoustic'. For the planar line the situation is reversed. The currents in the lower branch are now in antiphase, thus this is the one we should, strictly speaking, call 'optical' (Fig. 7.20).

Let us next consider parametric amplification and the problem of ensuring synchronism between the signal wave and the pump wave. We choose the signal frequency at $\omega_0/2\pi = 63.87$ MHz. The propagation constant may be chosen at $2\beta a = \pi/3$ and the propagation constant of

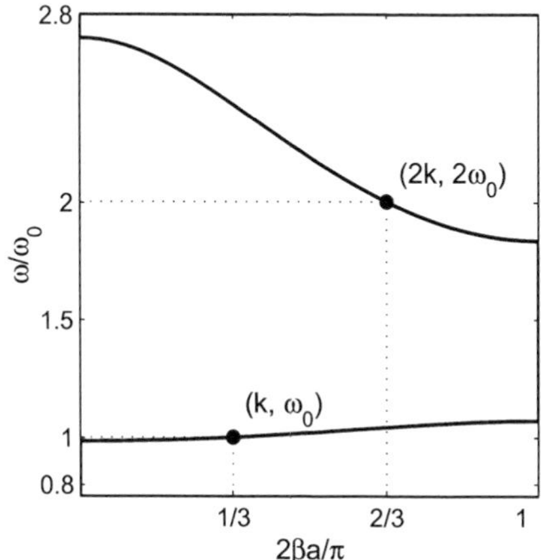

Fig. 7.21 Dispersion curve for the biperiodic array permitting the propagation of both signal (k, ω) and pump $(2k, 2\omega)$ waves required for parametric amplification. From Sydoruk *et al.* (2005). Copyright © 2005 American Institute of Physics

the pump wave as $2\beta a = 2\pi/3$ at the frequency $2\omega_0/2\pi = 127.74$ MHz. These requirements can be satisfied with an axial structure where the distance between the elements is $a = 0.5r_o$ $(M = 0.336L)$, $C_1 = 164$ pF, $C_2 = 56$ pF, $L = 33$ nH. The corresponding dispersion curve is shown in Fig. 7.21. It can be seen that the condition of synchronism is satisfied and both the signal and the pump waves can propagate in the system.

So far we have talked about an infinitely long biperiodic line. Does a terminal impedance exist that matches the line, that makes it possible to have a single travelling wave? Since the line consists now of two different kinds of elements it may be expected that we shall need two terminal impedances, one to be inserted into the last, and the other one into the last-but-one element. This may be shown to be the case. The values of the two terminal impedances may be found from a calculation similar to that in Section 7.3 as

$$Z_T(1,2) = -\frac{M_{1,2} Z_{01,02}}{M_{1,2} + M_{2,1}\, \mathrm{e}^{\mathrm{j}\, k(a_1 + a_2)}} \,. \tag{7.48}$$

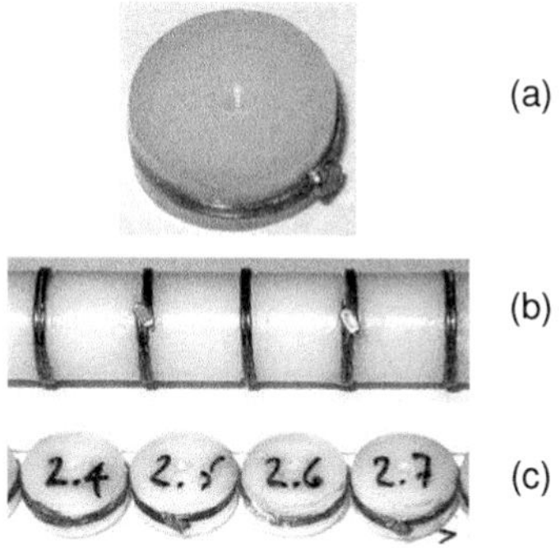

Fig. 7.22 Capacitively loaded loops. Photographs of single element (a) and fragments of an axial line (b) and a planar line (c). From Wiltshire *et al.* (2003*b*). Copyright © 2003 IEE

7.11 Experimental results

The first experimental results were obtained (Wiltshire *et al.*, 2003*b*) not long after the derivation of the dispersion equation (Shamonina *et al.*, 2002*a*). The basic element of the line, a capacitively loaded loop was realized by winding two turns of 1-mm diameter copper wire on a dielectric rod (diameter 9.6 mm), and tuned to the desired frequency of 60 MHz by inserting a capacitor (nominally 100 pF) between the ends of the wire, as shown in Fig. 7.22(a). Two lines were assembled from these elements, an axial line in which 32 elements were arranged along the axis of a dielectric rod, spaced by their diameter (Fig. 7.22(b)) and a planar line in which 15 elements were placed side-by-side (Fig. 7.22(c)). The resonant frequency and the quality factor were determined as $f_0 = (61.4 \pm 0.4)$ MHz and $Q = 48 \pm 5$.

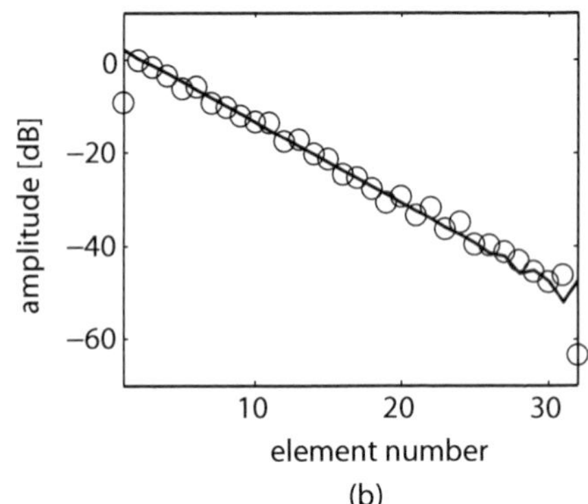

Fig. 7.23 Phase (a) and amplitude variation (b) along axial structure at 61 MHz. From Wiltshire *et al.* (2003*b*). Copyright © 2003 IEE

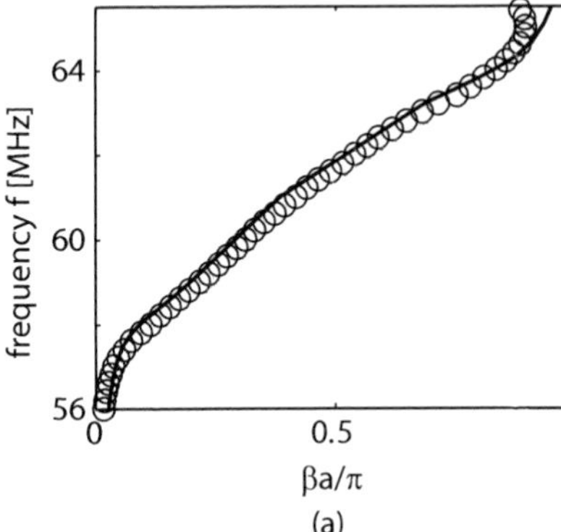

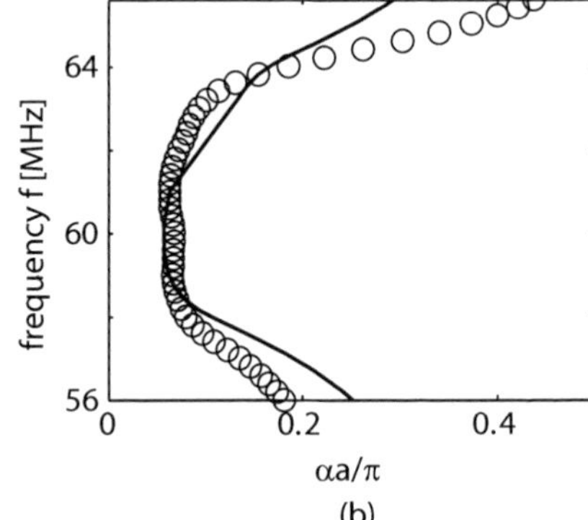

Fig. 7.24 Dispersion relationship for axial line. (a) Propagation constant and (b) attenuation constant versus frequency. From Wiltshire *et al.* (2003*b*). Copyright © 2003 IEE

The overall transmission characteristics were measured with a network analyzer at 401 frequency points in the range 50–70 MHz by placing the transmitter loop near to the first element and the receiver loop near to the last element. The measurements showed clearly the existence of a pass band flanked by two stop bands.

The dispersion characteristics were measured using the same setup, with the input loop next to the first element and the output loop over the rod adjacent to each successive element in turn, and the phase and amplitude of the output wave measured at each element for the same 401 frequencies. A typical behaviour of the phase and amplitude against element position at 61 MHz is shown by circles in Figs. 7.23(a) and (b). The phase variation may be seen to be substantially linear. The amplitude variation, plotted on a logarithmic scale, is also linear, apart from the end of the line where there is some variation due to a reflected wave (note that no attempt was made to match the line). The theoretical values are shown by solid lines. The agreement is quite good.

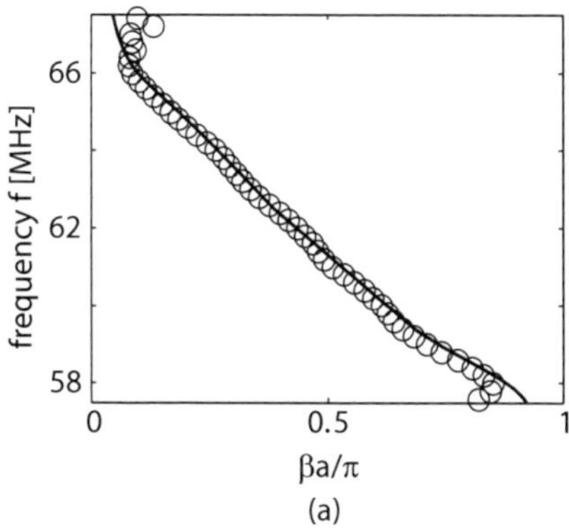

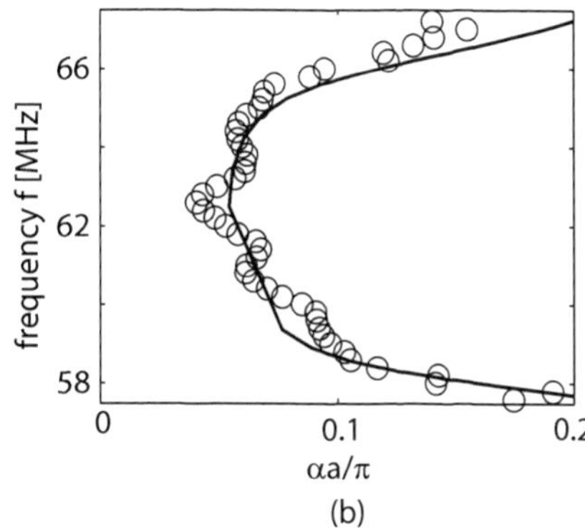

Fig. 7.25 Dispersion relationship for planar line. (a) Propagation constant and (b) attenuation constant versus frequency. From Wiltshire *et al.* (2003*b*). Copyright © 2003 IEE

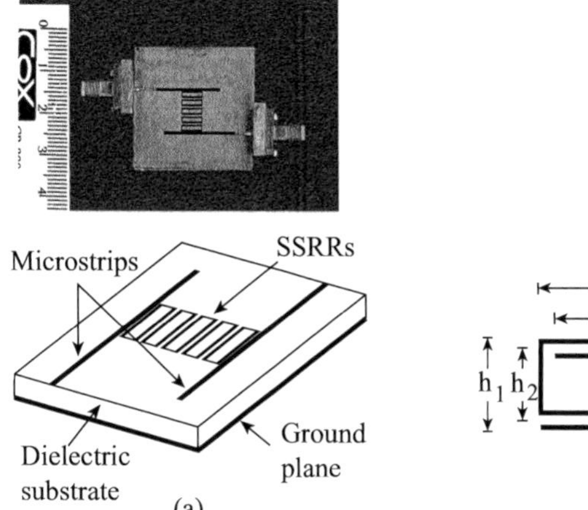

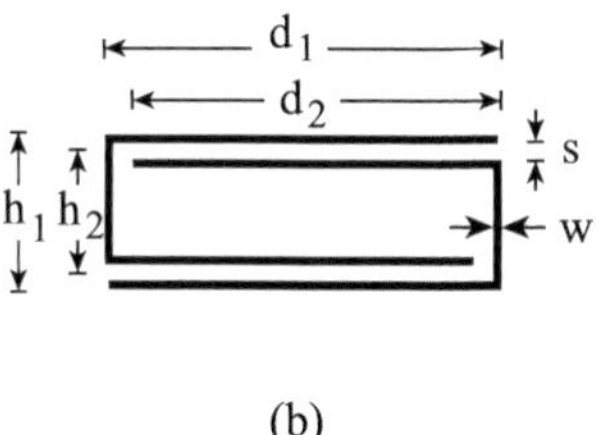

Fig. 7.26 (a) Photograph and sketch of the planar MI transducer. (b) sketch of the element. From Freire *et al.* (2004). Copyright © 2004 American Institute of Physics

Having obtained the phase and amplitude variation from element to element at all the frequencies we can now deduce the dispersion curves versus $\beta a/\pi$ and $\alpha a/\pi$. These are shown in Figs. 7.24 and 7.25 for the axial and planar configurations, respectively. The theoretical curves, solid lines, are obtained from the parameters of the two lines and the measured value of the quality factor. The agreement is very good for the phase variation and quite reasonable for the attenuation.

Propagation of MI waves at a much higher frequency (in the range of 3.5 to 5 GHz) were measured by Freire *et al.* (2004) using a MI transmission line between two microstrip lines as shown in Fig. 7.26(a). The shape of the individual elements is a variation on the split-ring resonator, as may be seen in Fig. 7.26(b). The high length to width ratio was chosen to strengthen the coupling between adjacent elements. They found a pass band centred around 4.5 GHz. The total transmission loss was close to 3 dB, i.e. the loss per element was 0.6 dB, in contrast to about 1.4 dB per element that follows from Fig. 7.23(b).

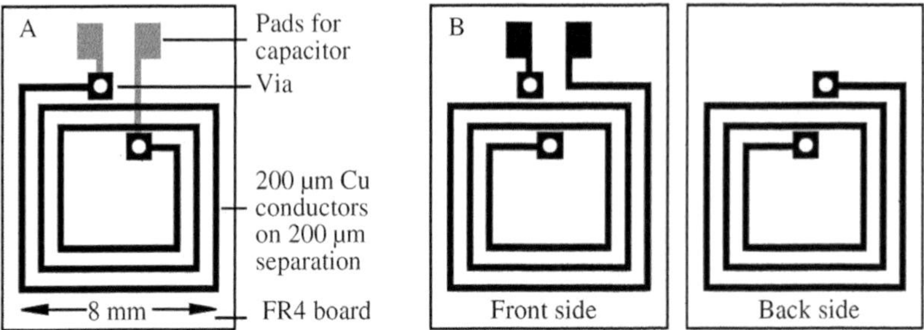

Fig. 7.27 Single-sided (a) and double-sided PCB element (b). From Syms *et al.* (2006*b*). Copyright © 2006 IOP Publishing Ltd

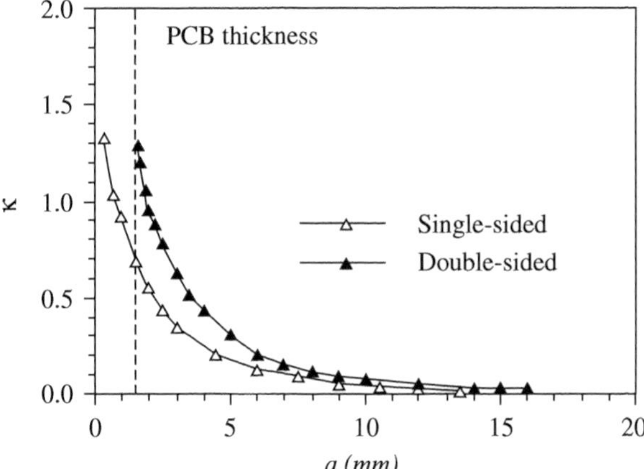

Fig. 7.28 Variation of the coupling coefficient κ with the element spacing a. From Syms *et al.* (2006*b*). Copyright © 2006 IOP Publishing Ltd

Since low attenuation is a necessary condition for any practical application for signal processing further attempts were made by Syms *et al.* (2006*b*) to lower the attenuation. The solutions chosen may be seen in Fig. 7.27. In Fig. 7.27(a) the turns made of copper are only on one side of a printed circuit board (PCB), whereas in Fig. 7.27(b) the inductor is double-sided. There are pads in both cases for adding capacitors to the inductors in order to bring the resonant frequency to the desired value.

As may be expected, the coupling coefficient is higher for the double-sided case as shown in Fig. 7.28. It is worth noting that by placing the elements close to each other in the axial configuration a coupling coefficient as high as 1.5 can be achieved, which makes wide-band operation possible.

The experimental setup using double-sided elements is shown in Fig. 7.29(a). The measured values of S_{21} as a function of frequency for a 30-element line and for six different element spacings may be seen in Fig. 7.29(b). For the smallest spacing of 2.5 mm between the elements the lowest value of S_{21} is about 4.4 dB. The coupling losses were measured as 0.4 dB, both at the input and at the output. This low value was achieved by introducing matching elements between the first (last) element of the line and the network analyzer. Deducting the coupling losses the

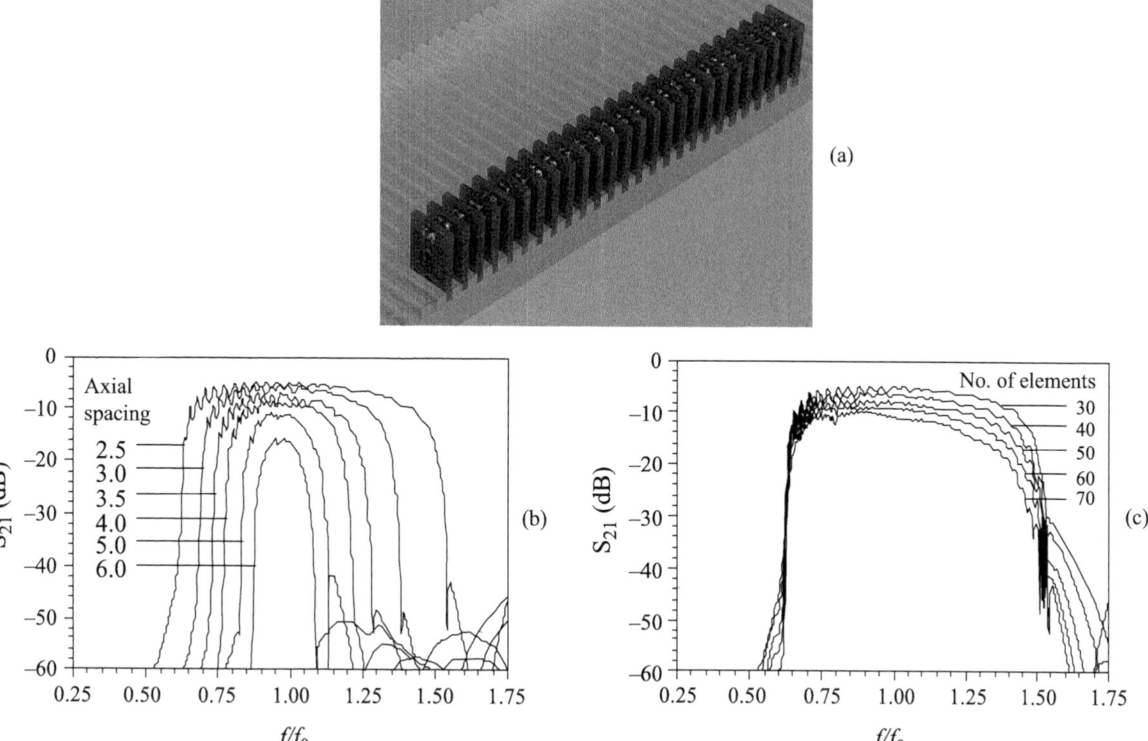

Fig. 7.29 (a) Experimental arrangement of MI waveguides; (b) experimental frequency variation of S_{21} for waveguides based on double-sided coils and a fixed number of elements (30) and different element spacing, and (c) corresponding result for a fixed element spacing (2.5 mm) and different numbers of elements. From Syms *et al.* (2006*b*). Copyright © 2006 IOP Publishing Ltd

attenuation per element comes to the low figure of 0.12 dB per element. The dependence on the number of elements of the S_{21} versus frequency curve is shown in Fig. 7.29(c). It may be seen that it is feasible to set up long lines.

All the experiments mentioned so far were on lines consisting of identical elements. We shall now report experimental results measured on biperiodic lines by Radkovskaya *et al.* (2007*a*). The loop used in the experiments, we shall call it a split pipe, is shown in Fig. 7.30(a). It became a resonant element at a desired frequency when loaded by a capacitor. Two sets of elements were used, one loaded by a capacitor of 330 pF, and the other one by a capacitor of 680 pF, yielding resonant frequencies of 46.21 MHz and 32.46 MHz. The two sets were interleaved to produce a biperiodic line as shown in Fig. 7.30(b) for the axial and in Fig. 7.30(c) for the planar configuration where the positions of the transmitter and receiver coils are also shown. The measurement technique was the same as described earlier in this section.

The measured and theoretical curves of ω versus βa are shown in Figs. 7.31(a) and (b) for the axial and planar configurations, respectively. The

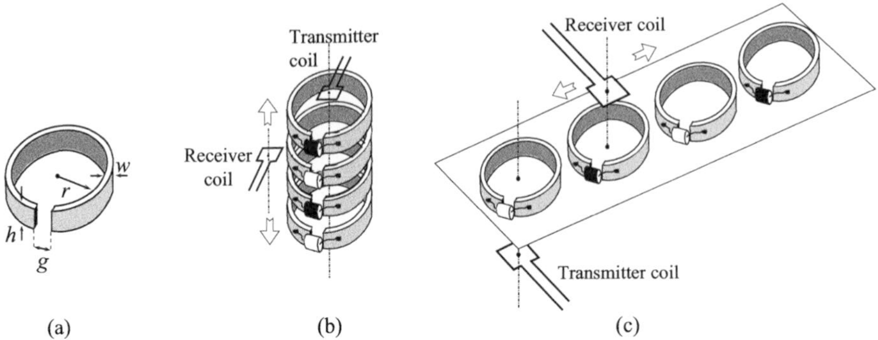

Fig. 7.30 Biperiodic arrays of capacitively loaded split pipes. Schematic representations of (a) element dimensions, (b) axial and (c) planar configurations with measuring coils. From Radkovskaya *et al.* (2007*a*). Copyright © 2007 IEE

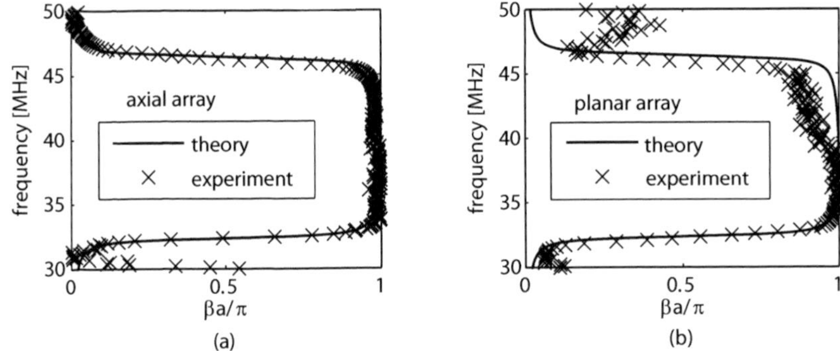

Fig. 7.31 Dispersion characteristics for the axial (a) and planar (b) biperiodic structure. From Radkovskaya *et al.* (2007*a*). Copyright © 2007 IEE

agreement is remarkably good for the planar line and quite good for the axial line. In particular, it should be emphasized that for the biperiodic line there is hardly any difference between the dispersion curves of the axial and planar configurations as predicted by eqn (7.47).

7.12 Higher-order interactions

We have so far considered nearest-neighbour interaction only. This is usually a good approximation when there is fast decay of the fields away from the element. By fast decay we mean cubic decay as would be the case for elements that can be regarded static magnetic dipoles. The question nevertheless arises: what kind of modifications in the properties of MI waves would be caused by taking higher-order interactions into account? The generalization to higher interactions is straightforward. In eqn (7.2) Kirchhoff's voltage law is applied to element n assuming nearest-neighbour interaction only. If elements further away may also induce voltages then eqn (7.2) will modify to

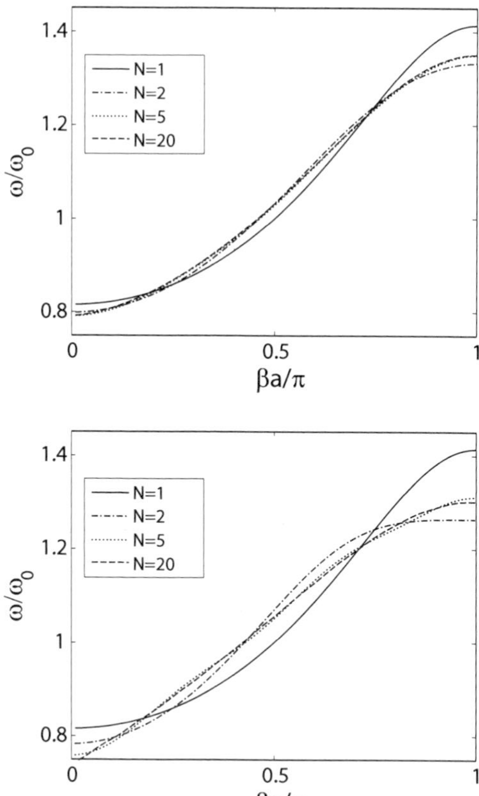

Fig. 7.32 Dispersion characteristic of a MI waveguide assuming a cubic decay of the coupling constant with distance and including interactions up to $N = 1$, 2, 5 and 20

Fig. 7.33 Same as in Fig. 7.32 but assuming a quadratic decay of the coupling constant with distance

$$Z_0 I_n + \mathrm{j}\omega \sum_{m=1}^{\infty} M_m (I_{n+m} + I_{n-m}) = 0\,, \tag{7.49}$$

where M_m is the mutual inductance between two elements a distance ma from each other. Assuming again a wave solution of the form of eqn (7.4) the dispersion equation can be derived as

$$1 - \frac{\omega_0^2}{\omega^2} + \sum_{n=1}^{\infty} \kappa_n \cos(nkd) = 0\,, \tag{7.50}$$

where $\kappa_n = 2M_n/L$ and only the lossless case is considered. How large is the influence of higher-order couplings? Assuming cubic decay of the mutual inductance with distance and $\kappa_1 = 0.5$ the dispersion equation is plotted in Fig. 7.32 for $N = 1$, 2, 5 and 20. The effect of higher interactions may be seen to be small. For finite-size elements the cubic decay is not a good approximation. The actual decay may be closer to a quadratic one. In Fig. 7.33 we plot again the dispersion curve for $N = 1$, 2, 5 and 20 and $\kappa_1 = 0.5$. For this lower decay, as expected, the effect of higher orders is more significant.

Can we say anything in more general terms about the dispersion curve? The answer is yes, when κ_n can be expanded into a series in

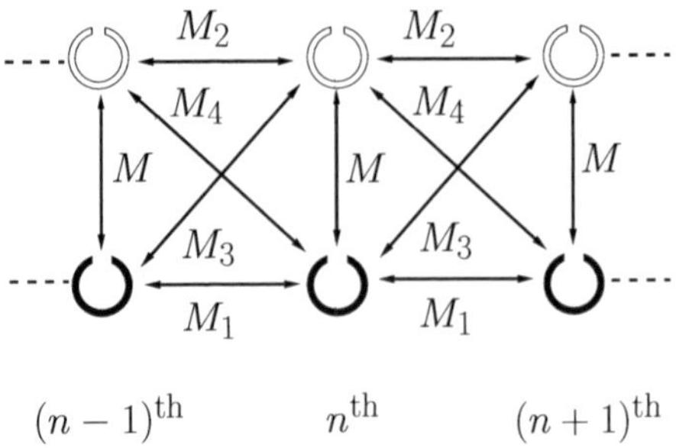

Fig. 7.34 Schematic representation of the coupling between two lines of resonant magnetic metamaterial elements. Mutual inductances M_1, M_2, M, M_3 and M_4 between the nearest neighbours are shown by arrows. From Sydoruk *et al.* (2006). Copyright © 2006 by the American Physical Society

inverse powers of the distance,

$$\kappa_n = \sum_{m=2}^{\infty} c_m (na)^{-m} , \tag{7.51}$$

where the coefficients c_m need to be determined. With the aid of eqn (7.51) we find the dispersion equation (7.50) reduces to the form

$$1 - \frac{\omega_0^2}{\omega^2} + \sum_{m=2}^{\infty} c_m a^{-m} \sum_{n=1}^{\infty} n^{-m} \cos(nka) = 0 . \tag{7.52}$$

Note that the summation over n can be expressed in terms of Li_m, a set of special functions called polylogarithms (Lewin, 1981). The relation is

$$\sum_{n=1}^{\infty} n^{-m} \cos(nka) = \frac{1}{2} \left[\mathrm{Li}_m \left(\mathrm{e}^{\mathrm{j}\,ka} \right) + \mathrm{Li}_m \left(\mathrm{e}^{-\mathrm{j}\,ka} \right) \right] . \tag{7.53}$$

For our purpose the most important property of this special function is that it is monotonic, from which it follows that the dispersion curve, provided the expansion of κ_n is possible in the form of eqn (7.51), is also monotonic.

The generalization to higher-order interactions is also straightforward for the case when there is excitation by a set of voltages. The mathematical form is still given by eqn (7.13) as

$$\mathbf{V} = Z\mathbf{I} , \tag{7.54}$$

but Z is no longer a tri-diagonal matrix. The main diagonal elements are still equal to Z_0 but the off-diagonal elements are now equal to

$$Z_{ij} = \mathrm{j}\,\omega M_{|i-j|}, \qquad i \neq j . \tag{7.55}$$

For a given set of voltage excitations we may still obtain the current distribution by inverting the relation in eqn (7.13).

7.13 Coupled one-dimensional lines

We shall now look at the case (Sydoruk *et al.*, 2006) when two lines, both capable of propagating MI waves, are coupled to each other, as shown schematically in Fig. 7.34.

We shall describe their properties by first assuming nearest-neighbour interaction. However, in contrast to a one-dimensional line, there are now not two but five nearest neighbours. The notations for the mutual inductances are shown in Fig. 7.34 as M_1 (between neighbouring elements in line 1), M_2 (between neighbouring elements in line 2), M (between the nth element in line 1 and the corresponding nth element in line 2), M_3 (between the $(n-1)$th element in line 1 and the nth element in line 2), and M_4 (between the nth element in line 1 and the $(n-1)$th element in line 2).

Kirchhoff's voltage equations written for element n in line 1 and the corresponding element n in line 2 are as follows,

$$Z_{01} I_n + \mathrm{j}\omega M_1 (I_{n-1} + I_{n+1}) + \mathrm{j}\omega M J_n + \mathrm{j}\omega M_4 J_{n-1} + \mathrm{j}\omega M_3 J_{n+1} = 0 \tag{7.56}$$

and

$$Z_{02} J_n + \mathrm{j}\omega M_2 (J_{n-1} + J_{n+1}) + j\omega M I_n + \mathrm{j}\omega M_3 I_{n-1} + \mathrm{j}\omega M_4 J_{n+1} = 0\,, \tag{7.57}$$

where I_n and J_n are the currents in lines 1 and 2, respectively. Assuming further wave solutions in both lines 1 and 2 eqns (7.56) and (7.57) yield the dispersion equation in analytic form (Sydoruk *et al.*, 2006). We shall show here only the results for two different coupling arrangements, as shown in Figs. 7.35(a) and (b), where the mutual inductances taken into account are also shown. The first one shows two planar lines placed above each other, a simple enough situation, but even then we find the unusual coupling relationship that the intraline mutual inductances are negative, whereas the interline mutual inductances are positive. Figure 7.35(b) shows a planar line coupled to an axial line. The interesting feature is now that the upper line carries a forward wave, the lower line a backward wave and the elements just above each other are not coupled.

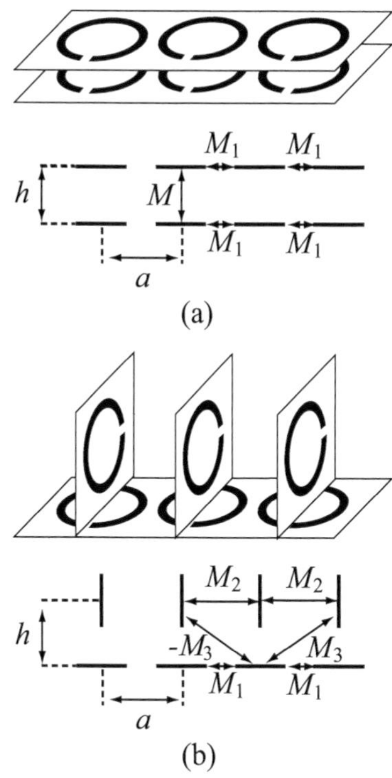

Fig. 7.35 (a) Configurations of bi-atomic metamaterial structures (a) planar lines above each other. (b) Planar and axial lines above each other. The non-zero mutual inductances are shown by arrows. From Sydoruk *et al.* (2006). Copyright © 2006 by the American Physical Society

What kind of dispersion equation curves would we expect for the two coupled planar lines? A single planar line supports backward waves. If the coupling between the two planar lines is small (the two lines are at a distance of 20 mm from each other) then we may expect a small split in the dispersion curve. If the distance between the lines is less (10 mm) then we may expect a bigger split.

Experiments on two coupled lines were conducted by Sydoruk *et al.* (2006) using the split-pipe elements shown in Fig. 7.30(a) used in the experiments of Radkovskaya *et al.* (2007*b*). The loading capacitor had a capacitance of $C = 330$ pF, which led to a resonant frequency of 46.21 MHz. For the planar lines (Fig. 7.35(a)) the experimental and theoretical results are shown in Figs. 7.36(a) and (b) for line separations of $h = 20$ mm and 10 mm. As expected, there is a small split for the large separation and a large split for the smaller separation.

The interesting result is that when the split is large we have effectively two pass bands with a stop band between them. The same elements were used for the coupled line shown in Fig. 7.35(b), where a backward wave is

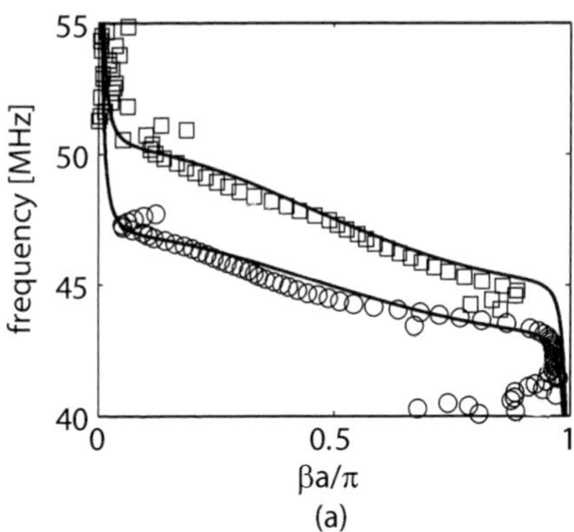

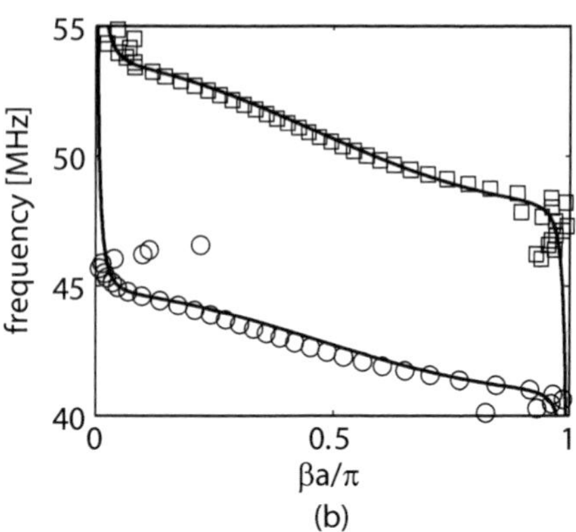

Fig. 7.36 Dispersion characteristics of coupled planar lines with h = 20 mm (a) and 10 mm (b). Theory (solid lines) and experiment (circles and squares). From Sydoruk *et al.* (2006). Copyright © 2006 by the American Physical Society

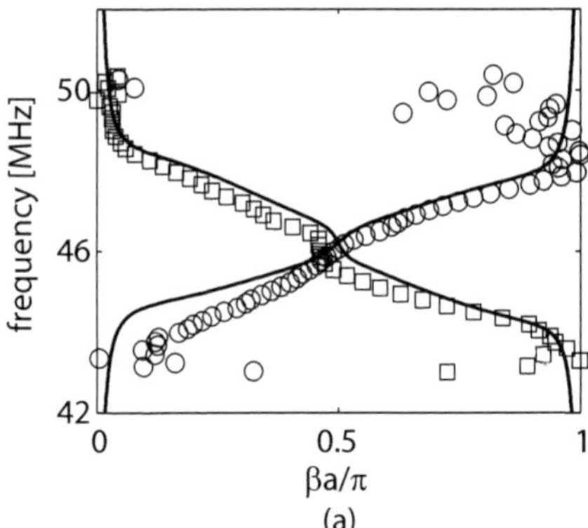

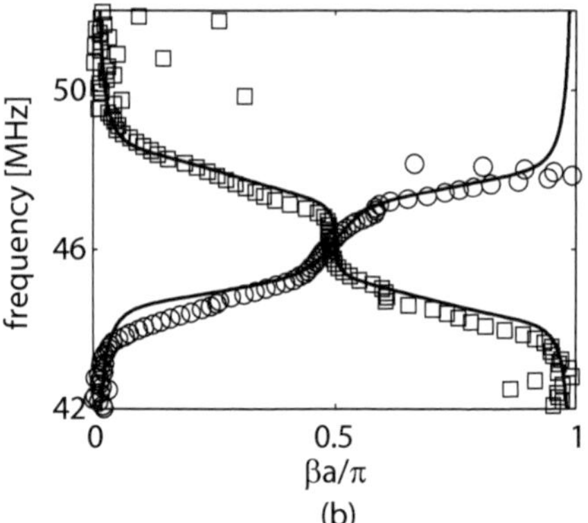

Fig. 7.37 Dispersion characteristics of coupled planar–axial lines with h = 30 mm (a) and 15 mm (b). Theory (solid lines) and experiment (circles and squares). From Sydoruk *et al.* (2006). Copyright © 2006 by the American Physical Society

coupled to a forward wave. The theoretical and experimental dispersion curves are shown in Figs. 7.37(a) and (b). As may be expected, we have a combination of forward and backward waves. Notice that there is a stop band around the resonant frequency. For smaller coupling ($h = 30$ mm) the stop band is smaller, for larger coupling ($h = 15$ mm) the stop band is larger. The agreement between theory and experiment is very good, although we have a few spurious experimental points that cannot be accounted for by the theory.

Having looked at the dispersion curves we shall return to the coupled lines with a different question in mind. How large is the transmission, i.e. what is the value of S_{21}, between the first element of line 1 and the last element of line 2 (see Fig. 7.38(a) for the experimental arrangement) and how does it vary as the two lines are shifted (Fig. 7.38(b)) relative to each other? The theoretical calculations have been done by Radkovskaya *et al.* (2007*b*) by assuming interaction between any two elements, i.e. relying on the generalized Ohm's law presented in eqn (7.13). This

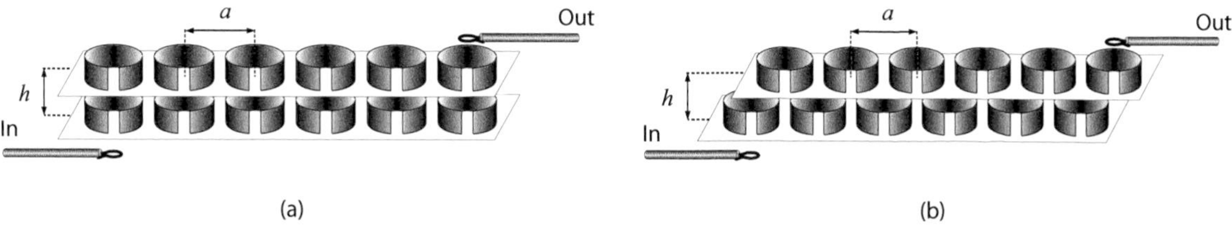

Fig. 7.38 Schematic representation of the (a) unshifted and (b) half-a-period shifted coupled lines. The first element of the lower array is excited by a transmitting coil and the signal in the last element of the upper array is measured by a receiving coil. From Radkovskaya *et al.* (2007*b*). Copyright © 2007 Wiley-VCH Verlag GmbH & Co. KGaA

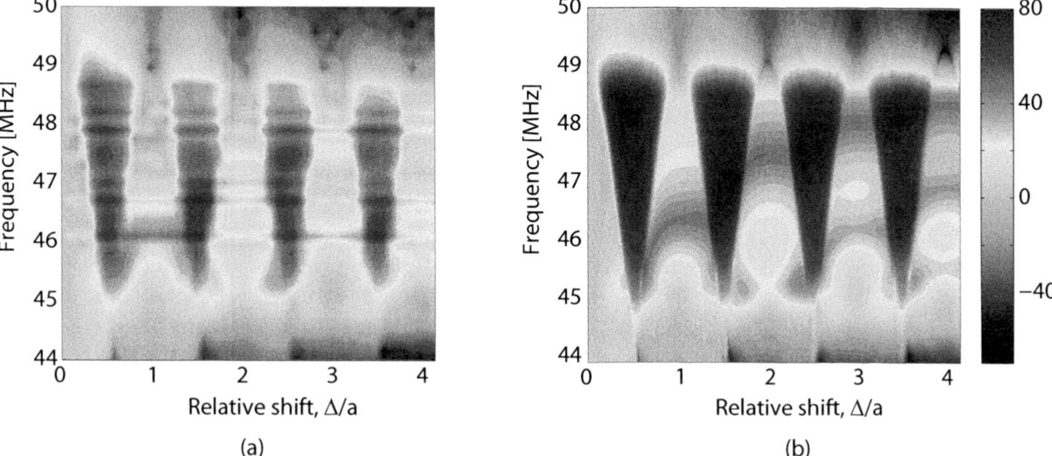

Fig. 7.39 Contour plots of transmission between the split-pipe arrays as a function of frequency and shift, (a) experiment, (b) theory. From Radkovskaya *et al.* (2007*b*). Copyright © 2007 Wiley-VCH Verlag GmbH & Co. KGaA. For coloured version see plate section

necessitated the determination of the mutual inductances between any two elements. It was done by assuming each element to be represented by a filamentary loop. If we know all the elements of the $N \times N$ Z matrix then the current distribution may be obtained by inverting numerically the Z matrix. The experiments were performed and comparisons with theoretical results were made by Radkovskaya *et al.* (2007*b*) as shown in Figs. 7.39(a) and (b). The amplitude of S_{21} is colour-coded. The variables are the shift between the elements (this is done for four full periods) on the horizontal axis and the frequency on the vertical axis. The agreement between theory and experiment may be regarded as very good. Note that maximum transmission occurs when the shift is a half-period and minimum transmission when the elements are just above each other. The results have an interesting implication for coupled lines. As far as we know this is the first time ever that by reducing the coupling between opposite elements in coupled lines the transfer of power actually increases. The physical reason can be attributed to the dispersion curve shown in Fig. 7.36(b). When the coupling between the two lines is strong a stop band appears and total transmission is minimum.

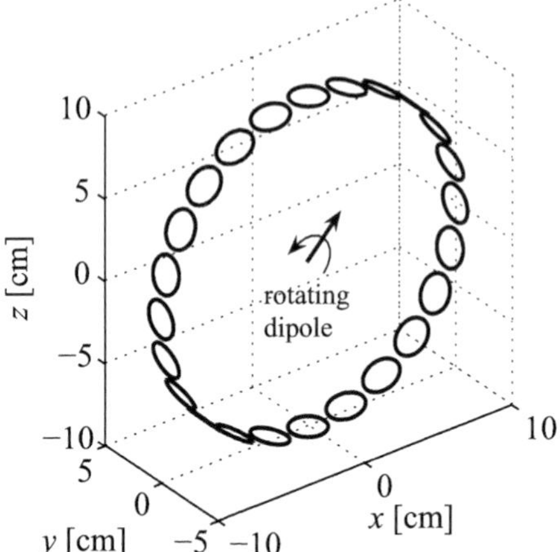

Fig. 7.40 Schematic view of rotational resonator composed of 24 capacitively loaded loops. From Solymar *et al.* (2006). Copyright © 2006 American Institute of Physics

7.14 Rotational resonance

We have discussed so far MI wave propagation along a line and analyzed what happens when two lines are coupled to each other. We shall now consider the case when the MI wave travels round and round on a ring structure of resonant elements, as shown in Fig. 7.40.

We call the resulting resonance a rotational resonance (Solymar *et al.*, 2006). Such a construction is known in microwaves as a strip-line ring resonator (Chang, 1996). It operates on the principle that the total phase shift that a wave accumulates round a closed path should be an integral multiple of 2π. An alternative description is that the resonance occurs when the circumference of the circle is equal to an integral number of wavelengths of the MI wave. Another analogy is with the cavity magnetron that is operated by coupling power from a circulating electron beam to an electromagnetic wave propagating round a ring of coupled cavity resonators.

The general dispersion equation, with all interactions taken into account, is given by eqn (7.50). The difference is that in the ring structure there are only a finite number of interactions, depending obviously on the number of elements. As it happens, there is a slight difference in the dispersion equation depending on whether the number of elements constituting the ring is odd or even. The dispersion equations can be found easily with the aid of a few algebraic operations. For even numbers it is

$$1 - \frac{\omega_0^2}{\omega^2} + \frac{\kappa_N}{2}\,\mathrm{e}^{-\mathrm{j}Nka} + \sum_{n=1}^{N-1} \kappa_n \cos(nka) = 0\,, \qquad (7.58)$$

where the number of elements is equal to $2N$, $\kappa_n = 2M_n/L$ and M_n is the mutual inductance between two elements n neighbours apart. For an odd number of elements the dispersion equation is

Table 7.1 Rotational resonances ω_n/ω_0 for a 9-element ring. From Solymar *et al.* (2006)

resonance number n	nearest-neighbour case	all-interaction case
0	1.069	1.085
1	1.052	1.049
2	1.011	1.005
3	0.970	0.969
4	0.946	0.949

$$1 - \frac{\omega_0^2}{\omega^2} + \sum_{n=1}^{N} \kappa_n \cos(nka) = 0\,, \tag{7.59}$$

where the number of elements is equal to $2N + 1$. As an example we shall choose a 9-element ring. The elements are assumed to be circular resonant loops nearly touching each other. The corresponding positions of the rotational resonances are given in Table 7.1 both in the nearest-neighbour and in the all-interaction cases for a quality factor, $Q = 100$. It may be seen that the more accurate calculations show a small shift in the resonant frequencies.

7.15 Applications

7.15.1 Introduction

Waves propagating on coupled resonant structures had applications in microwave tubes and in linear accelerators (Bevensee, 1964). In the metamaterial context at the time of writing we can only talk about potential applications. To that category belongs the early work of Wiltshire *et al.* (2001) who conducted, with the aid of swiss rolls, magnetic information from an MRI machine to a detector. An application as a delay line was envisaged by Freire *et al.* (2004). Their device was shown in Fig. 7.26(a). The information travelled from one strip line to the other strip line via a MI wave. The maximum delay time measured was about 6 ns. Phase shifters were designed by Nefedov and Tretyakov (2005). They made good use of the fact that lines with positive phase shift (those that support forward waves) and also with negative phase shift (those that support backward waves) are simultaneously available.[5] Near-field imaging was demonstrated by Freire and Marques (2005). Later, it was shown by the same group (Mesa *et al.*, 2005) that the imaging strongly depends on the characteristics of the receiver.

In this section we shall investigate three kinds of applications in a little more detail. Section 7.15.2 will describe the signal processing aspects of MI waves that require various waveguide components. Section 7.15.3 will discuss potential applications for imaging with some experimental results showing the quality of imaging obtained. Finally, Section 7.15.4

[5] That arrangement can lead to shorter phase shifters. If, for example, a phase shifter of -20° is required it can be a short one if negative phase shift is available. On the other hand, if one has to rely on a positive phase shift then the phase shifter must be 17 times larger to offer a phase shift of 340°.

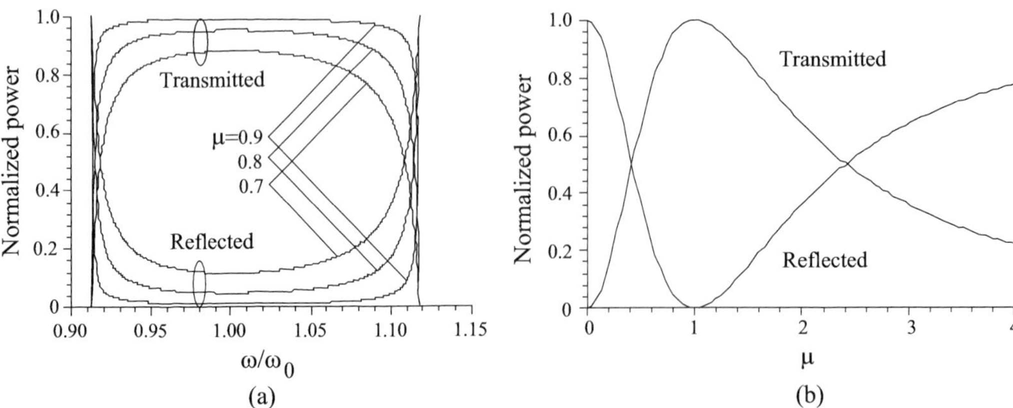

Fig. 7.41 Magnetoinductive waveguide mirror. (a) Variation of $|R|^2$ and $|T|^2$ with ω/ω_0 for $\kappa = 0.2$ and different values of μ. (b) Variation of $|R|^2$ and $|T|^2$ with μ for $\omega = \omega_0$. From Syms *et al.* (2006*a*). Copyright © 2006 IEE

will resurrect the rotational resonance of the previous section and discuss briefly its potential applications in magnetic resonance measurements.

7.15.2 Waveguide components

The design of various MI waveguide components was treated in Shamonina and Solymar 2004; Syms *et al.* 2005*c*; Syms *et al.* 2006*a*. We shall present here some of the components discussed by Syms *et al.* (2006*a*). The first one is a frequency-dependent reflector that we have already analyzed in Section 7.9 as a boundary-reflection problem. The setup may be seen in Fig. 7.15. Two one-dimensional lines may be seen to be coupled to each other by a mutual inductance M_b across the boundary. Let us now simplify the problem and say that the two lines are identical, so that the only new parameter is M_b. Then, eqns (7.35) and (7.36) lead to the power reflection and transmission coefficients

$$R^2 = \frac{(\mu^2-1)^2}{D} \qquad \text{and} \qquad T^2 = \frac{4\mu^2 \sin^2(ka)}{D}, \tag{7.60}$$

where

$$D = 1 + \mu^4 - 2\mu^2\cos(2ka) \qquad \text{and} \qquad \mu = \frac{M_\mathrm{b}}{M}. \tag{7.61}$$

The normalized power for $\kappa = 0.2$ and $\mu = 0.9$, 0.8 and 0.7 is shown in Fig. 7.41(a) as a function of frequency over the range corresponding to the pass band of the MI wave. The curves are slowly varying across the band. Towards the band edge, where $ka = 0$ or π, the power reflection coefficient tends to unity: the discontinuity reflects everything. Near the centre of the band the transmission is high but steadily reduces with μ. In this region we may obtain the approximation

$$R = \frac{\mu^2 - 1}{\mu^2+1} \qquad \text{and} \qquad T = \frac{2\mu}{\mu^2+1}. \tag{7.62}$$

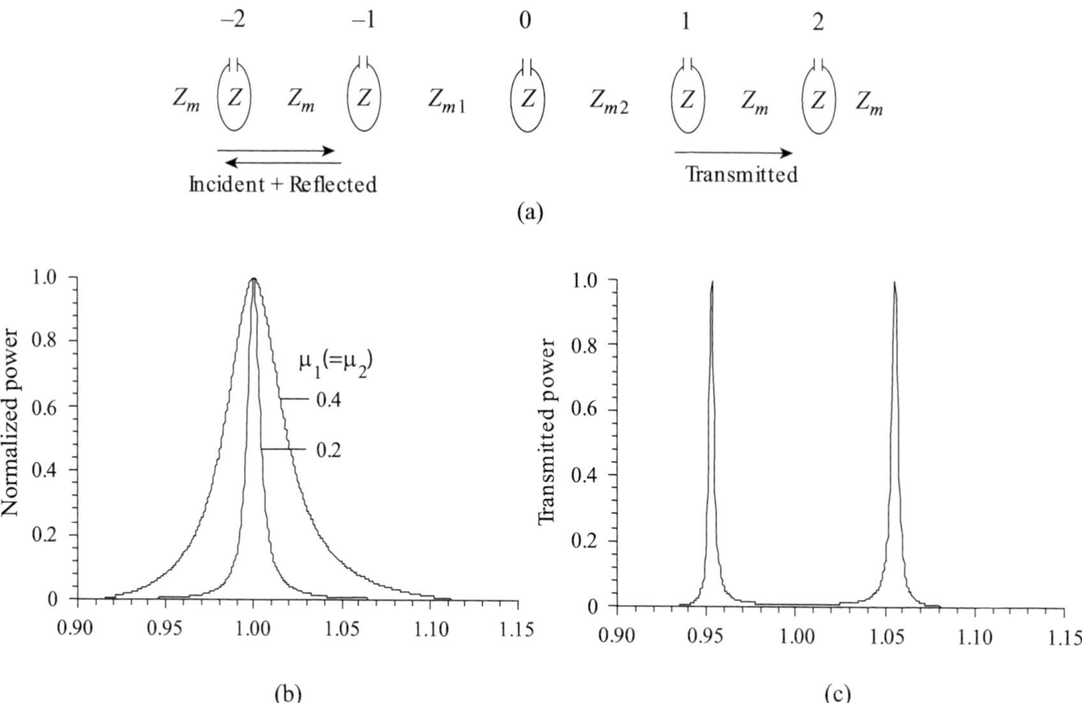

Fig. 7.42 (a) Magnetoinductive waveguide Fabry–Perot resonator. (b) Variation of $|T|^2$ with ω/ω_0 for $\kappa = 0.2$ and different values of μ. (c) Variation of $|T|^2$ with ω/ω_0 for two-loop cavity with $\kappa = 0.2$ and $\mu = 0.2$. From Syms *et al.* (2006*a*). Copyright © 2006 IEE

Figure 7.41(b) shows the variation of R^2 and T^2 with μ obtained from the above equation. These results are independent of ka and show that a reflector with reasonable broadband performance may be constructed by a slight variation in the spacing at the junction between two lines.

As known in optics, two collinear reflectors make up a Fabry–Perot resonator and that applies to MI waves as well. This is achieved by inserting an additional reflector as shown in Fig. 7.42(a). The mutual inductance between the elements is taken as M, with the exception of those between elements -1 and 0, and between 0 and $+1$, which are denoted by M_1 and M_2, respectively. We may then write Kirchhoff's equations for elements -1, 0 and 1 as follows

$$Z_0 I_{-1} + \mathrm{j}\omega M_1 I_0 + \mathrm{j}\omega M I_{-2} = 0, \tag{7.63}$$

$$Z_0 I_0 + \mathrm{j}\omega M_2 I_1 + \mathrm{j}\omega M_1 I_{-1} = 0, \tag{7.64}$$

$$Z_0 I_1 + \mathrm{j}\omega M I_2 + \mathrm{j}\omega M_2 I_0 = 0. \tag{7.65}$$

The current in the uniform line to the left of element -1 may be assumed in the form of an incident and a reflected wave, and as a traveling wave to the right of element 1. There are three unknowns then, the reflection coefficient, the transmission coefficient and the current in

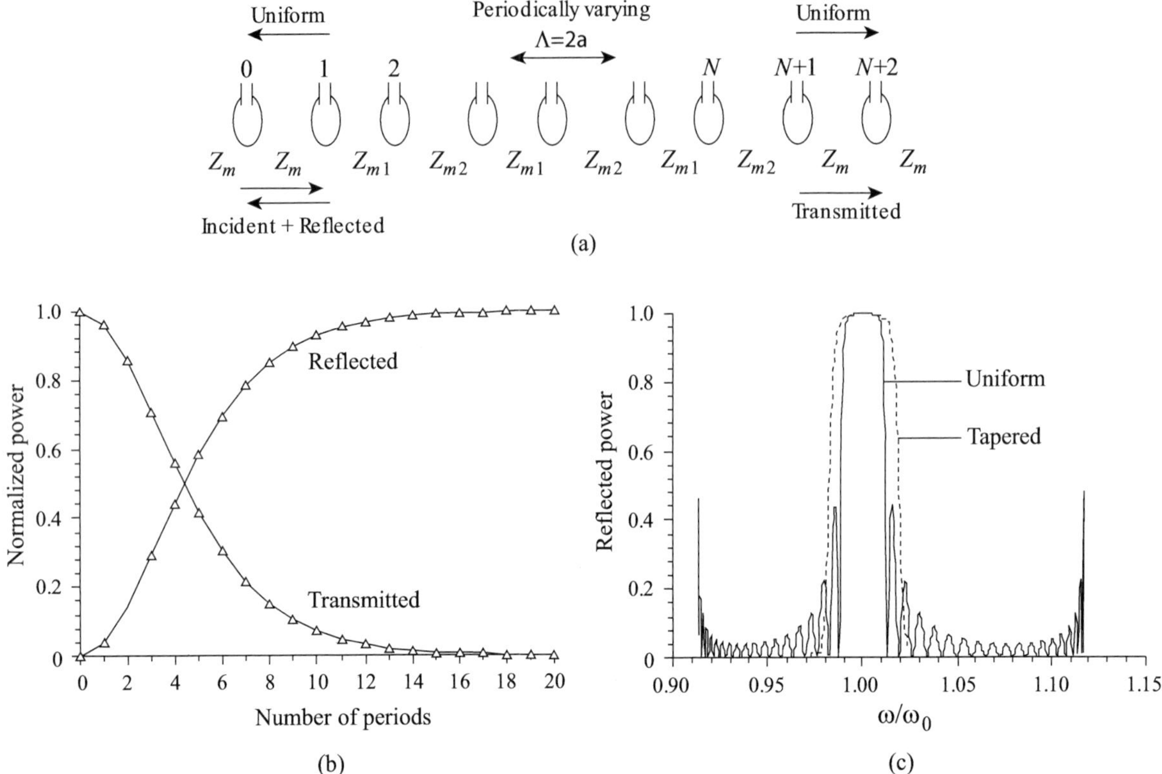

Fig. 7.43 (a) Magnetoinductive waveguide Bragg grating. (b) Variation of $|R|^2$ and $|T|^2$ at $\omega = \omega_0$, with the number of periods for a Bragg grating with $\kappa = 0.2$ and $\mu_1 = 1.1$, $\mu_2 = 0.9$. (c) Variation of $|R|^2$ with ω/ω_0 for a similar grating containing 20 periods (full line) and corresponding result for grating with raised cosine taper (dashed line). From Syms *et al.* (2006*a*). Copyright © 2006 IEE

element 0. They can be obtained from eqns (7.63)–(7.65) and the analytical solution may be found in Syms *et al.* (2006*a*). Choosing again the coupling coefficient between nearest neighbours in the uniform lines, $\kappa = 0.2$ and $\mu_1 = M_1/M = \mu_2 = M_2/M$ equal to 0.2 and 0.4 the normalized transmitted power is plotted in Fig. 7.42(b). Clearly, there is a transmission peak around the resonant frequency.

If we insert additional resonant loops between elements 0 and 1 in Fig. 7.42(a) then we have additional transmission peaks. The transmitted power as a function of frequency for a two-loop cavity is shown in Fig. 7.42(c) for the same set of parameters. There are now two narrow transmission peaks.

Following further the optical analogy we should be able to have large reflection in a certain frequency band if a large number of small reflections add coherently. These are called Bragg reflectors in optics. We can achieve a large number of small reflections if the mutual inductances undergo a small but periodic variation along the line (see Fig. 7.43(a), where only one period is shown). The period is $\Lambda = 2a$. How many peri-

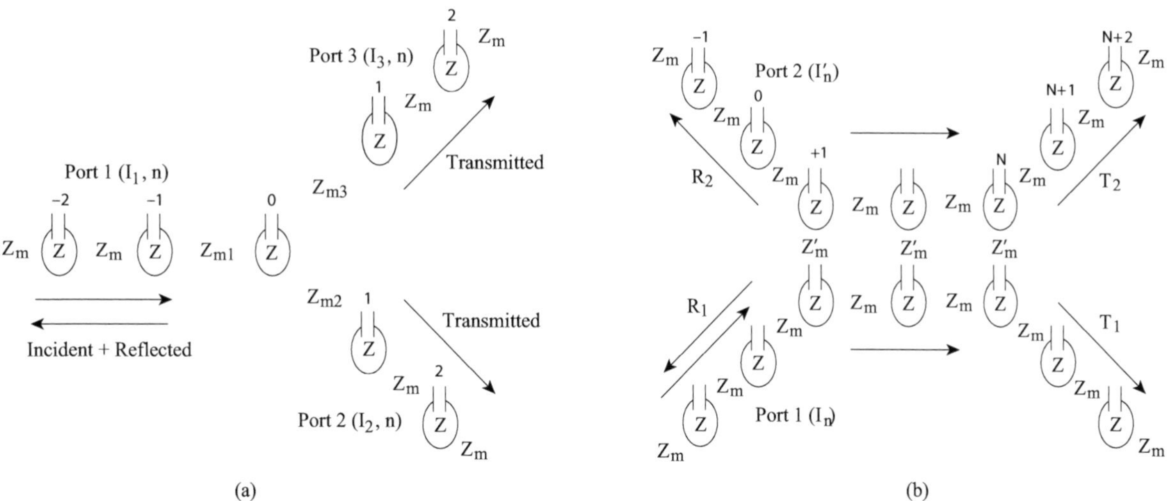

Fig. 7.44 (a) Magnetoinductive waveguide 3-port power splitter. (b) Magnetoinductive waveguide directional coupler. From Syms *et al.* (2006*a*). Copyright © 2006 IEE

ods do we need for large reflection? The answer is given in Fig. 7.43(b), where the dependence of the reflection and transmission coefficients is plotted as a function of the number of periods for $\mu_1 = 1.1$ and $\mu_2 = 0.9$. It may be seen that to reach saturation (that is 100% reflection) about 20 periods are needed. Next, we calculate the reflected power as a function of frequency for the same 20 periods and the same parameters. It may be seen in Fig. 7.43(c) that close to total reflection occurs within a narrow band centred on the resonant frequency of the element.

Have we seen similar things before? If we look at the alternating values of mutual inductance in Fig. 7.43(a) it should remind us of the biperiodic lines of Figs. 7.17(a)–(d). In Section 7.10 we looked at this problem and came to the conclusion (an infinite line was assumed there) that for a biperiodic line there is a stop band in the middle of the dispersion characteristics. In the present section we have a finite biperiodic line but the conclusion is the same. Instead of saying that there is a stop band we say now that we have nearly perfect reflection within a certain band due to Bragg reflection—and that's the same thing.

Another useful device may be seen in Fig. 7.44(a). It is shown in Syms *et al.* (2006*a*) that by judicious choice of M_1, M_2 and M_3 any desired power ratio between the two output lines can be achieved without any reflection in the input line.

Our final example is a four-port device shown schematically in Fig. 7.44(b). The free parameters are M' and N (note that Fig. 7.44(b) shows only one extra element in the coupling region), the mutual inductance and the number of elements in the coupling region. The aim is to direct desired powers to the two outputs. The analysis may be found in Syms *et al.* (2006*a*).

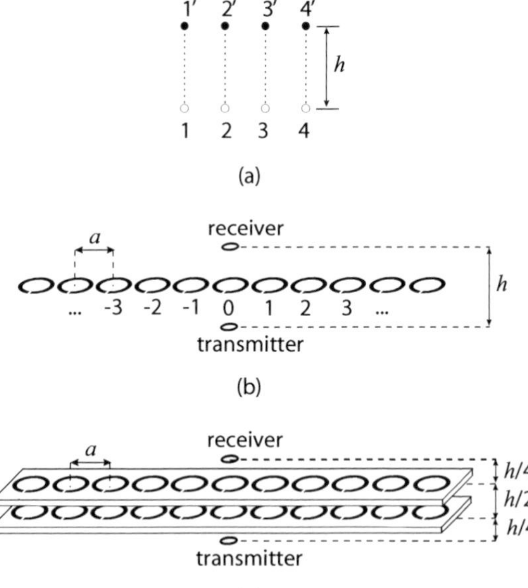

Fig. 7.45 (a) Near-field pixel-to-pixel imaging scheme. Schematic representation of (b) the single-layer and (c) the double-layer magnetoinductive lens. From Sydoruk *et al.* (2007*b*). Copyright © 2007 American Institute of Physics

7.15.3 Imaging

We shall discuss here a particular type of imaging under conditions when all the dimensions are small relative to the wavelength. It is a pixel-by-pixel imaging that simply translates the object. An example is shown in Fig. 7.45(a) where the object, consisting of points 1, 2, 3, 4 is translated along the dotted lines to $1'$, $2'$, $3'$, $4'$, a distance h away. In order to simplify the problem we shall look at one object point only and represent it by a small non-resonant transmitter coil at the point $x = 0$, $y = 0$. The imaging is tested by moving a small, non-resonant receiver coil along the line $y = h$. If the received power has a narrow maximum in the vicinity of the point $x = 0$, $y = h$ then we can regard it as an image. However, if all we have is a small transmitter coil and a small receiver coil then the maximum along the $y = h$ line will be wide, corresponding to the field distribution of the small transmitter coil. How could we make it sharper? Let's insert a MI waveguide between the transmitter and the receiver, as shown in Fig. 7.45(b). We may now claim that the field at the point $x = 0$, $y = h$ will be higher due to the coupling of element 0 both to the transmitter and to the receiver. This claim, however, is not necessarily correct. Inserting the MI waveguide will not, in general, make the field opposite the transmitter more concentrated because there will be a MI wave propagating in both directions away from element zero spreading the power in the x direction.

The remedy is to have a MI wave in the y direction but suppress it in the x direction. How can we do this? We have actually done so in Section 7.13. When we have two coupled lines and the elements are above each other then there is no power transfer along the coupled lines, provided the coupling is high enough. This conclusion can be

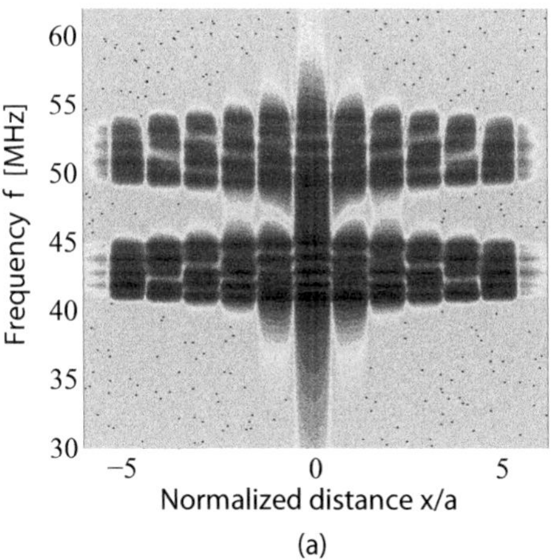

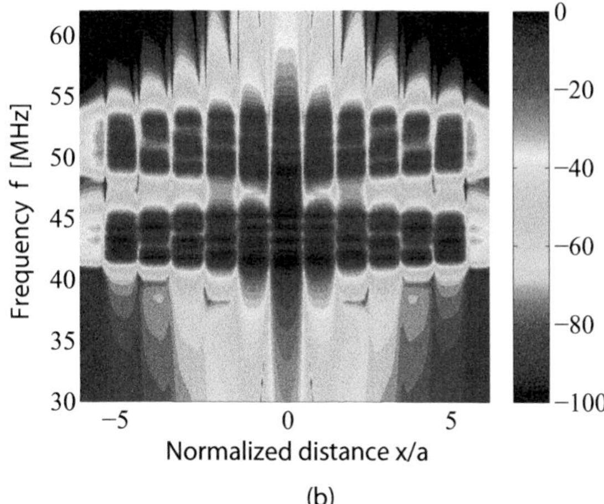

Fig. 7.46 Near-field imaging for the double lens with $h = 10$ mm. Magnetic-field distribution in the image plane versus frequency (contour plot). Experiment (a) and theory (b). From Sydoruk *et al.* (2007*b*). Copyright © 2007 American Institute of Physics. For coloured version see plate section

drawn from Fig. 7.36(b), which shows that there is a stop band at ω_0 for $h = 10$ mm, and also from Fig. 7.39(a), showing no power transfer for the unshifted case. Thus, power can be transferred in the y direction but now it cannot spread in the x direction.

The relevant experiment was done (Sydoruk *et al.*, 2007*b*) with two coupled lines consisting of split-pipes and separated from each other by $h/2$. The elements are arranged as in Fig. 7.38 but the measurement is done now by placing the transmitter below the element in the middle at a distance $h/4$, while the receiver moves above the upper line at a distance $h/4$, as shown schematically in Fig. 7.45(c). The experimental and theoretical results are shown in Figs. 7.46(a) and (b) for a distance of $h = 20$ mm. The horizontal axis shows the displacement of the receiver, whereas the frequency is on the vertical axis. The measured field strength is colour-coded. There is remarkable agreement between theory and experiment, both showing that an image exists in the vicinity of the resonant frequency at which sideways propagation of MI waves is prohibited. The image is translated in the present case by $2h$. Translation further away can be realized by inserting further lines. The essential criterion is that sideways propagation of MI waves must be prohibited.

7.15.4 Detection of nuclear magnetic resonance

We shall discuss here the application of MI waves to the detection of nuclear magnetic resonance. It needs to be remembered at this stage that there are three distinct resonance phenomena at play: the resonance of the element at ω_0, the rotational resonances occurring in the ring structure and the nuclear magnetic resonances of the various nuclei. The frequency of nuclear magnetic resonance is known to be proportional to

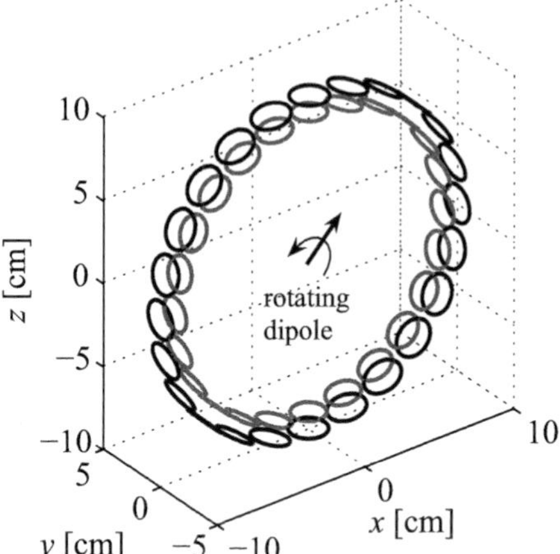

Fig. 7.47 Bilayered rotational resonator

the applied dc magnetic field. For a magnetic field of 1 T the resonant frequency of protons (the one most often examined in medical imaging) is 41 MHz, which is just in the range of frequencies at which MI waves can easily propagate and for which most of the experiments reported in this chapter have been done.

Under resonance conditions the nuclear dipoles precess, producing a rotating magnetic field. The simplest detecting mechanism consists of a single coil in which the rotating magnetic field induces a voltage. In order to have a more sensitive detecting mechanism the best chance is to create synchronism between the rotating magnetic field and the MI wave propagating round the ring structure of Fig. 7.40. This occurs when the phase velocity of the rotating field at a radius R is equal to the phase velocity of the MI wave at rotational resonance. It is easy to show that for the synchronism condition to be satisfied the frequency of rotational resonance must agree with the nuclear magnetic resonance.

Some aspects of this detecting mechanism were analyzed in Solymar *et al.* (2006). The power extracted from a single element was compared with that that can be extracted from an N-element synchronous ring by an optimized load impedance and by optimal matching. It was shown that in the latter case N times the power from a single ring can be obtained.

A further advantage of the synchronous detection scheme is that the circulating MI wave can be amplified by parametric amplification at the expense of a pump wave. In order to do so a second ring is needed (see Fig. 7.47) to couple magnetically to the first one, furthermore, the dispersion characteristics need to be tailored so that both the fundamental and the second temporal harmonic can propagate and, of course, a nonlinear element like a varactor diode is needed to couple the fundamental frequency to the pump wave.

For a practical realization see Syms *et al.* (2008) where experiments with a three-frequency parametric amplifier are shown.

Magnetoinductive waves II

8

8.1 MI waves in two dimensions

8.1.1 Introduction

A one-dimensional treatment was sufficient to present most of the properties of MI waves: how they propagate, how they attenuate, how they are reflected, how to find a matching impedance, what is the power density and how they are coupled to each other. However, for some other properties, e.g. refraction, a two-dimensional theory is needed. In the theory of metamaterials refraction occupies a central position. One might even say that the subject started with the discovery of Veselago's paper (1968) by Smith *et al.* (2000). The new physical concept introduced, as mentioned several times in this book, was negative refractive index and negative refraction. We may now ask the question whether MI waves can exhibit negative refraction? For that we need to have a formulation of MI wave propagation in a 2D medium and we have to find out what happens at the boundary of two periodic materials both of them capable of propagating MI waves.

Since metamaterial elements often have circular shapes, one of the popular 2D structures is a hexagonal one. We shall investigate their spatial resonances and imaging properties that were the subject of experimental work by Wiltshire *et al.* 2003*a*; Wiltshire *et al.* 2004*a*.

The theoretical treatment does not differ much from the one-dimensional one. The physics has not changed. The MI wave propagates in the same way due to the coupling between the elements. The nearest-neighbour approximation may still be used but the number of nearest neighbours is now four for the square configuration and six for the hexagonal one. The dispersion equations will look different and the frequency against wave-vector diagrams must also be presented in a different manner. However, if we consider the treatment in terms of currents, applied voltages and the mutual impedance matrix then there is no qualitative difference.

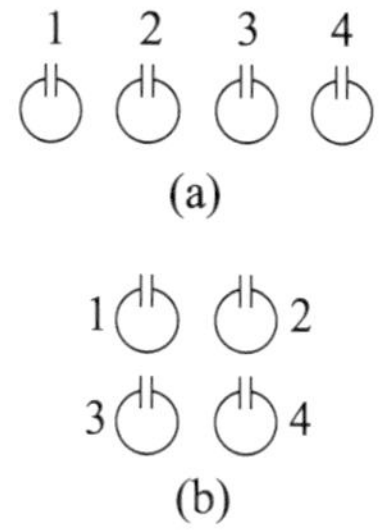

In eqn (7.16) the current flowing in the nth element is related to the voltage excitation in all the other elements. The relationship is via the inverse mutual impedance matrix

$$\mathbf{I} = \mathbf{Z}^{-1}\mathbf{V}. \tag{8.1}$$

The impedance matrix as presented in eqn (7.15) is a tri-diagonal matrix because the treatment there was restricted to nearest-neighbour

Fig. 8.1 Four capacitively loaded loops arranged (a) in a line, (b) in two rows

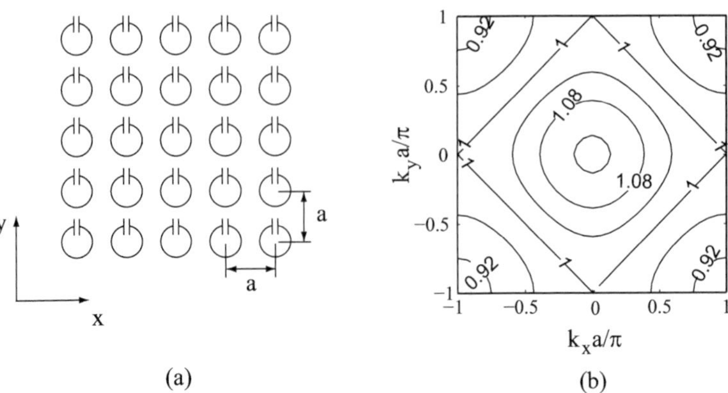

Fig. 8.2 Capacitively loaded loops in the planar configuration (a) geometry, (b) normalized dispersion (isofrequency) curves for $a = 2.25r_0$. From Syms *et al.* (2005*b*). Copyright © 2005 EDP Sciences

interaction in one dimension, but in the general case, when we consider every element coupled to every other element, the two-dimensional case looks just the same. This is illustrated in a simple example. Two sets of four elements are shown in Fig. 8.1, the first one is a one-dimensional array, the second one is a two-dimensional one. The impedance matrix for both of them may be written in the form

$$\begin{pmatrix} Z_{11} & Z_{12} & Z_{13} & Z_{14} \\ Z_{21} & Z_{22} & Z_{23} & Z_{24} \\ Z_{31} & Z_{32} & Z_{33} & Z_{34} \\ Z_{41} & Z_{42} & Z_{43} & Z_{44} \end{pmatrix}. \tag{8.2}$$

Of course the values of the mutual inductance between element 1 and 4 for example would be quite different for the two configurations but the mathematical formulation would be identical.

8.1.2 Dispersion equation, group velocity, power density

It is rarely necessary to consider interaction between all the particles. For most purposes nearest-neighbour interaction is sufficient. Kirchhoff's law for element (n, m) in a square lattice may be written as

$$Z(\omega)I_{n,m}+\mathrm{j}\,\omega M_x(I_{n+1,m}+I_{n-1,m})+\mathrm{j}\,\omega M_y(I_{n,m+1}+I_{n,m-1}) = 0, \tag{8.3}$$

where $I_{n,m}$ is the current in the element located at the nth row and mth column, M_x and M_y are the mutual inductances in the x (horizontal) and y (vertical) directions, respectively, and, as before, $Z(\omega)$ is the self-impedance of the elements. For the planar configuration (see Fig. 8.2(a)), $M_x = M_y$ and they are both negative. For the planar–axial case shown in Fig. 8.3(a) M_x is of course still negative, but M_y is positive and under the present conditions (element spacing being the same in both directions) its value is smaller than $|M_x|$.

Assuming the current in the form

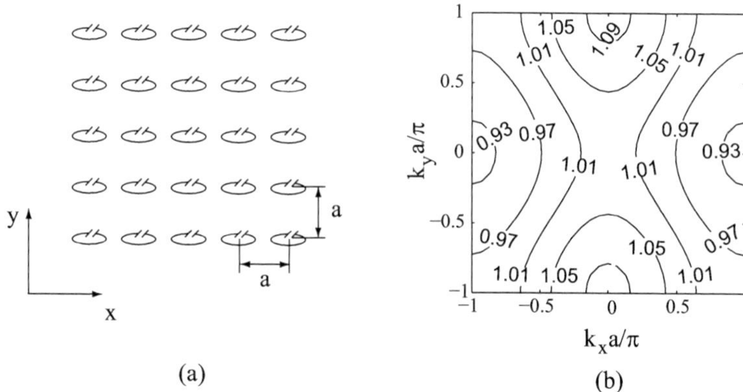

Fig. 8.3 Capacitively loaded loops in the planar–axial configuration (a) geometry, (b) normalized dispersion (isofrequency) curves for $a = 2.25r_0$. From Syms *et al.* (2005*b*). Copyright © 2005 EDP Sciences

$$I_{n,m} = I_0\,\mathrm{e}^{-\mathrm{j}\,(nk_xa + mk_ya)}, \tag{8.4}$$

where k_x and k_y are the x and y components of the wave vector, $\mathbf{k}$, and I_0 is a constant, and substituting eqn (8.4) into eqn (8.3) we obtain the dispersion equation in the form

$$\frac{\omega}{\omega_0} = A^{-1/2}\,; \quad A = 1 + \kappa_x \cos(k_xa) + \kappa_y \cos(k_ya), \tag{8.5}$$

where κ_x and κ_y are the coupling coefficients defined as $\kappa_{x,y} = 2M_{x,y}/L$. For our purpose the most useful presentation of the equation is in the form of the $\omega/\omega_0 =$ constant curves where $\omega_0 = (LC)^{-1/2}$ in Fig. 8.2(b) for the planar configuration ($\kappa_x = \kappa_y = -0.106$) and in Fig. 8.3(b) for the planar–axial configuration ($\kappa_x = -0.106$, $\kappa_y = 0.066$). In both cases the separation of the elements is $2.25r_0$, where r_0 is the external radius of the SRR. When $k_xa, k_ya \ll 1$ the curves in Fig. 8.2(b) may be shown to be circles, whereas those in Fig. 8.3(b) are hyperbolae.

In the 2D case the group velocity is given by the gradient of the ω versus $\mathbf{k}$ curves plotted in Figs. 8.2(b) and 8.3(b). Mathematically, they may be obtained from eqn (8.5) as

$$\mathbf{v}_\mathrm{g} = \frac{a\omega_0}{2} A^{-3/2} \left[\kappa_x \sin(k_xa)\mathbf{i}_x + \kappa_y \sin(k_ya)\mathbf{i}_y\right], \tag{8.6}$$

where $\mathbf{i}_x$ and $\mathbf{i}_y$ are unit vectors in the x and y directions, respectively.

For the planar case, when $\kappa_x = \kappa_y$ and the arguments of both sine functions are small, the group velocity may be seen to be in a direction opposite to the phase velocity—a clear case of a backward wave. The relationship is more complicated for the planar–axial configuration, as will be discussed in the next section.

The direction of power flow is given by the group velocity and we may again obtain the power density by multiplying the group velocity by the stored energy per unit surface

$$\mathbf{S} = \frac{1}{2}\mathbf{v}_\mathrm{g} E_\mathrm{s}. \tag{8.7}$$

The energy stored in any of the elements is given by the sum of the energies in the inductance, the capacitance and in the mutual inductances relating to nearest neighbours. In terms of a single element (m, n) it is given by

$$\begin{aligned} E_{\mathrm{s}} &= \frac{1}{2a^2}\left[L|I_{n,m}|^2 + \frac{|I_{n,m}|^2}{\omega C} + M_x I_{n,m}\left(I^*_{n-1,m} + I^*_{n+1,m}\right)\right. \\ &+ \left. M_y I_{n,m}\left(I^*_{n,m-1} + I^*_{n,m+1}\right)\right]. \end{aligned} \tag{8.8}$$

With our wave assumption for the current (eqn (8.4)), and using the expression of the group velocity given by eqn (8.6), the power density may be written as follows

$$S = \frac{1}{2}\omega|I_0|^2\left[M_x \sin(k_x a)\mathbf{i}_x + M_y \sin(k_y a)\mathbf{i}_y\right], \tag{8.9}$$

i.e. it is independent of the circuit parameters L and C. The condition for ω to be in the pass band must of course be satisfied.

8.1.3 Reflection and refraction

We shall now look at reflection and refraction of a MI wave at the boundary of two different media. Reflection and transmission at perpendicular incidence has already been considered in Section 7.9. We shall now assume that the MI wave is incident at an angle, so we shall be able to determine refraction as well.

Let us assume two semi-infinite 2D media lying on each side of the $x = 0$ line, which can support MI waves and have the same regular rectangular lattice with a lattice constant a. We shall consider here three examples. In the first case the elements in both media are in the planar configuration, medium 1 having the same parameters as in Fig. 8.2(b), whereas medium 2 differs from medium 1 by having a slightly different capacitance and, consequently, a slightly different resonant frequency that we shall take as $\omega_{02} = 1.03\,\omega_{01}$, see Fig. 8.4. Assuming further that $k_x a$ and $k_y a$ are small relative to unity the constant frequency curves are circles with good approximation. They are shown in Fig. 8.5(a) for $\omega/\omega_{01} = 1.11$.

It needs to be noted that we have here a rather unusual situation. Normally, positive refraction is due to forward waves propagating in both media. In the present case both media support backward waves. This does not lead to any complications but it means that if we want a wave incident at a positive angle (i.e. the group velocity to be in the first quarter) we have to choose k_{x1} and k_{y1} in medium 1, to be in the third quarter.

The boundary condition to be satisfied is that the phase velocities along the boundary must be the same on both sides (see discussion and the Ewald circle construction in Fig. 1.4) hence $k_{y2} = k_{y1}$ and k_{x2} is given by the construction shown in Fig. 8.5(a). The corresponding group

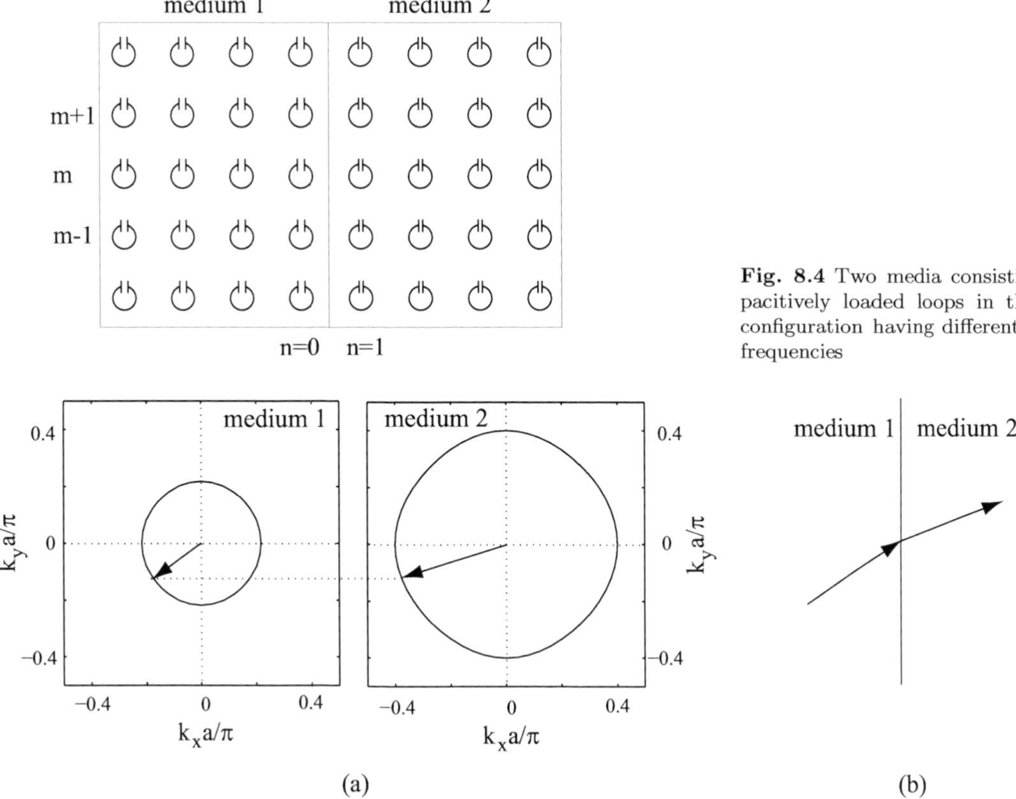

Fig. 8.4 Two media consisting of capacitively loaded loops in the planar configuration having different resonant frequencies

Fig. 8.5 Refraction at the boundary of two media. Planar configuration with $\omega_{02}/\omega_{01} = 1.03$. (a) The loci $\omega/\omega_{01} = 1.11$ and corresponding phase velocity vectors; (b) directions for the group velocity vectors, assuming $(k_{x1}a, k_{y1}a) = (-0.18, -0.12)\pi$ and $(k_{x2}a, k_{y2}a) = (-0.38, -0.12)\pi$. From Syms *et al.* (2005*b*). Copyright © 2005 EDP Sciences

velocities, very closely opposite to the chosen wave vectors, are shown in Fig. 8.5(b). It may be seen from Fig. 8.5(a) that the construction can be performed for all possible values of k_{x1}, i.e. at that particular frequency a refracted wave exists for any incident wave.

In our next example we shall take the same planar configuration on both sides of the boundary but now the resonant frequency is assumed to be smaller in medium 2. We take $\omega_{02} = 0.97\omega_{01}$. The ratio of the radii is then reversed as shown in Fig. 8.6(a). Using the same construction as before we find that $k_{x2} > k_{x1}$ and refraction is now pointing away from the perpendicular, as may be seen in Fig. 8.6(b), where the corresponding group velocities are shown at the boundary of the two media. It may also be seen in Fig. 8.6(a) that no refraction is possible for a range of incident angles. This is the case of total internal reflection.

In our third example we choose planar–axial configurations on both sides with all the parameters being the same but the orientation in medium 2 is at right angle to that in medium 1, as shown in Fig. 8.7(a). Note that the lattice in medium 2 is shifted by one half of the lattice constant relative to that in medium 1, which ensures increased magnetic

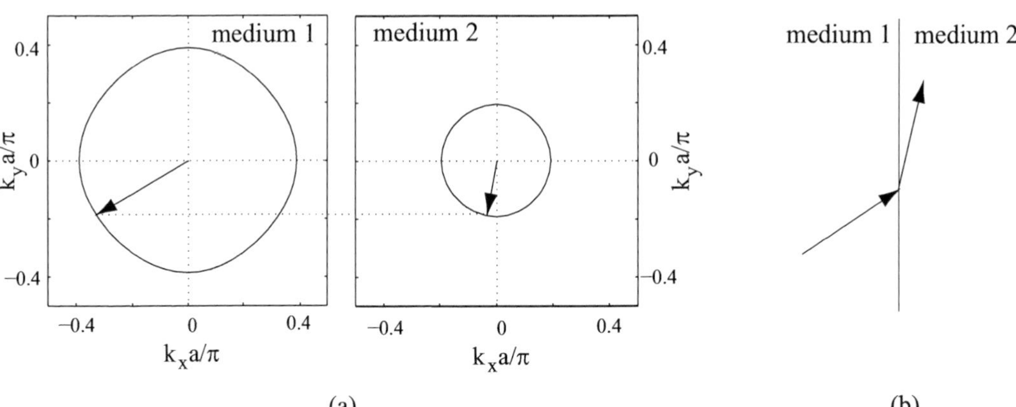

Fig. 8.6 Refraction at the boundary of two media. Planar configuration. $\omega_{02}/\omega_{01} = 0.97$. (a) the loci $\omega/\omega_{01} = 1.08$ and corresponding phase velocity vectors; (b) directions for the group velocity vectors, assuming $(k_{x1}a, k_{y1}a) = (-0.33, -0.19)\pi$ and $(k_{x2}a, k_{y2}a) = (-0.04, -0.12)\pi$. From Syms *et al.* (2005*b*). Copyright © 2005 EDP Sciences

coupling between the two media.

Our aim is now the same as in the previous examples: we wish to have an incident wave with group velocity in the first quarter. Let us remember that the group velocity is the gradient vector of the $\omega =$ constant curves. Hence, any $\omega =$ constant curve in the second quarter of the dispersion diagram, shown in Fig. 8.3(b) would qualify. Let us choose the $\omega/\omega_{01} = 1.01$ curve plotted in Fig. 8.7(b) with small arrows showing the direction of the group velocity. The relevant dispersion curve for the same value of ω/ω_{01} in medium 2 is rotated by 90 degrees and is therefore in the first quarter, as shown also in Fig. 8.7(b). It may be seen that there is only a very limited range of incident wave vectors for which refraction exists. The effect could be used for switching, modulation or spatial filtering, for example. Choosing a value of $k_y a = 0.4\pi$, and adhering again to the rule that the phase velocities must agree across the boundary, the group velocities in the two media are shown in Fig. 8.7(c). There is negative refraction. Note that in the present case the angle between phase and group velocities is slightly less than 90 degrees in both media, so the waves on both sides would qualify as forward waves. In fact, we could make the angles between phase and group velocity further decrease by choosing $\omega/\omega_0 = 1$ or increase the angle (and going thereby into the region where the waves on both sides are backward waves) by choosing $\omega/\omega_0 = 1.01$. This is similar to the conclusions reached by Luo *et al.* (2002*b*) in the sense that negative refraction may occur without the presence of backward waves but in our case the refraction angle is a strongly varying function of the incident angle.

We could also change the angle of negative refraction in medium 2 by leaving medium 1 unchanged and rotating the orientation of the loops in medium 2, relative to medium 1, by less than 90 degrees. In addition, we could considerably influence the dispersion characteristics of MI waves in medium 2 by changing the resonant frequency and the

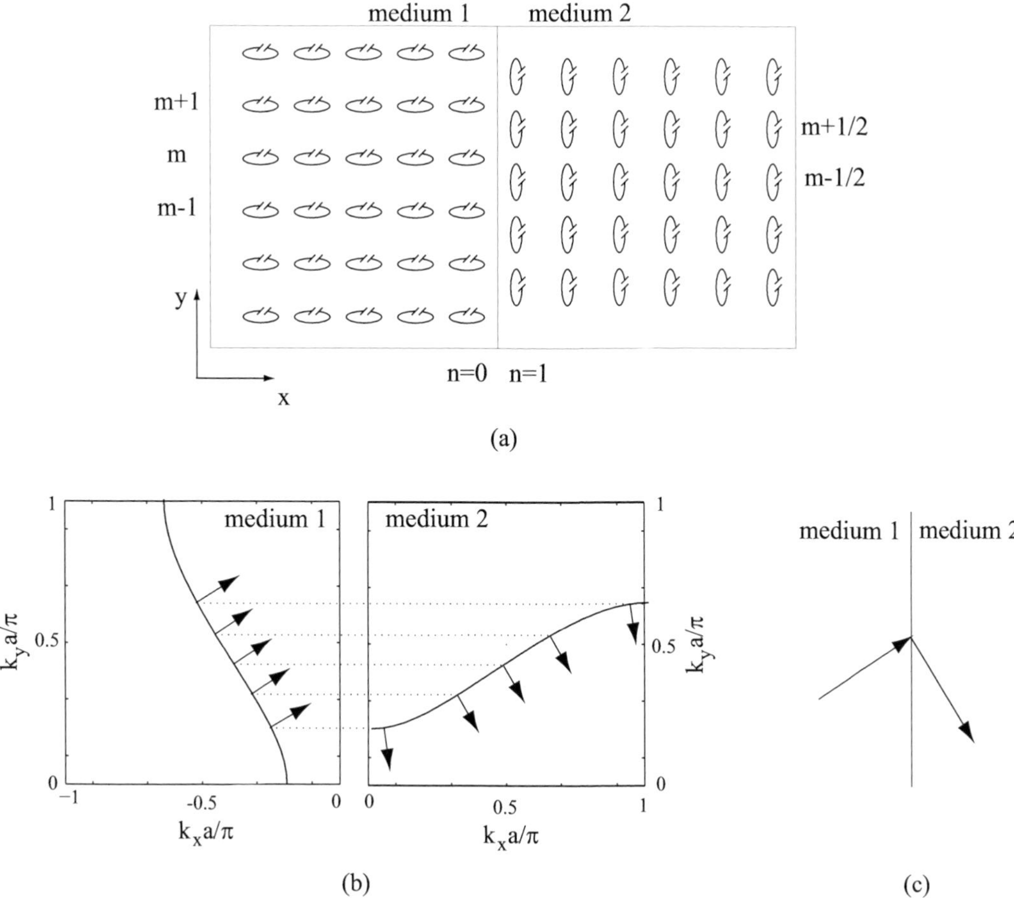

Fig. 8.7 Refraction at the boundary of two media. Planar–axial configuration with different orientations of the loops in medium 1 and medium 2. (a) The geometry of the loops; (b) the loci $\omega/\omega_{01} = 1.01$ and directions for group velocity vectors; (c) directions for the group velocity vectors for $(k_{x1}a, k_{y1}a) = (-0.39, -0.42)\pi$. From Syms *et al.* (2005*b*). Copyright © 2005 EDP Sciences

coupling coefficients. In fact, any of the four combinations may lead to negative refraction: (i) both waves forward, (ii) both waves backward, (iii) incident wave in medium 1 forward, refracted wave in medium 2 backward and (iv) incident wave in medium 1 backward, refracted wave in medium 2 forward.

Most efforts in the literature to find negative refraction have been aimed at electromagnetic waves incident from free space upon a periodic medium. In our case both media are periodic. The MI wave, quite obviously, could not be incident from free space because it can exist only in certain periodic media. It needs to be noted that we have a large amount of freedom choosing the dispersion characteristics in both media, thereby allowing a large variety of refractive angles, positive or negative, to be realized.

Next, we shall derive the reflection and transmission coefficients for

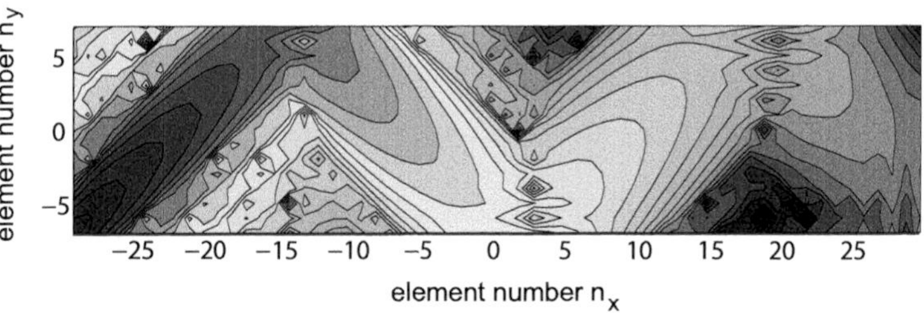

Fig. 8.8 Contour plot of currents in a 2D array of resonators showing reflection of a magnetoinductive wave. For coloured version see plate section

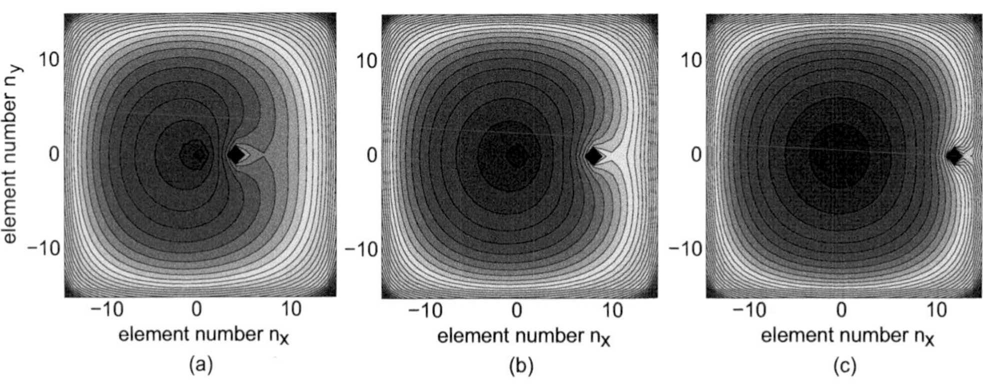

Fig. 8.9 Contour plot of currents in a 2D array of resonators showing diffraction on a defect. For coloured version see plate section

the wave incident at an interface between two media. The technique is the same as in Section 7.9. The current in medium 1 is written as the sum of the incident and reflected waves, and in medium 2 as a transmitted wave. We can then demand that the currents across the boundary satisfy the recursion equations as in Section 7.9. The equations turn out to be the same as eqns (7.35) to (7.39), we just need to replace k by k_x and the mutual inductances by their values in the x direction perpendicular to the boundary.

8.1.4 Excitation by a point source: reflection and diffraction

Up to now we have assumed that MI waves propagate in a 2D medium with plane wavefronts. We shall now look at the propagation of MI waves when one of the elements is excited by a temporally varying magnetic field. We wish to see how the MI wave originating in a single point is reflected by a boundary and diffracted by an obstacle. In other words we wish to show that MI waves are no different from other kinds of waves physicists love and cherish. The aim is to find the current distribution. The solution is provided by the matrix equation (8.1) in which the voltage vector has one non-zero element.

Our first example is a 2D sheet of 15 × 60 elements in a rectangular lattice. We assume that the element in the lower left-hand corner is excited. Having 900 elements necessitates the inversion of a 900 by 900 matrix that is well within the capabilities of the relevant Matlab program. The resultant distribution of the currents is shown in Fig. 8.8 by coloured contour plots. The excitation and reflection of the wave may be clearly seen.

In our second example a square sheet of 31×31 elements is considered that is excited at the centre. The regularity of the square lattice is broken by a missing element. Figures 8.9(a)–(c) show the wave pattern of the MI wave for three different positions of the missing element. The MI wave is clearly diffracted.

8.1.5 Spatial resonances in hexagonal lattices

Next, we shall look at the propagation of MI waves in hexagonal lattices and in particular at spatial resonances that were experimentally investigated by Wiltshire *et al.* (2004*a*). The hexagonal lattice and the corresponding co-ordinate system is shown in Figs. 8.10(a) and (b). Considering only nearest-neighbour interactions Kirchhoff's law for the voltage in element (n, m) may be written as

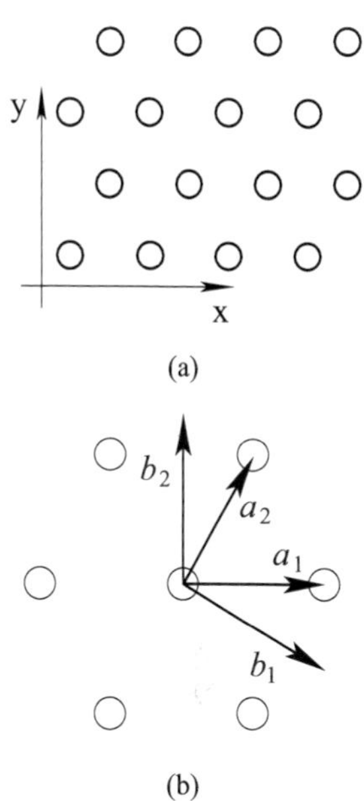

Fig. 8.10 (a) Lattice of resonant elements with hexagonal arrangement. (b) $(\mathbf{a}_1, \mathbf{a}_2)$ direct vectors of the hexagonal lattice and $(\mathbf{b}_1, \mathbf{b}_2)$ reciprocal vectors

$$\begin{aligned} &Z I_{n,m} + \mathrm{j}\omega M(I_{n,m-1} + I_{n,m+1} + I_{n-1,m} \\ &+ I_{n+1,m} + I_{n-1,m+1} + I_{n+1,m-1}) = 0, \end{aligned} \tag{8.10}$$

where $I_{m,n}$ is the current in the (m, n) element, Z is the self-impedance assumed again as lossless, M is the mutual inductance between nearest neighbours. The position of an element is given by the radius vector

$$\mathbf{r}_{n,m} = n\mathbf{a}_1 + m\mathbf{a}_2, \tag{8.11}$$

where $\mathbf{a}_1$ and $\mathbf{a}_2$ are the direct vectors of the lattice, shown in Fig. 8.10(b). In the x, y co-ordinate system they can be written as

$$\mathbf{a}_1 = \mathbf{i}_x a, \quad \mathbf{a}_2 = a(\mathbf{i}_x \cos 60^\circ + \mathbf{i}_y \sin 60^\circ), \tag{8.12}$$

where a is the spacing between the elements and $\mathbf{i}_x$ and $\mathbf{i}_y$ are unit vectors in the x and y directions, respectively. The corresponding vectors of the reciprocal lattice, $\mathbf{b}_1$ and $\mathbf{b}_2$ may be obtained by requiring the following conditions be satisfied

$$\begin{aligned} \mathbf{a}_1 \cdot \mathbf{b}_1 &= 1, & \mathbf{a}_1 \cdot \mathbf{b}_2 &= 0, \\ \mathbf{a}_2 \cdot \mathbf{b}_1 &= 0, & \mathbf{a}_2 \cdot \mathbf{b}_2 &= 1. \end{aligned} \tag{8.13}$$

We shall now look for the solution of eqn (8.10) in the form

$$I_{n,m} = I_{0,0}\, \mathrm{e}^{-\mathrm{j}\mathbf{k} \cdot \mathbf{r}_{n,m}}. \tag{8.14}$$

$I_{0,0}$ is a constant and $\mathbf{k}$ is the wave vector that may be expressed in terms of the reciprocal lattice vector as

$$\mathbf{k} = 2\pi(f_1\mathbf{b}_1 + f_2\mathbf{b}_2), \tag{8.15}$$

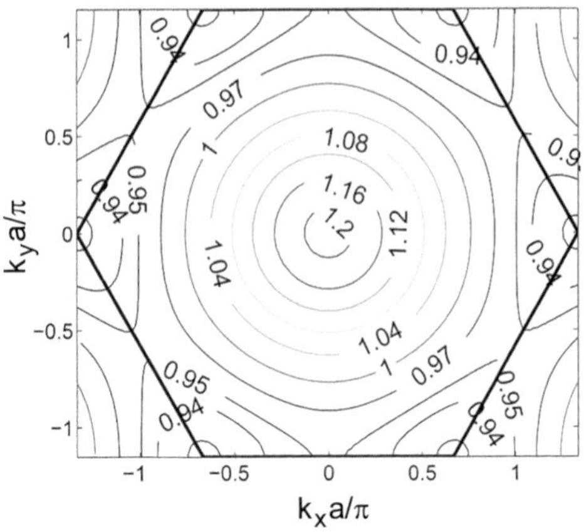

Fig. 8.11 2D normalized dispersion (isofrequency) curves for a hexagonal lattice. The boundaries of the first Brillouin zone are shown by bold lines. From Zhuromskyy *et al.* (2005*a*). Copyright © 2005 Optical Society of America

where f_1 and f_2 are constants.

Substituting eqn (8.14) into eqn (8.10) we obtain the dispersion equation in the form

$$\frac{\omega}{\omega_0} = \frac{1}{\sqrt{1 + \kappa[\cos(2\,\pi f_1) + \cos(2\,\pi f_2) + \cos(2\,\pi f_1 - 2\,\pi f_2)]}}, \tag{8.16}$$

where $\omega_0 = 1/\sqrt{LC}$ is the resonant frequency of the element and $\kappa = 2M/L$ is the coupling coefficient. The ω/ω_0 = constant curves as functions of the wave vector are plotted in Fig. 8.11. The pass band is found to extend from $0.93\omega_0$ to $1.21\omega_0$. Note that the waves are backward waves with phase and group velocities in opposite directions.

In the examples to follow in the presence of an excitation we shall use eqn (8.1) to determine the current distribution. First, however, we shall attempt to find an approximate solution that will lead us to a mathematically familiar territory and offer a clear physical picture. We may expect that the propagation of MI waves, similarly to the propagation of most other waves, can be mathematically described by a second-order partial differential equation. We shall therefore convert our difference equation (eqn (8.10)) into a differential equation. We can do that when the wave vectors are sufficiently small or in other words when the wavelength of the MI wave is much larger than the element spacing. Consequently we shall introduce the ν, μ co-ordinate system in the directions $\mathbf{a}_1$ and $\mathbf{a}_2$ with the continuous variables $\nu = na$ and $\mu = ma$, and replace the discrete function $I_{n,m}$ by the continuous function $I(\nu, \mu)$. A change in the subscript by unity would then be equivalent to a change of the continuous variable by a, which is then regarded as an elementary change. To convert all the terms in eqn (8.10) into continuous variables we need to expand the current into a Taylor series as follows

$$I(\nu + \Delta\nu, \mu + \Delta\mu) = \left\{1 + \Delta\nu\frac{\partial}{\partial\nu} + \Delta\mu\frac{\partial}{\partial\mu}\right.$$

$$+\frac{1}{2}\left[(\Delta\nu)^2\frac{\partial^2}{\partial\nu^2}+2\Delta\nu\Delta\mu\frac{\partial^2}{\partial\nu\partial\mu}\right.$$
$$\left.\left.+(\Delta\mu)^2\frac{\partial^2}{\partial\nu^2}\right]\right\}I(\nu,\mu), \tag{8.17}$$

where $\Delta\nu$ and $\Delta\mu$ are small deviations from ν and μ. With the aid of eqn (8.17) we may now convert eqn (8.10) into the differential equation

$$\frac{\partial^2 I}{\partial\nu^2}-\frac{\partial^2 I}{\partial\nu\partial\mu}+\frac{\partial^2 I}{\partial\nu^2}+\frac{1}{2d^2}\left(6+\frac{Z}{j\omega M}\right)I=0. \tag{8.18}$$

A further transformation to the x, y co-ordinate system yields the relationship

$$\frac{\partial^2}{\partial\nu^2}-\frac{\partial^2}{\partial\nu\partial\mu}+\frac{\partial^2}{\partial\nu^2}=\frac{3}{4}\left(\frac{\partial^2}{\partial x^2}+\frac{\partial^2}{\partial y^2}\right). \tag{8.19}$$

So, we end up with the familiar wave equation

$$\frac{\partial^2 I}{\partial x^2}+\frac{\partial^2 I}{\partial y^2}+k^2 I=0, \tag{8.20}$$

where

$$k^2=\frac{4}{a^2}\left[1+\frac{1}{3\kappa}\left(1-\frac{\omega^2}{\omega_0^2}\right)\right], \tag{8.21}$$

and $k=|\mathbf{k}|$. It may be easily shown that the dispersion equation (8.16) reduces to eqn (8.21) when $ka \ll 1$.

A clear advantage of having the wave equation is that, at least for certain geometries, we know the solutions from past experience. For rectangular boundaries, possible solutions are

$$I=I_0\sin(k_x x)\sin(k_y y), \tag{8.22}$$

where I_0 is a constant and k_x and k_y are the x and y components of the wave vector. For a circular boundary there are circularly symmetrical solutions of the form

$$I=I_0 J_0(kr), \tag{8.23}$$

where J_0 is the zero-order Bessel function of the first kind and r is the distance from the centre of the circular structure of the elements.

If we choose a frequency and know which elements are excited (our formulation allows all of them to be separately excited) we can determine the current distribution from eqn (8.1). The result might be in the form of odd-looking current distributions because several wave vectors may coexist at a particular frequency. It is only at the spatial resonance that no more than one wave vector survives. Spatial resonances are characteristic to all wave phenomena. They exhibit the same behaviour whether they occur in vibrating membranes, organ pipes or Fabry–Perot resonators. The boundary condition to be satisfied in the present case is that the current must vanish at the boundary.

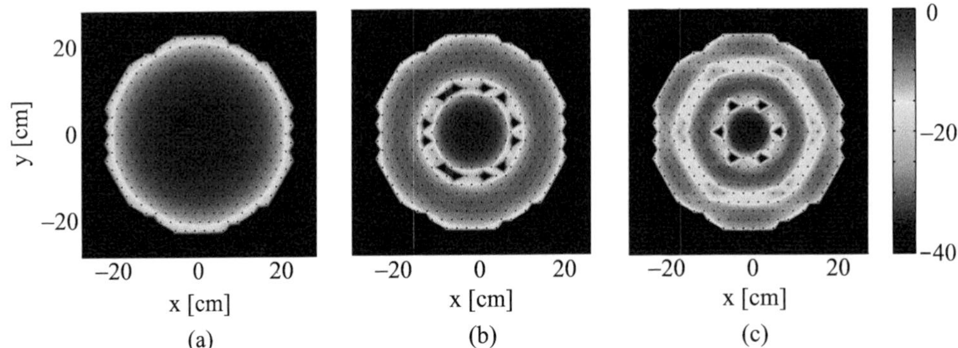

Fig. 8.12 Numerically obtained current distributions for circular boundary conditions at $\omega/\omega_0 =$ (a) 1.207, (b) 1.192, (c) 1.167. Black dots show positions of the elements. From Zhuromskyy *et al.* (2005*a*). Copyright © 2005 Optical Society of America. For coloured version see plate section

Array of circular shape. For a circle of radius R the condition for spatial resonance is

$$kR = \rho_i, \tag{8.24}$$

where ρ_i is the ith root of J_0.

We shall now look at the phenomenon of spatial resonance with the aid of a few examples. At this stage we need to commit ourselves as to the distance between the elements and the resonant frequency. The loop radius, the wire diameter and the distance between the elements are taken as 10 mm, 2 mm and 22.5 mm, and the resonant frequency as 21.5 MHz. The inductance of the loop may then be determined from standard formulae (Grover, 1981) that give $L = 33$ nH. The corresponding capacitance can be determined from the resonant frequency as 1.66 pF. The mutual inductance between two neighbouring elements is 1.75 nH, yielding $\kappa = -0.106$. Owing to the hexagonal arrangement it is not possible for all the elements to lie exactly on a circular boundary. With our choice of 361 elements the deviation from the circular boundary is quite small, as may be appreciated by looking at Figs. 8.12(a)–(c).

Knowing the geometry the frequencies of the first 3 spatial resonances may be determined from eqn (8.24) as $\omega/\omega_0 = 1.207, 1.187, 1.155$. Numerical calculations based on the known values of the mutual impedance matrix yield 1.207, 1.191, 1.165, a very good approximation. The analytical results are of interest because they give good approximation for low values of k and, of course, they give an immediate idea of what the current distribution looks like. For the general case, however, it is more accurate to rely on the exact solution of the discrete problem that is based on the inversion of the impedance matrix. Note that for the numerical determination of the current distribution we need an excitation. We assume that the central element out of the 361 is excited and then proceed with the numerical solution. The numerically determined current distributions are shown by a colour code in Figs. 8.12(a)–(c). Normalization in each figure is to the maximum value within the figure.

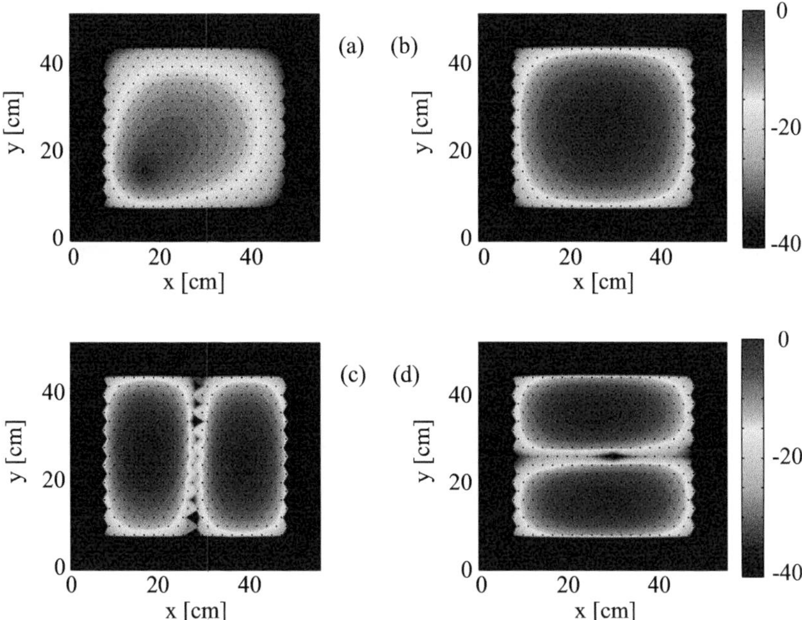

Fig. 8.13 Numerically obtained current distributions for rectangular boundary conditions, (a) non-resonant distribution for asymmetric excitation, (b)–(d) resonant excitation at $\omega/\omega_0 = 1.207$, 1.202 and 1.201, respectively. Black dots show positions of the elements. From Zhuromskyy *et al.* (2005*a*). Copyright © 2005 Optical Society of America. For coloured version see plate section

As may be expected, at the first spatial resonance there is zero current only at the boundary. For the second and third spatial resonances the currents are approximately zero at one and two radii, respectively. An interesting feature of Fig. 8.12(c) is that in spite of the close-to-circular boundary the hexagonal nature of the element geometry re-establishes itself further away from the centre.

Array of rectangular shape. We shall assume the array to extend from 0 to D_x in the x direction and from 0 to D_y in the x direction. In view of eqn (8.22) the wave vector components leading to spatial resonances are given by the relationships

$$k_x D_x = p_x \pi \quad \text{and} \quad k_y D_y = p_y \pi, \tag{8.25}$$

where p_x and p_y are integers.

We shall arrange now the hexagonal lattice in a 19×19 square geometry. As in the previous example we shall compare the analytical and numerical values for the frequencies of 3 spatial resonances (fundamental plus two of the second order) shown in Figs. 8.13(b)–(d). The resonant frequencies are $\omega/\omega_0 = 1.207$, 1.202 and 1.201 from the numerical solutions. There is some ambiguity in the analytical expression: due to the jagged boundaries D_x cannot be exactly defined. The values we obtain for the resonant frequencies from the analytical solution are 1.207, 1.202 and 1.202. The last two figures agree because in theory the two current

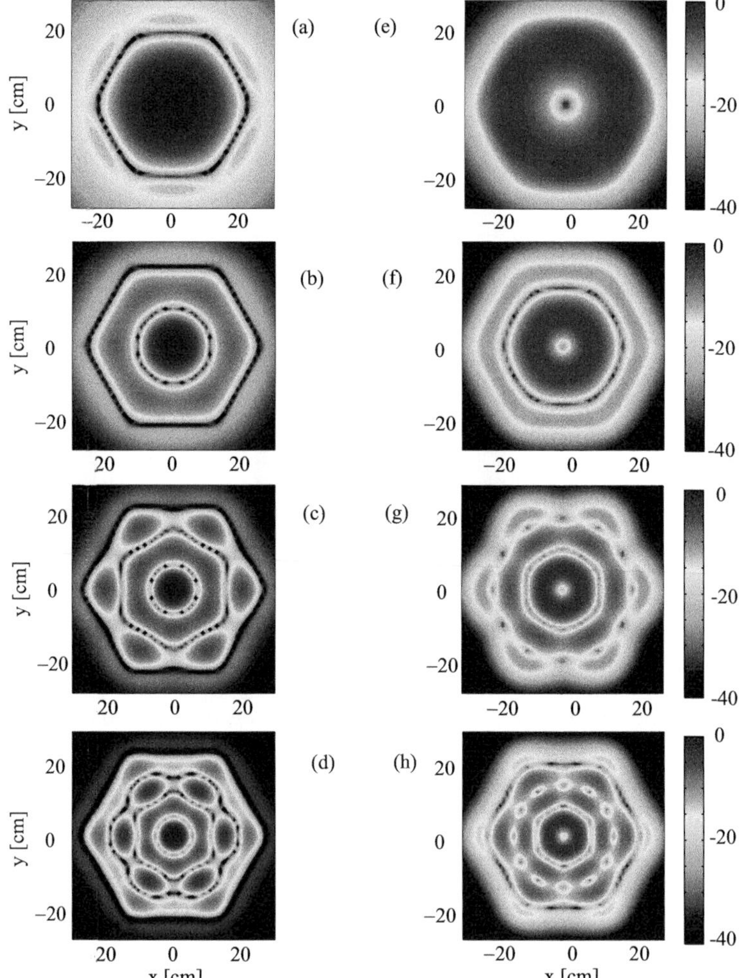

Fig. 8.14 Magnetic-field distribution at four resonant frequencies: (a) and (e) $\omega/\omega_0 = 1.206$, (b) and (f) $\omega/\omega_0 = 1.186$, (c) and (g) $\omega/\omega_0 = 1.154$, (d) and (h) $\omega/\omega_0 = 1.114$; normal component (a)–(d) and tangential component (e)–(h). From Zhuromskyy *et al.* (2005*a*). Copyright © 2005 Optical Society of America. For coloured version see plate section

distributions are identical, only rotated by 90 degrees. We may claim again good agreement between the analytical and numerical results.

Note that the establishment of the spatial resonances depends little on the exact position of the excitation as long as it is not at an expected current minimum. For the present rectangular case the excitation is taken one quarter of the way along one of the diagonals. We can obtain with this excitation all the spatial resonances, but the current pattern will of course depend on the position of the excitation if we are not at a spatial resonance. At a frequency of $\omega/\omega_0 = 1.208$ we do have an asymmetric pattern characteristic to the excitation, as may be seen in Fig. 8.13(a).

For hexagonal boundaries we have a chance to compare the theoretical results with a set of experimental ones measured by Wiltshire *et al.* (2004*a*). The experiments were performed on a 2D array of 271 'swiss rolls'. The array was centrally excited on one side of the structure and

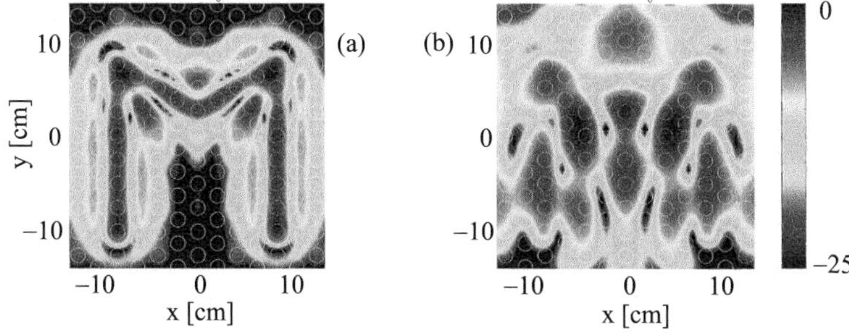

Fig. 8.15 The normal component of magnetic field at ω/ω_0 = (a) 0.98 and (b) 1.01. From Zhuromskyy *et al.* (2005*a*). Copyright © 2005 Optical Society of America. For coloured version see plate section

the axial (perpendicular to the plane of the elements) and radial components of the magnetic field were measured on the other side. The measured 2D distribution of these two components of the magnetic field are given in their Figs. 4–7 for the first four spatial resonances. Unfortunately, we have not been able to get hold of these figures. The reader might want to look at the original publication. Our theoretical results plotted in Fig. 8.14 display practically the same spatial variation for both components of the magnetic field as in the experiments.

8.1.6 Imaging

We have already discussed imaging with MI waves (Sydoruk *et al.*, 2007*b*), both experimentally and theoretically, in Section 7.15.3 where the 'lens' consisted of two coupled 1D lines made up by split pipes. It was a pixel-by-pixel imaging that could be best described as channelling the spatial information across the imaging device. The channelling occurred under conditions when the propagation of the MI wave was forbidden along the length of the coupled lines. Much earlier, Wiltshire *et al.* (2003*a*) successfully imaged an object with the aid of a hexagonally arranged array of swiss rolls consisting of 271 elements. The authors excited the 2D resonant structure by placing an M-shaped wire antenna below the structure and measuring the axial component of the magnetic field on the top. The experimentally obtained image is shown in their Fig. 4. The image we have calculated from our simple model (in which we consider capacitively loaded rings instead of swiss rolls) may be seen in Figs. 8.15(a) and (b) for $\omega/\omega_0 = 0.98$ and 1.01, respectively. The agreement with the experimental results of Wiltshire *et al.* (2003*a*) is good. However, as the frequency increases beyond the resonant frequency of the element the image quickly deteriorates. At $\omega/\omega_0 = 1.01$ the object is unrecognizable.

Imaging with MI waves using two parallel planes of broadside-coupled SRRs was carried out by Freire and Marques (2005). Their experimental setup is shown in Fig. 8.16(a). The two planes have areas of 7×7 cm^2. They are separated from each other by a foam slab of thickness $d = 4$ mm. The substrate thickness is $h = 0.254$ mm and the dielectric permittivity is $\varepsilon_r = 10$. The transmitting and receiving antennas are square loops of 1 cm^2. The field strength was measured by moving

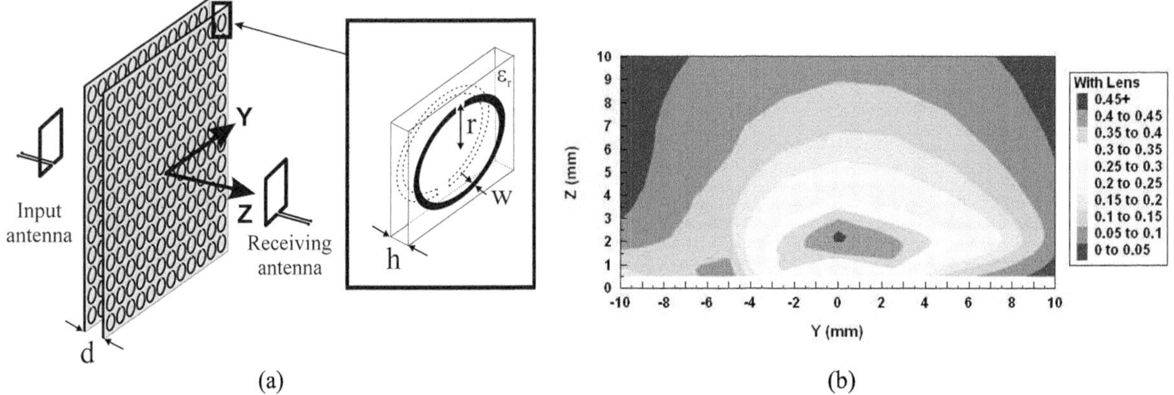

Fig. 8.16 (a) Experimental setup for imaging with magnetoinductive waves. (b) The magnitude of the transmission coefficient between the input and output antennas at a frequency of 3.23 GHz. From Freire and Marques (2005). Copyright © 2005 American Institute of Physics. For coloured version see plate section

the receiving antenna in the image plane. The experimental results are given in Fig. 8.16(b). A focal region may be clearly seen where the field strength is higher than in the surrounding region. The question whether it is a proper focus is discussed by Mesa *et al.* (2005). A proper focus is one that is measured by an ideal receiving device that does not interact with the elements of the lens. The authors conclude that the focus measured by Freire and Marques (2005) was due to that kind of interaction i.e. not a proper focus.[1] Further imaging work using MI waves in similar configuration was done later by Freire and Marques (2006), using this time low-impedance receivers in order to avoid the interaction. They showed both experimentally and theoretically (theory based on that of Maslovski *et al.* (2004)) that good imaging can be achieved. A further generalization of the problem was done by Marques and Freire (2005). They showed theoretically that subdiffraction imaging devices cannot produce focusing of power into three-dimensional spots of subdiffraction size.

[1] A generalization of the principle was presented by Sydoruk *et al.* (2007*d*). The authors showed that such 'improper' focus can be placed in any desired position for any transmitter–lens configuration by appropriate matching of the receiver.

8.2 MI waves retarded

8.2.1 Introduction

All the MI waves investigated so far could be termed low-frequency waves or perhaps quasi-static waves. In either case, one envisages waves that propagate on a line, which may attenuate due to ohmic losses but do not lose power by radiation. The dispersion characteristics were determined entirely by disregarding radiation. If the dispersion equation is to be found in the presence of radiation then all the assumptions made so far need to be reassessed. The most often used approximation was that of nearest-neighbour interaction. The tacit assumption is then that the field strength declines so rapidly away from an element that it is too small to matter at the position of a neighbour further away. In the presence of radiation this assumption cannot be valid. But what about our calculation of the magnetic field via the vector potential? Is that valid? Let us recall how we did that. We used eqn (1.14) that cannot cater for radiation effects, and we assumed a uniform current distribution around the loop. The magnetic field obtained in this way declined rapidly away from the loop (in the approximation of a small loop it can be regarded as a magnetic dipole and the field decays with the third power of the distance). If the free-space wavelength is comparable with some dimension of the MI waveguiding structure then we can no longer argue in favour of these approximations.

Let us start with the weakest constraint. What if the total length of the guiding structure is comparable with the wavelength? Can it happen? Not at the frequencies used for magnetic resonance imaging, one of the candidates for the application of MI waves. For a large magnetic field of 2 T the proton magnetic resonant frequency is still only about 100 MHz, corresponding to a wavelength of 3 m, which is still large enough to ignore, unless the line is very long. However, if we consider phenomena at microwaves with free-space wavelengths of less than 10 cm and element sizes of about 10 mm, then one feels radiation effects might matter. Then, the vector potential can no longer be calculated from the current as given by eqn (1.14). We need to take into account that the effect of the current is retarded. To reach point r_2 from r_1 a time $t = (r_2 - r_1)/c$ is required. The expression for the vector potential becomes

$$\mathbf{A}(\mathbf{r}_2, t) = \frac{\mu_0}{4\pi} \int \mathbf{J}(\mathbf{r}_1) \frac{\mathrm{e}^{-\mathrm{j}\, k_0 |\mathbf{r}_2 - \mathbf{r}_1|}}{|\mathbf{r}_2 - \mathbf{r}_1|} \mathrm{d}\tau, \tag{8.26}$$

where the current flows in point $\mathbf{r}_1$ and the vector potential is calculated in point $\mathbf{r}_2$. Hence, the mutual inductance between two loops must be calculated from the above equation, increasing considerably the numerical work.

If the size of the loop is comparable with the wavelength (say the diameter of the loop is equal to $\lambda/4$, that may very well happen in the optical region) then there are further complications. The current distribution can no longer be regarded as uniform and one needs to resort to the integral equations of antenna theory (see, e.g., King 1969) to find the current as a function of azimuthal angle. We shall certainly not go that far. In the present section we shall make a compromise: assume a uniform current in the element but include retardation in calculating the field quantities. As a result, the fields far away from the element will decay as the inverse of the distance, which means that for a long array it is necessary to include the effect of radiation when calculating the interaction between any two elements in the array. The concept of mutual inductance may still be used but it has to be calculated in a different manner and, due to the phase delay it turns out to be complex.

In Section 8.2.2 we shall derive the dispersion equation for an infinitely long line that now includes retardation, and compare the result with that obtained by the quasi-static approximation. In Section 8.2.3 we make an attempt to clarify the usefulness and limitations of the dispersion-equation approach. In Section 8.2.4 we shall return to the finite-line problem. Two excitations will be investigated for the full range of frequencies: (i) the first element excited, a situation we have examined before in the quasi-static approximation, and (ii) all elements excited with the value of k prescribed. Conclusions will be drawn in Section 8.2.5.

8.2.2 Dispersion equation

We shall now include retardation in the dispersion equation, but to simplify the matter, and to have a complete analytical formulation, we shall replace our loops by magnetic dipoles. We have done so before in Section 2.9 for the axial configuration, where we relied on the concept of polarizability to derive the dispersion equation in the form (see eqn (2.66))

$$\frac{1}{\alpha_{\mathrm{m}}} = I_{\mathrm{F}}, \tag{8.27}$$

where I_{F} is the interaction function. We shall now do the derivation for the planar case where the interaction between the elements is due to the H_θ component given by eqn (1.87) where we need to take $\theta = 90°$. The polarizability for the lossless case is still the same as given by eqn (2.41) but now we need to include radiation damping (see eqn (2.43)) in the expression for polarizability, and of course the interaction function must include the term declining with the inverse distance responsible for radiation. Formally, the dispersion equation is the same as in Section 2.9. It can be written in the form

$$1 = \alpha_{\mathrm{m}} \sum_{n=1}^{\infty} f(n) \cos(kna), \tag{8.28}$$

where now

$$\frac{1}{\alpha_{\mathrm{m}}} = \frac{L}{\mu_0^2 S^2}\left(1 - \frac{\omega_0^2}{\omega^2}\right) + \mathrm{j}\,\frac{k_0^3}{6\,\pi\mu_0}, \tag{8.29}$$

and

$$f(n) = \frac{k_0^3}{2\,\pi\mu_0}\left[\frac{1}{nk_0a} - \frac{\mathrm{j}}{(nk_0a)^2} - \frac{1}{(nk_0a)^3}\right]\mathrm{e}^{-\mathrm{j}\,k_0na}. \tag{8.30}$$

Note that the dispersion equation as given by eqns (8.28)–(8.30) is complex. It has a real part and an imaginary part. Unlikely as it seems, it may be shown (first realized by Simovski *et al.* (2005) on analyzing an infinite set of nanoparticles) that the imaginary parts cancel, provided $k > k_0$. Having assumed no ohmic losses the equation being real implies that there are no radiation losses either. Can one offer a physical

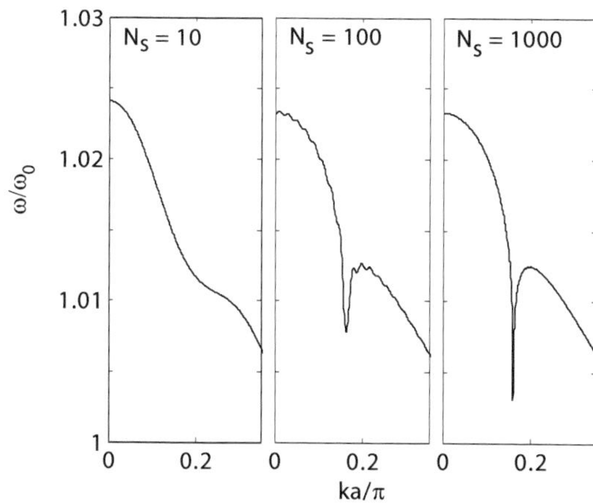

Fig. 8.17 Dispersion curves derived from eqns (8.28)–(8.30) summed for the first N_s terms. From Zhuromskyy (2008)

explanation? One may argue that if the line is infinitely long it cannot radiate for the reason that there is no space left into which it could radiate. However, this argument does not apply to the part of the dispersion characteristics to the left of the light line. The imaginary part does not cancel then: the dispersion equation predicts decay of amplitude due to radiation. These waves belong to the category of leaky waves (see, e.g., Hessel and Oliner 1965; Marcuvitz 1956), an interesting subject with a large literature, but beyond the scope of the present book.

Equation (8.28) is in an analytical form but the infinite series still needs to be summed numerically up to a certain number of elements. It can be shown that the series is convergent but converges rather slowly in the vicinity of the light line. To demonstrate that convergence numerically we need a few parameters, which we shall choose as $L = 33$ nH, $\omega_0/(2\,\pi) = 0.96$ GHz, $a = 25$ mm, $r_0 = 10$ mm. The dispersion curves corresponding to $N = 10$, 100 and 1000 elements are shown in Fig. 8.17. Taking now $N = 1000$ we can next examine the difference between the quasi-static and the full dispersion equation for the lossless case. For the previous set of parameters it is shown in Fig. 8.18. The main difference is that for the quasi-static case there is no dip at the light line. For larger values of k there is hardly any difference.

8.2.3 The nature of the dispersion equation

As we have said many times before the dispersion equation relates the frequency to the wave number. If we know one of them we can find the other one from the dispersion equation. In the derivation above we managed to show that as long as we take into account all the elements up to infinity and $k > k_0$, all the terms in the dispersion equation are real: there is no radiation loss. What about ohmic losses? There are no difficulties formally to include losses. The wave then declines as it propagates, which is taken into account by making k complex. But then the concept of an infinite number of elements becomes a little unrealistic. Surely, there is no power left infinitely far away from the excitation,

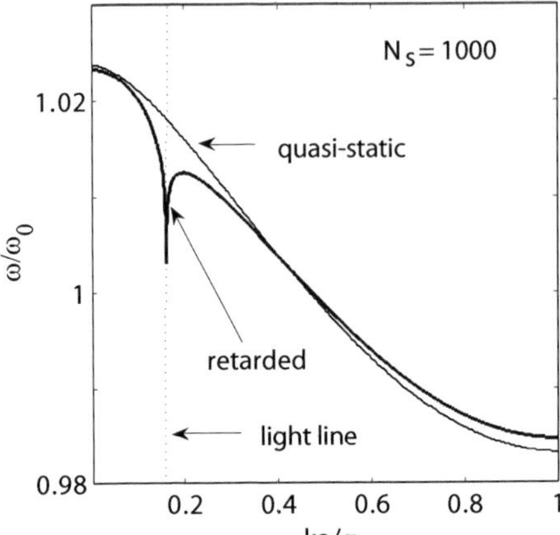

Fig. 8.18 Comparison between the retarded and quasi-static dispersion curves for $N_s = 1000$. From Zhuromskyy (2008)

and, anyway, where is the line excited? At minus infinity? At zero? The dispersion equation does not know about it. So we are in trouble with the infinitely long line if we want to include losses. Is there a way out? In principle there is. Instead of specifying the frequency we could specify the wave number, i.e. impose a periodic spatial variation upon the infinitely long line. For the lossless case the physical picture does not change. To the chosen k there will be a corresponding ω, meaning that the wave amplitude will vary harmonically at every point in the line. The physical picture is still OK for the lossy case. After being excited at every element of the infinite line the wave amplitude declines as a function of time everywhere in space while it oscillates at the frequency ω. Mathematically, the decay appears by assigning an imaginary ω to the real k. For a given set of parameters and for a given real k we can solve the dispersion equation and obtain from it both the real and the imaginary part of ω. So the physical picture is consistent. But this approach also has some drawbacks. First, in any practical situation we are interested in $\mathrm{Im}(k)$ (we want to know how the wave declines as it propagates), and not in $\mathrm{Im}(\omega)$, and also, in practice, we impose the frequency and not the wave number. The third drawback is the infinite number of elements in the model: it is difficult experimentally to realize it. So neither of the models discussed above are perfect. The question arises: has it been a worthwhile exercise to derive the dispersion equation? The answer is: certainly, yes. The dispersion equation gives a good idea of what is going on and what can be expected, but if we want to know what happens in a practical situation we need to look at a finite line and do the analysis for the actual excitation. We shall do this in the next section.

8.2.4 A 500-element line

Our aim is to find the current distribution for particular excitations, and in addition, to see how a kind of dispersion equation can be deduced from the data obtained and how close the values of ω and k are to those predicted by the dispersion equation derived in Section 8.2.2. We also need to choose the length of the line. The longer the line the easier it is to find the Fourier components of the current. In a paper concerned with the effect of retardation on magnetoinductive waves (Zhuromskyy *et al.*, 2005*b*) a 100-element line was chosen and analyzed. Below we shall repeat the exercise for a larger number of elements. A good compromise between accuracy and the ease of a numerical solution resulted in 500 elements.

The equation that will yield the current distribution for a given excitation is eqn (8.1), the generalized Ohm's law. There is no difficulty calculating the mutual inductance between any two elements. It has been done a number of times before in this book. However, in this section we chose to simplify the problem by assuming magnetic dipoles instead of small loops. Consequently, the mutual inductance will be given by the simple expression (see Appendix J)

$$M = \frac{\mu_0 \pi r_0^4}{4a^3}(1 + \mathrm{j}\, k_0 a - k_0^2 a^2)\, \mathrm{e}^{-\mathrm{j}\, k_0 a} \,. \tag{8.31}$$

A further problem is the choice of resistances. We have decided to regard the ohmic resistance as a parameter that follows from the assumed value of Q, which we shall often vary within a wide range to illustrate the effect of losses. What about the radiation resistance? Haven't we proven that the line does not radiate so we need not bother about radiation resistance? Yes, that is true to the right of the light line and for an infinite number of elements. Once the number of elements is finite we are not entitled to disregard radiation. Hence, we shall include the radiation resistance of a small loop that was given by eqn (1.96) and repeated below

$$R_{\mathrm{s}} = \frac{\pi}{6}\eta_0 \left(\frac{2\,\pi r_0}{\lambda}\right)^4 . \tag{8.32}$$

For the parameters given above this radiation resistance turns out to be 0.32 ohm. It must be added to the ohmic resistance so the effective Q will decline.

Having got all the parameters of the impedance matrix we are ready to determine the current distribution by inverting the impedance matrix.

Excitation by the first element. This means that in the voltage vector V the first element is taken as 1 V and the further 499 elements as zero. The current distribution may now be calculated for a given value of frequency that we shall choose to be $f = 1.01\, f_0$.

The modulus and phase of the current flowing in the nth element are plotted in Figs. 8.19(a) and (b) for $Q = 100$. The current, as may be expected, declines roughly exponentially. The phase variation looks rather odd, quite different from phase variations one usually encounters. It may be approximated by two straight lines: up to the first 15 elements the phase increases linearly along the line and then the phase variation reverses and declines linearly up to the end of the line. The increasing variation is a sign of a backward wave, whereas the decreasing phase

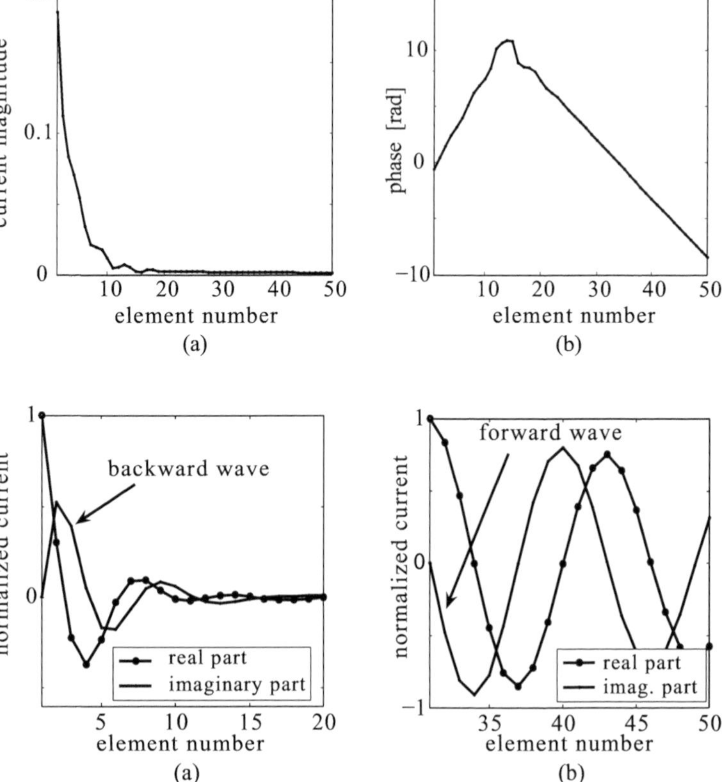

Fig. 8.19 (a) The magnitude of the current and (b) its phase as a function of element number for a quality factor of 100. From Zhuromskyy (2008)

Fig. 8.20 (a) The real and (b) the imaginary part of the current as a function of element number showing whether a wave is forward or backward. From Zhuromskyy (2008)

variation indicates a forward wave. The explanation for this rather odd behaviour is that two waves are present: a backward wave and a forward wave, both excited at element 1. The backward wave is excited with a higher amplitude, hence it dominates for the first 15 elements. However, this backward wave is lossy and has a higher attenuation than the forward wave. After propagating for 15 elements the two waves are of about equal amplitude. After the 15th element the forward wave dominates, and hence the decreasing phase variation. A rough calculation would give for ka (i.e. for the phase change per element) $-0.22\,\pi$ for the backward wave and $0.16\,\pi$ for the forward wave. Note that the power moves from the excited first element to the end of the line corresponding to a positive group velocity, which applies both to the backward and to the forward wave. But, as the name implies, the backward wave has a phase velocity pointing towards the source.

An alternative way to gain an intuitive feel for the periodic variation of the current is to plot its real and imaginary part. These are shown in Figs. 8.20(a) and (b) for $n = 1$ to 20 and for $n = 30$ to 50, respectively. These figures confirm the conclusions drawn from Fig. 8.19(b). At the beginning of the line one can see only the backward wave. By element 15 it declines sufficiently so that the forward wave, which is less attenuated, can be seen. Figures 8.20(a) and (b) may, clearly, provide the value of

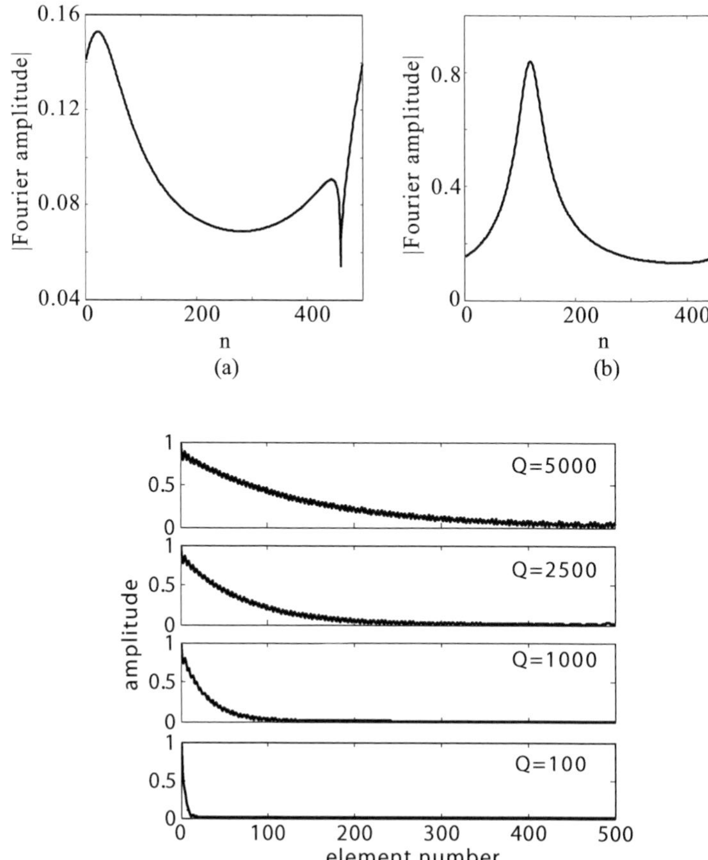

Fig. 8.21 Fourier spectrum as a function of the components for (a) $f/f_0 = 1.03$ and (b) $f/f_0 = 1.00$. From Zhuromskyy (2008)

Fig. 8.22 Decay of current amplitude as a function of element number. From Zhuromskyy (2008)

ka from the periodic behaviour of the curves.

The above two methods of extracting the information have been shown for their intuitive value. There is of course a more rigorous way of obtaining the spatial spectrum of the waves. Take the complex value of the current for each element from $n = 1$ to 500 and find its Fourier components. The results of that exercise (using Matlab program *fft*) are shown in Figs. 8.21(a) and (b) for $f/f_0 = 1.00$ and 1.03. The value of ka can be obtained from these diagrams by the following rule. $ka = -2\pi(n-1)/499$ when $n > 250.5$ and $ka = 2\pi(501-n)/499$ otherwise. The peaks both for the backward wave and for the forward wave may be clearly seen. The spectra are broad due to losses.

We can repeat the exercise for a number of Q values. As Q increases losses decrease and the amplitude decline is more gradual. This is shown in Fig. 8.22 for $Q = 100$, 1000, 2500, 5000. The phase against element number also depends strongly on the value of Q, as shown in Fig. 8.23. We can see again that as losses are lower the backward wave dominates further away in the line.

Next, we shall find the Fourier coefficients of the current distribution (Figs. 8.24(a)–(d)) presented by a contour code for four values of Q:

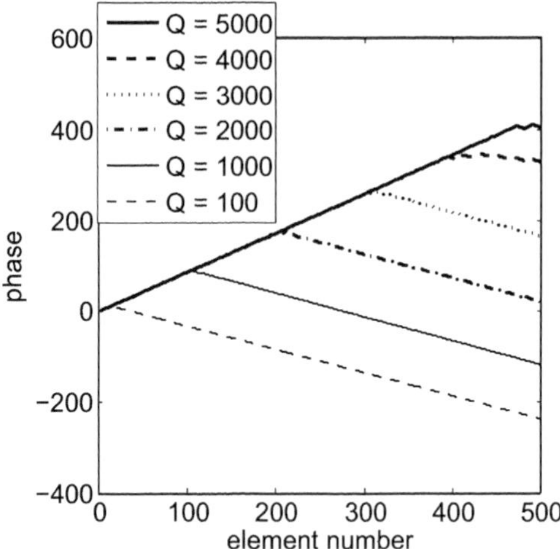

Fig. 8.23 Variation of the phase of the current as a function of element number. From Zhuromskyy (2008)

100, 1000, 10 000 and infinity. For a given value of Q we find a spread in the corresponding k values. The amount of the spread depends, as may be expected, on the value of Q. For $Q = 10\,000$ and infinity the lines representing the dispersion curve are quite narrow so they may be taken as the 'true' dispersion curve. For $Q = 10\,000$ there is a just discernible curve in the range $1 < ka/\pi < 2$. For $Q = \infty$ the curve is clearly visible. The reason is that without a matched load there is a nearly perfect reflection from the end of the array, hence, in the presence of low ohmic losses, the k values are about the same in both directions.

The spread in the wave vector was shown in order to appreciate how narrow (or wide) is the Fourier spectrum. Taking the maxima of the spectrum, an unambiguous dispersion curve is obtained for each value of Q as shown in Figs. 8.25(a)–(d). The new feature shown is the convergence of the curve to the light-line for $Q = 10\,000$ and ∞, which means that another solution exists (represented by a local maximum) as well. It could not be seen in Figs. 8.24(c) and (d) because their values were too small. This is in line with the conclusions arrived at earlier in this section. There are two waves excited, a backward wave and a forward wave, the latter one propagating close to the velocity of light.

The dispersion curves shown in Figs. 8.25(a)–(d) have been found by exciting the first element. Would we find the same curves if more elements were excited? We tried exciting the first few elements (up to 5); the dispersion curves found were no different.

Imposing the wave number. In the example given above the temporal frequency ω was imposed and we looked at the spread in the wave number. There is another obvious possibility: to impose the wave number and look at the spread in the temporal frequency. We found similar results to those of Figs. 8.24(a)–(d). We shall not show the spread here. We shall give only one value of ω for one value of k. Our criterion for finding that single value of ω is based on the concept of the input power

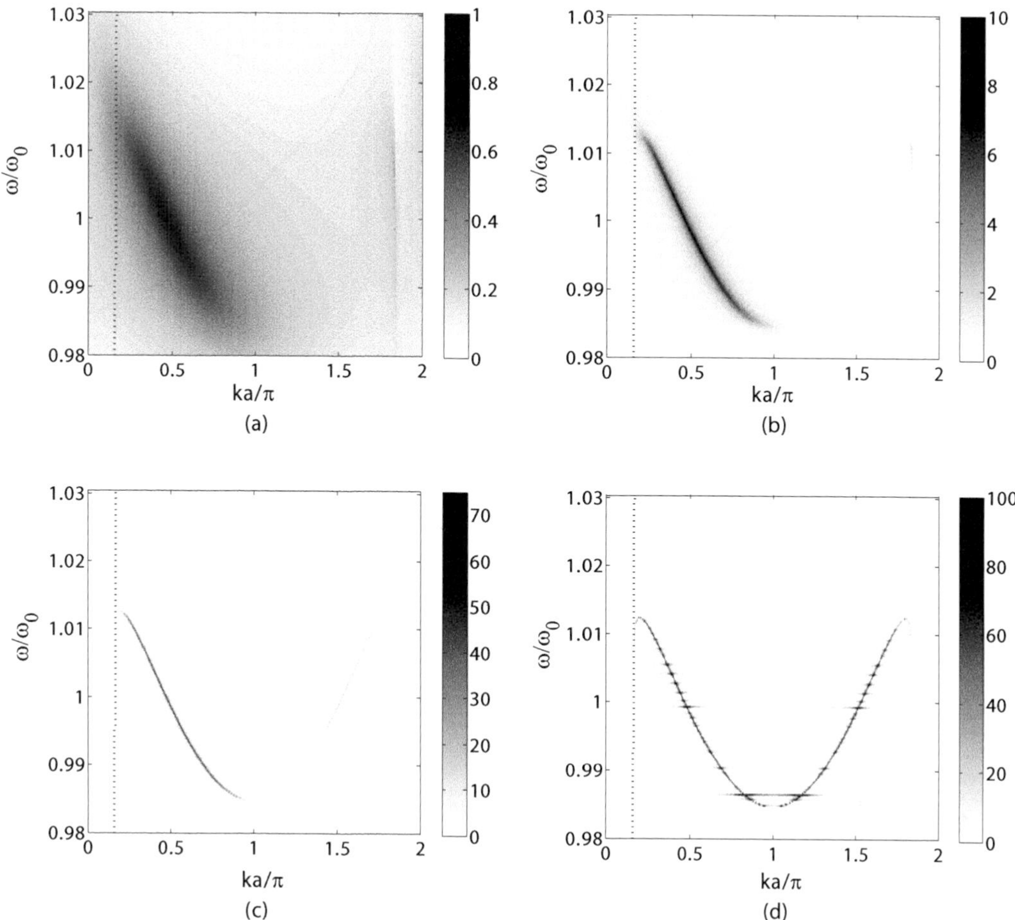

Fig. 8.24 Magnitude and distribution of the Fourier component plotted in the normalized $\omega - ka$ plane for various values of the quality factor, (a)–(d) $Q = 100$, 1000, 10 000 and ∞. From Zhuromskyy (2008)

$$P_{\text{in}} = \sum_{i=1}^{500} V_i I_i^* \,. \tag{8.33}$$

The frequency at which the input power is maximum is then regarded as a point on the dispersion curve. For $Q = 1000$ the dispersion curve is plotted in Fig. 8.26. Curves are not shown for other values of Q because they are very close to that in Fig. 8.26. Apparently, losses have little influence on the ω–ka relationship, presumably because the excitation occurs all along the line.

One would hope that exciting the first element or exciting all of them in a periodic manner will lead more or less to the same dispersion curve for the same value of Q. We can check this assertion by comparing, say, the curves in Fig. 8.25(c) with that of Fig. 8.26. This is shown in Fig. 8.27 where both are plotted. The agreement may be seen to be very

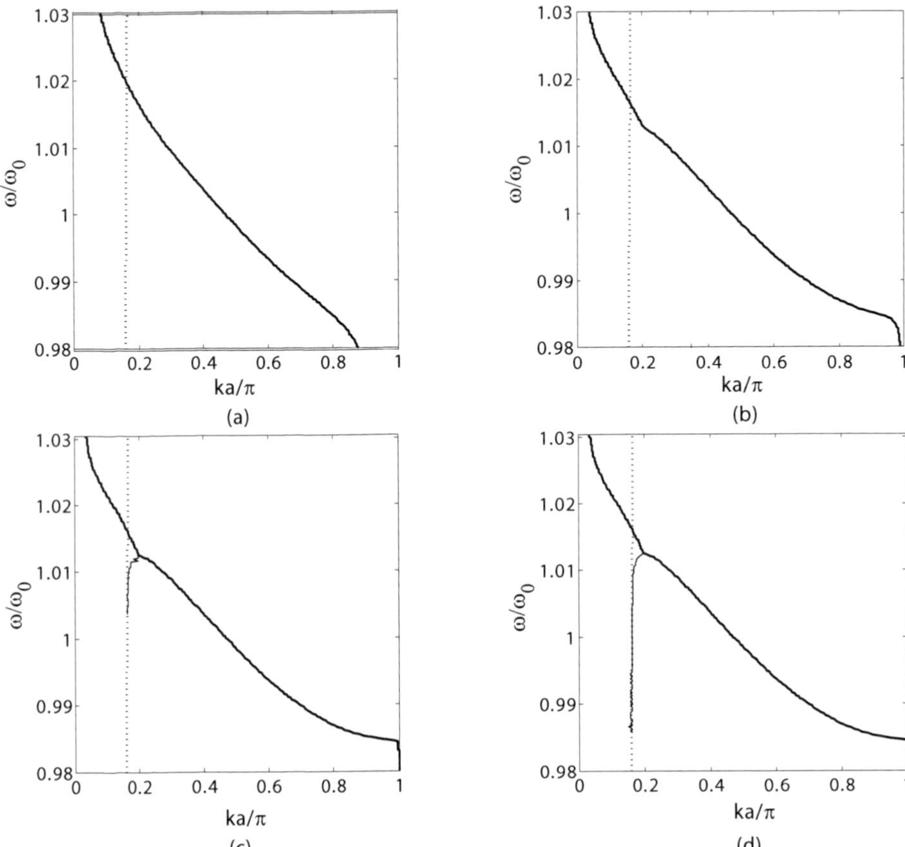

Fig. 8.25 Dispersion curves derived from the maximum of the Fourier spectrum, (a)–(d) $Q = 100$, 1000, 10 000 and ∞. From Zhuromskyy (2008)

good to the right of the light line. The two curves coincide within the thickness of the line. To the left of the light line they differ considerably. We cannot say which is correct, they are just the results of different excitations but there is no doubt that the one obtained from imposed wave number is closer to the dispersion curve of the infinite line. The information about the solution at the light line comes from the first element excitation. The reason is technical. When ω is imposed we can find a value of k close to the light line but we cannot impose k values infinitely close to each other.

8.2.5 Conclusions

We cannot claim to be able to draw clear conclusions. When the line is excited in a certain manner then we can predict what happens whether it is ω or k that is imposed. We have shown that in the planar configuration two waves are excited: a high attenuation backward wave and a low-attenuation forward wave that travels practically at the velocity of light.

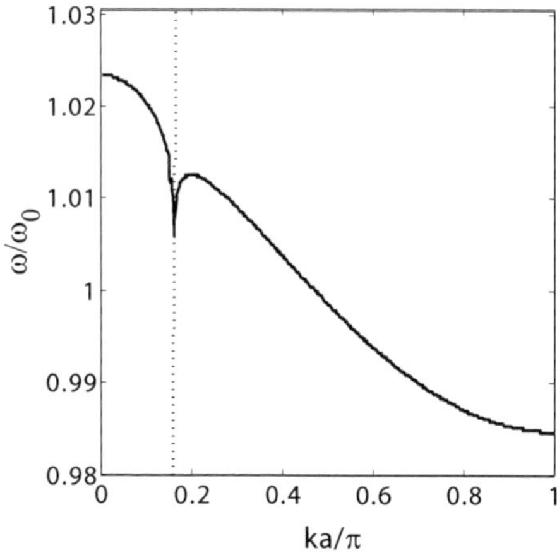

Fig. 8.26 Dispersion curve for $Q = 1000$ derived from the criterion of maximum input power. From Zhuromskyy (2008)

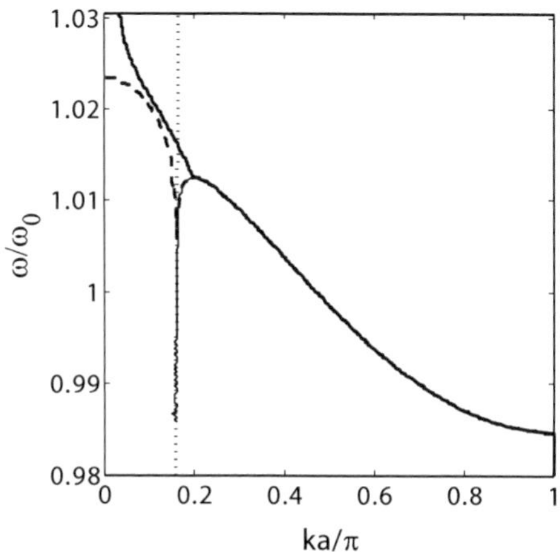

Fig. 8.27 Dispersion curves for $Q = 1000$ derived from the criterion of Fourier spectrum maximum (solid lines) and input power maximum (dotted lines). From Zhuromskyy (2008)

We can find current distributions and from these, if needed, we can find the radiation patterns. The ambiguity comes in when we consider the dispersion curve for infinite number of elements. It is not something that could ever be measured and for the lossy case its significance is rather limited.

8.3 Non-linear effects in magnetoinductive waves

8.3.1 Introduction

We have looked at a number of properties of magnetoinductive waves: we know about the interaction between resonant elements, the dispersion curve, propagation and attenuation, stop bands and pass bands, the interaction between arrays, and various applications. So far they were all linear. Can we extend our analysis to the non-linear case? Experimentally it is quite easy. Since in most cases the basic building block was the capacitively loaded loop all we need to do is to insert an element whose capacity varies with the voltage. Such an element is a varactor diode. As an approximation we shall assume that the variable part of the capacitance is a linear function of the voltage

$$C = C_0^{(\mathrm{nl})}(1 + \gamma V)\,, \tag{8.34}$$

where C_0 and γ are constants, and the second term is assumed to be small relative to the first term. What do we want to do with this non-linear diode? The application we have in mind is parametric amplification. It could serve for amplifying a signal with little noise added, and it could, in general, be used for loss compensation.

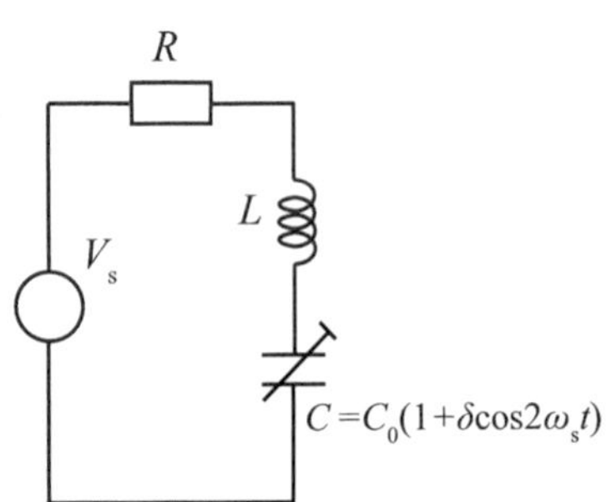

Fig. 8.28 *RLC* circuit in which the value of the capacitance is changing at twice the signal frequency of ω_s

Parametric amplifiers were invented in the 1930s. For a long time they were no more than scientific curiosities but came into their own when, owing to their low-loss properties, they were used as pre-amplifiers in satellite communications. With the advent of coherent optics they also played a role in producing practical oscillators (see, e.g., Solymar and Walsh 2004). The basic circuit is shown in Fig. 8.28. We have an *RLC* circuit in which the value of the capacitance is changing at twice the signal frequency of ω_s. In general, there are three frequencies the signal at ω_s, the pump at ω_p and the idler at ω_i. For our purpose it is perfectly adequate to omit the idler and to use the pump at $2\,\omega_\mathrm{s}$.

The aim is to amplify waves that propagate in a line. In our case they happen to be magnetoinductive waves, but whatever type the waves to be amplified are they need a pump wave, with which they can interact and which travels along with the same phase velocity. Thus, besides the condition of $\omega_\mathrm{p} = 2\,\omega_\mathrm{s}$, we also need to satisfy the condition $\beta_\mathrm{p} = 2\,\beta_\mathrm{s}$. In that case the signal velocity $v_\mathrm{s} = \omega_\mathrm{s}/\beta_\mathrm{s}$ will be equal to the pump velocity $v_\mathrm{p} = \omega_\mathrm{p}/\beta_\mathrm{p}$. This is the so-called phase-matching condition, the subject of Section 8.3.2.

8.3.2 Phase matching

In most of the examples we have seen so far the magnetoinductive wave was confined to a narrow band of frequencies. The corresponding values of the coupling coefficient were not much larger than $|\kappa| = 0.1$. Such a band is clearly too narrow for any chance of satisfying the phase-matching condition. We need a much larger $|\kappa|$. How large? We shall find it by substituting the phase-matching conditions into the dispersion equation of magnetoinductive waves, which yield

$$\frac{\omega_\mathrm{s}^2}{\omega_0^2} = \frac{1}{1 + \kappa\cos(\beta_\mathrm{s} a)} \tag{8.35}$$

for the signal wave and

$$\frac{4\omega_\mathrm{s}^2}{\omega_0^2} = \frac{1}{1 + \kappa\cos(2\,\beta_\mathrm{s} a)} \tag{8.36}$$

for the pump wave. We shall choose $\omega_\mathrm{s}/(2\,\pi) = 63.87$ MHz, and the mutual inductance as $M = 0.45L$ (i.e. $\kappa = 0.9$). We can then solve

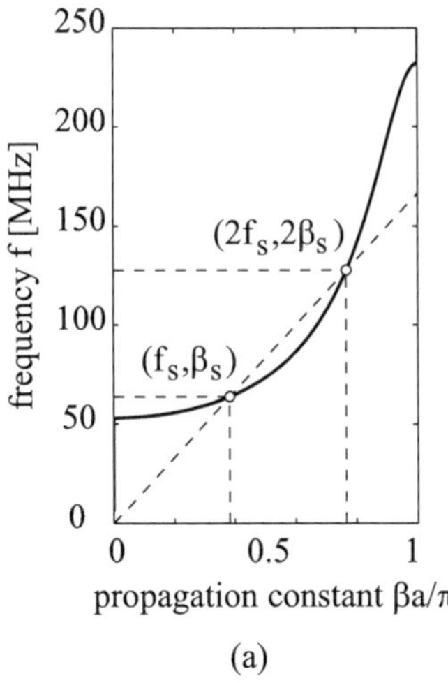

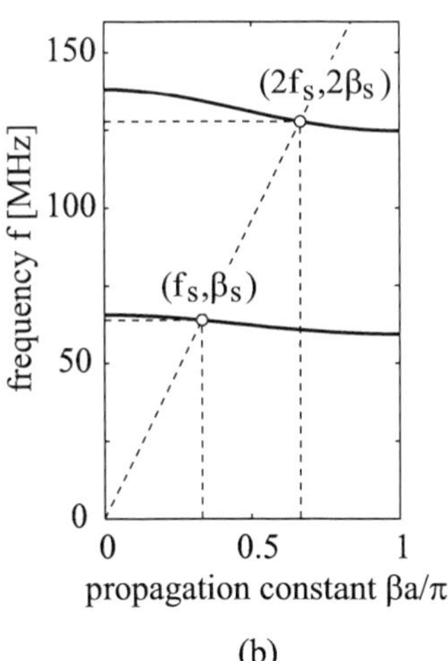

Fig. 8.29 Phase matching in (a) single axial and (b) two coupled planar arrays supporting MI waves. From Sydoruk *et al.* (2007*a*), copyright © 2007 Wiley-VCH Verlag GmbH & Co. KGaA, and Sydoruk *et al.* (2007*c*), copyright © 2007 IOP Publishing Ltd

eqns (8.35) and (8.36) for the resonant frequency and propagation constant, giving, respectively, $\omega_0/(2\,\pi) = 73.44$ MHz and $\beta_s a = 0.38\,\pi$. For elements with inductance $L = 50$ nH this value of the resonant frequency can be achieved by loading the elements by capacitances of $C_0 = 94$ pF.

The question arises whether we can have such a high coupling coefficient. According to recent experimental work (Syms *et al.*, 2006*b*; Syms *et al.*, 2007*a*) this is indeed possible, provided the elements in the axial configuration are sufficiently close[2] to each other. So, the possibility is there. If the sole purpose is to amplify a magnetoinductive wave then such an axial array could do it. For the elements given above the dispersion curve is shown in Fig. 8.29(a). It can be seen that both the signal and the pump waves can propagate and are phase matched.

[2]In fact, coupling coefficients as high as $\kappa = 1.5$ have been reported (Syms *et al.*, 2006*b*).

An alternative solution is to rely on two coupled arrays (Sydoruk *et al.*, 2006). The dispersion curve for such coupled arrays has already been shown in Fig. 7.36 for the planar configuration. It consists of two branches, hence the phase-matching condition can be satisfied with small values of the coupling coefficient. We shall choose again $\omega_s/(2\,\pi) = 63.87$ MHz and $L = 50$ nH for the signal frequency and the inductance of the loop. We take two identical arrays with separation of $h = 10$ mm between them (see Fig. 7.35(a)). This time the upper array is tuned to the signal frequency and the lower array to the pump frequency. The distance between the elements within the same array is $a = 24$ mm. Hence, the mutual inductances can be calculated[3] as $M = 0.15L$ and $M_1 = -0.05L$. These values are close to those investigated experimentally in Section 7.13. For choosing the propagation coefficients we have quite a lot of freedom. We shall take them as $\beta_s = \pi/(3a)$ and $\beta_p = 2\,\pi/(3a)$. Two parameters that are left undefined are the capacitances of the elements $C_{1,2}$. They are to be determined from the condition that the two phase-matched waves satisfy the dispersion relation. For the two coupled waveguides it is given (Sydoruk *et al.*, 2006) as

[3]Note that in Section 7.13 several mutual inductances were considered. In the present section, for simplicity, we use the approximation that only nearest-neighbour coupling needs to be taken into account.

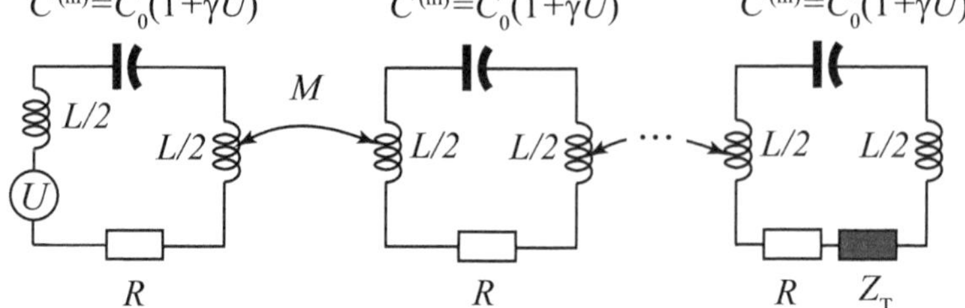

Fig. 8.30 An equivalent circuit of a single array. The capacitance is voltage-dependent, leading to parametric interaction between the signal and the pump

$$\begin{aligned}\left(\mathrm{j}\,\omega L + \frac{1}{\mathrm{j}\,\omega C_1} + R_1 + 2\mathrm{j}\,\omega M_1 \cos ka\right) & \\ \times \left(\mathrm{j}\,\omega L + \frac{1}{\mathrm{j}\,\omega C_2} + R_2 + 2\mathrm{j}\,\omega M_1 \cos ka\right) &= -\omega^2 M^2 . \quad (8.37)\end{aligned}$$

The above equation must be satisfied both for ω_s, β_s and for $2\,\omega_s$, $2\,\beta_s$. From the solution we find then that $C_1 = 130$ pF and $C_2 = 30$ pF. The corresponding resonant frequencies are $\omega_{01}/(2\,\pi) = 62.49$ MHz for the upper array and $\omega_{02}/(2\,\pi) = 129.15$ MHz for the lower one. The two branches of the dispersion curve showing phase matching are plotted in Fig. 8.29(b).

8.3.3 Theoretical formulation of amplification for the single array

The equivalent circuit of a non-linear magnetoinductive waveguide is shown in Fig. 8.30. There is a voltage source in the first element and a matching impedance, both for the signal and pump waves (see Section 7.3), in the last element.

The voltage in the nth element across the non-linear capacitor is taken as

$$U_s(n) = \frac{1}{2}\left[u_s(n)\,\mathrm{e}^{\mathrm{j}\,\nu_s(n)} + \mathrm{c.c.}\right], \quad (8.38)$$

$$U_p(n) = \frac{1}{2}\left[u_p(n)\,\mathrm{e}^{2\mathrm{j}\,\nu_s(n)} + \mathrm{c.c.}\right], \quad (8.39)$$

where

$$\nu_s(n) = \omega_s t - n\beta_s a\,, \quad (8.40)$$

and c.c. denotes complex conjugate. Similarly, we can write for the current in the nth element. Note that when relations are non-linear we need to exercise great care with complex notations. The intermediate variables u_s, u_p can be complex, but we ensure, by taking the complex conjugate, that the physically significant quantities, the signal and pump voltages are real.

The current flowing through the capacitor is the temporal derivative of the charge across the capacitor, hence it may be written in the general form

$$I_n = \frac{\mathrm{d}(CU)}{\mathrm{d}t}. \tag{8.41}$$

The problem is now that the capacitor, being voltage-dependent (see eqn (8.34)), the current will have a number of harmonics, even if we can assume (as we did in eqns (8.39) and (8.40)) that only the signal and pump voltages will be present across the capacitor. We shall follow here the time-honoured method and simplify the problem by further assuming that the currents will also have only those two frequencies:

$$I_\mathrm{s}(n) = \frac{1}{2}\left[i_\mathrm{s}(n)\,\mathrm{e}^{\mathrm{j}\,\nu_\mathrm{s}(n)} + \mathrm{c.c.}\right] \tag{8.42}$$

and

$$I_\mathrm{p}(n) = \frac{1}{2}\left[i_\mathrm{p}(n)\,\mathrm{e}^{2\mathrm{j}\,\nu_\mathrm{s}(n)} + \mathrm{c.c.}\right]. \tag{8.43}$$

The relationship between the voltages and currents will be determined by eqn (8.41). The linear part of the problem is trivial but the non-linear part is less so. Let us evaluate the term γV^2 when the signal and pump voltages are simultaneously present. We find

$$\gamma V^2 = \frac{\gamma}{4}\left[u_\mathrm{s}(n)\,\mathrm{e}^{\mathrm{j}\,\nu_\mathrm{s}(n)} + u_\mathrm{p}(n)\,\mathrm{e}^{2\mathrm{j}\,\nu_\mathrm{s}(n)} + \mathrm{c.c.}\right]^2. \tag{8.44}$$

It can be immediately seen that the term varying at the pump frequency will take the form

$$\frac{\gamma}{4}u_\mathrm{s}^2(n)\,\mathrm{e}^{2\mathrm{j}\,\nu_\mathrm{s}(n)}. \tag{8.45}$$

It is a little more difficult to recognize the term varying at the signal frequency. It is

$$\frac{\gamma}{2}u_\mathrm{p}u_\mathrm{s}^*(n)\,\mathrm{e}^{\mathrm{j}\,\nu_\mathrm{s}(n)}, \tag{8.46}$$

where the star is our other notation for the complex conjugate. There are 13 more terms in eqn (8.44) but none of them are at these frequencies. Adding now the current due to the linear term in eqn (8.41) to those in eqns (8.45) and (8.46) we find the signal and pump currents as

$$i_\mathrm{s}(n) = \mathrm{j}\,\omega_\mathrm{s}C_0\left[u_\mathrm{s}(n) + \gamma u_\mathrm{p}(n)u_\mathrm{s}^*(n)\right] \tag{8.47}$$

and

$$i_\mathrm{p}(n) = 2\mathrm{j}\,\omega_\mathrm{s}C_0\left[u_\mathrm{p}(n) + \frac{\gamma}{2}u_\mathrm{p}^2(n)\right]. \tag{8.48}$$

One more relationship that needs to be satisfied is of course Kirchhoff's voltage law that we have quoted a number of times and shall write down below but this time we need to keep the temporal derivatives,

$$M\frac{\mathrm{d}}{\mathrm{d}t}\left[I(n+1)+I(n-1)\right]+L\frac{\mathrm{d}I(n)}{\mathrm{d}t}+RI(n)+U(n)=0\,. \tag{8.49}$$

The above equation relates nearest neighbours to each other involving the $(n-1)$th, the nth and the $(n+1)$th element. The equation needs to be modified when it is written for the first and the last element in the array that have only one neighbour. The technique is described in Section 7.3 for finding the matching impedance. The present case is a little more complicated because the line needs to be matched at both the signal and the pump frequency so that

$$\begin{aligned} Z_{\mathrm{T}}(\omega_{\mathrm{s}}) &= \mathrm{j}\,\omega_{\mathrm{s}}M\,\mathrm{e}^{-\mathrm{j}\,\beta_{\mathrm{s}}a} \\ Z_{\mathrm{T}}(2\omega_{\mathrm{s}}) &= 2\mathrm{j}\,\omega_{\mathrm{s}}M\,\mathrm{e}^{-2\mathrm{j}\,\beta_{\mathrm{s}}a} \end{aligned}\,. \tag{8.50}$$

Performing the rather tedious algebraic operations the final result may be presented in matrix form (Sydoruk, 2007) as follows

$$\left\{\begin{aligned} &\left(\mathbf{Z}^{(\mathrm{s})}-\mathrm{j}\,\frac{1}{\omega_{\mathrm{s}}C_0}\mathbf{E}\right)\mathbf{u}_{\mathrm{s}}+\gamma\mathbf{Z}^{(\mathrm{s})}\mathbf{q}_{\mathrm{s}}=\mathbf{u}_{\mathrm{s}0} \\ &\left(\mathbf{Z}^{(\mathrm{p})}-\mathrm{j}\,\frac{1}{2\omega_{\mathrm{s}}C_0}\mathbf{E}\right)\mathbf{u}_{\mathrm{p}}+\frac{\gamma}{2}\mathbf{Z}^{(\mathrm{p})}\mathbf{q}_{\mathrm{p}}=\mathbf{u}_{\mathrm{p}0} \end{aligned}\right. . \tag{8.51}$$

Here, $\mathbf{E}$ is an $N\times N$ identity matrix,

$$\mathbf{Z}^{(\mathrm{s})}=\mathrm{j}\,\omega_{\mathrm{s}}\begin{pmatrix} L+\dfrac{R}{\mathrm{j}\,\omega_{\mathrm{s}}} & M\,\mathrm{e}^{-\mathrm{j}\,\beta_{\mathrm{s}}a} & \cdots & 0 \\ M\,\mathrm{e}^{\mathrm{j}\,\beta_{\mathrm{s}}a} & L+\dfrac{R}{\mathrm{j}\,\omega_{\mathrm{s}}} & \cdots & \vdots \\ \vdots & \ddots & \ddots & 0 \\ \vdots & \ddots & \ddots & M\,\mathrm{e}^{-\mathrm{j}\,\beta_{\mathrm{s}}a} \\ 0 & \cdots & M\,\mathrm{e}^{\mathrm{j}\,\beta_{\mathrm{s}}a} & L+\dfrac{R+Z_{\mathrm{T}}(\omega_{\mathrm{s}})}{\mathrm{j}\,\omega_{\mathrm{s}}} \end{pmatrix} \tag{8.52}$$

and

$$\mathbf{Z}^{(\mathrm{p})}=2\mathrm{j}\,\omega_{\mathrm{s}}\begin{pmatrix} L+\dfrac{R}{2\mathrm{j}\,\omega_{\mathrm{s}}} & M\,\mathrm{e}^{-2\mathrm{j}\,\beta_{\mathrm{s}}a} & \cdots & 0 \\ M\,\mathrm{e}^{2\mathrm{j}\,\beta_{\mathrm{s}}a} & L+\dfrac{R}{2\mathrm{j}\,\omega_{\mathrm{s}}} & \cdots & \vdots \\ \vdots & \ddots & \ddots & 0 \\ \vdots & \ddots & \ddots & M\,\mathrm{e}^{-2\mathrm{j}\,\beta_{\mathrm{s}}a} \\ 0 & \cdots & M\,\mathrm{e}^{2\mathrm{j}\,\beta_{\mathrm{s}}a} & L+\dfrac{R+Z_{\mathrm{T}}(2\omega_{\mathrm{s}})}{2\mathrm{j}\,\omega_{\mathrm{s}}} \end{pmatrix}. \tag{8.53}$$

$\mathbf{u}_{s0} = (u_{s0}, 0, \ldots, 0)^T$ and $\mathbf{u}_{p0} = (u_{p0}, 0, \ldots, 0)^T$ (the superscript T means transpose of a vector). Here, u_{s0} and u_{p0} characterize the amplitudes of, respectively, the signal and pump sources in the first element of the array. The vector $\mathbf{q}_s$ has elements $q_{sn} = u_p(n)u_s^*(n)$, and the vector $\mathbf{q}_p$, the elements $q_{pn} = u_s^2(n)$.

Equation (8.49) constitutes a system of difference equations in space that can be solved numerically for the signal and pump voltages for a given input. For an array of N elements a system of $2N$ equations should be solved. It is rather tiresome to do so but if the elements are different that is the only way to solve the problem. For identical elements it is usually preferable to convert the problem into a continuous one and turn the difference equations into differential equations. Provided the change of variables from element to element is small, this can be done. In fact, we have already done so in Section 8.1.5. In the present section the technique will be somewhat different because we don't want to impose the restriction that the phase change must be small from element to element. We only wish to assume that $u_s(n)$ will have a value close to $u_s(n-1)$ and to $u_s(n+1)$. In that case we can convert the discrete problem into a continuous one by using the following relationships

$$z = (n-1)a \qquad \text{and} \qquad u_{s,p}(n \pm 1) = u_{s,p}(n) \pm a\frac{du_{s,p}(n)}{dz} . \tag{8.54}$$

As an illustration, let us convert the expression for the signal wave

$$\mathrm{j}\,\omega_s M u_s(n+1)\,\mathrm{e}^{-\mathrm{j}\,\beta_s a} + \mathrm{j}\,\omega_s M u_s(n-1)\,\mathrm{e}^{\mathrm{j}\,\beta_s a} \tag{8.55}$$

into continuous form. We find that it comes to

$$\begin{aligned} &\mathrm{j}\,\omega_s M \left\{ \left[u_s(z) + a\frac{du_s}{dt} \right] \mathrm{e}^{-\mathrm{j}\,\beta_s a} + \left[u_s(z) - a\frac{du_s}{dt} \right] \mathrm{e}^{\mathrm{j}\,\beta_s a} \right\} \\ = \;\; &\mathrm{j}\,\omega_s M \left\{ u_s(z) + a\frac{du_s}{dt} \sin(\beta_s a) \right\} . \end{aligned} \tag{8.56}$$

Following the same technique and assuming that losses are weak and the non-linearity of the capacitance γ is small we end up (Sydoruk, 2007) with the differential equations

$$\begin{aligned} \frac{du_s(z)}{dz} &= -\mathrm{j}\, g_s u_p(z) u_s^*(z) - \alpha_s u_s(z) \\ &, \\ \frac{du_p(z)}{dz} &= -\mathrm{j}\, g_p u_s^2(z) - \alpha_p u_p(z) \end{aligned} \qquad , \tag{8.57}$$

where

$$g_s = \frac{\gamma}{2a\omega_s^2 C_0 M \sin\beta_s a} \qquad \text{and} \qquad g_p = \frac{\gamma}{16a\omega_s^2 C_0 M \sin 2\beta_s a} \tag{8.58}$$

characterize non-linearity and

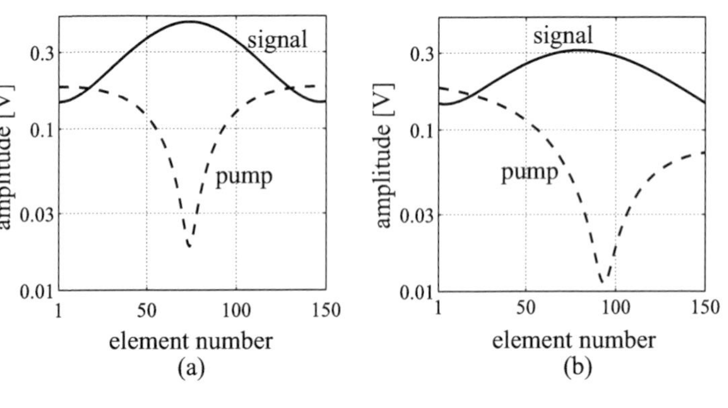

Fig. 8.31 Parametric amplification in a single array: (a) without and (b) with losses. From Sydoruk *et al.* (2007*a*). Copyright © 2007 Wiley-VCH Verlag GmbH & Co. KGaA

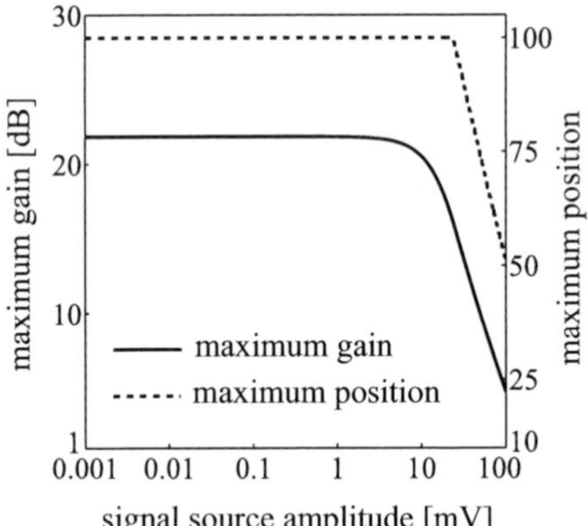

Fig. 8.32 Maximum gain achieved and position of the signal maximum for different amplitudes of the signal source. From Sydoruk *et al.* (2007*a*). Copyright © 2007 Wiley-VCH Verlag GmbH & Co. KGaA

$$\alpha_\mathrm{s} = \frac{R}{2a\omega_\mathrm{s} M \sin\beta_\mathrm{s} a} \quad \text{and} \quad \alpha_\mathrm{p} = \frac{R}{4a\omega_\mathrm{s} M \sin 2\beta_\mathrm{s} a} \tag{8.59}$$

are the attenuation coefficients. Note that eqns (8.57) are analogous to the corresponding equations for parametric amplification and second-harmonic generation well known in non-linear optics (Shen, 1984; Bloembergen, 2005).

For a numerical example we shall take the parameters given above and $\gamma = 0.1\,\mathrm{V}^{-1}$, $u_{\mathrm{p}0} = 0.25$ V and $u_{\mathrm{s}0} = 0.5$ mV. In the absence of losses there is a periodic exchange of power between the signal and pump waves, as shown in Fig. 8.31(a). Both waves decline in the presence of loss as shown in Fig. 8.31(b) for $Q = 250$. In a practical case of course the pump wave is much stronger than the signal wave. The amplification against signal input amplitude for that case is shown in Fig. 8.32 for $u_{\mathrm{s}0} = 1\,\mu\mathrm{V}$–100 mV. It may be seen that the amplification is constant up to a certain signal amplitude and then starts to decline.

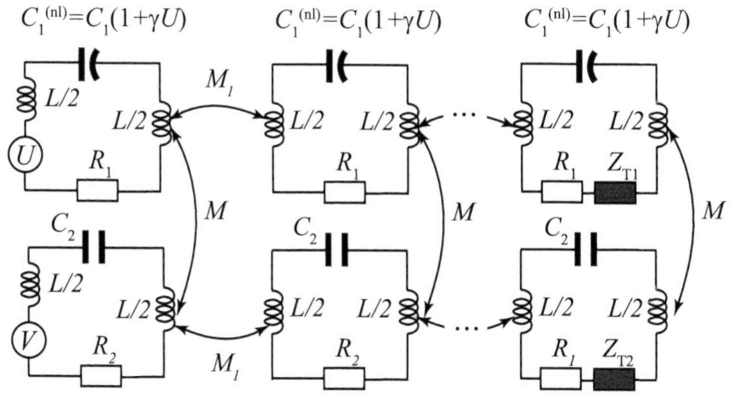

Fig. 8.33 Equivalent circuit of two coupled arrays. The capacitance of the top array is voltage-dependent, leading to parametric interaction between the signal and the pump. From Sydoruk *et al.* (2007*c*). Copyright © 2007 IOP Publishing Ltd

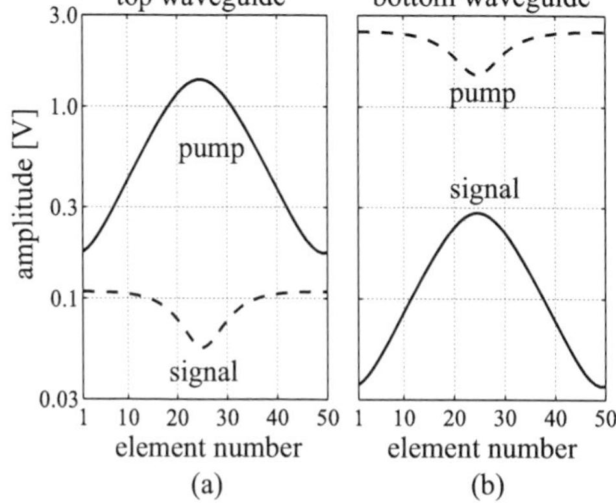

Fig. 8.34 Periodic exchange of power between the signal and pump in the general case, top (a) and bottom (b) waveguides. $U_0 = 10\,\text{mV}$, $V_0 = 125\,\text{mV}$, $\gamma = 0.1\,\text{V}^{-1}$. Losses are not included. From Sydoruk *et al.* (2007*c*). Copyright © 2007 IOP Publishing Ltd

8.3.4 Theoretical formulation for the coupled arrays

As indicated in Section 8.3.2 an alternative solution for achieving phase matching is to use two coupled arrays. This has two main advantages: firstly, more parameters give more freedom to design and a solution is possible in terms of planar arrays, which give a chance of further interaction with external agents. The equivalent circuit is shown in Fig. 8.33. The signal input is in the upper array and that of the pump in the lower array. However, due to the coupling the signal and the pump propagate in both arrays in spite of the fact that (remember from Section 8.3.2) the upper array is tuned to the signal frequency and the lower array to the pump frequency. A non-linear element is clearly necessary for parametric amplification, but it turns out that it is sufficient to insert a non-linear element in the upper array only.

The mathematical formulation is entirely analogous to that of a single array. We can derive a discrete set of equations in matrix form and it is also possible to assume continuity of the variables and formulate the problem in terms of partial differential equations. For the details of the derivations see Sydoruk *et al.* (2007*c*). Here, we shall only formulate the

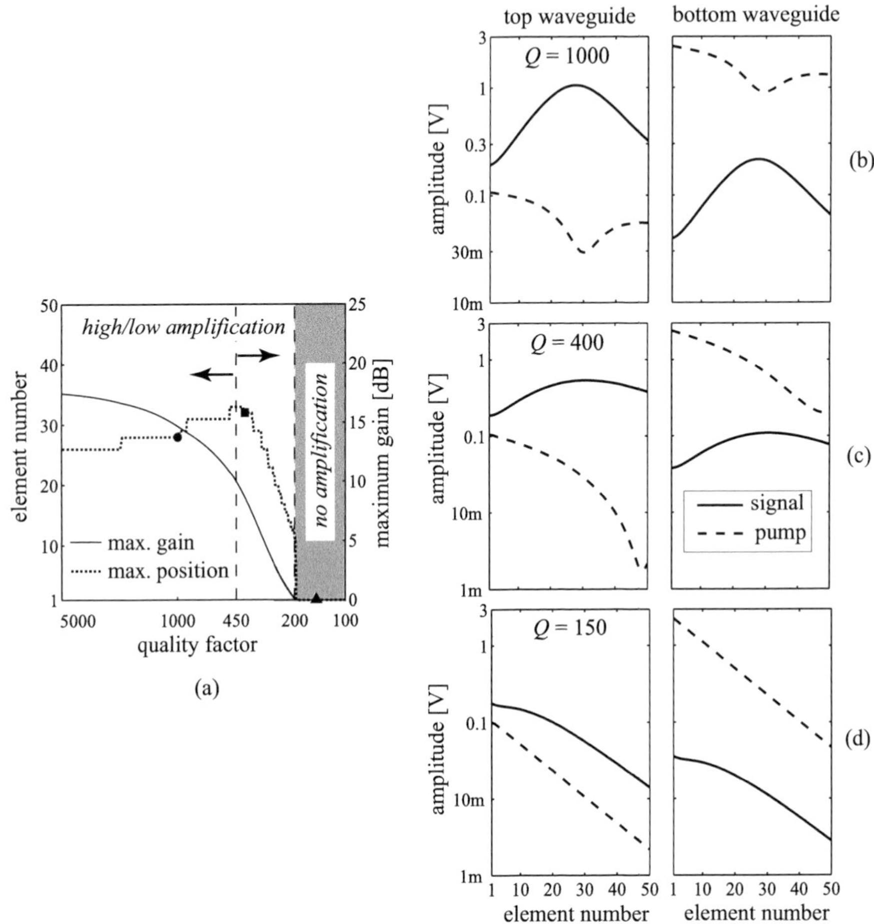

Fig. 8.35 Maximum signal gain achieved (dotted line) and the position of the signal maximum (solid line) depending on the amount of loss (a) and the distribution of the signal (solid lines) and pump (dashed lines) amplitudes along the waveguides for (b) $Q = 1000$, (c) $Q = 400$, and (d) $Q = 150$. Other parameters are as in Fig. 8.34. From Sydoruk *et al.* (2007*c*). Copyright © 2007 IOP Publishing Ltd

problem and show some results.

It is convenient to write the equations in terms of the voltages across the capacitors of the nth elements in both arrays

$$\begin{aligned} U_{\rm s}(n) &= \frac{1}{2}\left[u_{\rm s}(n)\, {\rm e}^{{\rm j}\,(\omega_{\rm s}t - n\beta_{\rm s}a)} + {\rm c.c.}\right] \\ U_{\rm p}(n) &= \frac{1}{2}\left[u_{\rm p}(n)\, {\rm e}^{2{\rm j}\,(\omega_{\rm s}t - n\beta_{\rm s}a)} + {\rm c.c.}\right] \\ V_{\rm s}(n) &= \frac{1}{2}\left[v_{\rm s}(n)\, {\rm e}^{{\rm j}\,(\omega_{\rm s}t - n\beta_{\rm s}a)} + {\rm c.c.}\right] \\ V_{\rm p}(n) &= \frac{1}{2}\left[v_{\rm p}(n)\, {\rm e}^{2{\rm j}\,(\omega_{\rm s}t - n\beta_{\rm s}a)} + {\rm c.c.}\right] \end{aligned} \quad . \qquad (8.60)$$

Here, $u_s(n)$ and $u_p(n)$ are the amplitudes of, respectively, the signal and pump waves in the upper array, and $v_s(n)$ and $v_p(n)$ are their amplitudes in the lower array. The corresponding currents flowing through the elements of the upper and lower arrays are, respectively,

$$
\begin{aligned}
I_n(n) &= \frac{d[C_1^{(\mathrm{nl})}\,(U_s(n)+U_p(n))]}{dt} \\
J_n(n) &= C_2\frac{d[V_s(n)+V_p(n)]}{dt}
\end{aligned} \quad . \tag{8.61}
$$

Neglecting higher harmonics the currents can be written separately at the signal and pump frequencies as

$$
\begin{aligned}
I_s(n) &= \frac{\mathrm{j}\,\omega_s C_1}{2}\left[u_s(n)+\gamma u_p(n)u_s^*(n)\right]\,\mathrm{e}^{\mathrm{j}\,(\omega_s t-n\beta_s a)}+\mathrm{c.c.} \\
I_p(n) &= \frac{\mathrm{j}\,\omega_s C_1}{2}\left[2u_p(n)+\gamma u_s^2(n)\right]\,\mathrm{e}^{2\mathrm{j}\,(\omega_s t-n\beta_s a)}+\mathrm{c.c.} \\
J_s(n) &= \frac{\mathrm{j}\,\omega_s C_2}{2}v_s(n)\,\mathrm{e}^{\mathrm{j}\,(\omega_s t-n\beta_s a)}+\mathrm{c.c.} \\
J_p(n) &= \mathrm{j}\,\omega_s C_2 v_p(n)\,\mathrm{e}^{2\mathrm{j}\,(\omega_s t-n\beta_s a)}+\mathrm{c.c.}
\end{aligned} \quad . \tag{8.62}
$$

The next steps are to substitute the currents and voltages into the difference equations of the magnetoinductive waves taking into account the excitation and matched terminal impedances, turn them into differential equations and solve the differential equations numerically.

A set of results showing the periodic exchange of energy between the signal and pump waves is shown in Fig. 8.34. The gain against signal amplitude is plotted in Fig. 8.35(a), whereas Figs. 8.35(b)–(d) show the variation of signal and pump amplitudes as a function of element number. An interesting conclusion is that in the upper array the signal amplitude may well exceed that of the pump. The pump amplitude is of course still high in the lower array.

A possible use of parametric amplification is for loss compensation. Taking $Q = 500$ and an appropriate choice of parameters the signal amplitude in the presence and absence of loss is shown in Fig. 8.36. It may be seen that the amplitude of the signal can be kept constant with good approximation.

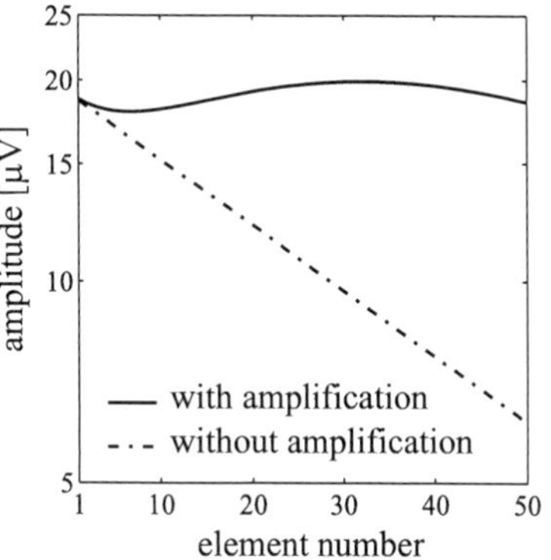

Fig. 8.36 Loss compensation by means of parametric amplification. The amplitude of the signal in the amplified case (solid line) is almost constant along the waveguide, whereas it declines exponentially in the non-amplified case (dash-dotted) line. $U_0 = 1\ \mu\mathrm{V}$, $V_0 = 50$ mV, $Q = 500$, $\gamma = 0.1\ \mathrm{V}^{-1}$. From Sydoruk *et al.* (2007*c*). Copyright © 2007 IOP Publishing Ltd

8.3.5 MRI detector

At the time of writing experimental work is going on in Syms' group at Imperial College on a detector (see Fig. 8.37) based on parametric amplification of MI waves (Syms *et al.*, 2007*b*; Syms *et al.*, 2008). The elements are spirals loaded by lumped capacitors to make them resonant at a required frequency. The signal wave is excited by rotating nuclear dipoles and travels (on the lower set of elements) in synchronism with

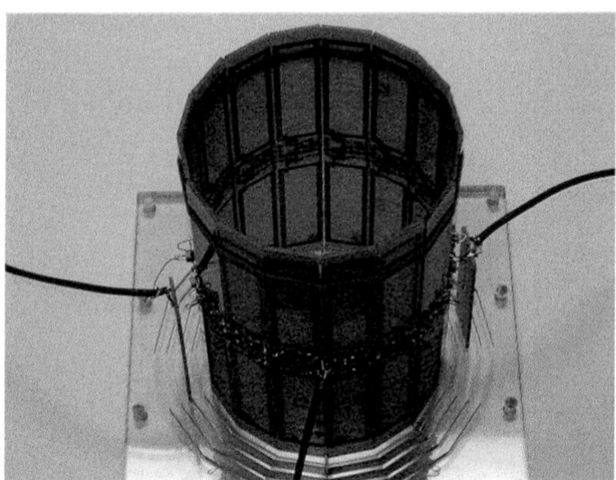

Fig. 8.37 Experimental setup of a magnetoinductive ring resonator. From Syms *et al.* (2008). For coloured version see plate section

them under conditions of rotational resonance. In contrast to the two-wave amplifiers discussed in the previous sections, this is a three-wave amplifier. The signal and pump waves both propagate in the planar configuration. The idler wave is realized by lumped circuits.

Seven topics in search of a chapter

9

9.1 Introduction

We have tried to be as comprehensive as possible in our treatment of the various phenomena of metamaterials. It is, however, inevitable that some topics just do not fit into the general framework. We introduced imaging by negative-index materials in Chapter 2 and discussed it later in Chapter 5 in considerable detail. Then, there was imaging by MI waves both in Chapters 7 and 8. Nevertheless, there are still some aspects of imaging that are important and need to be included. They will be discussed in Section 9.2.

Although multilayers were discussed in Section 5.5 and have also been treated in a different context in Section 8.1.2 there is still a combination of layers (negative permittivity and negative permeability in neighbouring layers) that did not seem to fit into any of the discussions. We describe them briefly in Section 9.3.

Focusing by indefinite media is an odd phenomenon because it happens only under specific circumstances when both a certain type of anisotropy and a periodic structure are present. The physics is similar to that of MI waves (see construction of the output beam in Fig. 8.7) but there we were concerned with only a single plane wave incident at a particular angle, whereas for focusing with indefinite media the incident beam is taken as divergent. The mathematical treatments are also quite different so we decided to treat indefinite media separately in Section 9.4.

Another interesting phenomenon is the Goos–Hanchen shift that stipulates that at reflection a finite beam appears to be shifted laterally. It is well known for positive-index materials. We shall give in Section 9.5 a summary of what happens when one of the media has negative index, and discuss briefly how the beam is guided when a negative-index material is surrounded by two positive-index materials.

Section 9.6 is on waves in nanoparticles. It is a little doubtful whether it should be included here because nanoparticles-cum-nanophotonics is a vast subject on its own. We have decided to include here those aspects that are akin to magnetoinductive waves.

Section 9.7 is devoted to phenomena occurring when the permittivity is close to zero.[1] The main effect is high phase velocity, well above the velocity of light. The consequences can be described by saying that those

[1] Similar phenomena will of course occur when the permeability is near to zero.

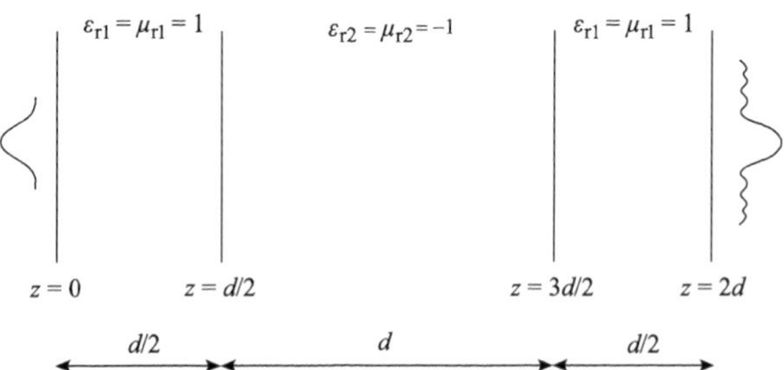

Fig. 9.1 The classical imaging arrangement of the 'perfect lens'. The region between $z = d/2$ and $z = 3d/2$ is usually a negative-index material. However, this figure may serve to illustrate the presence of two plates that should be resonant and coupled

in a hurry do not pay attention to small obstacles. They just sweep past them. Finally, we shall come to a late entry to the field: cloaking and invisibility. The basic principles will be discussed and some of the proposed solutions illustrated in Section 9.8.

On the whole, this chapter differs from the others in the sense that there will be very little mathematics. The emphasis will always be on the multifarious physical phenomena, a hallmark of the subject of metamaterials.

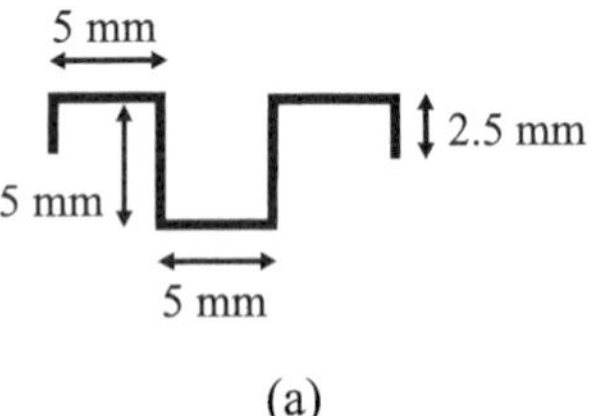

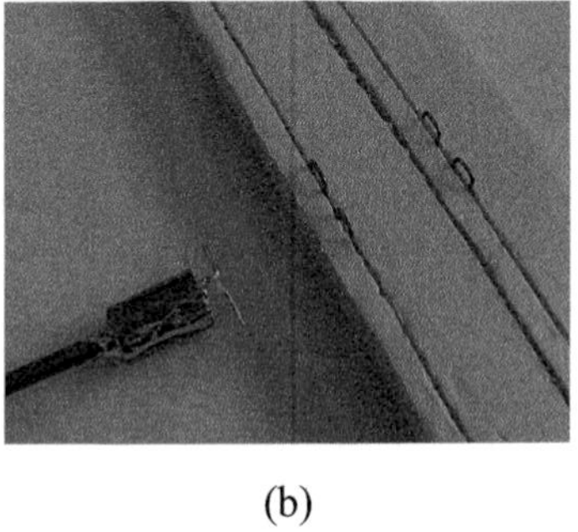

Fig. 9.2 (a) Resonant element made of thin wire of a length close to half-wavelength. (b) Two small elements in foam holders on each side with a dipole antenna as a source. From Maslovski *et al.* (2004). Copyright © 2004 American Institute of Physics

9.2 Further imaging mechanisms

9.2.1 Parallel sheets consisting of resonant elements

We shall reproduce here in Fig. 9.1 once more the basic configuration of the flat lens suitable for subwavelength imaging (Pendry, 2000). The lens consisting of a material with $\varepsilon_r = \mu_r = -1$ is between $z = d/2$ and $z = 3d/2$, the object is at $z = 0$ and the image is obtained at $z = 2d$. Now let us imagine that the two parallel lines in Fig. 9.1 are not the boundaries of a negative-index material but represent two plates, each one consisting of two-dimensional arrays of resonators. Would we still have subwavelength imaging? In Sections 5.3 and 5.4 we repeatedly called attention to the relationship between the growing evanescent waves and the SPP resonances. Is it possible that the negative-index material is only there to provide two surfaces that can propagate SPPs? Rao and Ong (2003) concluded on the basis of an FDTD (finite-difference time domain) numerical analysis that their results could be explained by a simple model that relies on coupling between two resonant entities a certain distance apart. Maslovski *et al.* (2004) went further by proposing that for subwavelength imaging it is a sufficient condition to have two coupled resonant sheets. For simplicity, they chose resonant lines instead of resonant sheets in their experiments. It was accomplished by placing the experimental apparatus between two conducting sheets a distance of 25 mm apart.

The element chosen by Maslovski *et al.* (2004) is shown in Fig. 9.2(a). It is made of thin wire. Its total length is 25 mm. It was found to be resonant at a frequency of 5 GHz, which is about right considering that the half-wavelength is 30 mm. Two such elements were placed in foam

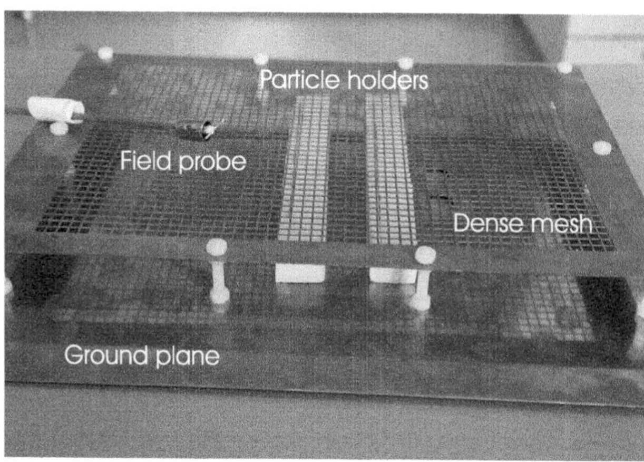

Fig. 9.3 Photograph of the experimental setup. The foam holders are between two planes made by a dense mesh of conducting strips. From Maslovski *et al.* (2004). Copyright © 2004 American Institute of Physics

holders at $x = -105$ mm and $x = -65$ mm, a distance of 40 mm apart, shown in Fig. 9.2(b). The object was a horizontal electric dipole placed at $x = -125$ mm, i.e. 20 mm behind the first set of elements. A photograph of the measuring apparatus may be seen in Fig. 9.3. The two conducting planes are realized by dense meshes. The upper mesh was made weakly penetrable to the fields in order to allow the measurement of the field distribution by a probe positioned at the top of the mesh. Note that the electric dipole is parallel with the meshes, and the distance between the meshes is less than half-wavelength. Consequently, the dipole can excite only evanescent waves. The wave amplitudes measured just above the mesh are shown in Fig. 9.4. They may be seen to follow the expected pattern. The field declines from the source to the position of the first line, grows afterwards until it reaches the second line, and declines again beyond the second line. Its value at the image position, 20 mm behind the second line, is the same as at the source. Further experiments with more elements yielded similar conclusions.

The essential requirement is, apparently, the presence of two coupled resonant lines. In the experiments of Freire and Marques (2005, 2008), discussed in Section 8.1.6, the resonant sheets supported MI waves, hence the mechanism was very likely the same as that of Maslovski *et al.* (2004), with channelling effects of the type discussed in Section 7.15.3 playing a role.

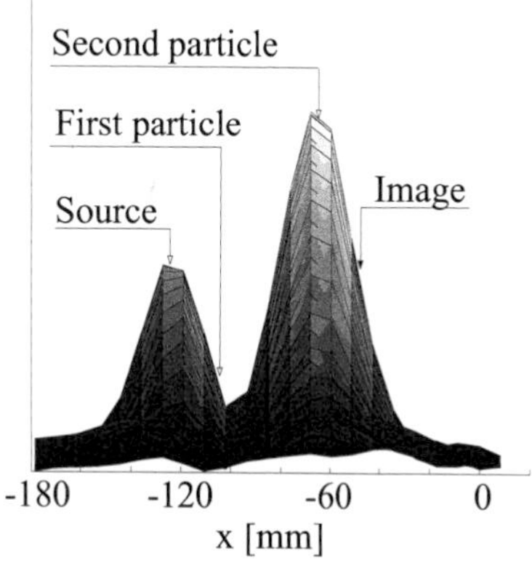

Fig. 9.4 Field amplitude variation along the axis. Key positions on the plot are indicated by arrows. From Maslovski *et al.* (2004). Copyright © 2004 American Institute of Physics

9.2.2 Channelling by wire structures

The interaction between wire structures and electromagnetic waves have been known for a long time. See the early papers of Kock 1964; Macfarlane 1946; Cohn 1946; Brown 1950; Brown 1953; Lewis and Casey 1952; Casey and Lewis 1952, the still influential paper of Rotman (1962) and those of Pendry *et al.* 1996; Pendry *et al.* 1998 in which the wires' plasma properties were rediscovered. Wires were of course used in the

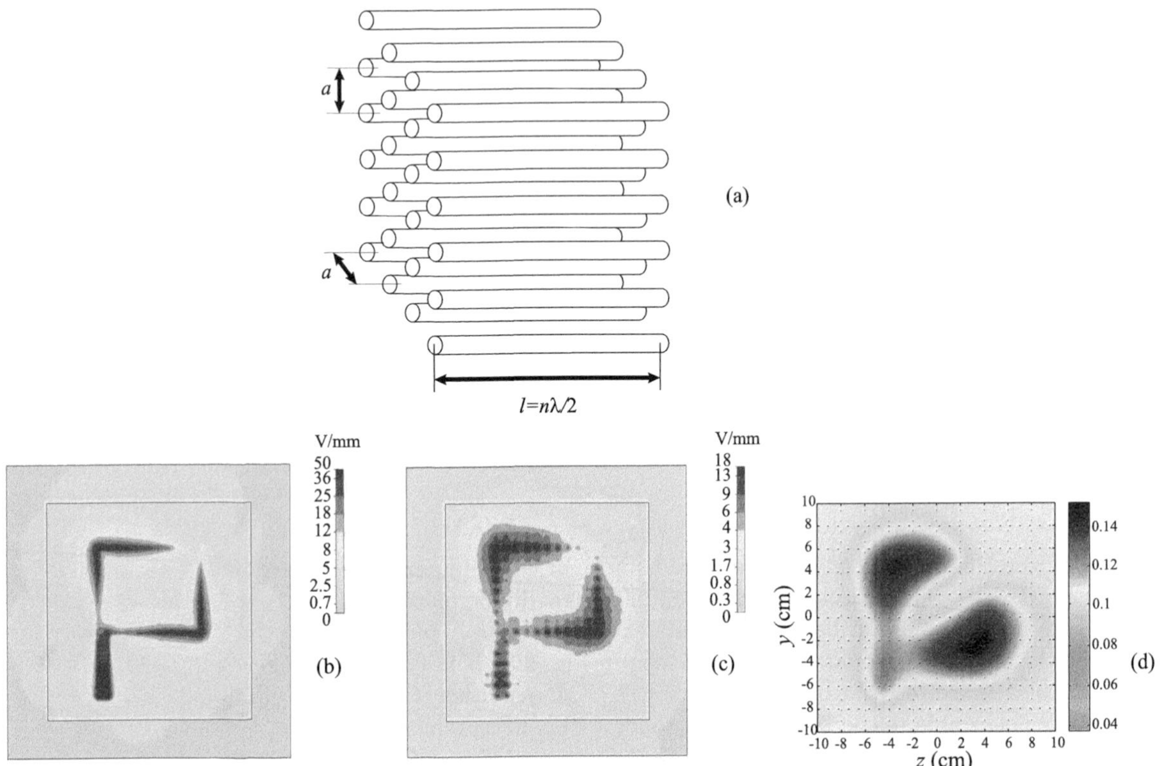

Fig. 9.5 (a) Horizontal stack of a square lattice of thin wires. Length chosen so as to satisfy the Fabry–Perot condition of resonance. (b) Image of letter P 2.5 mm in front of the object obtained by numerical simulation. (c) Image 2.5 mm behind the lens obtained by numerical simulation. (d) Electric-field distribution in the image plane measured by near-field scanning. From Belov *et al.* (2006*a*). Copyright © 2006 by the American Physical Society. For coloured version see plate section

experiments of Shelby *et al.* (2001*a*) to produce the negative permittivity, while the negative permeability was provided by SRRs.

In this section we shall discuss how to use a wire structure for channelling the spatial information from the object to the image. In other words this is pixel-by-pixel imaging. The basic idea is quite simple (Belov *et al.*, 2006*a*; Zhao *et al.*, 2006; Zhao *et al.*, 2007). Wires arranged as in Fig. 9.5(a) will carry a wave with p polarization along the wires. The resolution is limited by the period. The main problem is reflection from the surfaces, which is quite significant for thicker slabs. The solution is to choose thicknesses (integral multiples of the half-wavelength) that give rise to Fabry–Perot resonance. The wire structure chosen both for numerical simulation and for experiment had wires of 1 mm radius, the period was $a = 10$ mm, there were 21×21 wires, and the slab thickness was 15 cm, a half-wavelength at the frequency of 1 GHz. The object was in the form of the letter P, placed at 5 mm in front of the lens.

The images at 2.5 mm in front of the lens (Fig. 9.5(b)) and 2.5 mm behind the lens (Fig. 9.5(c)) were obtained by the CST MICROWAVE STUDIO package. It may be seen that transmission though the lens does not lead to significant distortion. The results were further confirmed by experiments. The object was an antenna in the form of the letter P fed by a coaxial cable and located 3 mm away from the wires. The receiver was a short piece of wire connected to a coaxial cable. The electric field was scanned at 5 mm behind the lens. At the best frequency of 0.98 GHz the measured image is shown in Fig. 9.5(d).

Similar experiments by a linearly expanding set of wires were conducted by Ikonen *et al.* (2007*b*) using the letter M as the object. The pattern was found to be distorted at the output but still recognizable. Thanks to the radial expansion a magnification by a factor of 3 was achieved.

Wire structure experiments with a meander line as the object were performed by Belov *et al.* (2006*b*). They showed (in agreement with the theoretical predictions of Belov and Silveirinha (2006)) that a resolution of $\lambda/15$ can be achieved irrespective of the shape and complexity of the object.

Similar but much smaller wire structures suitable for the infra-red and optical range were proposed by Ono *et al.* (2005) and Silveirinha and Engheta (2007). For a non-local homogenization model see Silveirinha and Engheta (2006).

A quite different idea for imaging was conceived by Belov *et al.* (2005). The wire structure consisting of 14 layers is now perpendicular to the direction of propagation as shown in Figs. 9.6(a) and (b). Note in the inset that the vertical rods are capacitively loaded (analyzed earlier by Belov *et al.* (2002)). The isofrequency curves corresponding to the orientation of the wire structure shown in Fig. 9.6(b) may be seen in Fig. 9.7. Note that at a certain value of the longitudinal wave vector the contour is flat within a wide range of the transverse wave vector. The incident wave vectors are represented by the half-circle on the left of the diagram. As usual, the criterion of refraction is that tangential components of the wave vectors must agree outside and inside the material. This means that the transverse component of $\mathbf{k}^{(1)}$ (incident wave vector) is identical with the transverse component of $\mathbf{q}^{(1)}$ (wave vector of the propagating wave inside the wire structure). If the isofrequency curve is flat then all the propagating waves inside the wire structure will have identical group velocities $\mathbf{v}_{\mathrm{g}}^{(1)}$ pointing in the same (horizontal) direction—and that is the condition of perfect channelling. Interestingly, the group velocity will be roughly the same for a range of evanescent waves as well because $\mathbf{v}_{\mathrm{g}}^{(2)}$ corresponding to the input evanescent wave of $\mathbf{k}^{(2)}$ differs very little from $\mathbf{v}_{\mathrm{g}}^{(1)}$. Hence, the evanescent waves in air turn into propagating waves in the wire structure. At the output of the structure the inverse phenomenon takes place: the propagating wave turns into an evanescent wave. Thus, the evanescent waves representing subwavelength information can be reproduced. It is again an advantage to have thicknesses corresponding to Fabry–Perot resonance. They are shown by vertical lines in Fig. 9.7.

The results of a numerical simulation may be seen in Fig. 9.8. The object is a point course located at a distance of d in front of the structure.

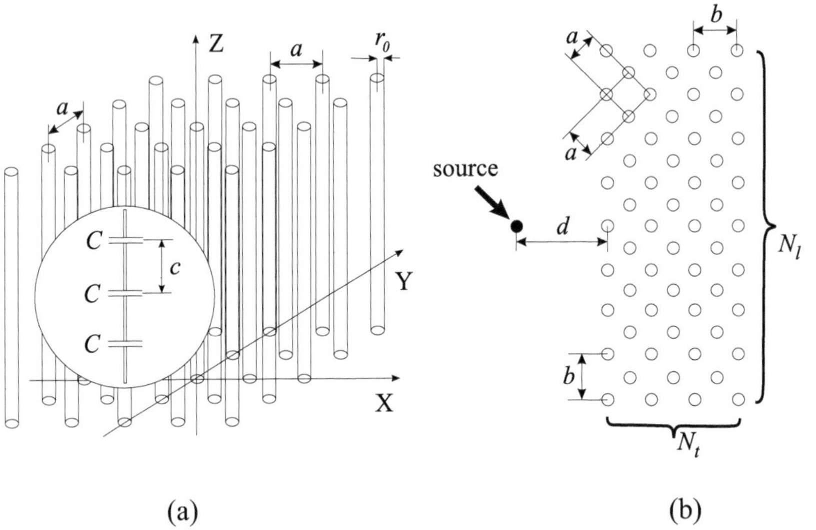

Fig. 9.6 (a) Square lattice of capacitively loaded vertical wire structure, (b) view from above. From Belov *et al.* (2005). Copyright © 2005 by the American Physical Society

The focal spot was calculated as equal to $\lambda/6$.

It needs to be noted that there are no growing evanescent waves. Thus, if the object is farther away from the input surface then it will decline before it turns into a propagating wave inside the structure. There is no mechanism to restore the evanescent waves to their original amplitude. There are no growing waves and there are no SPPs either. The mechanism of subwavelength imaging in this case is entirely different from that proposed by Pendry (2000).

The 1D metal–dielectric structure was already discussed in Section 5.5. It was claimed there that the structure is, under certain circumstances, equivalent to an anisotropic structure with permittivities zero in the transverse direction and infinitely large in the axial direction (see Appendix I). Its imaging properties were discussed in more detail by Belov and Zhao (2006) and Li *et al.* (2007). The qualitative mechanism invoked was channelling, which applied both to the propagating and evanescent components. The length of the device could be arbitrarily long in the lossless case. High transmission amplitude was ensured by making the length of the device equal to an integral multiple of the half-wavelength, i.e. by constructing a Fabry–Perot resonator. In one of their designs, Belov and Zhao (2006) used an anisotropic high positive permittivity metamaterial that was proposed earlier by Shen *et al.* (2005).

Finally, we wish to mention that the wire structures discussed in this section (in particular the one where the wires are parallel to the interface) could also be regarded as examples of photonic crystals. We included them here owing to their significance for metamaterials. Photonic crystals themselves are beyond the scope of the present book. We shall make only a few brief comments on them in the next section.

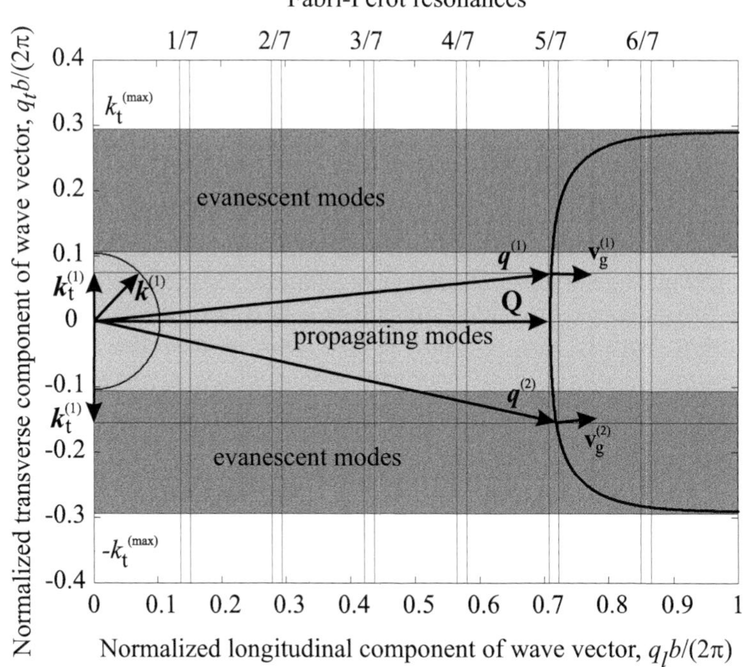

Fig. 9.7 Iso-frequency curves in the wave-vector plane. Lines of Fabry–Perot resonances also shown. From Belov *et al.* (2005). Copyright © 2005 by the American Physical Society

9.2.3 Imaging by photonic crystals

Photonic crystals rose to fame as periodic (usually dielectric) structures capable of reflecting electromagnetic waves coming from any directions, hence capable of trapping electromagnetic waves. Their operation is based on a generalization of the principles upon which reflection holograms works: total reflection is built up from elementary reflections satisfying the Bragg conditions. Hence, their period is comparable with the wavelength. Unfortunately, their design is much more complicated because the periodic structures of photonic crystals are not minor perturbations of the background. The contrast must be high, e.g. the background material for a periodic set of holes may have a relative dielectric

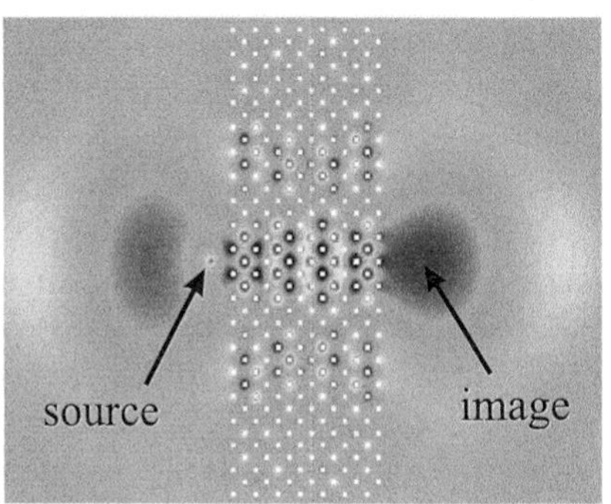

Fig. 9.8 Distribution of the electric field showing the source and the image. From Belov *et al.* (2005). Copyright © 2005 by the American Physical Society. For coloured version see plate section

constant as high as 10.

As a response to the advent of metamaterials those working on photonic crystals were keen to show that they can be made suitable both for negative refraction and imaging. Whether the subject has thereby become part of metamaterials is disputable, largely a matter of definition. In a recent book edited by Engheta and Ziolkowski (2006) entitled metamaterials about one half of the book is devoted to electromagnetic bandgap materials which is just another phrase for photonic crystals. As explained in the Preface we believe that both the physical description and the mathematical formulation of photonic crystals differ so much from metamaterials that it is difficult to give a unified description that applies to both of them. Hence, in this book we have not so far mentioned photonic crystals. Nonetheless, we shall say a few words about them here and give a few references.

Notomi (2000) pointed out that at frequencies close to the bandgap it is possible to define an index of refraction for waves propagating in photonic crystals. Subsequently, it was shown theoretically that other configurations are also possible and the concept of negative index is not necessary (Luo *et al.*, 2002*a*; Luo *et al.*, 2002*b*; Luo *et al.*, 2002*c*). Experimental confirmation of imaging was provided by Parimi *et al.* 2003; Parimi *et al.* 2004.

9.3 Combinations of negative-permittivity and negative-permeability layers

We have mentioned several times multilayer structures in different contexts. In each case the aim was to produce properties superior to those of single layers. The choice in each case was a metal–dielectric structure, or in other words combinations of layers with negative and positive permittivities. A different combination of layers, namely an $\varepsilon < 0$, $\mu > 0$ layer combined with an $\varepsilon > 0$, $\mu < 0$ layer, was proposed by Fredkin and Ron (2002). They showed that such a sandwich may behave as a negative-index material with phase and group velocities in opposite directions and may also exhibit negative refraction. The authors also found solutions with positive effective index with bandgaps associated with the period of the structure. A more detailed analysis of the same problem is due to Alu and Engheta (2003) who showed that the combination had a number of remarkable properties, like resonance and perfect tunnelling. To show the physics of the perfect tunnelling mechanism they plotted Poynting vector streamlines. The streamlines were highly curved in both media but the joint effect was a 100% transmission of the electromagnetic wave incident at a particular angle as if the bilayer was not there. A similar analysis with similar conclusions was performed by Wang *et al.* (2006). A version that uses indefinite media anisotropic layers (see Section 9.4) and is suitable for imaging was proposed by Schurig and Smith (2005). Experimental realization of imaging with the aid of loaded anisotropic transmission lines was achieved by Feng *et al.* (2007). They also concluded that imaging by the combination of negative-permittivity and negative-permeability layers was less sensitive to losses than Pendry's 'perfect lens'.

9.4 Indefinite media

In a search for media exhibiting negative refraction, Lindell *et al.* (2001) came to the conclusion that uniaxially anisotropic media are suitable for the purpose, as could be deduced from the dispersion equation. They showed that under certain circumstances it is sufficient to have just one of the material parameters as negative in order to achieve negative refraction. The concept was further developed by Smith and Schurig 2003; Schurig and Smith 2005 and Smith *et al.* 2004*b*. They referred to the media, in which the diagonal elements of the material parameter tensors may be of opposite sign, as indefinite media. These media have some interesting properties that we shall investigate in this section.

The assumption is that the material parameters are given by diagonal tensors, the electric field has only a component in the y direction and the direction of propagation is the z direction. The corresponding wave equation was obtained in Section 1.17. We may obtain the dispersion equation from this by assuming the electric field in the form

$$E_y = E_0 \, \mathrm{e}^{-\mathrm{j}\,(k_z z + k_x x)}\,, \tag{9.1}$$

and substituting it into the wave equation. The dispersion equation may then be obtained as

$$k_z^2 = \mu_{xx}\left(\varepsilon_{yy}k_0^2 - \frac{k_x^2}{\mu_{zz}}\right). \tag{9.2}$$

The question of interest is whether the wave is propagating ($k_z^2 > 0$) or evanescent ($k_z^2 < 0$). The boundary between the two cases may be easily obtained from eqn (9.2). It occurs at k_c, a critical value of the spatial frequency, when

$$k_x = k_\mathrm{c} = k_0\sqrt{\varepsilon_{yy}\mu_{zz}}\,. \tag{9.3}$$

In eqn (9.2) there are two parameters, $\mu_{xx}\varepsilon_{yy}$ and μ_{xx}/μ_{zz}. We can distinguish four cases:

$$\text{(i)} \quad \mu_{xx}\varepsilon_{yy} > 0 \quad \mu_{xx}/\mu_{zz} > 0\,, \tag{9.4}$$

$$\text{(ii)} \quad \mu_{xx}\varepsilon_{yy} > 0 \quad \mu_{xx}/\mu_{zz} < 0\,, \tag{9.5}$$

$$\text{(iii)} \quad \mu_{xx}\varepsilon_{yy} < 0 \quad \mu_{xx}/\mu_{zz} > 0\,, \tag{9.6}$$

$$\text{(iv)} \quad \mu_{xx}\varepsilon_{yy} < 0 \quad \mu_{xx}/\mu_{zz} < 0\,. \tag{9.7}$$

It follows from eqn (9.2) that in case (i) there is propagation below k_c, in case (ii) there is propagation above k_c, in case (iii) there is propagation for all values of the spatial frequency, and in case (iv) there is no propagation for any value of the spatial frequency.[2]

[2] The authors called these four cases cutoff, anti-cutoff, never cutoff and always cutoff (a terminology shared by Li *et al.* (2007)). Although the terminology is entirely appropriate we did not adopt it because we did not want to introduce new definitions.

Case (iii) is particularly interesting. Propagation for all values of the spatial frequency means that waves that are evanescent in air (i.e. for which $k_x > k_c$) can propagate in this medium. The evanescent waves do not grow but do not decline either if the medium is lossless. Does such a medium exist? It does. The material parameters of a set of SRRs satisfy eqn (9.5). The longitudinal component of the permeability tensor is negative, whereas μ_{xx} and ε_{yy} are positive. If, in addition, $\mu_{xx} = 1$, $\varepsilon_{yy} = 1$ and $\mu_{zz} = -1$, then we obtain the nice hyperbola

$$k_z^2 - k_x^2 = k^2 , \tag{9.8}$$

plotted in Fig. 9.9.

What is this medium good for? Smith *et al.* (2004*b*) proposed that it would bring a diverging beam to a partial focus. Why? Let us draw an input wave vector from the origin to the hyperbola. When it is incident upon our indefinite medium the direction of the group velocity is obtained by drawing perpendiculars to the ω = const line, as has already been explained in Section 8.1. The different focusing properties of Veselago's lens and that relying on the properties of indefinite media are shown in Fig. 9.10. Figure 9.10(a) is the well-known ray diagram for Veselago's flat lens, whereas Fig. 9.10(b) was calculated by the authors from the dispersion curves. It may be seen that the rays converge upon a reasonably good focal point at the back of the slab. The electric-field distribution in the slab given by the numerical package HFSS is given in Fig. 9.11(a).

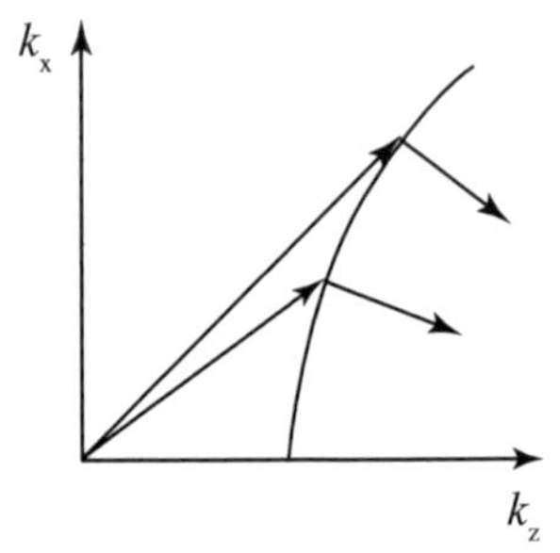

Fig. 9.9 A hyperbolic isofrequency curve in the 2D wave vector plane. Arrows show the incident and refracted 'rays'

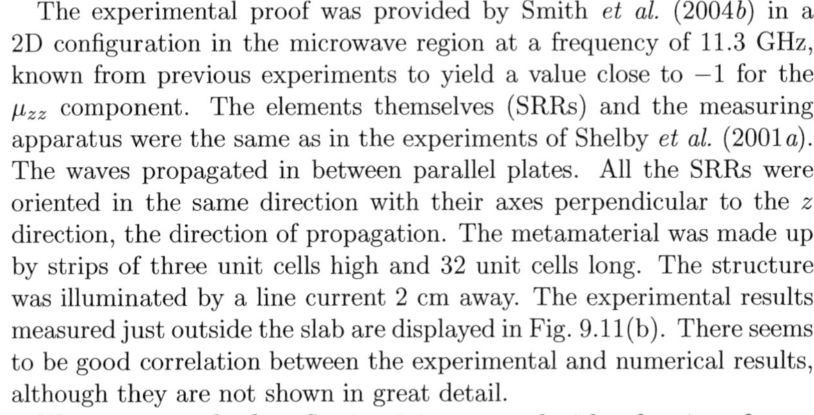

The experimental proof was provided by Smith *et al.* (2004*b*) in a 2D configuration in the microwave region at a frequency of 11.3 GHz, known from previous experiments to yield a value close to -1 for the μ_{zz} component. The elements themselves (SRRs) and the measuring apparatus were the same as in the experiments of Shelby *et al.* (2001*a*). The waves propagated in between parallel plates. All the SRRs were oriented in the same direction with their axes perpendicular to the z direction, the direction of propagation. The metamaterial was made up by strips of three unit cells high and 32 unit cells long. The structure was illuminated by a line current 2 cm away. The experimental results measured just outside the slab are displayed in Fig. 9.11(b). There seems to be good correlation between the experimental and numerical results, although they are not shown in great detail.

We can now go back to Section 8.1 concerned with refraction of magnetoinductive waves. The configuration of interest is the one illustrated in Fig. 8.3(b). It is the dispersion characteristics of the planar–axial configuration of capacitively loaded loops shown in Fig. 8.3(a). The similarity with the indefinite media dispersion comes from the fact that the mutual inductance is positive in the axial direction and negative in the planar direction. The dispersion curves for the magnetoinductive waves are hyperbolic allowing the same construction as shown above in Fig. 9.9. Thus, the whole argument about possible focusing could have been presented in Chapter 8. We postponed the discussion to the present section for the reason that we can show here the experimental results

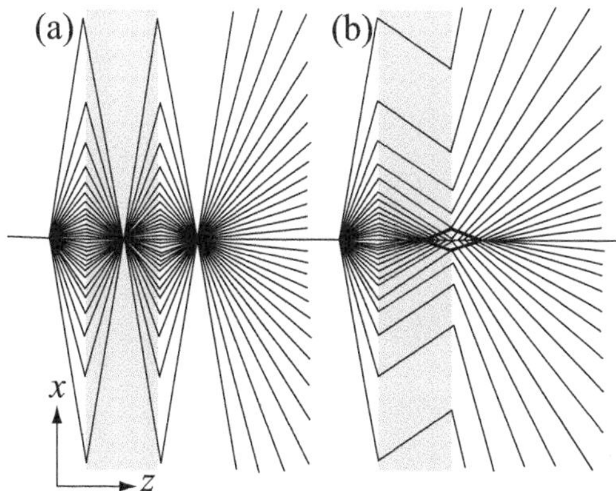

Fig. 9.10 A ray diagram showing the trajectories of rays emerging from a point source for a slab of (a) negative-index and (b) indefinite medium. From Smith *et al.* (2004*b*). Copyright © 2004 American Institute of Physics

of Smith *et al.* (2004*b*) thus lending more credence to the qualitative argument.

Note that the arguments have been conducted in terms of rays that have limited validity when the dimensions of the lens are comparable with the wavelength. However, for this particular arrangement of SRRs it is worth considering an alternative approach relaying on magnetoinductive waves and on the generalized Ohm's law

$$\mathbf{V} = \mathbf{ZI}\,, \tag{9.9}$$

discussed several times in this book. Each element is then excited by the current source and the currents in the elements may be determined by assuming that each element is coupled to every other element. Such an approach was attempted by Kozyrev *et al.* (2007). They were able to find the focus in a narrow frequency band but no regular pattern at different frequencies.

The general question of interest is when to use effective-medium theory and when to resort to solution in terms of eigenwaves. The latter is more accurate, but feasible only when the number of elements are not too large (the limit posed by computational difficulties is around 10^4). For magnetically coupled elements like SRRs or capacitively loaded loops these waves are magnetoinductive waves. There are of course other eigenwaves as well, like the electroinductive waves of Beruete *et al.* (2006) and waves propagating on an array of dipoles to be discussed in Section 9.6 concerned with nanoparticles.

9.5 Gaussian beams and the Goos–Hanchen shift

Gaussian beams propagating through a material with negative ε have already been discussed in Chapter 5, where it was shown how an image degrades (Shamonina *et al.*, 2001). In negative-index materials the re-

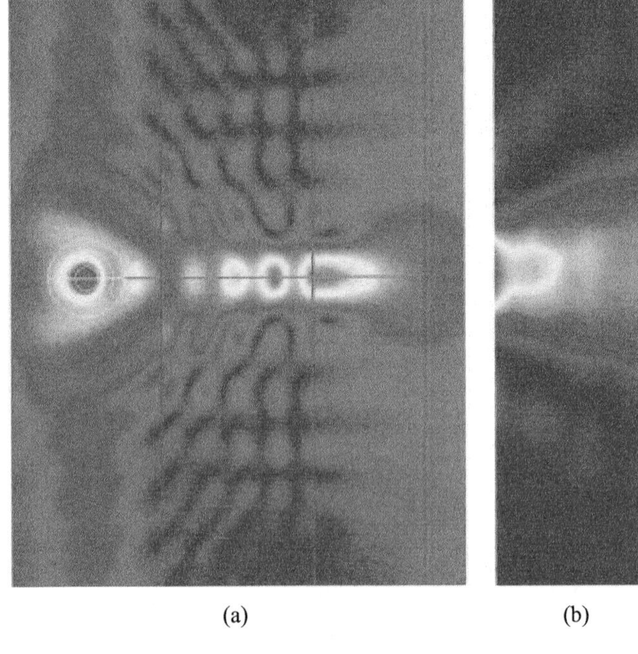

Fig. 9.11 (a) Numerical simulation of the distribution of the electric field. The slab indicated by the solid lines has $\varepsilon_r = -1$ and a diagonal permeability tensor for which the longitudinal component $\mu_z = -1$ and $\mu_x = \mu_y = 1$. The slab is 16 cm long, with a line source placed 2 cm from the slab. The slab thickness is 4 cm. (b) Experimentally obtained field distribution in the plane starting 4 cm away from the output of the slab. From Smith *et al.* (2004*b*). Copyright © 2004 American Institute of Physics. For coloured version see plate section

$n_1>0$

$0\leq n_2 \leq n_1$ positive lateral shift

(a)

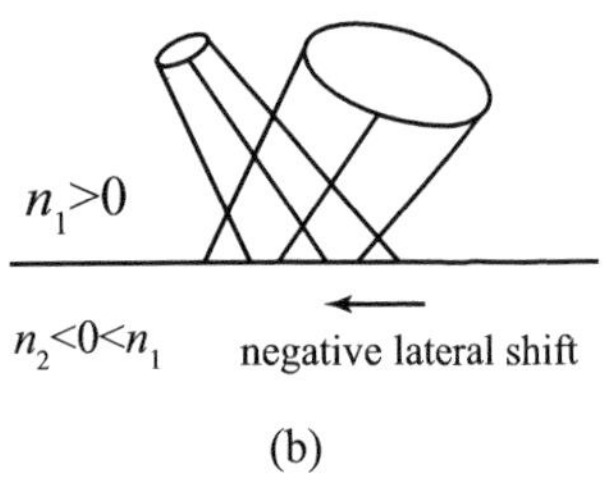

Fig. 9.12 Schematic representation of (a) a positive and (b) a negative lateral shift of an incident beam upon reflection from (a) a positive- and (b) a negative-index material. From Ziolkowski (2003*a*). Copyright © 2003 Optical Society of America

fraction and reflection of a Gaussian beam was treated by Kong *et al.* 2002*a*; Kong *et al.* 2002*b*.

A particularly interesting reflection of a finite beam under conditions of total internal reflection is known as the Goos–Hanchen shift (Goos and Hanchen, 1947). The reflected beam suffers a positive shift, as shown schematically in Fig. 9.12(a).

For an incident plane wave in the TM configuration we derived the reflection coefficient in Section 1.6 in the form

$$R = \frac{1-\zeta_e}{1+\zeta_e}, \tag{9.10}$$

where ζ_e is dependent on the propagation coefficients on both sides of the boundary. Since all finite beams may be constructed by the superposition of infinite beams this means that the reflection coefficient is different for each plane-wave component. When the reflected beam is reconstructed from its elements it turns out that the beam still keeps its shape but it is laterally shifted in the positive direction. This applies to the normal case, that is when the refractive index on both sides of the boundary is positive.

When the beam is incident from a positive-index material upon a negative-index material then the same model predicts a negative shift (Fig. 9.12(b)). This was shown practically simultaneously by Berman (2002) and Lakhtakia (2003) both of them formulating the problem analytically. A similar analytical approach was used by Ziolkowski 2003*a*; Ziolkowski 2003*b* who also did numerical calculations with the aid of the FDTD (finite-difference time domain) package. The frequency depen-

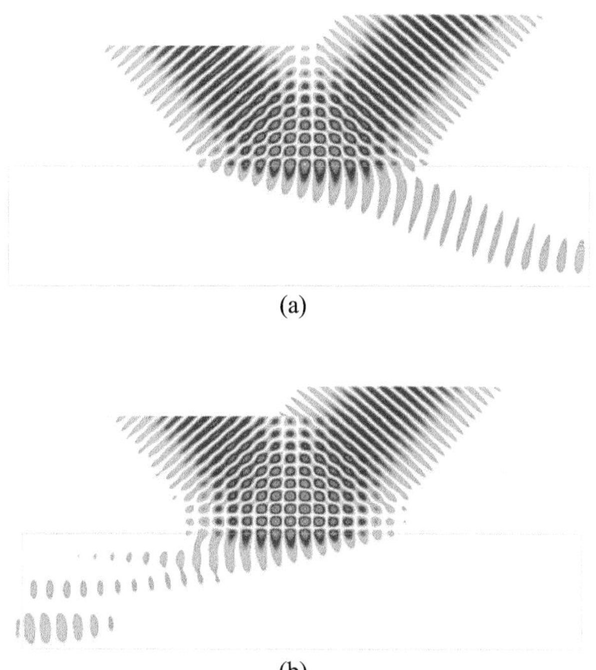

Fig. 9.13 Intensity distribution for a Gaussian beam incident at 40° from a medium with $\varepsilon_r = 9$, $\mu_r = 1$ upon a slab of (a) $\varepsilon_r = 3$, $\mu_r = 1$ and (b) $\varepsilon_r = -3$, $\mu_r = -1$. From Ziolkowski (2003*a*). Copyright © 2003 Optical Society of America. For coloured version see plate section

dence of the permittivity and permeability, needed for this package, was taken in the functional form of the Drude model.[3] A Gaussian beam of 1-cm waist is incident at a frequency of 30 GHz from medium 1 upon medium 2. The material parameters of medium 1 are $\varepsilon_r = 9$ and $\mu_r = 1$ giving a refractive index of $n_1 = 3$. Medium 2 is assumed to be a positive index medium with $\varepsilon_r = 3$ and $\mu_r = 1$ in the first case, and a negative index medium with $\varepsilon_r = -3$ and $\mu_r = -1$ in the second case. The incident angle is chosen to be 40° above the critical angle of 35.26°, hence the beam in both cases will suffer total internal reflection. The reflected and scattered beams yielded by the numerical package are shown in Figs. 9.13(a)) and (b), respectively. The direction of scattering may be immediately seen. For the positive-index material it is in the positive direction and for the negative-index material it is in the negative direction. The actual shift is too small to appreciate it just by looking at the figures. Ziolkowski (2003*a*) gives values of 32 cells for the positive shift and 33 cells for the negative shift, both very close to the predictions of his analytical model. The simulation space was 520 × 1040 cells. The size of a cell was taken as 0.1 mm.

[3]To become negative at a threshold frequency is a good approximation for the permittivity. However, the known negative-permeability materials do not follow this frequency dependence. The permeability is negative within a band, and not below a threshold value, as discussed many times in this book (see, e.g., Section 2.8). Nevertheless, it seems likely that the model is good enough to find the main features of the incident, scattered and reflected beams.

A large Goos–Hanchen shift was predicted theoretically by Shadrivov *et al.* (2003) by having an extra layer between the positive- and negative-index materials that can bring about the resonant excitation of surface polaritons.

An interesting consequence of the negative Goos–Hanchen shift is that in a negative-index waveguide, depending on the parameters of the three media, propagation may be forward or backward, as shown in Figs.

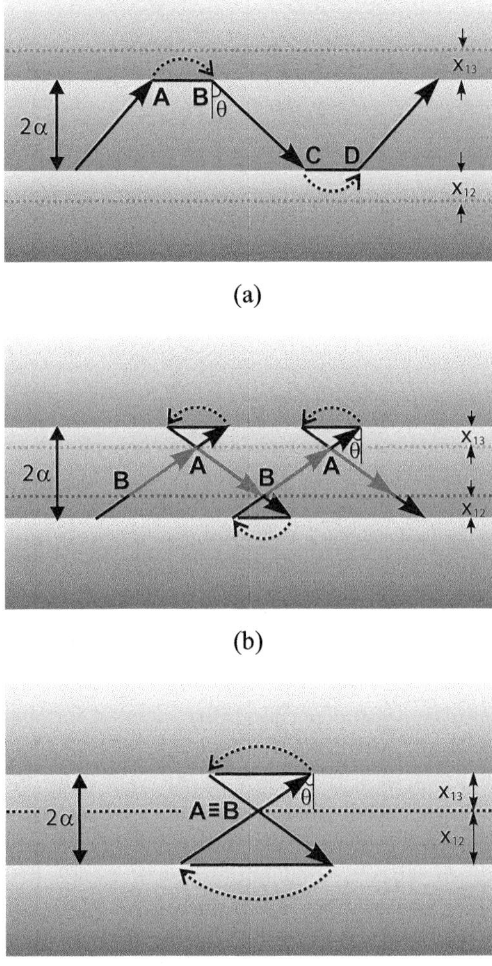

Fig. 9.14 A waveguide consisting of a negative-index material bounded by two positive-index materials. Parameters are chosen so that (a) the wave is forward, (b) the wave is backward, and (c) the wave is stationary. From Tsakmakidis *et al.* (2007). Copyright © 2008 Nature Publishing Group. For coloured version see plate section

9.14(a) and (b) on the basis of a ray picture. It follows then (Tsakmakidis *et al.*, 2006; Tsakmakidis *et al.*, 2007) that for some parameters in between, the rays may bite their own tails: the propagation is neither forward nor backward, it is stationary,[4] as shown in Fig. 9.14(c). The conclusion may then be drawn that such a device may stop and store light.

[4]The authors called the resulting shape an optical 'clepsydra', which according to the Oxford Dictionary is an ancient time-measuring device presumably similar to an hourglass. The pattern could also be described as biconical.

9.6 Waves on nanoparticles

It is doubtful whether the present section should be part of a book on waves in metamaterials. Nanoelectronics-cum-nanophotonics is a new subject that has made enormous strides in the last few years. The idea is to manipulate surface plasmons with a view to optical signal processing. Obviously we shall not be able to do more than scratch the surface. We wish to scratch the surface at that particular point where analogies emerge. In spite of the enormous difference in dimensions and in the shape of the elements it turns out that waves on nanoparticles are very similar to magnetoinductive waves. They are both based on coupling between elements, they can both be treated by a method in which only the near field and only nearest-neighbour interactions are taken into account, and for both cases the theory can be formulated by taking into account interactions between any two elements and determining that interaction by including radiation terms as well. In fact, the ease with which experiments on magnetoinductive waves can be realized might serve as a testing ground for ideas emerging in the field of nanophotonics.

By nanoparticles we mean small metallic spheres, although other shapes are not excluded either. We actually mentioned small metallic triangles by Kottmann *et al.* 2000*b*; Kottmann *et al.* 2000*a*; Kottmann and Martin 2001; Kottmann *et al.* 2001 and full rings by Aizpurua *et al.* (2003) of the order of tens of nanometers in Chapter 4 among small resonant elements. A chain of spherical nanoparticles might consist of 50 elements, the diameter may be 50 nm and the distance between the elements 75 nm. These were actually the parameters used by Quinten *et al.* (1998) when they first investigated waves propagating along these elements. Their motivation was to prove the potential of waveguides at optical frequencies that are small relative to the wavelength and thus they may be instrumental in founding a new type of integrated optics. Other aspects of these waves, like switching (Brongersma *et al.*, 2000), splitting between longitudinal and transverse modes (Maier *et al.*, 2002), pulse propagation (Maier *et al.*, 2003*a*), detection of electromagnetic energy (Maier *et al.*, 2003*b*), multipoles (Park and Stroud, 2004), effect on a channel waveguide (Quidant *et al.*, 2004), coupling and 2D representations (Maier *et al.*, 2005) have also been discussed.

Why can waves propagate on nanoparticles at all? The reason is the presence of free electrons in the metal that can form electric dipoles,[5] and two dipoles can interact in the same manner as two loops. We have actually referred to experiments on dipole arrays (see Fig. 2.8) conducted half a century ago (Shefer, 1963). In that case the wave on the dipoles was excited by a horn antenna and it was detected by another horn. Those dipoles could carry waves, although they were not resonant. To have them resonant is even better and nanoparticles are resonant indeed, belonging to the family of plasma resonances known also as Mie resonances (see, e.g., Stratton 1941; Kreibig and Vollmer 1995).

[5] Multipoles as well but they only matter when the elements are extremely close to each other so they rarely come into consideration.

Let us first work out their resonant frequency. It may be obtained from a combination of the quasi-static model of Clausius–Mossotti with the Drude model. The polarizibility of a metallic sphere is given by

$$\alpha_{\mathrm{e}} = d^3 \frac{\varepsilon_{\mathrm{r}}(\omega) - 1}{\varepsilon_{\mathrm{r}}(\omega) + 2} . \tag{9.11}$$

At resonance, the polarizability tends to infinity, i.e. it occurs when $\varepsilon_r = -2$. Then, using the lossless Drude model

$$\varepsilon_r(\omega) = 1 - \frac{\omega_p^2}{\omega^2}, \tag{9.12}$$

we find that the resonant frequency is

$$\omega_0 = \frac{\omega_p}{\sqrt{3}}. \tag{9.13}$$

The polarizability of the metallic sphere is then obtained by substituting eqn (9.13) into eqn (9.11). We find

$$\frac{1}{\alpha_e} = \frac{1}{d^3}\left(1 - \frac{\omega^2}{\omega_0^2}\right). \tag{9.14}$$

Since in this section we intend to include radiation effects we should add to the above equation the radiation damping term derived in Section 2.7 for electric dipoles

$$\mathrm{Im}\left(\frac{1}{\alpha_e}\right) = \frac{k_0^3}{6\,\pi\varepsilon_0}. \tag{9.15}$$

How will this metallic sphere respond to an electric field? It will set up an electric dipole. By definition, the dipole moment of the electric dipole (see Section 1.16) is related to the electric field as

$$p = \alpha_e E. \tag{9.16}$$

Let us now consider a linear chain consisting of N metallic nanospheres. The aim is now the same as in Sections 2.9 and 8.2.2, where we derived the dispersion characteristics of magnetoinductive waves based on the concept of polarizability. We need two equations, the effect of the electric field on the polarization of the nth element, and the electric field at the nth element produced by all the other elements known as I_F, the interaction function. The electric field due to an electric dipole has already been given in closed form in Section 1.12. Assuming that we have N dipoles with an interelement spacing of a the field at the nth dipole is given by summing up the individual contributions. From the requirement that $\alpha_e I_F = 1$ the dispersion equation was obtained by Weber and Ford (2004) as

$$\frac{1}{\alpha_e} = \frac{k_0^3}{2\,\pi\varepsilon_0}\sum\left[\frac{1}{nk_0a} - \frac{\mathrm{j}}{(nk_0a)^2} - \frac{1}{(nk_0a)^3}\right]\mathrm{e}^{-\mathrm{j}\,k_0na} \tag{9.17}$$

for the electric dipoles transverse to the direction of propagation.[6] It may be seen that eqn (9.17) is entirely analogous to eqns (8.28)–(8.30) only the magnetic polarizability should be changed to the electric polarizability and μ_0 to ε_0. This is not really surprising because in Section 8.2.2 we used the magnetic dipole approximation to MI waves and in this section the metallic nanospheres are regarded as electric dipoles.

[6] In eqn (9.17) we used our notation instead of that of Weber and Ford and also changed to SI units. The dispersion equation of Simovski *et al.* (2005) is identical but, as mentioned in Section 8.2.2, they show that the imaginary part of the equation cancels, and they also sum up some of the infinite series.

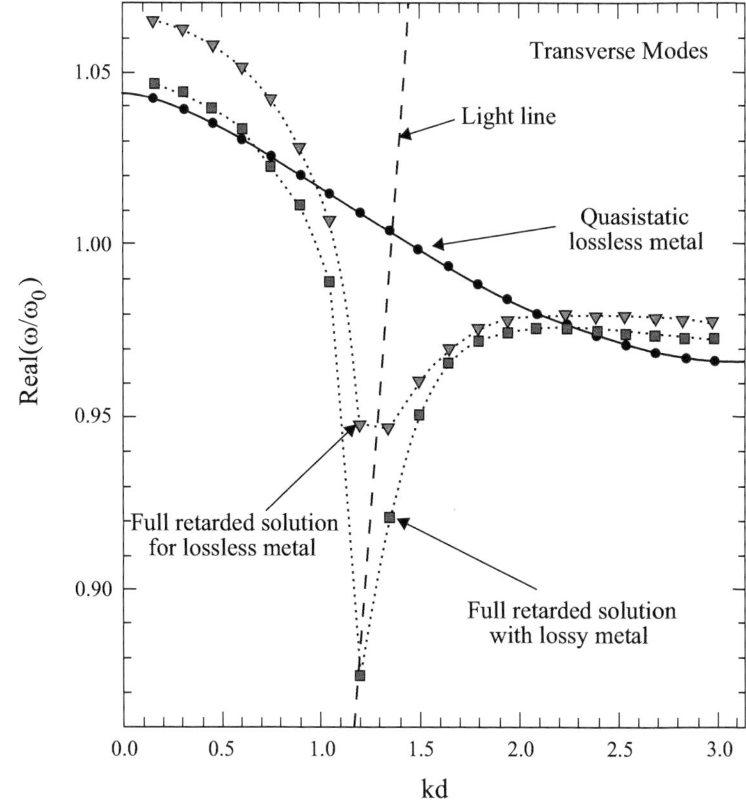

Fig. 9.15 Dispersion curves of nanoparticle chains with and without retardation in the transverse mode. From Weber and Ford (2004). Copyright © 2004 by the American Physical Society

Weber and Ford also proceeded to compare the quasi-static case with that containing full retardation. This is shown in Fig. 9.15, which is analogous to Fig. 8.18. It needs to be noted that Weber and Ford find the retarded solution not from the infinite series but from a finite number of elements with an imposed wave vector and assuming complex ω. For the transverse case, when dipoles are perpendicular to the direction of propagation, the dispersion curves have a dip at the light line not present in the longitudinal case, as we have already noted in the previous chapter.

Another attempt at finding the dispersion equation was made by Koenderink and Polman (2006). They also solved the dispersion equation for k imposed and working in the complex ω region. Their calculations took into account both ohmic and radiation losses. Their results differ considerably from those of Weber and Ford in the vicinity of the light line as shown in Fig. 9.16. The two branches, to the right and to the left of the light line do not cross, which is, apparently, not unusual for polaritons. Their dispersion curve has zeros only at the band edges, in contrast to those of Simovski *et al.* (2005) that yield zero group velocity at a particular value of ka within the Brillouin zone. Since in both models there is some numerical work involved it is too early to say whether the two models contradict each other.

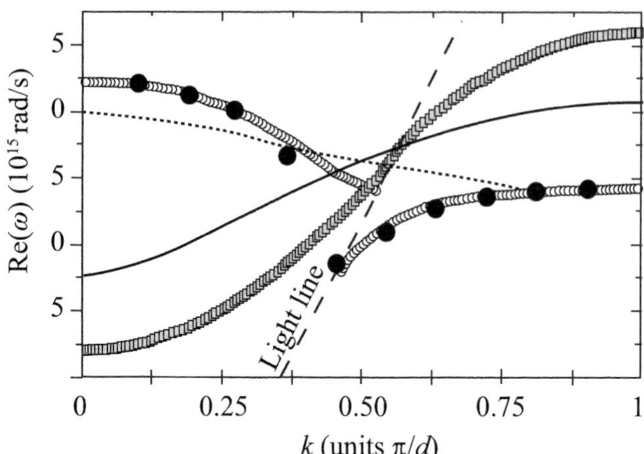

Fig. 9.16 Dispersion curves of nanoparticle chains by Koenderink and Polman (2006). Copyright © 2006 by the American Physical Society. Thin solid (dotted) line, quasi-static dispersion curve of the longitudinal (transverse) mode. Squares (circles), infinite chain dispersion curves of the longitudinal (transverse) mode. Full circles, transverse mode dispersion curve from ten elements according to Weber and Ford (2004). Copyright © 2004 by the American Physical Society

Finally, we wish to mention a perturbation analysis by Citrin (2006) who solved the dispersion equation with the aid of Clausen functions to which the infinite sums reduce. He also considered the possibility of reducing attenuation by embedding the nanoparticle chain in a gain medium.

9.7 Refractive index close to zero

9.7.1 Introduction

We know that the refractive index is given by

$$n = \sqrt{\varepsilon_r \mu_r}\,. \tag{9.18}$$

We have also seen that both ε and μ may have very small values so we may end up with a very small refractive index. The consequences were explored in a theoretical paper by Ziolkowski (2004). He assumed that both ε and μ are zero but in such a manner that the characteristic impedance $\eta = (\mu/\varepsilon)^{1/2}$ remains the same as for the adjoining medium. He made the point that a zero-index medium has a static character in space but the field magnitudes vary in time. He performed an FDTD numerical study of a zero-index material flanked by two media of the same characteristic impedance. He showed that during the rather long transients the electric field does vary as a function of space. But when the steady state is reached the field distribution in the adjoining finite-index materials is the usual sinusoidal variation as a function of space at a given moment in time but in the zero-index material there is no spatial variation at all.

It would be difficult to produce a zero-index material, particularly with the right characteristic impedance, but a low-index metamaterial is well within the practical possibilities. The prominent effect would then be on the refraction properties of the material. Let us remember Snell's law and the relationship between propagation constant and frequency. They are

$$n_1 \sin\theta_1 = n_2 \sin\theta_2 \quad \text{and} \quad k = n\frac{\omega}{c}\,. \tag{9.19}$$

Let us explore the implications of both relationships. We have already had a good look at refraction in Section 2.11. We found that for an incident angle of θ_1 from medium 1 the refracted ray can propagate in any direction from $+90°$ to $-90°$, depending on the refractive index of medium 2 (see Fig. 2.23), and of course for certain values of n_2 there can be total internal reflection. Now, the more interesting case arises when medium 2 has a refractive index of unity and that of medium 1 is close to zero, say, 0.05. Then, for an incident angle of θ_1, the refracted angle is given by

$$\sin\theta_2 = 0.05\,\sin\theta_1\,. \tag{9.20}$$

Plotting θ_2 in Fig. 9.17(a) as a function of θ_1 we can see that θ_2 varies only between 0 and 3°, while θ_1 covers the whole angular range from 0 to 90°. We shall discuss a potential application in Section 9.7.2.

Let us now look at the propagation coefficient. If the index of refraction is close to zero then k tends to zero as well and the wavelength in the medium tends to infinity. A low index means that the phase of the electromagnetic field in this material changes very slowly. This is a major departure from normal practice. We shall discuss some of the potential applications in Section 9.7.3.

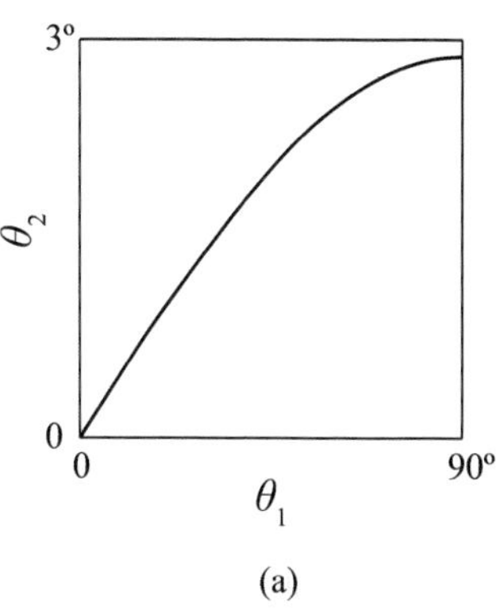

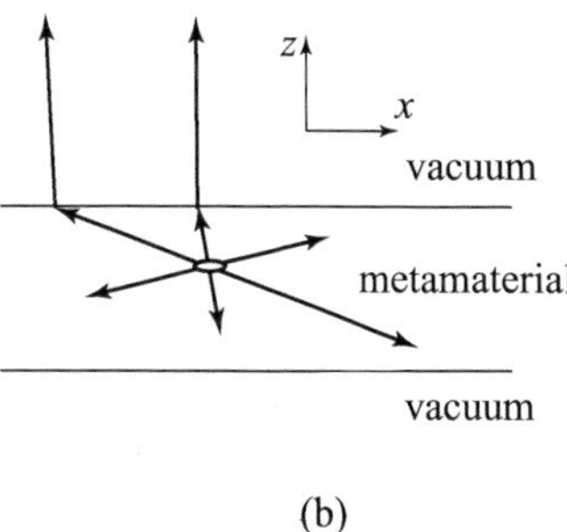

Fig. 9.17 (a) Angle of refraction versus angle of incidence for waves incident from a metamaterial with $n = 0.05$ incident upon vacuum with $n = 1$. (b) Refraction of rays for the same two media

9.7.2 Wavefront conversion

Let us assume now a source of electromagnetic waves embedded in a near-zero-index material close to the surface of a material with refraction index of 1. Figure 9.17(a) has shown the variation of θ_2 with θ_1 when $n_2 = 1$ and $n_1 = 0.05$. In Fig. 9.17(b) the corresponding ray picture is shown. The output rays are very close to being parallel, which implies a good directional radiation pattern.

The effect was demonstrated experimentally by Enoch *et al.* (2002) at a frequency of 14.5 GHz using six sheets of a metallic grid as the metamaterial. The source was a monopole antenna, the period in the lattice was 5.8 mm and the total size 226 mm. The authors achieved a directivity of 372.

The effect is interesting. Its suitability for producing a directional antenna has been proven. It is unlikely, however, that it would be competitive with existing antennas for two reasons. One is the problem of matching. There is a large mismatch in the impedance of the two media that is not easy to overcome. It is also doubtful that the directivity achieved is particularly good by the standards of an antenna engineer. The size of this antenna is 10.91 wavelengths in both directions. According to the well-known expression for the directivity of an antenna with

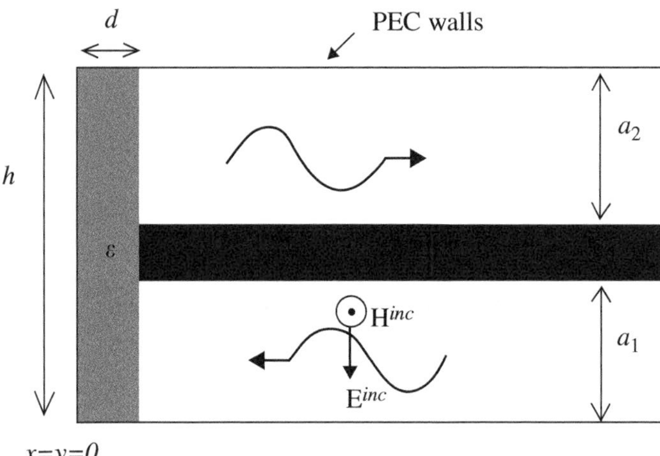

Fig. 9.18 Geometry of a 2D parallel-plate waveguide with walls of perfect electrical conductors and a 180° bend, filled with a permittivity near zero material. From Silveirinha and Engheta (2006). Copyright © 2006 by the American Physical Society

a plane wavefront, $D = 4\pi A/\lambda^2$, where A is the area, this directivity is 1497, way above 372.

The experiments of Bulu *et al.* (2005*b*) may or may not belong to this category. They aimed to produce a directional antenna by a 3D array of SRRs excited by a monopole antenna inside the structure. They report increased directivity in the vicinity of the resonant frequency but the corresponding value of μ_{eff} (needed if the mechanism is the same) is not known. They compare the measured beamwidth (18°) in the H plane with that of a rectangular aperture of the same size ($3.8\,\lambda^2$). They find that the measured beamwidth is smaller than the beamwidth obtained by the uniform aperture. This part of their result may be controversial because one needs a superdirective distribution in order to beat the uniform one. It is also difficult to explain their measured radiation pattern in the E plane. Probably further work is required to clarify the radiation mechanism.

9.7.3 Effect of low phase variation

If the phase varies very slowly then it will ignore obstacles and discontinuities. An example, given by Silveirinha and Engheta (2006) is shown in Fig. 9.18. Two 2D parallel plate waveguide are joined by an abrupt 180° bend that is filled with an ε-near-zero material. Under normal conditions a wave incident from either of the waveguides would suffer high reflection. The authors showed using the numerical package of CST MICROWAVE STUDIO that at certain frequencies near-perfect transmission may be obtained.

In another example, Silveirinha and Engheta (2007) investigated the transmission through a double bend filled with split-pipe resonators in which the slots are filled with a dielectric (see Fig. 9.19) so arranged that the resulting effective index is near zero at a particular frequency. The authors have shown that near-perfect transmission may be achieved.[7]

[7]Note the similarity between these simulations and the experimental results on propagation in waveguides conducted by Marques *et al.* (2002*b*) and Hrabar *et al.* (2005), discussed in Chapter 6.

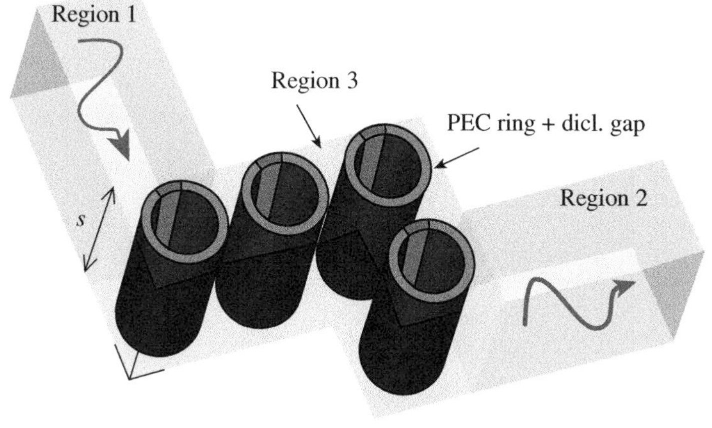

Fig. 9.19 Geometry of the 3D rectangular metallic waveguide. The H-plane width, s, is chosen so that region 3 behaves as a metamaterial with index of refraction close to zero. Regions 1 and 3 have unity indices of refraction. From Silveirinha and Engheta (2007). Copyright © 2007 by the American Physical Society

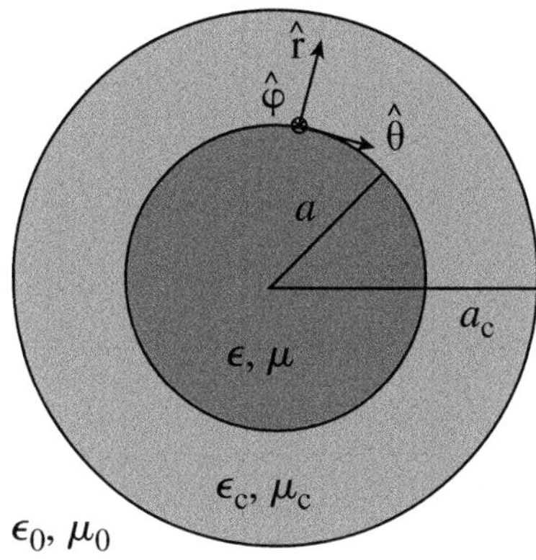

Fig. 9.20 Cross-section of a spherical scatterer composed of two concentric layers of different isotropic materials. From Alu and Engheta (2005). Copyright © 2005 by the American Physical Society

9.8 Invisibility and cloaking

Is it science fiction? The invisible man of H. G. Wells goes back over a century. In more modern times Harry Potter, a budding magician, was also able to acquire an invisibility cloak. Can it really be realized? If it can, one could expect quite substantial orders from the various Ministries of Defence. No general would want to be caught without some kind of invisibility gadget.

In science, in contrast to science fiction, the problem was broached recently by Alu and Engheta (2005) followed by a plethora of articles by popular science writers (a good title among many others was: 'Invisibility cloaks are in sight'). The news even reached the general public via many of the daily papers. A number of scientific publications followed (Lee *et al.*, 2005; Pendry *et al.*, 2006; Cummer *et al.*, 2006; Leonhardt, 2006; Leonhardt and Philbin, 2006; Leonhardt and Philbin, 2007; Schurig *et al.*, 2006; Silveirinha and Engheta, 2006; Alu and Engheta, 2006; Alu and Engheta, 2007*a*; Alu and Engheta, 2007*b*; Cai *et al.*, 2007; Uslenghi, 2007). So what are the chances? It is early days; too early to

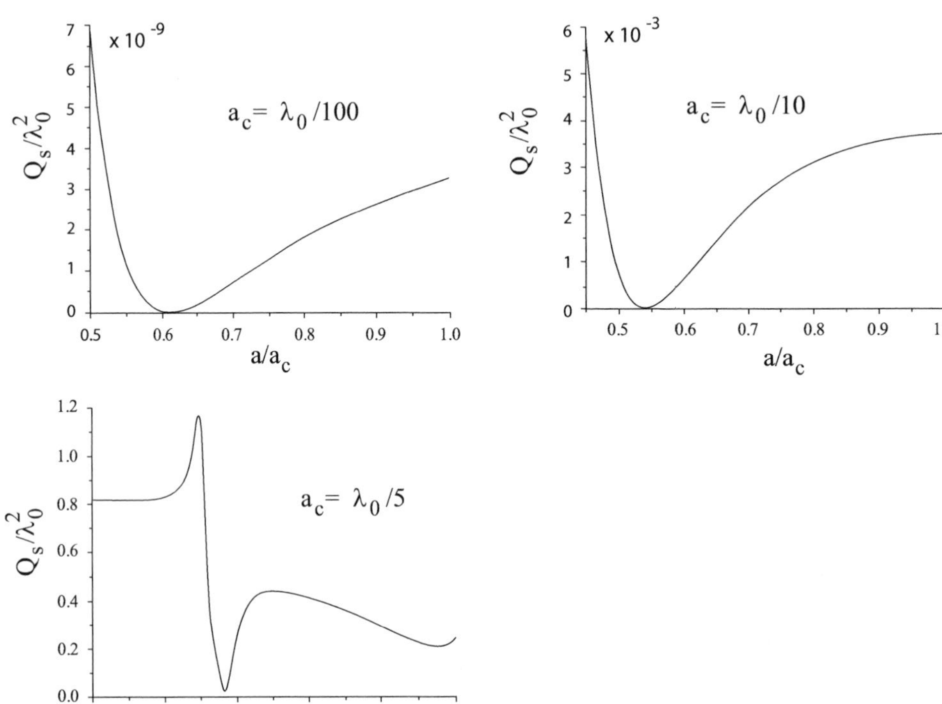

Fig. 9.21 Normalized scattering cross-section for a spherical element with three different sizes of the outer radius of the cover. From Alu and Engheta (2005). Copyright © 2005 by the American Physical Society

say whether anything will come out of it.[8]

There is no doubt that the questions asked are interesting and the arguments, that such effects are not impossible, quite convincing. We shall summarize here briefly the basic principles and the state of affairs.

The basic idea of Alu and Engheta (2005) follows from studies on scattering. How do we see an object? We know it is there because the light scattered by it enters our eyes. If that scatter could be cancelled we

[8]The desired effect might be qualitatively similar to that of superdirectivity in the theory of antennas, which predicts that arbitrarily high directivities are possible in principle. It was shown that superdirectivity is compatible with Maxwell's equations (Oseen, 1922), but the practical difficulties of realizing those antennas have been enormous. They work in a very narrow frequency band, and their performance is extremely sensitive to tolerances in realizing the required current distribution. Similarly, perfect (or near-perfect) cloaking may be possible in principle but it will work only in a very narrow frequency band and tolerance sensitivity could turn out to be prohibitive. At the time of writing there is not much information about the effect of non-ideal conditions (see though Cummer *et al.* (2006) later in this section). If we do not quite succeed in making things invisible, what then? Can we measure the degree of invisibility? If we say an object has become half as visible as it was before we put on the cloak, what do we mean by this? In this respect superdirectivity comes out better. We know what it means. Any increase in directivity over the classical limit is welcome, whereas invisibility has to be very close to 100% to be of any value.

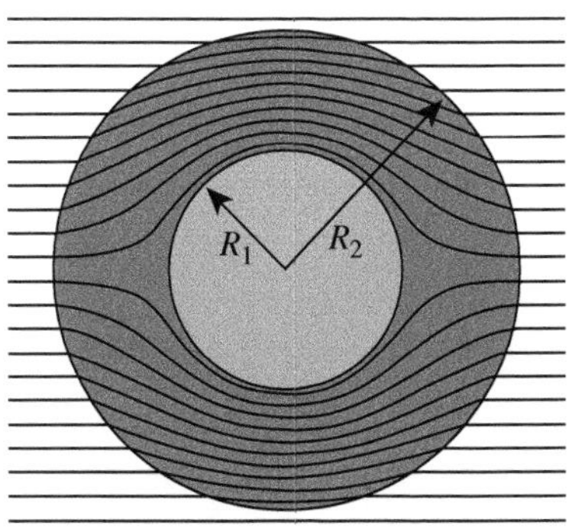

Fig. 9.22 Poynting vector streamlines for an incident plane wave in a 2D configuration. Cloak is designed so that the electromagnetic power moves round it. From Pendry *et al.* (2006). Copyright © 2005 AAAS

would not be able to see the object. Theoretical proof for this hypothesis was provided by considering a spherical scatterer having materials constants ε and μ. The scatter from this sphere is to be cancelled by a coating with material constants ε_c and μ_c, shown in Fig. 9.20. The mathematics is rather complicated: the scattered field must be written as a sum of discrete spherical harmonics and it turns out that the largest term (the dipolar one) can be cancelled. The final results can be presented in a simple manner. In Fig. 9.21 the normalized scattering cross-section is plotted as a function of a/a_c, where a is the radius of the sphere to be hidden and a_c is the radius of the sphere with the coating added. The parameters chosen are $\varepsilon = 4\varepsilon_0$, $\varepsilon_c = -3\varepsilon_0$, $\mu = \mu_c = \mu_0$. Note that the coating must have negative permittivity. It may be seen from Fig. 9.21 that there is very good cancellation at a particular value of the coating thickness when the object is well in the subwavelength region ($a_c = \lambda_0/100$ and $\lambda_0/10$, where λ_0 is the free-space wavelength) but much worse when the external radius is a fifth of the wavelength. Thus, the idea works for subwavelength structures but, apparently, loses its validity for larger objects. It is also true that, with scattering being shape-dependent, each object requires a separate design for scattering cancellation.

A different idea was proposed by Leonhardt (2006) and Pendry *et al.* (2006). An object is made invisible because the rays of light do not go through it but go around it, as shown schematically in Fig. 9.22 for a simple case. Leonhardt (2006) starts with a disclaimer. He quotes Nachman (1988) who proved that the inverse scattering problem has a unique solution. If we measure the scattered field in all directions we can uniquely determine the spatial variation of ε and μ that caused the scattering. Hence, from our measurements we should always be able to determine the distribution of the material parameters whether there is a cloak there or not. Hence, there cannot be perfect invisibility. If our measurements show that there is nothing there then the only possibility is that there is indeed no more than empty space there. However, if we are allowed to talk about rays instead of waves then the situation changes. Waves diffract, rays are willing to bend if the index of refraction varies spatially but they do not diffract. This is, of course, the geometrical optics approximation. In that approximation, perfect

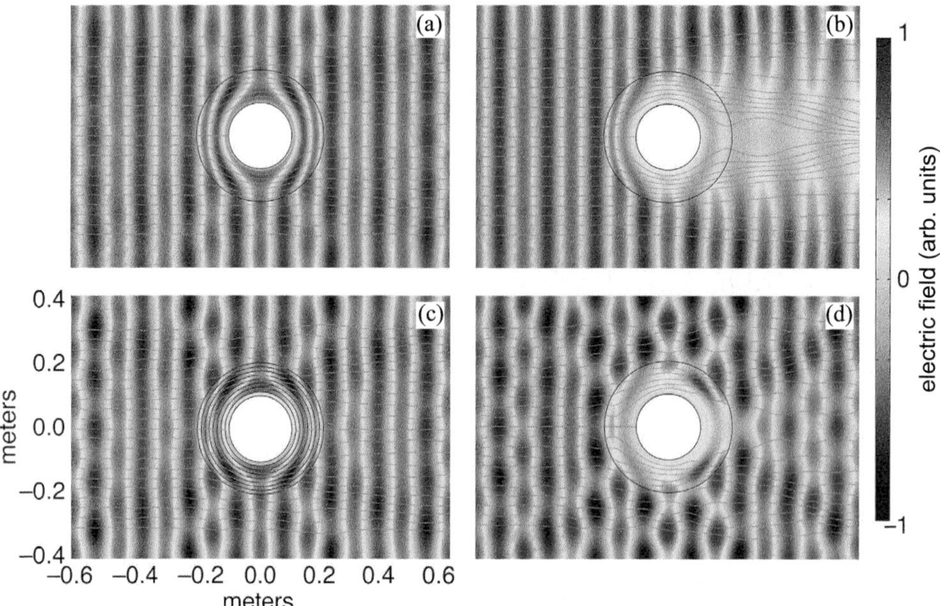

Fig. 9.23 Electric-field distribution for an incident plane wave in the vicinity of a perfectly conducting shell and a cloak. (a) ideal parameters, (b) with a loss tangent of 0.1, (c) with an 8-layer approximation to the desired distribution of the material parameters, (d) with a simplified cloak in which only μ_r is varying spatially. From Cummer *et al.* (2006). Copyright © 2006 by the American Physical Society. For coloured version see plate section

invisibility is possible in principle. If we know the desired trajectories the required variation of the index of refraction may be obtained by a conformal transformation in the 2D case. The bandwidth would be wide if the index of refraction were to be independent of frequency. But it is not independent, and that would limit the bandwidth. In fact, in order to go round an object, the rays that do that must propagate faster than the velocity of light (phase velocity only!). That is easy to realize; all we need is a refractive index less than unity. But then there is strong dispersion, hence a limitation of the bandwidth.

A full wave numerical simulation[9] of the 2D cylindrical problem, aimed at investigating deviations from the ideal case (Fig. 9.22) was performed by Cummer *et al.* (2006). The wave incident was a TE wave at a frequency of 2 GHz in a 2D configuration. The object that the rays are supposed to circumnavigate is a thin shell of 0.2 m diameter made of a perfect electric conductor (PEC). It was surrounded by the cloak with outer diameter of 0.4 m in which the permittivity and permeability tensors varied in a prescribed manner. The simulations were performed with the shell present but the authors claim that similar results were obtained in the absence of the shell.[10] The results are displayed in Fig. 9.23 where the electric-field distribution is shown at a particular moment in time. The ideal situation may be seen in Fig. 9.23(a). The field distribution appears to be unperturbed even in the vicinity of the cloak. Figure 9.23(b) shows the effect of losses. For a loss tangent of 0.1 there

[9]The COMSOL Multiphysics finite-element-based electromagnetics solver was used since it allows the specification of both anisotropy and inhomogeneity in the material parameters.

[10]One feels, however, that it is much easier for a ray to propagate round the shell when the shell is impenetrable anyway. A more testing experiment would use a dielectric cylinder as the object to be hidden.

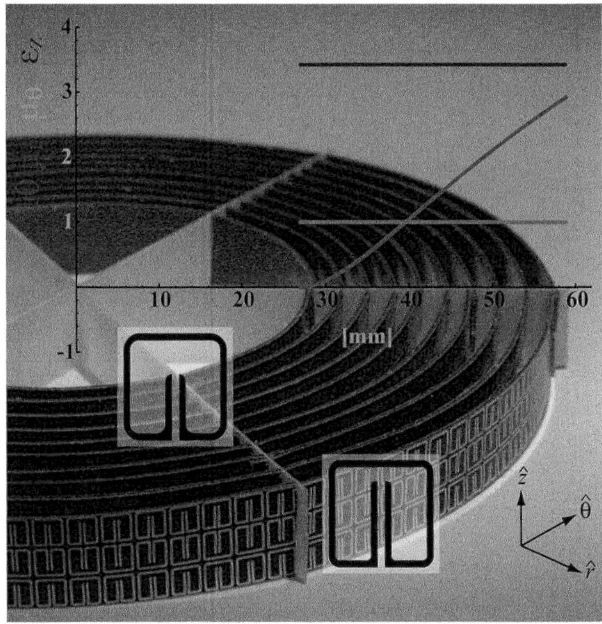

Fig. 9.24 2D microwave cloaking structure with a plot of the material parameters implemented. μ_r (red line) is multiplied by a factor of 10 for clarity. μ_θ(green line) $= 1$, $\varepsilon_z = 3.423$. The SRRs of cylinder 1 (inner) and cylinder 10 (outer) are shown in expanded schematic form. From Schurig *et al.* (2006). Copyright © 2006 AAAS. For coloured version see plate section

is quite a lot of distortion of the forward scattered wave, indicating that loss may be the greatest obstacle to be overcome. For a layered solution consisting of eight discrete homogeneous layers (Fig. 9.23(c)) the field distribution looks good again. Figure 9.23(d) shows the case when instead of the ideal materials distribution a simpler one, which is easier to realize, is chosen. Again, the distortion of the field distribution does not look bad. As said before, these are early times. We have no objective criterion for the measure of invisibility but it is gratifying to know that the results look encouraging.

An experimental investigation of cloaking was carried out by Schurig *et al.* (2006) in a 2D geometry (the waves were confined between parallel metallic plates) at a frequency of 8.5 GHz. The inner shell is made of copper and the cloak by a set of thin cylindrical metamaterial shells. The element used is shown in the inset of Fig. 9.24. It is the same as that of Fig. 4.24(b) apart from a little rounding of the corners. There are 10 discrete layers. The design value of the permittivity is $\varepsilon_z = 3.423\,\varepsilon_0$. The permeabilities μ_r and μ_θ are functions of the radius. The experimental results in the absence and in the presence of cloaking are shown in Figs. 9.25(a) and (b). It may be seen that the cloak helps to hide the copper shell. The constant electric field lines on the right are straighter in the presence of the cloak, hence the object is then less visible.

A similar cloak for a 2D geometry was proposed by Cai *et al.* (2007) for TM polarization in which only the permittivity needed to be varied. In their design the cloak was made of a dielectric host into which metal wires were embedded, as seen in Fig. 9.26.

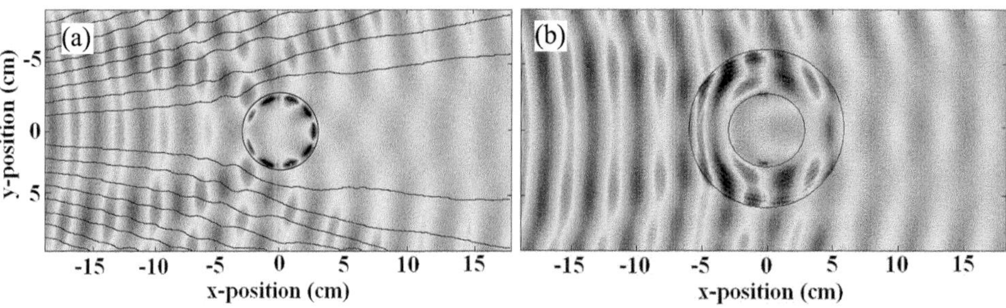

Fig. 9.25 Experimental field distribution for a copper cylinder (to be hidden) and a ten-layer cloak made up by resonating elements. (a) in the absence and (b) in the presence of the cloak. From Schurig *et al.* (2006). Copyright © 2006 AAAS. For coloured version see plate section

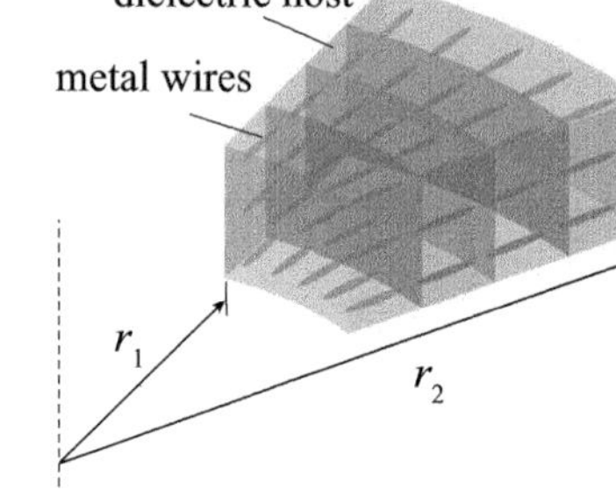

Fig. 9.26 Section of the cloak containing metallic wires for the radial change of the dielectric constant, suitable for the optical region. From Cai *et al.* (2007). Copyright © 2007 Nature Publishing Group

In conclusion, we can only repeat what was said before. This is an interesting research topic; whether it will lead to substantial degrees of invisibility remains to be seen (well, as much as one can ever see invisibility).

A historical review

10

It is probably too early, the subject is too young and not very well defined, for writing an all-embracing review of the history of metamaterials. Our aim here is much more modest. We shall make am attempt at finding the various antecedents leading to the birth of the subject and enumerate the seminal papers instrumental in establishing the initial momentum. A large number of closely related topics will be described that contributed to the process of cross-fertilization and led to sustained interest and further rapid development.

10.1 Introduction

Some births are more difficult than others. Pallas Athene is reputed to have sprung out of Zeus' head, and Aphrodite from the foam of the sea. Other births have been known to be less instantaneous and to have required the simultaneous efforts of a number of parents. Metamaterials belong to that category. It had a long gestation period and many contributors. Before going into specifics and enumerating the various prior influences it might be worth having a broader look at Physics that has often been accorded a privileged position in the ranks of natural science. We may even go further and quote Ernest Rutherford saying that 'Physics is the only science: the rest is stamp collecting.' This is not quite true in the twenty-first century. Many other disciplines have been using rigorous criteria for proving their theses. Nonetheless, we could claim with good conscience that Physics is still the discipline farthest away from the art of stamp collecting.

When we come to physics-in-the-making it is no longer a rational subject. We do not know which hypotheses are useful and which are useless. We do not know whether any particular approach to solving a problem is feasible or not. We do not know the direction that further research should take. We can only guess. It is necessary of course to say nowadays that the new research we have in mind will benefit society but we have no criteria to determine what benefits society and what does not. Let us quote Rutherford again. In 1933, in his speech to the British Association for the Advancement of Science, he claimed: 'We cannot control atomic energy to an extent which would be of any value commercially, and I believe we are not likely ever be able to do so.' Atomic energy provides as much as three quarter of French electrical energy generation at the moment so, at least in the opinion of some French authorities, the continuing research on atomic energy bore fruits.

According to some other opinions, coming with hindsight, research on atomic energy should have been banned from the very beginning. We just do not know.

The sole purpose of the arguments presented so far is to emphasize how difficult is to know the direction research should take. So how does a new research topic arise? What will determine whether it fails or flourishes? Coming now to our specific question: why is there a new research direction in Physics that is known under the generic term of metamaterials (or electromagnetic metamaterials if we wish to exclude their association with other branches of Physics)? How has it come about? Has it been suggested by those in high administrative positions? Has it come about by grass-root pressure? Have there been some random factors at play? Have there been some seminal papers opening the floodgates? Yes, we can say with some certainty that there were a number of seminal papers. How many? Four, we think. The first one that needs to be included is the one by Viktor Veselago (1968) who asked the speculative question: what happens in a material when both the electric permittivity and the magnetic permeability are negative? It would lead both to a backward wave and to a negative refractive index, he suggested. The latter was, no doubt, a suggestion that should have grabbed the headlines. For centuries people believed in a positive index of refraction and suddenly, it was suggested that it might be negative. In addition, Veselago claimed that radiation emerging from a point source on one side of a negative-index slab could be brought to a focus on the other side. He postulated a flat lens, a possibility never considered before. His paper was published in 1967 in Russian, and a year later in English. Did it open the floodgates? It did not. Strangely enough it lay dormant until discovered by Smith *et al.* (2000) three decades later.[1] They designed an artificial material that, in the same frequency band, could produce both negative permittivity and negative permeability. They showed experimentally that electromagnetic waves cannot propagate in a medium in which one of the materials constant is negative but propagation is restored when they are both negative. But if they are both negative then, according to Veselago, the refractive index is negative and hence they should be able to observe negative refraction. And that was indeed what they did next (Shelby *et al.*, 2001*a*). They sent an electromagnetic wave at a frequency of 10.5 GHz upon a negative-index prism and found that refraction was in the expected but highly unusual direction. It was on the other side of the perpendicular to the boundary. The angle of refraction obeyed Snell's law, provided the index of refraction was taken negative.

[1]There was actually an earlier paper written by Sivukhin, published in 1957, which could have been discovered instead. Sivukhin considers the case when both materials constants are negative and concludes that such medium would propagate a backward wave.

Would the papers Smith *et al.* 2000 and Shelby *et al.* 2001*a* have been sufficient to open the floodgates? We do not know, but probably not. The existence of negative refractive index was no doubt sensational (much more sensational than talking about negative permittivity and permeability) but it lacked practical applications. That missing requirement was provided by the fourth seminal paper, Pendry's perfect lens (2000). This was not about beating the classical limit. There have been many publications on that. This was about the possibility of perfect imaging, of being able to reproduce not only the travelling waves associated with the object but all the evanescent waves as well. The four papers taken together acted as a catalyst. Papers started to pour in. Metamaterials was not an accepted term yet but everyone knew which the fundamental papers were. People agreed and disagreed. The large majority agreed, with a few dissenting voices.

Our aim in this chapter is not to review the birth of the subject in any detail. It is rather to point out a number of crucial springboards that kept up the momentum after the initial upsurge in popularity. The seminal papers were a necessary condition for launching the subject, the springboards were responsible for the continued interest. In the next section we shall enumerate a selection of topics that were alive and flourishing at the time of the upsurge and could be regarded as various forerunners of the subject of metamaterials.

10.2 Forerunners

10.2.1 Effective-medium theory

An early success of materials science was to be able to describe in terms of macroscopic quantities the response of a real material (one that consists of atoms and molecules) to electric and magnetic fields. Every textbook in solid state physics devotes some attention to early attempts at homogenization and provides an expression for the permittivity (or permeability) of a material, known as the Clausius–Mossotti equation. These attempts, although made in the nineteenth century (Mossotti, 1850; Clausius, 1879), still give inspiration today. The theory of Gorkunov *et al.* (2002) relies on similar arguments. Another example is the paper by Belov and Simovski (2005*a*) that discusses homogenization in metamaterials including the radiation term. The arguments are presented as generalizations of the Clausius–Mossotti equation.

It is also worth mentioning an early example by Lewin (1947) of the calculation of the effective permittivity and permeability of a medium loaded with spherical particles. The approach was taken up in the metamaterial context by Holloway *et al.* (2003).

10.2.2 Negative permittivity

The plasma frequency was defined, and plasma oscillations were found experimentally by Tonks and Langmuir in 1929. Considering then the Drude model, it was shown (see, e.g., Jackson 1967) that the permittivity associated with a plasma can be negative below the plasma frequency. This was a simple calculation based on Maxwell's equations and on the equation of motion. The aim was to simplify the treatment by introducing a macroscopic quantity that could explain a host of physical phenomena, the transparency of alkali metals for example at ultraviolet wavelengths.

The need of radar technology for higher-permittivity low-loss materials resulted in the development of artificial dielectrics (for a comprehensive survey see Collin (1991)). One of the structures studied was an array of thin wires that were shown to have an effective plasma frequency by Brown (1953). Later motivation came from the desire to simulate plasmas in order to have more insights into problems like the effect of rocket exhaust upon the radiation of re-entry vehicle antennas. The comprehensive paper of Rotman (1962) was a result of that investigation. Similar calculations reaching similar conclusions were done by Pendry *et al.* (1996) a quarter of a century later. The choice of thin wires for producing negative permittivity by Smith *et al.* (2000) and Shelby *et al.* (2001*a*) was inspired by Rotman (1962) and Pendry *et al.* (1996).

10.2.3 Negative permeability

In his search for negative refractive index materials Veselago (1968) was thinking about gaseous or solid-state plasmas (Chynoweth and Buchsbaum, 1965) (apparently, unaware of the potential of thin wires) for negative permittivity. He also proposed anisotropic gyroscopic substances in which, due to off-diagonal elements in the permeability tensor, the effective permeability may be negative for one of the two circular polarizations of an electromagnetic wave. Interestingly, unknown to Veselago, the existence of negative permeability was already shown in such a material (a ferrite with an applied dc magnetic field placed in a circular hollow metallic waveguide) by Thompson 1955; Thompson 1963 a decade earlier. More recent experiments on a metamaterial array in a hollow waveguide by Marques *et al.* (2002*a*) yielded similar results.

10.2.4 Plasmon–polaritons

The interaction of plasmas with electromagnetic waves led both to a new term, plasmon–polaritons and to a variety of phenomena both in the bulk and on surfaces (see, e.g., Cottam and Tilley 1988; Mahan and Obermair 1969). In fact, the so-called amplification of evanescent waves described by Pendry (2000) is due to the excitation of a surface plasmon–polariton on the far surface of the subwavelength lens. Recent studies by Kempa *et al.* (2005) and by Belov and Simovski (2006) that have their inspiration in the field of metamaterials are based on the earlier work of Mahan and Obermair (1969).

10.2.5 Backward waves

Backward waves were discussed as early as the beginning of the twentieth century. Schuster (1904) showed that negative refraction takes place at the boundary of two media, one supporting a forward wave and the other one a backward wave. Four decades later, Mandelshtam (1945) discussed the same idea but the theory and practice of backward waves remained on the whole of little interest to physicists. Most of the work on backward waves was done after the Second World War by electronic engineers concerned with device applications. A number of successful devices like backward-wave oscillators, amplifiers (Beck, 1958; Hutter, 1960) and antennas (Walter, 1965) made their appearance.

Veselago (1968) noted in his original paper the backward-wave character of the electromagnetic waves in negative refractive index materials, which he called left-handed media. Lindell *et al.* (2001) proposed that to avoid confusion with chiral materials it would be more logical to introduce the term backward-wave media. Unfortunately, their proposal was not generally accepted. At the time of writing a plethora of terms prevail.

10.2.6 Theory of periodic structures

We may say that the theories of Mossotti (1850) and Clausius (1879) were already part of the theory of periodic structures. The subject was just becoming popular towards the end of the nineteenth century (see, for example, Lord Rayleigh's article (1892) on the properties of a regular array). Man-made periodical structures also made their appearance. The first one was probably that of Lippmann (1894) who produced a standing-wave structure by interference techniques in a photographic emulsion, and used it for producing colour photographs. The major part of the theory was developed in the first three decades of the twentieth century related to X-ray diffraction by crystals. Some of the pioneering papers were written by Friedrich *et al.* (1912), Darwin (1914) and Ewald (1916). A comprehensive treatment devoted entirely to the theory of wave propagation in periodic structures was formulated somewhat later by Brillouin (1953). Part of this theory, that of periodically loaded transmission lines, was found particularly useful for describing wave propagation in metamaterials (see, e.g., Eleftheriades *et al.* 2002; Grbic and Eleftheriades 2003*a*; Caloz and Itoh 2004; Lai *et al.* 2004; Syms *et al.* 2005*a*; Engheta *et al.* 2005).

10.2.7 Resonant elements small relative to the wavelength

Metallic boxes of the order of the wavelength were well known as microwave resonators. The problem of producing smaller resonators arose when space was at a premium and there were specific requirements like providing a narrow interaction region for electrons. The first such element was probably the re-entrant cavity used in klystrons (Hansen, 1939) but many others were developed later, examples being the split-ring resonator of Hardy and Whitehead (1981) and the loop-gap resonator of Froncisz and Hyde (1982). The split-ring resonator of Pendry *et al.* (1999) used in the experiments of Smith *et al.* (2000) and Shelby *et al.* (2001*a*) was a variation on the same theme.

10.2.8 Chiral materials

They affect the polarization of the incident electromagnetic wave. The first artificially produced chiral material was probably that of Bose (1898). The first report of negative permeability by Thompson (1955) was based on experiments involving ferrites. A microstructured chiral element was proposed by Svirko *et al.* (2001). A more general proposal for a chiral route to metamaterials was made by Pendry (2004). Another chiral element of interest was studied recently by Bai *et al.* (2007).

10.2.9 Faster than light

The first lens based on rays going faster than the velocity of light was realized by Kock (1964). He used hollow metal waveguides that can propagate waves at any phase velocity up to infinity, obtaining lenses that were concave for converging beams and convex for diverging beams. Metamaterials can also reach such phase velocities by having a dielectric constant less (or much less) than unity. Silveirinha and Engheta (2006) showed how the waves in such a material can go round obstacles, while Pendry *et al.* (2006) laid down design rules for low dielectric constant, inhomogeneous metamaterial cloaks that can make an object inside invisible.

10.2.10 Frequency filters made of periodically arranged resonant elements

These represented a class of frequency filters particularly suitable for microwaves. They are briefly described by Atabekov (1965) in his textbook of 1965. In their more modern form they were reported by Hong and Lancaster 1996*a*; Hong and Lancaster 2000. The experimental work of Martin *et al.* (2003*b*) was also directed at producing filters. They used split-ring resonators inserted into coplanar waveguides.

10.2.11 Slow-wave structures

This was the term employed for a particular set of periodic structures used in microwave tubes. The structures slowed down the electromagnetic wave so that they could interact with electron beams drifting at the same velocity (Beck, 1958; Hutter, 1960; Bevensee, 1964; Silin and Sazonov, 1966). Similar kinds of structures were also used in particle accelerators (Knapp *et al.*, 1965) and in magnetic resonance imaging, as realized by bird-cage resonators (Leifer, 1997). In the metamaterial context these waves were resurrected by Shamonina *et al.* 2002*a*; Shamonina *et al.* 2002*b* by introducing magnetoinductive waves. For an application of the interaction between precessing dipoles and magnetoinductive waves see Solymar *et al.* (2006).

10.2.12 Waves arising from nearest-neighbour interactions

The classical example of such waves, acoustic waves including the optical branch, have been discussed in most textbooks on solid state physics and periodic structures (see, e.g., Brillouin 1953). They were derived for coupled waveguides by Syms 1986; Syms 1987, for coupled optical resonators by Yariv *et al.* (1999), for coupled nanoparticles by Brongersma *et al.* (2000), another set of forerunners of magnetoinductive waves.

10.2.13 Superdirectivity, superresolution, subwavelength focusing and imaging

It was believed for a long time that for the radiation of a sharp beam it is necessary to have a large aperture. Similarly, both a large aperture and high illumination angle were regarded as necessary conditions for producing a sharp focus by a lens. The latter was (and still is) known as the Rayleigh criterion, giving the classical limit that the focal region must be of the order of the wavelength. The theorem that claims that an arbitrarily sharp beam can be produced by a finite aperture is due to Oseen (1922). The resulting radiation pattern is known as an example of superdirectivity. The term superresolution was introduced by Toraldo di Francia (1952) who offered practical methods for tailoring the aperture distribution in order to beat the classical limit. High-resolution near-field imaging was first proposed by Ash and Nicholls (1972) who relied on the near-field leaking from a microwave cavity through a small hole. Near-field imaging in the optical region is now a major subject running under the acronym of SNOM (scanning near-field optical microscopy), see, e.g., Paesler and Mayer (1996). In the metamaterials context Pendry's proposal for a 'perfect' lens (Pendry, 2000) also relied on the near field but the configuration was quite different and no scanning was needed. There were though objections from people who disliked one or other aspect of the proposal but those doubts were quickly disposed of. The idea of subwavelength imaging was already in the air, and Pendry's proposal could very soon become part of conventional wisdom.

10.2.14 Inverse scattering

This is the problem of finding the scattering medium or the source when the scattered fields are known. For an early paper see Imbriale and Mittra (1970). For a more detailed account see Colton and Kress (1992). In the metamaterials context the problem was to find the real and imaginary parts of the index of refraction by measuring the complex reflection and transmission coefficients of a slab of negative-index material, see, e.g., Smith *et al.* (2002). One of the difficulties is that for the same scattering coefficients there may be multiple solutions. There are also questions whether the method can work when the number of elements per wavelength is not sufficiently large (Efros, 2004; Simovski and

Tretyakov, 2007).

10.2.15 Bianisotropy

The main questions arising are the properties of the elements and their collective effect upon the propagation of electromagnetic waves; exactly the same as those arising in the study of metamaterials. We could actually argue that the subject of metamaterials simply swallowed that of bianisotropy and related topics like chiral and bi-isotropic media. The introduction of the omega particle by Saadoun and Engheta (1992), treatment of chiral scatterers by Tretyakov *et al.* (1996) or more general books like that of Lindell *et al.* (1994) have been very specific forerunners of later studies. Examples are the study of the bianisotropic character of split-ring resonators by Marques *et al.* (2002*c*) and the consideration of chiral media for negative refraction by Pendry (2004).

10.2.16 Photonic bandgap materials

The subject was founded in the late 1980s (Yablonovitch, 1987; John, 1987) concerned with confining electromagnetic radiation within a certain bandwidth inside an artificially produced material. The term bandgap came from analogy with electron confinement to specific bands in a solid. The structure of the material had to be such as to produce Bragg reflection in every direction. A change towards concepts in metamaterials came with the work of Notomi (2000) who showed that refraction-like behaviour is possible close to the bandgap. Further moves towards metamaterials came later with the aim of showing negative refraction and imaging (see, for example, Luo *et al.* 2002*a*; Lu *et al.* 2005).

10.2.17 Waves on nanoparticles

This topic may be regarded as another example of wave propagation in slow-wave structures because resonant elements placed closely to each other relative to the wavelength are involved. The element in this case (Quinten *et al.*, 1998; Brongersma *et al.*, 2000) is a small metallic particle of the order of tens of nanometers whose resonant frequency is equal to $\omega_\mathrm{p}/\sqrt{3}$, where ω_p is the plasma frequency of the metal. The motivation is to produce optical circuits. In fact, this nanoparticle array is quite analogous to the array of capacitively loaded loops (Shamonina *et al.*, 2002*a*; Shamonina *et al.*, 2002*b*) if retardation terms are included in the latter. The equations are practically the same, although the frequencies of operation differ by nine orders of magnitude. In the metamaterials context Zhuromskyy *et al.* (2005*b*) determined the dispersion characteristics of magnetoinductive waves including the effect of radiation while chains of nanoparticles were studied, e.g., by Alu and Engheta (2006) and Alitalo *et al.* (2006).

10.3 ... and the subject went on and flourished...

For those interested in any of the seventeen topics mentioned above (and a few more we must have missed) the advent of negative refractive index, negative refraction and 'perfect' imaging provided an immediate challenge. Will these physical phenomena lead to new physics and possibly to some new applications? That was the question asked by many. The beauty of this new field was that it had so many different aspects and a very low barrier to entry. Reading the four seminal papers Veselago 1968; Smith *et al.* 2000; Shelby *et al.* 2001*a*; Pendry 2000 anyone doing research (or just having an interest) in any of the topics mentioned must have been tempted to become a metamaterialist (a term that has not been coined yet but would be quite an appropriate one).

We know what happened. Scores of people entered the field, contributions poured in leading to an exponential increase in the usual measures of research activities (scientific papers and books (Tretyakov, 2003; Eleftheriades and Balmain, 2005; Itoh and Caloz, 2005; Engheta and Ziolkowski, 2006; Marques *et al.*, 2008; Markos and Soukoulis, 2008), number of groups involved, number of citations, grants received, articles in popular science journals, etc.) bringing forth the necessity of baptism (Sihvola, 2007). By now there is a consensus that all the phenomena associated somehow with wave propagation on resonant elements and with negative refraction should be known under the generic term of metamaterials.

And the subject went on and flourished.

Acronyms

AANR (all-angle negative refraction)
BC-SRR (broadside-coupled split-ring resonator)
CDE (coupled dipole equations)
CLS (capacitively loaded strips)
CMM (composite metamaterial)
CRLH (composite right/left-handed)
CRR (closed-ring resonator)
CSRR (complementary split-ring resonator)
DNG (double negative)
DNM (double-negative material)
DPS (double positive)
DSDR (doubly split double ring)
DSR (double-spiral resonator)
DSRR (deformed split-ring resonator)
D SRR (double-slit split-ring resonator)
EC-SRR (edge-coupled split-ring resonator)
EEMR (electric excitation of magnetic resonance)
EFC (equal-frequency contour)
ENG (epsilon negative)
ENZ (epsilon near zero)
EVL (epsilon very large)
LH (left handed)
LHH (left-handed heterostructure)
LHM (left-handed media)
MI (magnetoinductive)
MIW (magnetoinductive wave)
MM (metamaterial)
MNG (mu negative)
MNP (metal nanoparticle)
MPP (magnetic plasmon–polariton)
MTM (metamaterial)
NB SRR (non-bianisotropic split-ring resonator)
NFPL (near-field perfect lens)
NIM (negative-index material)
NIR (negative-index refraction)
NMPM (negative-magnetic-permeability medium)
NNA (nearest-neighbour approximation)
NR (negative refractive)
NRI (negative refractive index)

NRIM (negative refractive index material)
NRM (negative-refraction medium)
OSRR (open split-ring resonator)
PLH (purely left handed)
PRH (purely right handed)
PRI (positive refractive index)
RCE (resonant conducting element)
RH (right handed)
RHM (right-handed medium)
SLPW (square-lattice photonic waveguide)
SNG (single negative)
SRR (split-ring resonator)
SSDR (singly split double ring)
SSSR (singly split single ring)
TLPW (triangular-lattice photonic waveguide)
VM (Veselago material)

Field at the centre of a cubical lattice of identical dipoles

B

In the usual derivation of the Clausius–Mossotti formula (given in Section 2.3) it is claimed within the validity of the quasi-static approximation that the field (electric or magnetic) due to a cubic lattice of dipoles (electric or magnetic) within a spherical volume is zero at the centre of the sphere. A similar problem arises in Section 2.8 where the mutual inductances need to be summed.

We show here the derivation for a magnetic dipole starting with the expression (see Section 1.12)

$$\mathbf{H} = \frac{1}{4\pi\mu_0 r^5}\left[3(\mathbf{r}\cdot\mathbf{m})\mathbf{r} - r^2\mathbf{m}\right], \tag{B.1}$$

which is the magnetic equivalent of the electric dipole expression given in eqn (1.79) valid for the static case. The vector $\mathbf{r}$ connects the point where the magnetic dipole is, to the point of observation, which is the $(0,0,0)$ point. $\mathbf{m}$ is the magnetic dipole moment, which is assumed to have only a z component (Fig. B.1).

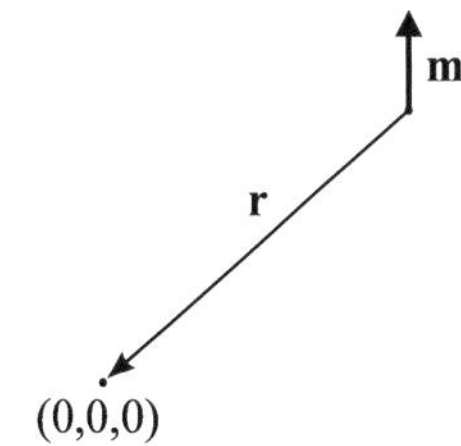

Fig. B.1 Geometry for calculating the magnetic field at the point $(0,0,0)$ due to a magnetic dipole at the general point (r,θ,φ)

Next, we shall find from eqn (B.1) the Cartesian components of the magnetic field at the $(0,0,0)$ point in the r, θ, φ spherical co-ordinate system, where r is the radial, θ is the elevation and φ is the azimuthal co-ordinate. We obtain

$$H_x = \frac{3m_z}{4\pi\mu_0 r^3}\sin\theta\cos\theta\cos\varphi\,, \tag{B.2}$$

$$H_y = \frac{3m_z}{4\pi\mu_0 r^3}\sin\theta\cos\theta\sin\varphi\,, \tag{B.3}$$

$$H_z = \frac{m_z}{4\pi\mu_0 r^3}(3\cos^2\theta - 1)\,. \tag{B.4}$$

We wish to find the magnetic field at $(0,0,0)$ due to all the magnetic dipoles within a sphere of radius R. For each component we need then to sum the contribution from each element. In a number of textbooks (see, e.g., Jackson 1967) the field at $(0,0,0)$ is summed in a rectangular co-ordinate system, which is the logical choice for a cubic lattice. We shall, in a less rigorous manner, keep the spherical co-ordinates and instead of summation integrate over the spherical volume. This method will easily lead to the answer that all three components are zero. The integration is of the form

$$\langle H_i \rangle = \frac{1}{V} \int_0^R \int_0^\pi \int_0^{2\pi} H_i \mathrm{d}V \,, \tag{B.5}$$

where $i = x,\ y,\ z$, V is the volume of the sphere and

$$\mathrm{d}V = r^2 \sin\theta \mathrm{d}r \mathrm{d}\theta \mathrm{d}\varphi \,. \tag{B.6}$$

It may be immediately seen that $\langle H_x \rangle$ and $\langle H_y \rangle$ must be zero because $\sin\varphi$ and $\cos\varphi$ integrate out to zero. For $\langle H_z \rangle$ an elementary integration in θ needs to be performed that shows again that the total field is zero in the z direction as well.

If the total field in the z direction is zero at the position of the dipole at $(0, 0, 0)$ the sum of all the mutual inductances within the sphere must be zero as well. Hence, the summation term in eqn (2.53) is zero for the cubic lattice. It turns out to be finite for other lattice configurations.

Derivation of material parameters from reflection and transmission coefficients

C

It was claimed by Koschny *et al.* (2004*d*) that both the real and the imaginary parts of the material parameters ε and μ can be deduced from measurements of the complex S parameters, S_{11} and S_{21}. They further pointed out that when the permeability has a resonant behaviour (e.g. for SRRs) then the permittivity will have an antiresonant behaviour: the signs of ε'' and μ'' will be of the opposite sign, and vice versa when the permittivity is resonant. The authors argued that although the wrong sign of the imaginary part of a material parameter leads to negative loss, the total loss, when both ε'' and μ'' are taken into account will still be positive, so there is no contradiction.

Opposite signs of ε'' and μ'' did not seem to bother many of the authors. It is found acceptable in a number of publications (Smith and Schurig, 2003; O'Brien and Pendry, 2002; O'Brien *et al.*, 2004; Smith *et al.*, 2004*a*; Huang *et al.*, 2004; Katsarakis *et al.*, 2004; Katsarakis *et al.*, 2005). However, the approach was criticized by Depine and Lakhtakia (2004) and Efros (2004) to which Koschny *et al.* (2004*c*) replied. For a reworking and refining of the retrieval method see Smith *et al.* (2005).

The procedure will always give the correct material parameters in the sense that a plane wave incident upon that lattice from that particular direction and having those material parameters would yield the same scattering coefficients. But that is all. If the angle of incidence is different or the lattice has more or fewer elements then those material parameters no longer lead to the correct result.

The difficulties due to the insufficient number of elements per wavelength were known to Drude (1959). The approximation of sharp boundaries is then no longer tenable because the phase shift of the fields over the lattice period cannot be neglected, the position of the boundary cannot be unambiguously defined. Hence, we do not know the plane at which the values of the material parameters suddenly change from that of one medium to that of the other. Drude's solution was to introduce a thin transition layer in which the material parameters gradually change. The Drude theory was further developed by Simovski *et al.* 2000*a*; Simovski *et al.* 2000*b*. In the paper by Simovski and Tretyakov (2007) the problem of homogenization for insufficient number of elements (but still far from the Bragg condition) is resolved by introducing a refined method for extracting the material parameters and introducing two transition layers. The values of parameters derived in this way are valid in whatever form (propagating, evanescent or wave packet) and at whatever angle the waves are incident. Examples of this procedure for interacting loaded wires have been presented by Ikonen *et al.* (2007*a*).

How does surface charge appear in the boundary conditions?

Let the value of the dielectric displacement be D_1 and D_2 in the adjoining media 1 and 2. The boundary condition to satisfy is then

$$D_{\mathrm{n}1} - D_{\mathrm{n}2} = 0\,, \tag{D.1}$$

where the subscript n refers to the component normal to the boundary. We have shown in Section 1.9 that in a plasma we can define an effective dielectric constant in the form

$$\varepsilon_{\mathrm{eff}} = \varepsilon_0 \left(1 - \frac{\omega_{\mathrm{p}}^2}{\omega^2}\right)\,. \tag{D.2}$$

If medium 1 is a plasma and medium 2 is a dielectric of dielectric constant $\varepsilon_{\mathrm{r}2}$ then eqn (D.1) may be written as

$$\varepsilon_0 \left(1 - \frac{\omega_{\mathrm{p}}^2}{\omega^2}\right) E_{\mathrm{n}1} + \varepsilon_0 \varepsilon_{\mathrm{r}2} E_{\mathrm{n}2} = 0\,, \tag{D.3}$$

where $E_{\mathrm{n}1}$ and $E_{\mathrm{n}2}$ are the normal components of the electric field in the two media. The above equation may be rewritten in the form

$$\varepsilon_0 E_{\mathrm{n}1} + \varepsilon_0 \varepsilon_{\mathrm{r}2} E_{\mathrm{n}2} = \varepsilon_0 \frac{\omega_{\mathrm{p}}^2}{\omega^2} E_{\mathrm{n}1}\,. \tag{D.4}$$

The aim of this Appendix is to point out the different physical interpretations of eqns (D.3) and (D.4). According to eqn (D.4) the difference of the normal components of the dielectric displacement in the two media is equal to the surface charge density which may be shown to be equal to $\varepsilon_0(\omega_{\mathrm{p}}^2/\omega^2)E_{\mathrm{n}1}$. Equation (D.3) tells a different story. On the left-hand side the dielectric constant used is the effective dielectric constant and there is nothing on the right-hand side. Hence, we no longer need to talk of surface charge.

Thus, all depends on the formulation of the problem. If we say that the plasma's relative dielectric constant is unity then we need to introduce the surface charge in the boundary condition. If we use the effective dielectric constant (as in eqn (D.2)) then we don't need to worry about surface charge at all.

The Brewster wave

E

Taking the positive sign in eqn (3.15) we find the dispersion curve of the Brewster (Boardman, 1982; Welford, 1991) wave plotted in Fig. E.1(a) together with the surface plasmon dispersion curve that extends from 0 to $\omega_s = \omega_p/\sqrt{2}$. The upper branch looks similar but not the same as the bulk plasmon dispersion curve. The curve intersects the y axis at the same point, ω_p, but the asymptote is different. As $\omega \to \infty$ we find that in the present case $k_x \to k_0/\sqrt{2}$, in contrast to $k_x \to k_0$ for the bulk wave.

The properties of this wave stem from the condition that it needs to be incident at the Brewster angle. There is then no reflected wave and we could look at the wave as moving along the dielectric–metal interface with a propagation coefficient k_x. Note that this is not a proper surface wave. It radiates. It may be called a radiative wave or a leaky surface wave. It refracts into the metal and propagates in the metal above the plasma frequency. The relative dielectric constant of the metal is between 0 and 1.

The condition for no reflection may be obtained from eqn (1.48) as

$$\zeta_e = \frac{k_{z2}\varepsilon_1}{k_{z1}\varepsilon_2} = 1 . \tag{E.1}$$

This is an unusual formulation of the condition of no reflection. Textbooks give it as

$$\tan\varphi_1 = \sqrt{\frac{\varepsilon_2}{\varepsilon_1}} , \tag{E.2}$$

where φ_1 is the angle of incidence. Since it is not obvious that the two expressions are identical we shall show it below. Using the relationships

$$k_{z1} = k_1 \cos\varphi_1 \qquad \text{and} \qquad k_{z2} = k_2 \cos\varphi_2 , \tag{E.3}$$

we find

$$\zeta_e = \frac{k_{z2}}{k_{z1}}\frac{\varepsilon_1}{\varepsilon_2} = \frac{\cos\varphi_2}{\sin\varphi_1} , \tag{E.4}$$

where φ_2 is the angle of refraction. But from Snell's law

$$\cos\varphi_2 = \sqrt{1 - \frac{\varepsilon_1}{\varepsilon_2}\sin^2\varphi_1} = \sin\varphi_1 , \tag{E.5}$$

with which $\zeta_e = 1$.

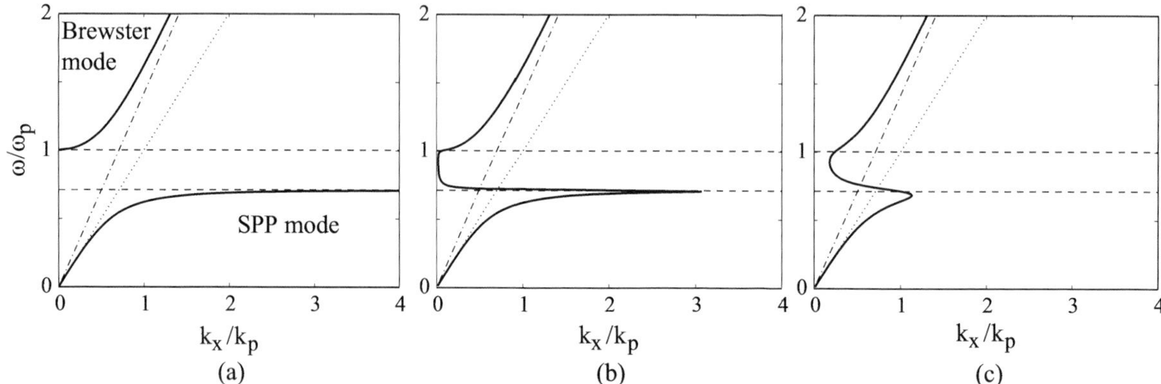

Fig. E.1 Dispersion curve of the surface wave. (a) no loss; Brewster mode for $\omega > \omega_{\mathrm{p}}$ and SPP mode for $\omega < \omega_{\mathrm{s}}$. (b) $\gamma_{\mathrm{p}} = 0.01$. (c) $\gamma_{\mathrm{p}} = 0.1$. Asymptotes: the light line $\omega/k = c$ (dotted line) and $\omega/k = c\sqrt{2}$ (dashed-dotted line)

It needs to be noted that k_{z1} and k_{z2} are now both positive and real, and the same applies to the dielectric constants. Can the Brewster wave be regarded as a surface wave? The justification is that the same pattern moves along the surface undistorted by reflection. Alternatively we may argue that the two different types of waves leading to the same dispersion equation may be construed as one being obtained from the condition that there is no incident wave (Section 1.10), and the other one (in the present appendix) demanding that the reflected wave should vanish.

Losses of surface plasmons have already been considered in Section 3.3.2 and plotted in Fig. 3.6(a) and (b). We shall replot them in Figs. E.1(b) and (c) for $\gamma_{\mathrm{p}} = 0.01\omega_{\mathrm{p}}$ and $0.1\omega_{\mathrm{p}}$ including this time the upper branch. It may be seen that the effect of losses is to connect the lower branch and the upper branch. The stop band between ω_{s} and ω_{p} has now become a rather odd-shaped pass band (Arakawa *et al.*, 1973; Alexander *et al.*, 1974; Kovener *et al.*, 1976) showing a backward wave between points where the group velocity gives the (physically impossible) value of infinity. The mode has very high attenuation, so it has no practical significance. The upper branch lying to the left of the light line is not much affected by loss but, as mentioned above, it is radiative. We shall not consider it further since our interest is in guided waves.

The electrostatic limit

F

The electrostatic limit is sometimes invoked when a proper (meaning all of Maxwell's equations) formulation seems too daunting or one is after a simple analytical solution. But very often it is part of a gradualist approach. One may argue: An electrostatic approach is easier than a proper one. If it leads to a solution that makes good sense physically and if, in addition, it gives reasonable agreement with experimental results then one might call off the chase for better results and has every right to be contented. In the absence of experimental results there is, however, a danger that what makes good physical sense is not necessarily true. An example is Pendry's contention that for the 'perfect lens' to operate there is no need to have a negative permeability; it is sufficient to have $\varepsilon_r = -1$. This is discussed in more detail in Chapter 5.

What do we mean by the electrostatic limit? It is a kind of hotch-potch. We say that things vary with time but not a lot, so we are entitled to ignore the time derivatives of the electric and magnetic fields. Having neglected time derivatives the electric and magnetic fields got uncoupled so we can rely on electric quantities alone. Our starting equations are

$$\nabla \times \mathbf{E} = 0 \,, \quad \nabla \cdot \mathbf{D} = 0 \,, \quad \mathbf{D} = \varepsilon \mathbf{E} \,. \tag{F.1}$$

If the curl of the electric field is zero then it can be expressed as the gradient of a scalar function

$$E = -\nabla \varphi \,. \tag{F.2}$$

Then, if the dielectric constant is independent of space the equation to solve is Laplace's equation

$$\nabla^2 \varphi = 0 \,. \tag{F.3}$$

F.1 Single interface

We shall now attempt the solution under the above approximations of the problem of surface waves propagating along the interface of two different media (Fig. F.1). The solutions in media 1 and 2 are assumed in the form

$$\varphi_1 = A \, \mathrm{e}^{-\mathrm{j} k_x x + \kappa_1 z} \tag{F.4}$$

and

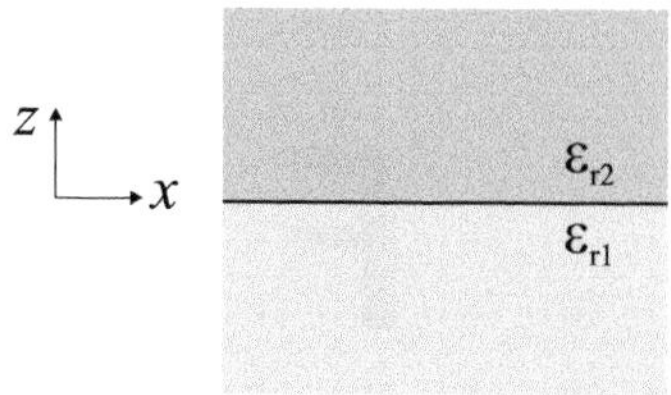

Fig. F.1 Two semi-infinite dielectrics

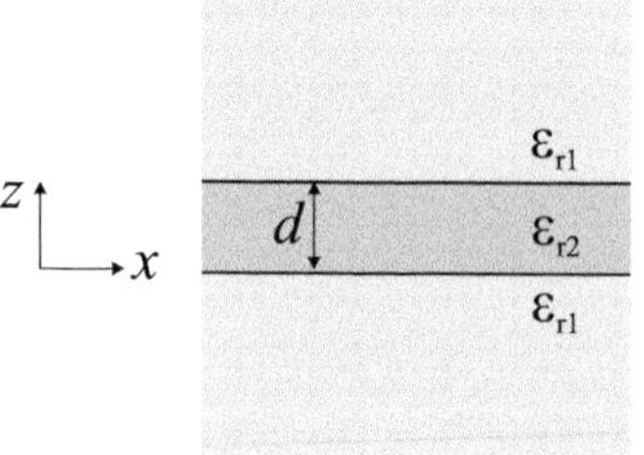

Fig. F.2 Thin dielectric slab between two identical semi-infinite dielectrics

$$\varphi_2 = B\,\mathrm{e}^{-\mathrm{j}\,k_x x - \kappa_2 z}\,. \tag{F.5}$$

The assumptions are the same sort as in Section 1.10 where we discussed surface waves. The waves are declining away from the interface in both media. The difference is that instead of solving the wave equation we now solve Laplace's equation. The time variation in the form of $\exp(\mathrm{j}\,\omega t)$ is tacitly assumed but made no use of.

Substituting our assumed solution into Laplace's equation we readily obtain

$$k_x^2 = \kappa_1^2 = \kappa_2^2\,. \tag{F.6}$$

Next, we need to satisfy the boundary conditions at the interface that both the function φ and its derivative in the x direction must be continuous across the boundary, and that, in addition,

$$\varepsilon_{\mathrm{r}1}\,\frac{\mathrm{d}\varphi_1}{\mathrm{d}z}\bigg|_{z=0} = \varepsilon_{\mathrm{r}2}\,\frac{\mathrm{d}\varphi_2}{\mathrm{d}z}\bigg|_{z=0}\,. \tag{F.7}$$

The result is that the electric fields decay nicely away from the interface at the same rate and the condition

$$\varepsilon_{\mathrm{r}1} = -\varepsilon_{\mathrm{r}2} \tag{F.8}$$

needs to be satisfied. This last condition may be seen to be much more restrictive than those given in Section 3.3.1.

In order to find the dispersion equation $\omega(k_x)$ we have to substitute the frequency dependence of $\varepsilon_{\mathrm{r}2}$ into eqn (F.8), which is[1] equal to

$$\varepsilon_{\mathrm{r}2} = 1 - \frac{\omega_\mathrm{p}^2}{\omega^2} \tag{F.9}$$

into eqn (F.8) leading to

$$\omega = \frac{\omega_\mathrm{p}}{\sqrt{1+\varepsilon_{\mathrm{r}1}}}\,, \tag{F.10}$$

the equation (3.17) quoted in Section 3.3.1. It is a straight horizontal line that gives good approximation, provided k_x is large enough.

[1] This is, of course, a contradiction because this particular dependence on the frequency comes about by taking into account the time derivative of the electric field that we were supposed to have neglected. Nevertheless, these approximations very often lead to good results.

F.2 Symmetric slab

We shall now go one step further and find the dispersion equation of surface waves when medium 2 with dielectric constant $\varepsilon_{\mathrm{r}2}$ is sandwiched between two semi-infinite media of dielectric constant $\varepsilon_{\mathrm{r}1}$ (Fig. F.2). The solutions in the three media are assumed in the form

$$\varphi_1 = B\,\mathrm{e}^{-\mathrm{j}\,k_x x + \kappa z}\,, \tag{F.11}$$

$$\varphi_2 = C\,\mathrm{e}^{-\mathrm{j}\,k_x x - \kappa z} + D\,\mathrm{e}^{-\mathrm{j}\,k_x x + \kappa(z-d)}\,, \tag{F.12}$$

$$\varphi_3 = F\,\mathrm{e}^{-\mathrm{j}\,k_x x - \kappa(z-d)}\,. \tag{F.13}$$

Again, Laplace's equation is there to be solved. The boundary conditions to be satisfied are the same as in the previous case. The φ functions, their derivatives and the normal component of the dielectric displacement must be continuous across the boundaries. It is a rather long slog to find the condition under which a solution exists but eventually it is found as

$$\frac{\varepsilon_{\mathrm{r}1}+\varepsilon_{\mathrm{r}2}}{\varepsilon_{\mathrm{r}1}-\varepsilon_{\mathrm{r}2}} = \pm\,\mathrm{e}^{-k_x d}\,. \tag{F.14}$$

Substituting again for $\varepsilon_{\mathrm{r}2}$ from eqn (F.9) we obtain the dispersion equation as

$$\omega^2 = \omega_{\mathrm{p}}^2 \frac{1 \pm \mathrm{e}^{-k_x d}}{(\varepsilon_{\mathrm{r}1}+1) \pm (\varepsilon_{\mathrm{r}1}-1)\,\mathrm{e}^{-k_x d}}\,, \tag{F.15}$$

which in air simplifies to

$$\omega^2 = \omega_{\mathrm{p}}^2 \left(1 \pm \mathrm{e}^{-k_x d}\right), \tag{F.16}$$

the expression quoted in eqn (3.37). The approximation is very good for large k_x. How good it is otherwise is discussed in Section 3.4.1 and shown in Figs. 3.18 and 3.19.

Alternative derivation of the dispersion equation for SPPs for a dielectric–metal–dielectric structure: presence of a surface charge

G

We derived the dispersion equation of a metal (plasma) slab by assuming infinitely large transmission, i.e. taking the denominator of eqn (1.76) equal to zero. In this appendix we derive the same dispersion equation in a somewhat different manner. It is not a radically different approach: it just puts the emphasis on the fields instead on the reflection and transmission coefficients as it was done before. In addition, this approach will easily lead to an analytic expression for the surface charge.

We have already given the field quantities both in the dielectric and in the metal in Section 1.10. That was for a single interface. We shall write them below for the case when the dielectric extends from $z = -\infty$ to $z = -d/2$, and again from $z = d/2$ to $z = \infty$, the metal being between then in the range $z = -d/2$ to $z = d/2$ (see Fig. G.1). We shall restrict here generality and assume a solution in which the transverse electric field is a symmetric function of z. The field quantities, as solutions of Maxwell's equations, may then be obtained for the TM mode as follows:

In the dielectric for $z < -d/2$

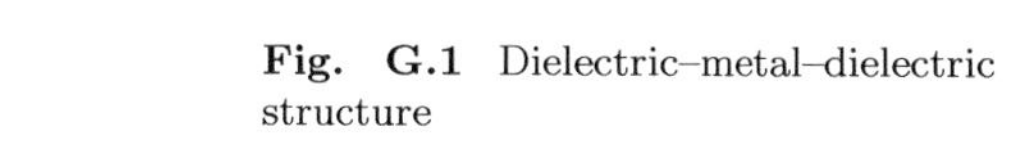

Fig. G.1 Dielectric–metal–dielectric structure

$$H_{y1} = H_{\mathrm{d}}\, \mathrm{e}^{\kappa_1 \left(z + \frac{d}{2}\right)}\, \mathrm{e}^{-\mathrm{j}\, k_x x}, \tag{G.1}$$

$$E_{x1} = -\frac{\kappa_1}{\mathrm{j}\,\omega\varepsilon_1} H_{\mathrm{d}}\, \mathrm{e}^{\kappa_1 \left(z + \frac{d}{2}\right)}\, \mathrm{e}^{-\mathrm{j}\, k_x x}, \tag{G.2}$$

$$E_{z1} = -\frac{k_x}{\omega\varepsilon_1} H_{\mathrm{d}}\, \mathrm{e}^{\kappa_1 \left(z + \frac{d}{2}\right)}\, \mathrm{e}^{-\mathrm{j}\, k_x x}, \tag{G.3}$$

where H_{d} is a constant. In the metal for $-d/2 < z < d/2$

$$H_{y2} = H_{\mathrm{m}} \cosh(\kappa_2 z)\,\mathrm{e}^{-\mathrm{j}\,k_x x}\,, \tag{G.4}$$

$$E_{x2} = -\frac{\kappa_2}{\mathrm{j}\,\omega\varepsilon_2} H_{\mathrm{m}} \sinh(\kappa_2 z)\,\mathrm{e}^{-\mathrm{j}\,k_x x}\,, \tag{G.5}$$

$$E_{z2} = -\frac{k_x}{\omega\varepsilon_2} H_{\mathrm{m}} \cosh(\kappa_2 z)\,\mathrm{e}^{-\mathrm{j}\,k_x x}\,, \tag{G.6}$$

where H_{m} is a constant. Note that we have adhered here to the notations of Section 1.10 where the dielectric is referred to as medium 1 and the metal as medium 2. In eqns (G.5) and (G.6) ε_2 is the effective dielectric constant of the plasma which for completeness we shall give again below. It is

$$\varepsilon_2 = \varepsilon_0 \left(1 - \frac{\omega_{\mathrm{p}}^2}{\omega^2}\right). \tag{G.7}$$

Next, let us satisfy the boundary conditions for the tangential components of the electric and magnetic fields, i.e. match H_y and E_x at $z = -d/2$. We obtain

$$H_{\mathrm{d}} = H_{\mathrm{m}} \cosh\left(\frac{\kappa_2 d}{2}\right) \tag{G.8}$$

and

$$-\frac{\kappa_1 H_{\mathrm{d}}}{\varepsilon_1} = \frac{\kappa_2}{\varepsilon_2} H_{\mathrm{m}} \sinh\left(\frac{\kappa_2 d}{2}\right), \tag{G.9}$$

whence the dispersion equation is

$$\coth\left(\frac{\kappa_2 d}{2}\right) + \frac{\varepsilon_1}{\varepsilon_2}\frac{\kappa_2}{\kappa_1} = 0\,, \tag{G.10}$$

in agreement with eqn (3.36).

It was already pointed out in Appendix D that when we introduce an effective dielectric constant there is no need for taking into account the surface charge. All we need is to satisfy the boundary conditions using the effective value of the dielectric constant. However, if we wish to determine the surface charge it can be easily obtained from the equation

$$\varrho_{\mathrm{s}} = \varepsilon_2 E_{z2} - \varepsilon_1 E_{z1}\,, \tag{G.11}$$

which with the aid of eqns (G.3) and (G.6) comes to

$$\varrho_{\mathrm{s}} = \frac{k_z}{\omega} H_{\mathrm{m}} \frac{\omega_{\mathrm{p}}^2}{\omega_{\mathrm{p}}^2 - \omega^2} \cosh\left(\frac{\kappa_2 d}{2}\right). \tag{G.12}$$

The surface charge may also be obtained in a less formal manner by invoking the underlining physics. Clearly, the surface charge will wax when the transverse current feeds it and will wane when the transverse current leads it away. Hence, the temporal rate of change of the surface current density must be equal to the transverse current, or

$$j\omega\varrho_s + J_z = 0\,. \tag{G.13}$$

The relationship between the current density and the electric field was one of the first few equations derived (see Section 1.2). Substituting that into eqn (G.13) and solving it for ϱ_s we find that it agrees with that obtained from eqn (G.12).

Talking of surface charge it is worth mentioning that if a transverse dc voltage is applied across a metal–dielectric sandwich then a dc surface charge will reside at the boundary. As a consequence there will be an ac surface current as well which will alter the boundary conditions for the magnetic field. For experiments and a discussion see Batke and Heitmann 1984; Tilley 1988.

Electric dipole moment induced by a magnetic field perpendicular to the plane of the SRR

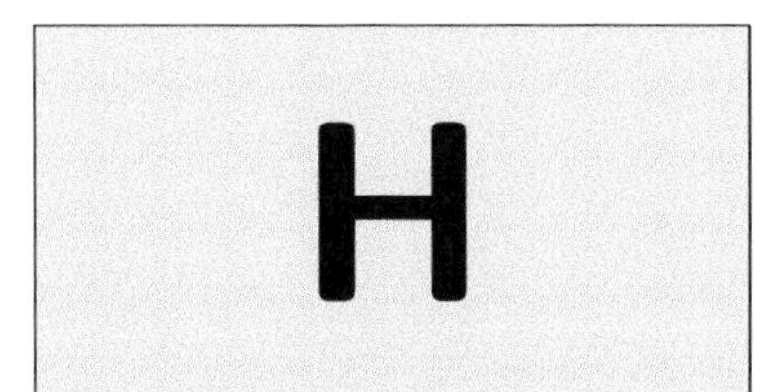

A magnetic field will set up a current in the SRR, more correctly it will set up currents both in the inner and outer rings that vary as a function of angle. Consequently, there will be a voltage distribution between the two rings that will lead to charges on the inner and outer surfaces. Alternatively, we may argue that a necessary consequence of the inter-ring capacitance are charges on the outer surface of the inner ring and on the inner surface of the outer ring, as shown in Fig. H.1. The dipole moment at a point φ is equal to qd, where q is the charge and d is the separation of the rings. Due to symmetry only the x component of the dipole moment will be different from zero, which is given by

$$p_x = qd \sin\varphi \,. \tag{H.1}$$

Hence, P_x, the total dipole moment, is its integral between 0 and π that needs to be multiplied by two because the other side of the ring contributes the same amount (see Fig. H.1). Thus

$$P_x = 2 \int_0^\pi p_x \mathrm{d}\varphi \,. \tag{H.2}$$

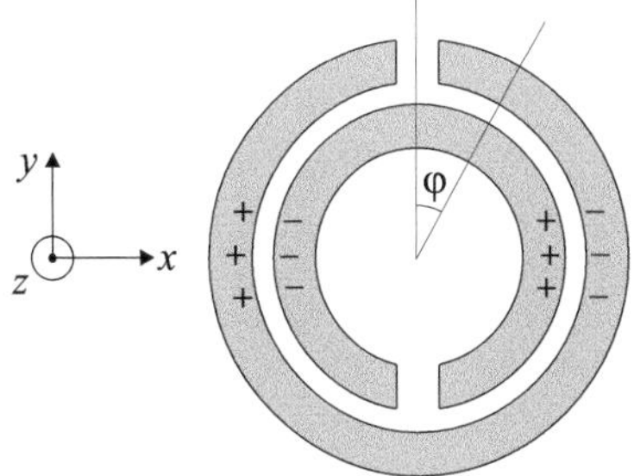

Fig. H.1 Charge distribution in a SRR

Using further the $q = CV$ relationship (C is the inter-ring capacitance) and knowing the variation of the voltage between the rings as a function of angle (the trigonometric functions discussed in Section 4.4) the integration can be performed. We also know the variation of the voltage along the rings that is related to the value of H_z that induces the currents. Thus, at the end we can relate the input magnetic field to the electric dipole moment created. After completing all the algebraic operations we obtain for the relevant element of the polarizability tensor

$$\alpha_{yz}^{\mathrm{em}} = -2\mathrm{j}\,\omega C \pi r_0^3 \left(1 - \frac{\omega_0^2}{\omega^2}\right)^{-1} \,. \tag{H.3}$$

Let us recall what the above notation means: m and z stand for the magnetic field in the z direction, and e and y mean that an electric dipole is created in the y direction.

Calculations by Marques *et al.* (2002*c*), based on a simplified physical picture, yield an expression very similar to eqn (H.3). The only difference is that ω is replaced by ω_0^2/ω, which is practically the same as eqn (H.3) because we are usually talking of a narrow-band device.

Average dielectric constants of a multilayer structure

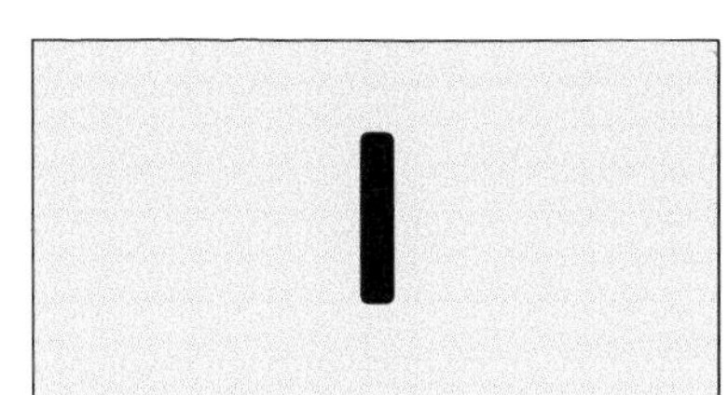

Let us assume a 1D multilayer structure (as in Fig. 5.27) consisting of thin layers of two materials alternating with dielectric constants of ε_1 and ε_2. Provided the layers are thin enough, we can regard this multilayer structure as a single layer of a homogeneous but anisotropic material. What will be then the values of the corresponding dielectric constants in the two distinguishable directions, the axial and the transverse? In both cases we shall rely on the boundary conditions to find some average values.

In the axial, z, direction it is the normal component of the dielectric displacement, D_z, that is conserved across the boundaries. The corresponding values of E_{1z} and E_{2z} may be obtained as

$$E_{1z} = \frac{D_z}{\varepsilon_1} \qquad \text{and} \qquad E_{2z} = \frac{D_z}{\varepsilon_2} \,. \tag{I.1}$$

And the average electric field is

$$\langle E_{1z} \rangle = \frac{1}{2} \left(\varepsilon_1^{-1} + \varepsilon_2^{-1} \right) D_z \,, \tag{I.2}$$

whence we can define the axial component of the permittivity tensor as

$$\varepsilon_z^{-1} = \frac{1}{2} \left(\varepsilon_1^{-1} + \varepsilon_2^{-1} \right) . \tag{I.3}$$

For the transverse case it is the tangential component of the electric field that is conserved across the boundary. Hence, we may obtain the values of the dielectric displacement as

$$D_{1\mathrm{t}} = \varepsilon_1 E_\mathrm{t} \qquad \text{and} \qquad D_{2\mathrm{t}} = \varepsilon_2 E_\mathrm{t} \,. \tag{I.4}$$

Then, the average value of the dielectric displacement is

$$\langle D_\mathrm{t} \rangle = \frac{1}{2} (\varepsilon_1 + \varepsilon_2) E_\mathrm{t} \,, \tag{I.5}$$

whence the transverse component of the permittivity tensor is

$$\varepsilon_\mathrm{t} = \frac{1}{2} (\varepsilon_1 + \varepsilon_2) \,. \tag{I.6}$$

Derivation of mutual inductance between two magnetic dipoles in the presence of retardation

In Section 2.9 we determined the mutual inductance between two magnetic dipoles for the static case in the axial configuration. We shall consider here the planar configuration and include retardation effects. The starting point is the H_θ component as given by eqn (1.87)

$$H_\theta = \frac{m \sin\theta}{4\pi\mu_0 r^3}(1 + \mathrm{j}\, k_0 r - k_0^2 r^2)\, \mathrm{e}^{-\mathrm{j}\, k_0 r}\,. \tag{J.1}$$

To find the magnetic field of dipole 1 at the position of dipole 2 at a distance a away we need to take $\theta = 90°$. The flux exciting the second dipole (regarded now as a small loop) is

$$\Phi = \pi r_0^2 \mu_0 H_\theta\,. \tag{J.2}$$

By definition

$$M = \frac{\Phi}{I} \quad \text{and} \quad m = \mu_0 \pi r_0^2 I\,, \tag{J.3}$$

where I is the current flowing in the loop. Hence, the mutual inductance is

$$M = \frac{(\pi r_0^2 \mu_0)^2 H_\theta}{m} = \frac{\pi\mu_0 r_0^4}{4r^3}(1 + \mathrm{j}\, k_0 r - k_0^2 r^2)\, \mathrm{e}^{-\mathrm{j}\, k_0 r}\,, \tag{J.4}$$

a complex quantity.

References

Abdalla, M. A. Y., Phang, K., and Eleftheriades, G. V. (2005). A 0.13-micron CMOS phase shifter using tunable positive/negative refractive index transmission lines. *IEEE Microw. Wireless Comp. Lett.*, **16**, 705–707.

Abeles, F. (1986). *Surface plasmon (SEW) phenomena*, Volume Electromagnetic surface excitations of *Springer series on wave phenomena*. Springer Verlag.

Aizpurua, J., Hanarp, P., Sutherland, D. S., Kall, M., Bryant, G. W., and Garcia de Abajo, F. J. (2003). Optical properties of gold nanorings. *Phys. Rev. Lett.*, **90**, 057401–1–4.

Al-Bader, S. J. (2004). Optical transmission on metallic wires-fundamental modes. *IEEE J. Quant. Electron.*, **40**, 325–329.

Al-Bader, S. J. and Imtar, M. (1992). Azimuthally uniform surface-plasma modes in thin metallic cylindrical shells. *IEEE J. Quant. Electron.*, **28**, 525–532.

Alexander, R. W., Kovener, G. S., and Bell, R. J. (1974). Dispersion curves for surface electromagnetic curves with damping. *Phys. Rev. Lett.*, **32**, 154–157.

Alitalo, P., Simovski, C., Viitanen, A., and Tretyakov, S. (2006). Near-field enhancement and subwavelength imaging in the optical region using a pair of two-dimensional arrays of metal nanospheres. *Phys. Rev. B*, **74**, 235425–1–6.

Alu, A. and Engheta, N. (2003). Pairing an epsilon-negative slab with a mu-negative slab: resonance, tunneling and transparency. *IEEE Trans. Ant. Prop.*, **51**, 2558–2571.

Alu, A. and Engheta, N. (2005). Achieving transparency with plasmonic and metamaterial coatings. *Phys. Rev. E*, **72**, 016623–1–9.

Alu, A. and Engheta, N. (2006). Theory of linear chains of metamaterial/plasmonic particles as subdiffraction optical nanotransmission lines. *Phys. Rev. B*, **74**, 205436–1–18.

Alu, A. and Engheta, N. (2007*a*). Dispersion characteristics of metamaterial cloaking structures. In *Proc. 1st Int. Congr. on Advanced Electromagnetic Materials in Microwaves and Optics (Metamaterials 2007)*, Rome, pp. 482–485.

Alu, A. and Engheta, N. (2007*b*). Plasmonic materials in transparency and cloaking problems: mechanism, robustness, and physical insight. *Opt. Exp.*, **15**, 3318–3332.

Antoniades, M. A. and Eleftheriades, G. V. (2003). Compact linear lead/lag metamaterial phase shifters for broadband applications. *IEEE Ant. Wireless Prop. Lett.*, **2**, 103–106.

Antoniades, M. A. and Eleftheriades, G. V. (2005). A broadband series power divider using zero-degree metamaterial phase-shifting lines. *IEEE Microw. Guided Wave Lett.*, **15**, 808–810.

Arakawa, E. T., Williams, M. W., Hamm, R. N., and Ritchie, R. H. (1973). Effect of damping on surface plasmon dispersion. *Phys. Rev. Lett.*, **31**, 1127–1129.

Ash, E. A. and Nicholls, G. (1972). Super-resolution aperture scanning microscope. *Nature*, **237**, 510–512.

Atabekov, G. I. (1965). *Linear network theory.* Pergamon Press, Oxford.

Aydin, K., Bulu, I., Guven, K., Kafesaki, M., Soukoulis, C. M., and Ozbay, E. (2005). Investigation of magnetic resonances for different split-ring resonator parameters and designs. *New J. Phys.*, **7**, 168–1–15.

Aydin, K., Bulu, I., and Ozbay, E. (2007). Subwavelength resolution with a negative-index metamaterial superlens. *Appl. Phys. Lett.*, **90**, 254102–1–3.

Aydin, K. and Ozbay, E. (2006). Identifying magnetic response of split-ring resonators at microwave frequencies. *Opto-Electron. Rev.*, **714**, 193–199.

Aydin, K. and Ozbay, E. (2007*a*). Capacitor-loaded split ring resonators as tunable metamaterial components. *J. Appl. Phys.*, **101**, 024911–1–5.

Aydin, K. and Ozbay, E. (2007*b*). Experimental and numerical analyses of the resonances of split ring resonators. *phys. stat. sol. (b)*, **244**, 1197–2001.

Baena, J. D., Bonache, J., Martin, F., Marques, R., Falcone, F., Lopetegi, T., Laso, M. A. G., Garcia-Garcia, J., Portillo, M. F., and Sorolla, M. (2005*a*). Equivalent-circuit models for split-ring resonators and complementary split-ring resonators coupled to planar transmission lines. *IEEE Trans. Microw. Theory Tech.*, **53**, 1451–1461.

Baena, J. D., Jelinek, L., and Marques, R. (2005*b*). Reducing losses and dispersion effects in multilayer metamaterial tunnelling devices. *New J. Phys.*, **7**, 166–1–13.

Baena, J. D., Jelinek, L., Marques, R., and Medina, F. (2005*c*). Near-perfect tunnelling and amplification of evanescent electromagnetic waves in a waveguide filled by a metamaterial: Theory and experiment. *Phys. Rev. B*, **72**, 075116–1–8.

Baena, J. D., Jelinek, L., Marques, R., and Zehentner, J. (2006). Electrically small isotropic three-dimensional magnetic resonators for metamaterial design. *Appl. Phys. Lett.*, **88**, 134108–1–3.

Baena, J. D., Marques, R., Medina, F., and Martel, J. (2004). Artificial magnetic metamaterial design by using spiral resonators. *Phys. Rev. B*, **69**, 014402–1–5.

Bahr, A. J. and Clausing, K. (1994). An approximate model for artificial chiral media. *IEEE Trans. Ant. Prop.*, **42**, 1592–1599.

Bai, B., Svirko, Y., Turunen, J., and Vallius, T. (2007). Optical activity in planar chiral metamaterials: Theoretical study. *Phys. Rev. A*, **76**, 023811–1–12.

Barnes, W. L. (1998). Fluorescence near interfaces: The role of photonic mode density. *J. Mod. Opt.*, **45**, 661–669.

Barnes, W. L. (2006). Surface plasmon-polariton length scales: a route to sub-wavelength optics. *J. Opt. A: Pure Appl. Opt.*, **8**, S87–S93.

Barnes, W. L., Dereux, A., and Ebbesen, T.W. (2003). Surface plasmon subwavelength optics. *Nature*, **424**, 824–830.

Batke, E. and Heitmann, D. (1984). Rapid-scan Fourier transform spectroscopy of 2-D space charge layers in semiconductors. *Infrared Phys.*, **24**, 189–197.

Beck, A. H. W. (1958). *Space charge waves and slow electromagnetic waves.* Pergamon Press, New York.

Belov, P. A., Hao, Y., and Sudhakaran, S. (2006*a*). Subwavelength microwave imaging using an array of parallel conducting wires as a lens. *Phys. Rev. B*, **73**, 033108–1–4.

Belov, P. A., Hao, Y., Sudhakaran, S., Alomainy, A., and Hao, Y. (2006*b*). Experimental study of the subwavelength imaging by a wire medium slab. *Appl. Phys. Lett.*, **89**, 262109–1–3.

Belov, P. A. and Silveirinha, M. G. (2006). Resolution of subwavelength transmission devices formed by a wire medium. *Phys. Rev. E*, **73**, 056607–1–9.

Belov, P. A. and Simovski, C. R. (2005*a*). On homogenization of electromagnetic crystals formed by uniaxial resonant scatterers. *Phys. Rev. E*, **72**, 026615–1–15.

Belov, P. A. and Simovski, C. R. (2005*b*). Subwavelength metallic waveguides loaded by uniaxial resonant scatterers. *Phys. Rev. E*, **72**, 036618–1–11.

Belov, P. A. and Simovski, C. R. (2006). Boundary conditions for interfaces of electromagnetic crystals and the generalized ewald-oseen extinction principle. *Phys. Rev. B*, **73**, 045102–1–14.

Belov, P. A., Simovski, C. R., and Ikonen, P. (2005). Canalization of subwavelength images by electromagnetic crystals. *Phys. Rev. B*, **71**, 193105–1–4.

Belov, P. A., Simovski, C. R., and Tretyakov, S. A. (2002). Two-dimensional electromagnetic crystals formed by reactively loaded wires. *Phys. Rev. E*, **66**, 036610–1–7.

Belov, P. A. and Zhao, Y. (2006). Subwavelength imaging at optical frequencies using a transmission device formed by a periodic layered metal-dielectric structure operating in the canalization regime. *Phys. Rev. B*, **73**, 113110–1–4.

Bergman, D. J. (1978). The dielectric constant of a composite material - a problem in classical physics. *Phys. Rep.*, **43**, 377–407.

Berini, P. (1999). Plasmon-polariton modes guided by a metal film of finite width. *Opt. Lett.*, **24**, 1011–1013.

Berini, P. (2000*a*). Plasmon-polariton modes guided by a metal film of finite width bounded by different dielectrics. *Opt. Exp.*, **7**, 329–335.

Berini, P. (2000*b*). Plasmon-polariton waves guided by thin lossy metal films of finite width: Bound modes of symmetric structures. *Phys. Rev. B*, **61**, 10484–10503.

Berini, P. (2001). Plasmon-polariton waves guided by thin, lossy metal films of finite width: bound modes of asymmetric structures. *Phys. Rev. B*, **63**, 125417–1–15.

Berini, P. (2006). Figures of merit for surface plasmon waveguides. *Opt. Exp.*, **14**, 13030–13042.

Berman, P. R. (2002). Goos-Hanchen shift in negatively refractive media. *Phys. Rev. E*, **66**, 067603–1–3.

Beruete, M., Freire, M. J., Marques, R., Falcone, F., and Baena, J. D. (2006). Electroinductive chains of complementary metamaterial elements. *Appl. Phys. Lett.*, **88**, 083503–1–3.

Bevensee, R. M. (1964). *Electromagnetic slow wave systems.* John Wiley and Sons Inc., New York.

Blaikie, R. J. and McNab, S. J. (2002). Simulation study of 'perfect lenses' for nearfield optical nanolithography. *Microelectron. Eng.*, **61-62**, 97–103.

Bloembergen, N. (2005). *Nonlinear optics.* World Scientific Publishing, Singapore.

Boardman, A. D. (ed.) (1982). *Electromagnetic surface modes.* Wiley, Chichester.

Boot, H. A. H. and Randall, J. T. (1976). Historical notes on the cavity magnetron. *IEEE Trans. ED*, **2**, 724–729.

Born, M. and Wolf, E. (1975). *Principles of optics* (5th edn). Pergamon Press, Oxford.

Bose, J. C. (1898). On the rotation of plane of polarization of electric waves by a twisted structure. *Proc. Royal Soc.*, **63**, 146–152.

Brillouin, L. (1953). *Wave propagation in periodic structures.* Dover, New York.

Brongersma, M. L., Hartman, J. W., and Atwater, H. A. (2000). Electromagnetic energy transfer and switching in nanoparticle chain arrays below the diffraction limit. *Phys. Rev. B*, **62**, R16356–R16359.

Brown, J. (1950). The design of metallic delay dielectric. *Proc. IEE*, **97**(Part III), 45–48.

Brown, J. (1953). Artificial dielectrics having refractive indices less than unity. *Proc. IEE*, **100**(pt. 4), Monograph No. 62R, 51–625.

Bulu, I., Caglayan, H., Aydin, K., and Ozbay, E. (2005*b*). Compact size highly directive antennas based on the SRR metamaterial medium. *New J. Phys.*, **7**, 223–1–10.

Bulu, I., Caglayan, H., and Ozbay, E. (2005*a*). Experimental demonstration of subwavelength focusing of electromagnetic waves by labyrinth-based two-dimensional metamaterials. *Opt. Lett.*, **31**, 814–817.

Burke, J.J., Stegeman, G. I., and Tamir, T. (1986). Surface-polariton-like waves guided by thin, lossy metal films. *Phys. Rev. B*, **33**, 5186–5201.

Cai, W., Chettiar, U. K., Kildishev, A. V., and Shalaev, V. M. (2007). Optical cloaking with Metamaterials. *Nature Photonics*, **1**, 224–227.

Caloz, C. and Itoh, T. (2004). Transmission line approach of left-handed (LH) materials and microstrip implementation of an artificial LH transmission line. *IEEE Trans. Ant. Prop.*, **52**, 1159–1166.

Caloz, C. and Itoh, T. (2006). *Electromagnetic metamaterials: Transmission line theory and microwave applications.* Wiley-Interscience, Hoboken, NJ.

Caloz, C., Sanada, A., and Itoh, T. (2004). A novel composite right-/left-handed coupled-line directional coupler with arbitrary coupling level and broad bandwidth. *IEEE Trans. Microw. Theory Tech.*, **52**, 980–992.

Cartlidge, E. (2002). Negative reaction to negative refraction. *Phys. World*, **8**, 8–9.

Casey, J. P. and Lewis, E. A. (1952). Interferometer action of a parallel pair of wire gratings. *J. Opt. Soc. Am.*, **42**, 971–977.

Chang, K. (1996). *Microwave ring circuits and antennas.* Wiley, New York.

Chen, H., Ran, L., Huangfu, J., Grzegorczyk, T. M., and Kong, J. A. (2006). Equivalent circuit model for left-handed metamaterials. *J. Appl. Phys.*, **100**, 024915–1–6.

Chen, J. J., Grzegorczyk, T.M., Wu, B.-I., and Kong, J.A. (2005). Limitations of FDTD in simulation of a perfect lens imaging system. *Opt. Exp.*, **13**, 10840–10845.

Chynoweth, A. G. and Buchsbaum, S. J. (1965). Solid-state plasma (Characteristics of solid state plasma conditions necessary for existence of plasma in semiconductor or semimetal). *Phys. Today*, **18**, 26–30, 32–34, 36–37.

Citrin, D. S. (2006). Plasmon-polariton transport in metal-nanoparticle chains embedded in a gain medium. *Opt. Lett.*, **31**, 98–100.

Clausius, R. J. E. (1879). *Die mechanische Behandlung der Electricitat*, Volume Abschnitt III. F. Vieweg, Braunschweig.

Cohn, S. B. (1946). Analysis of the metal strip delay structure. *J. Appl. Phys.*, **20**, 257–262.

Collin, R. E. (1991). *Field theory of guided waves.* Oxford University Press, Oxford.

Colton, D. and Kress, R. (1992). *Inverse acoustic and electromagnetic scattering theory.* Springer Verlag, Berlin.

Cottam, M. G. and Tilley, D. R. (1988). *Introduction to surface and superlattice excitations.* Cambridge University Press, Cambridge.

Craven, G. (1972). Slim guide microwave components. *Electrical Communication*, **47**, 245–258.

Cui, T. J., Cheng, Q., Lu, W. B., Jiang, Q., and Kong, J.A. (2005). Localization of electromagnetic energy using a left-handed-medium slab. *Phys. Rev. B*, **71**, 045114–1–11.

Cummer, S. A., Popa, B.-I., Schurig, D., Smith, D. R., and Pendry, J. (2006). Full wave simulations of electromagnetic cloaking structures. *Phys. Rev. E*, **74**, 036621–1–5.

Darmanyan, S. A., Neviere, M., and Zakhidov, A. A. (2003). Surface modes at the interface of conventional and left-handed media. *Opt. Commun.*, **225**, 233–240.

Darwin, C. G. (1914). The theory of X-ray reflection. *Philos. Mag.*, **27**, Part I (6th Series). 315–333, Part 2. 675–690.

Dekker, A. J. (1965). *Solid state physics.* Macmillan, London.

Depine, R. A. and Lakhtakia, A. (2004). Comment on ‘Resonant and antiresonant frequency dependence of the effective parameters of metamaterials’. *Phys. Rev. E*, **70**, 048601–1.

Dolling, G., Enkrich, C., Wegener, M., Zhou, J. F., Soukoulis, C. M., and Linden, S. (2005). Cut-wire pairs and plate pairs as magnetic atoms for optical metamaterials. *Opt. Lett.*, **30**, 3198–3200.

Dorofeenko, V., Lisyansky, A. A., Merzlikin, A. M., and Vinogradov, A. P. (2006). Full-wave analysis of imaging by the Pendry–Ramakrishna stackable lens. *Phys. Rev. B*, **73**, 235126–1–4.

Drude, P. (1959). *The theory of optics* (3rd edn). Dover, London.

Economou, E. N. (1969). Surface plasmons in thin films. *Phys. Rev.*, **182**, 539–554.

Efros, A. L. (2004). Comment II on ‘Resonant and antiresonant frequency dependence of the effective parameters of metamaterials’. *Phys. Rev. E*, **70**, 048602–1.

Eleftheriades, G. V. and Balmain, K. G. (2005). *Negative refraction metamaterials: Fundamental principles and applications.* John Wiley and Sons Inc., New York.

Eleftheriades, G. V. and Islam, R. (2007). Minaturized microwave components and antennas using negative-refractive index transmission-line (nri-tl) metamaterials. *Metamaterials*, **1**, 53–61.

Eleftheriades, G. V., Iyer, A. K., and Kremer, P. C. (2002). Planar negative refractive index media using periodically LC loaded transmission lines. *IEEE Trans. Microw. Theory Tech.*, **50**, 2702–2712.

Eleftheriades, G. V., Siddiqui, O., and Iyer, A. K. (2003). Transmission line models for negative refractive index media and associated implementation without excess resonators. *IEEE Microw. Wireless Comp. Lett.*, **13**, 31–33.

Engheta, N. (2002). An idea for thin subwavelength cavity resonators using metamaterials with negative permittivity and permeability. *IEEE Ant and Wireless Prop. Lett.*, **1**, 10–13.

Engheta, N., Nelatury, S. R., and Hoorfar, A. (2002). The role of geometry of inclusions in forming metamaterials with negative permittivity and permeability. In *XXVII URSI-GA*, Maastricht, Netherlands.

Engheta, N., Salandrino, A., and Alu, A. (2005). Circuit elements at optical frequencies: Nanoinductors, nanocapacitors, and nanoresistors. *Phys. Rev. Lett.*, **95**, 095504–1–3.

Engheta, N. and Ziolkowski, R. W. (ed.) (2006). *Metamaterials: Physics and engineering explorations.* John Wiley and Sons Inc., New York.

Enkrich, C., Perez-Willard, F., Gerthsen, D., Zhou, J., Koschny, T., Soukoulis, C. M., Wegener, M., and Linden, S. (2005*a*). Focused-ion-beam nano-fabrication of near infrared magnetic metamaterials. *Adv. Mater.*, **17**, 2547–2549.

Enkrich, C., Wegener, M., Linden, S., Burger, S., Zschiedrich, L., Schmidt, F., Zhou, J. F., Koschny, T., and Soukoulis, C. M. (2005*b*). Magnetic metamaterials at telecommunication and visible frequencies. *Phys. Rev. Lett.*, **95**, 203901–1–4.

Enoch, S., Tayeb, G., Sabouroux, P., Guerin, N., P., and Vincent (2002). A metamaterial for directive emission. *Phys. Rev. Lett.*, **89**, 213902–1–4.

Ewald, P. P. (1916). Zur Begründung der Kristalloptik. *Ann. d. Phys.*, **49**, 1–38 (Part 1). 1916; 49: 117–143 (Part 2). 1917; 54: 519–597 (Part 3).

Falcone, F., Lopetegi, T., Baena, J. D., Martin, F., Marques, R., and Sorolla, M. (2004*a*). Effective negative ε stopband microstrip lines based on complementary split ring resonators. *IEEE Microw. Wireless Comp. Lett.*, **14**, 280–282.

Falcone, F., Martin, F., Bonache, J., Laso, M. A. G., Garcia-Garcia, J., Baena, J. D., Marques, R., and Sorolla, M. (2004*b*). Stop-band and band-pass characteristics in coplanar waveguides coupled to spiral resonators. *Microw. Opt. Technol. Lett.*, **42**, 386–388.

Falcone, F., Martin, F., Bonache, J., Marques, R., Lopetegi, T., and Sorolla, M. (2004*d*). Left handed coplanar waveguide band pass filters based on bi-layer split ring resonators. *IEEE Microw. Wireless Comp. Lett.*, **14**, 10–12.

Falcone, F., Martin, F., Bonache, J., Marques, R., and Sorolla, M. (2004*c*). Coplanar waveguide structures loaded with split-ring resonators. *Microw. Opt. Technol. Lett.*, **40**, 3–6.

Fang, N., Lee, H., Sun, C., and Zhang, X. (2005). Sub-diffraction-limited optical imaging with a silver superlens. *Science*, **308**, 534–537.

Feng, S. and Elson, J. M. (2006). Diffraction-suppressed high-resolution imaging through metallodielectric nanofilms. *Opt. Exp.*, **14**, 216–221.

Feng, S., Elson, J. M., and Overfelt, P. L. (2005). Optical properties of multilayer metal-dielectric nanofilms with all-evanescent modes. *Opt. Exp.*, **13**, 4113–4124.

Feng, Y., Zhao, J., Teng, X., Chen, Y., and Jiang, T. (2007). Subwavelength imaging with compensated anisotropic bilayers realized by transmission-line metamaterials. *Phys. Rev. B*, **75**, 155107–1–6.

Fox, M. (2001). *Optical properties of solids.* Oxford University Press, Oxford.

Fredkin, D. R. and Ron, A. (2002). Effectively left-handed (negative index) composite material. *Appl. Phys. Lett.*, **70**, 1753–1755.

Freire, M. J. and Marques, R. (2005). A planar magnetoinductive lens for 3D subwavelength imaging. *Appl. Phys. Lett.*, **86**, 182505.

Freire, M. J. and Marques, R. (2006). Near-field imaging in the megahertz range by strongly coupled magnetoinductive surfaces: Experiment and ab initio analysis. *J. Appl. Phys.*, **100**, 063105–1–9.

Freire, M. J. and Marques, R. (2008). Optimizing the magnetoinductive lens: Improvement, limits, and possible applications. *J. Appl. Phys.*, **103**, 013115–1–7.

Freire, M. J., Marques, R., Medina, F., Laso, M. A. G., and Martin, F. (2004). Planar magnetoinductive wave transducers: theory and applications. *Appl. Phys. Lett.*, **85**, 4439–4441.

French, O. E., Hopcraft, K. I., and Jakeman, E. (2006). Perturbation on the perfect lens: the near-perfect lens. *New J. Phys.*, **8**, 271–1–12.

Friedrich, W., Knipping, P., and von Laue, M. (1912). Interferenz-Erscheinungen bei Rontgenstrahlen. *Sitzungsberichte der (Kgl.) Akad. der Wiss. Bayer*, **42**, 303–322.

Froncisz, W. and Hyde, J. S. (1982). The loop-gap resonator: A new microwave lumped circuit ESR sample structure. *J. Magn. Res.*, **47**, 515–521.

Gao, L. and Tang, C. J. (2004). Near-field imaging by a multi-layer structure consisting of alternate right-handed and left-handed materials. *Phys. Lett. A*, **322**, 390–395.

Garcia, N. and Nieto-Vesperinas, M. (2002). Left-handed materials do not make a perfect lens. *Phys. Rev. Lett.*, **88**, 207403–1–4.

Garcia-Garcia, J., Martin, F., Falcone, F., Bonache, J., Gil, I., Lopetegi, T., Laso, M. A. G., Sorolla, M., and Marques, R. (2004). Spurious passband suppression in microstrip coupled line band pass filters by means of split ring resonators. *IEEE Microw. Wireless Comp. Lett.*, **14**, 416–418.

Garwe, F., Rockstuhl, C., Etrich, C., Hubner, U., Bauerschafer, U., Setzpfandt, F., Augustin, M., Pertsch, T., Tunnermann, A., and Lederer, F. (2006). Evaluation of gold nanorod pairs as a potential negative index material. *Appl. Phys. B*, **84**, 139–148.

Gay-Balmaz, P., Maccio, C., and Martin, O. J. F. (2002). Microwire arrays with plasmonic response at microwave frequencies. *Appl. Phys. Lett.*, **81**, 2896–2898.

Gay-Balmaz, P. and Martin, O. J. F. (2002). Efficient isotropic magnetic resonators. *Appl. Phys. Lett.*, **81**, 939–941.

Ghim, B. T., Rinard, G. A., Quine, R. W., Eaton, S. S., and Eaton, G. R. (1996). Design and fabrication of copper-film loop-gap resonators. *J. Magn. Res. Ser. A*, **120**, 72–76.

Gil, I., Bonache, J., Garcia-Garcia, J., and Martin, F. (2006). Tunable metamaterial transmission lines based on varactor-loaded split-ring resonators. *IEEE Trans. Microw. Theory Tech.*, **54**, 2665–2674.

Gil, I., Garcia-Garcia, J., Bonache, J., Martin, F., Sorolla, M., and Marques, R. (2004). Varactor-loaded split ring resonators for tunable notch filters at microwave frequencies. *Electron. Lett.*, **40**, 1347–1348.

Gomez-Santos, G. (2003). Universal features of the time evolution of evanescent modes in a left-handed perfect lens. *Phys. Rev. Lett.*, **70**, 077401–1–4.

Goos, F. and Hanchen, H. (1947). Ein neuer and fundamentaler Versuch zur Totalreflexion. *Ann. Phys. Lpz.*, **1**, 333–346.

Gorkunov, M., Lapine, M., Shamonina, E., and Ringhofer, K. H. (2002). Effective magnetic properties of a composite material with circular conductive elements. *Eur. Phys. J. B*, **28**, 263–269.

Govyadinov, A. A. and Podolskiy, V. A. (2006). Metamaterial photonic funnels for subdiffraction light compression and propagation. *Phys. Rev. B*, **73**, 155108–1–5.

Gray, S. K. and Kupka, T. (2003). Propagation of light in metallic nanowire arrays: Finite-difference time-domain studies of silver cylinders. *Phys. Rev. B*, **68**, 045415–1–11.

Grbic, A. and Eleftheriades, G. V. (2002). Experimental verification of backward wave radiation from a negative refractive index metamaterial. *J. Appl. Phys.*, **92**, 5930–5935.

Grbic, A. and Eleftheriades, G. V. (2003*a*). Dispersion analysis of a micro-strip-based negative refractive index periodic structure. *IEEE Microw. Wireless Comp. Lett.*, **13**, 155–157.

Grbic, A. and Eleftheriades, G. V. (2003*b*). Growing evanescent waves in negative-refractive-index transmission-line media. *Appl. Phys. Lett.*, **82**, 1815–1817.

Grbic, A. and Eleftheriades, G. V. (2003*c*). Negative refraction, growing evanescent, and sub-diffraction imaging in loaded transmission-line metamaterials. *IEEE Trans Microw. Theory Tech.*, **51**, 2297–2305.

Grbic, A. and Eleftheriades, G. V. (2003*d*). Periodic analysis of a 2-d negative refractive index transmission line structure. *IEEE Trans. Ant. Prop.*, **51**, 2604–2611.

Grbic, A. and Eleftheriades, G. V. (2004). Overcoming the diffraction limit with a planar left-handed transmission-line lens. *Phys. Rev. Lett.*, **92**, 117403–1–4.

Grigorenko, A. N., Geim, A. K., Gleeson, H. F., Zhang, Y., Firsov, A. A., Khruschev, I. Y., and Petrovic, J. (2005). Nanofabricated media with negative permeability at visible frequencies. *Nature*, **438**, 335–338.

Grover, F. W. (1981). *Inductance calculations: Working formulas and tables.* Instrument Society of America, Research Triangle Park, N.C.

Gundogdu, T. F., Tsiapa, I., Kostopoulos, A., Konstantinidis, G., Katsarakis, N., Penciu, R. S., Kafesaki, M., Economou, E. N., Koschny, T., and Soukoulis, C. M. (2006). Experimental demonstration of negative magnetic permeability in the far infrared frequency region. *Appl. Phys. Lett.*, **89**, 084103–1–3.

Guo, Y., Goussetis, G., Feresidis, A. P., and Vardaxoglu, J. C. (2005). Efficient modelling of novel uniplanar left-handed metamaterials. *IEEE Trans. Microw. Theory Tech.*, **53**, 1462–1468.

Haldane, F. D. M. (2002). Electromagnetic surface modes at interfaces with negative refractive index make a 'not-quite-perfect' lens. arxiv:cond-mat/0206420.

Hammond, P. and Sykulski, J. (1994). *Engineering electromagnetism: Physical processes and computation.* Oxford University Press, Oxford.

Hansen, W. W. (1939). On the resonant frequency of closed concentric lines. *J. Appl. Phys.*, **10**, 39–45.

Hansen, W. W. and Richtmeyer, R. D. (1939). On resonators suitable for klystron oscillators. *J. Appl. Phys.*, **10**, 189–199.

Hao, T., Stevens, C. J., and Edwards, D. J. (2005*a*). Simulations and measurements for 1D metamaterial elements. In *Proc. SPIE-COO Conf.*, Warsaw, Poland.

Hao, T., Stevens, C. J., and J.Edwards, D. (2005*b*). Optimisation of metamaterials by Q factor. *Electron. Lett.*, **41**, 653–654.

Hardy, W. H. and Whitehead, L. A. (1981). Split-ring resonator for use in magnetic resonance from 20-2000 MHz. *Rev. Sci. Instrum.*, **52**, 213–216.

Hesmer, F. (2008). unpublished.

Hessel, A. and Oliner, A. A. (1965). A new theory of Wood's anomalies on optical gratings. *Appl. Opt.*, **4**, 1275–1297.

Hohenau, A., Krenn, J. R., Schider, G., Ditlbacher, H., Leitner, A., Aussenegg, F. R., and Schaich, W. L. (2005). Optical near-field of multipolar plasmons of rod-shaped gold nanoparticles. *Europhys. Lett.*, **69**, 538–543.

Holloway, C. L., Kuester, E. F., Baker-Jarvis, J., and Kabos, P. (2003). A double negative (DNG) composite medium composed of magnetodielectric spherical particles embedded in a matrix. *IEEE Trans. Ant. Prop.*, **51**, 2596–2603.

Homola, J. (2003). Present and future of surface plasmon resonance biosensors. *Anal. Bioanal. Chem.*, **377**, 528–239.

Hong, J. S. (2000). Couplings of asynchronously tuned coupled microwave resonators. *IEE Proc. Microw. Ant. Prop.*, **147**, 354–358.

Hong, J. S. and Lancaster, M. J. (1996*a*). Couplings of microstrip square open loop resonators for cross-coupled planar microwave filters. *IEEE Trans. Microw. Theory Tech.*, **44**, 2099–2109.

Hong, J. S. and Lancaster, M. J. (1996*b*). Theory and experiment of novel microstrip slow-wave open-loop resonator filters. *IEEE Trans. Microw. Theory Tech.*, **45**, 2358–2365.

Hong, J. S. and Lancaster, M. J. (1998). Cross-coupled microstrip hairpin resonator filters. *IEEE Trans. Microw. Theory Tech.*, **46**, 118–122.

Hong, J. S. and Lancaster, M. J. (1999). Aperture-coupled microstrip open-loop resonators and their applications to the design of novel microstrip bandpass filters. *IEEE Trans. Microw. Theory Tech.*, **47**, 1848–1855.

Hong, J. S. and Lancaster, M. J. (2000). *Microstrip filters for RF/microwave applications.* Wiley-Interscience, New York.

Hrabar, S., Bartolic, J., and Sipus, Z. (2005). Waveguide miniaturization using uniaxial negative permeability metamaterial. *IEEE Trans. Ant. Prop.*, **53**, 110–119.

Hsu, Y.-J., Huang, Y.-C., Lih, J.-S., and Chern, J.-L. (2004). Electromagnetic resonance in deformed split ring resonatrs of left-handed meta-materials. *J. Appl. Phys.*, **96**, 1979–1982.

Huang, K. C., Povinelli, M. L., and Joannopoulos, J. D. (2004). Negative effective permeability in polaritonic photonic crystals. *Appl. Phys. Lett.*, **85**, 543–545.

Hutter, R. G. E. (1960). *Beam and wave electronics in microwave tubes.* Van Nostrand, Princeton, NJ.

Ikonen, P., Saenz, E., Gonzalo, R., Simovski, C., and Tretyakov, S. (2007*a*). Mesoscopic effective material parameters for thin layers modeled as single and double grids of interacting loaded wires. *Metamaterials*, **1**, 89–105.

Ikonen, P., Simovski, C., Tretyakov, S., Belov, P., and Hao, Y. (2007*b*). Magnification of subwavelength field distributions at microwave frequencies using a wire medium slab operating in the canalization regime. *Appl. Phys. Lett.*, **91**, 104102–1–3.

Imbriale, W. A. and Mittra, R. (1970). The two-dimensional inverse scattering problem. *IEEE Trans. Ant. Prop.*, **18**, 633–642.

Ishikawa, A., Tanaka, T., and Kawata, S. (2005). Negative magnetic permeability in the visible light region. *Phys. Rev. Lett.*, **95**, 273401–1–4.

Islam, R. and Eleftheriades, G. V. (2004). Phase-agile branch-line couplers using metamateril lines. *IEEE Microw. Wireless Comp. Lett.*, **14**, 340–342.

Itoh, T. and Caloz, C. (2005). *Electromagnetic metamaterials: Transmission line theory and microwave applications.* John Wiley and Sons Inc., New York.

Iyer, K. A., Kremer, P. C., and Eleftheriades, G. V. (2003). Experimental and theoretical verification of focusing in a large periodically loaded transmission line negative refractive index metamaterial. *Opt. Exp.*, **11**, 696–708.

Jackson, J. D. (1967). *Classical electromagnetics.* John Wiley and Sons, New York.

Jacob, Z., Alekseyev, L. V., and Narimanov, E. (2006). Optical hyperlens: far-field imaging beyond the diffraction limit. *Opt. Exp.*, **14**, 8247–8256.

Jiang, S. H. and Pike, R. (2005). A full electromagnetic simulation study of near-field imaging using silver films. *New J. Phys.*, **7**, 169–1–20.

John, S. (1987). Strong localization of photons in certain disordered dielectric superlattices. *Phys. Rev. Lett.*, **58**, 2486–2489.

Johnson, P. B. and Christy, R. W. (1972). Optical constants of the nobel metals. *Phys. Rev. B*, **6**, 4370–4379.

Kafesaki, M., Koschny, T., Penciu, R. S., Gundogdu, T. F., Economou, E. N., and Soukoulis, C. M. (2005). Experimental demonstration of negative magnetic permeability in the far infrared frequency region. *J. Opt. A: Pure Appl. Opt.*, **7**, S12–S22.

Katsarakis, N., Konstantinides, G., Kostopoulos, A., Penciu, R. S., Gundogdu, T. F., Kafesaki, M., and Economou, E. N. (2005). Magnetic response of split-ring resonators in the far infrared frequency region. *Opt. Lett.*, **30**, 1348–1350.

Katsarakis, N., Koschny, T., Kafesaki, M., Economou, E. N., and Soukoulis, C. M. (2004). Electric coupling to the magnetic resonance of split ring resonators. *Appl. Phys. Lett.*, **84**, 2943–2945.

Kempa, K., Ruppin, R., and Pendry, J. B. (2005). Electromagnetic response of a point-dipole crystal. *Phys. Rev. B*, **72**, 205103–1–6.

Kik, P. G., Maier, S. A., and Atwater, H. A. (2002). The perfect lens in a non-perfect world. In *Proc. Progress in Electromagnetics Research Symposium (PIERS)*, Cambridge, MA, USA.

Kildishev, A. V., Cai, E., Chettiar, U. K., Yuan, H.-K., Sarychev, A. K., Drachev, V. P., and Shalaev, V. M. (2006). Negative refractive index in optics of metal-dielectric composites. *J. Opt. Soc. Am. B*, **23**, 423–433.

King, R. W. P. (1969). *The loop antenna for transmission and reception*, Volume Antenna Theory - Part I. McGraw Hill Co., New York.

Kittel, C. (1953). *Introduction to solid state physics.* John Wiley, New York.

Kittel, C. (1963). *Quantum theory of solids.* John Wiley, New York.

Klein, M. W., Enkrich, C., Wegener, M., Soukoulis, C. M., and Linden, S. (2006). Single-split split-ring resonators at optical frequencies: limits of size scaling. *Opt. Lett.*, **31**, 1259–1261.

Knapp, B. C., Knapp, E. A., Lucas, G. J., and Potter, J. M. (1965). Resonantly coupled accelerating structures for high-current proton Linacs. *IEEE Trans. Nuclear Science*, **12**, 159–165.

Kock, W. E. (1964). Metal lens antennas. *Proc. IRE*, **34**, 828–836.

Koenderink, A. F. and Polman, A. (2006). Complex response and polariton-like dispersion splitting in periodic metal nanoparticle chains. *Phys Rev. B*, **74**, 033402–1–4.

Kolinko, P. and Smith, D. R. (2003). Numerical study of electromagnetic waves interacting with negative index materials. *Opt. Exp.*, **11**, 640–648.

Kondrat'ev, I. G. and Smirnov, A. I. (2003). Comment on 'Left-handed media simulation and transmission of electromagnetic waves in subwavelength SRR loaded waveguide'. *Phys. Rev. Lett.*, **91**, 29401–1.

Kong, J. A., Wu, B.-I., and Zhang, Y. (2002*a*). Lateral displacement of a Gaussian beam reflected from a grounded slab with negative permittivity and permeability. *Appl. Phys. Lett.*, **80**, 2084–2086.

Kong, J. A., Wu, B.-I., and Zhang, Y. (2002*b*). A unique lateral displacement of a Gaussian beam transmitted through a slab with negative permittivity and permeability. *Microw. Opt. Technol. Lett.*, **33**, 136–139.

Korobkin, D., Urzhumov, Y., and Shvets, G. (2006*a*). Enhanced near-field resolution in midinfrared using metamaterials. *J. Opt. Soc. Am. B*, **23**, 468–478.

Korobkin, D., Urzhumov, Y., Zorman, C., and Shvets, G. (2006*b*). Far-field detection of the super-lensing effect in the mid-infrared: theory and experiment. *J. Mod. Phys.*, **52**, 2351–2364.

Koschny, T., Kafesaki, M., Economou, E. N., and Soukoulis, C. M. (2004*a*). Effective medium theory of left-handed materials. *Phys. Rev. Lett.*, **93**, 107402–1–4.

Koschny, T., Kafesaki, M., Penciu, R., Katsarakis, N., Soukoulis, C. M., and Economou, E. N. (2004*b*). Qualitative behaviour of left-handed materials and related metamaterials. In *Proc. Progress in Electromagnetics Research Symposium (PIERS)*, Pisa, Italy.

Koschny, T., Markos, P., Smith, D. R., and Soukoulis, C. M. (2004*c*). Reply to Comment on 'Resonant and antiresonant frequency dependence of the effective parameters of metamaterials. *Phys. Rev. E*, **70**, 048603–1.

Koschny, T., Markos, P., Smith, D. R., and Soukoulis, C. M. (2004*d*). Resonant and antiresonant frequency dependence of the effective parameters of metamaterials. *Phys. Rev. E*, **68**, 065602–1–4.

Koschny, Th., Moussa, R., and Soukoulis, C. M. (2006). Limits on the amplification of evanescent waves of left-handed materials. *J. Opt. Soc. Am. B*, **23**, 485–489.

Kostin, M. V. and Shevchenko, V. V. (1993). Theory of artificial magnetic substances based on ring currents. *Sov. J. Commun. Techn. Electron.*, **38**, 78–83.

Kostin, M. V. and Shevchenko, V. V. (1994). Artificial magnetics based on double circular elements. In *Proc. Bianisotropics'94, CHIRAL '94 3rd Int. Workshop on Chiral, Bi-isotropic and Bi-anisotropic Media,* (ed. F. Mariotte and J. P. Parneix), Perigueux, France, pp. 49–56.

Kottmann, J. P. and Martin, O. J. F. (2001). Plasmon resonant coupling in metallic nanowires. *Opt. Exp.*, **8**, 655–663.

Kottmann, J. P., Martin, O. J. F., Smith, D. R., and Schultz, S. (2000*a*). Field polarization and polarization charge distributions in plasmon resonant nanoparticles. *New J. Phys.*, **2**, 27.1–27.9.

Kottmann, J. P., Martin, O. J. F., Smith, D. R., and Schultz, S. (2000*b*). Spectral response of planar resonant nanoparticle with a non-regular shape. *Opt. Exp.*, **6**, 213–219.

Kottmann, J. P., Martin, O. J. F., Smith, D. R., and Schultz, S. (2001). Non-regularly shaped plasmon resonant nanoparticle as localized light source for near field microscopy. *J. Microscopy*, **202**, 60–65.

Kovener, G. S., Alexander, R. W., and Bell, R. J. (1976). Surface elecromagnetic waves with damping, I. Isotropic media. *Phys. Rev. B*, **14**, 1458–1464.

Kozyrev, A. B., Qin, C., Shadrivov, I. V., Kivshar, Y. S., Chuang, I. L., and van der Weide, D. W. (2007). Wave scattering and splitting by magnetic metamaterials. *Opt. Exp.*, **15**, 11714–11722.

Kreibig, U. and Vollmer, M. (1995). *Optical properties of metal clusters.* Springer, Berlin.

Kretschmann, E. and Raether, H. (1968). Radiative decay of non radiative surface plasmons excited by light (surface plasma waves excitation by light and decay into photons applied to nonradiative modes). *Z. Naturf. A*, **23**, 2135–2136.

Lagarkov, A. N. and Kissel, V. N. (2001). Electrodynamic properties of simple bodies made of materials with negative permeability and negative permittivity. *Doklady Phys.*, **46**, 163–165.

Lagarkov, A. N. and Kissel, V. N. (2005). Near-perfect imaging in a focusing system based on a left-handed-material plate. *Phys. Rev. Lett.*, **92**, 077401–1–4.

Lagarkov, A. N., Matytsin, S. M., Rozanov, K. N., and Sarychev, A. K. (1998). Dielectric properties of fiber-filled composites. *J. Appl. Phys.*, **84**, 3806–3814.

Lagarkov, A. N. and Sarychev, A. K. (1996). Electromagnetic properties of composites containing elongated conducting inclusions. *Phys. Rev. B*, **53**, 6318–6336.

Lai, A., Caloz, C., and Itoh, T. (2004). Composite right/left-handed transmission line metamaterials. *IEEE Microw. Mag.*, **5**, 34–50.

Lakhtakia, A. (2003). On planewave remittances and Goos-Hanchen shifts of planar slabs with negative real permittivity and permeability. *Electromagnetics*, **23**, 71–75.

Landau, L. D. and Lifschitz, E. M. (1984). *Electrodynamics of continuous media.* Pergamon Press, Oxford.

Lee, H., Xiong, Y., Fang, N., Srituravanich, W., Durant, S., Ambati, M., Sun, C., and Zhang, X. (2005). Realization of optical superlens imaging below the diffraction limit. *New J. Phys.*, **7**, 255–1–16.

Leifer, M. C. (1997). Resonant modes of a birdcage coil. *J. Magn. Res.*, **124**, 51–60.

Leonhardt, U. (2006). Optical conformal mapping. *Science*, **312**, 1777–1780.

Leonhardt, U. and Philbin, T. G. (2006). General relativity in electrical engineering. *New J. Phys.*, **8**, 247–1–18.

Leonhardt, U. and Philbin, T. G. (2007). General relativity in electrical engineering. In *Proc. 1st Int. Congr. on Advanced Electromagnetic Materials in Microwaves and Optics (Metamaterials 2007)*, Rome, pp. 478–481.

Lewin, L. (1947). The electrical constants of a material loaded with spherical particles. *Proc. IEE*, **94**, 65–68.

Lewin, L. (1981). *Polylogarithms and associated functions.* North Holland, New York.

Lewis, E. A. and Casey, J. P. (1952). Electromagnetic reflection and transmission by gratings of resistive wires. *J. Appl. Phys.*, **23**, 605–608.

Li, K., McLean, S. J., Greegor, R. B., Parazzoli, C. G., and Tanielian, M. H. (2003). Free-space focused-beam characterisation of left-handed materials. *Appl. Phys. Lett.*, **82**, 2535–2537.

Li, X., He, S., and Jin, Y. (2007). Subwavelength focusing with a multilayered Fabry–Perot structure at optical frequencies. *Phys. Rev. B*, **75**, 045103–1–7.

Lindell, I. V., Sihvola, A. H., Tretyakov, S. A., and Viitanen, A. J. (1994). *Electromagnetic waves in chiral and bi-isotropic media.* Artech House, Norwood, MA.

Lindell, I. V., Tretyakov, S. A., Nikoskinen, K. I., and Ilvonen, S. (2001). BW media - media with negative parameters, capable of supporting backward waves. *Microw. Opt. Technol. Lett.*, **31**, 129–133.

Linden, S., Enkrich, C., Wegener, M., Zhou, J., Koschny, T., and Soukoulis, C. M. (2004). Magnetic response of metamaterials at 100 THz. *Science*, **306**, 1351–1353.

Lippmann, G. (1894). Sur la theories de la photographie des couleurs simples et composee par la methode interferentielle. *J. Physique*, **3**, 97–107.

Liu, L., Caloz, C., Chang, C.-C., and Itoh, T. (2002). Forward coupling phenomena between artificial left-handed transmission lines. *J. Appl. Phys.*, **92**, 5560–5565.

Liu, Z., Durant, S., Lee, H., Picus, Y., Fang, N., Xiong, Y., Sun, C., and Zhang, X. (2007*a*). Far-field optical superlens. *Nano Lett.*, **7**, 403–408.

Liu, Z., Lee, H., Xiong, Y., Sun, C., and Zhang, X. (2007*b*). Far-field optical hyperlens magnifying sub-diffraction-limited objects. *Science*, **315**, 1686–1686.

Lu, Z., Murakowski, J. A., Schuetz, C. A., Shi, S., Schneider, G. J., and Prather, D. W. (2005). Three-dimensional subwavelength imaging by a photonic-crystal flat lens using negative refraction at microwave frequencies. *Phys. Rev. Lett.*, **95**, 153901–1–4.

Luo, C., Johnson, S. G., and Joannopoulos, J. D. (2002*a*). All-angle negative refraction in a three-dimensionally periodic photonic crystal. *Appl. Phys. Lett.*, **81**, 2352–2354.

Luo, C., Johnson, S. G., Joannopoulos, J. D., and Pendry, J. B. (2002*b*). All-angle negative refraction without negative effective index. *Phys. Rev. B*, **65**, 201104–1–4.

Luo, C., Johnson, S. G., Joannopoulos, J. D., and Pendry, J. B. (2002*c*). Subwavelength imaging in photonic crystals. *Phys. Rev. B*, **68**, 045115–1–15.

Macfarlane, G. G. (1946). Surface impedance of an infinite parallel wire grid at oblique angles of incidence. *J. IEE Part IIIA*, **93**, 1523–1527.

Mahan, G. D. and Obermair, G. (1969). Polaritons at surfaces. *Phys. Rev.*, **183**, 834–841.

Maier, S. A. (2008). The best of both worlds. *Nature Photonics*, **2**, 460–461.

Maier, S. A., Friedman, M. D., Barclay, P. E., and Painter, O. (2005). Experimental demonstration of fiber-accessible metal nanoparticle plasmon waveguides for planar energy guiding and sensing. *Appl. Phys. Lett.*, **86**, 071103–1–3.

Maier, S. A., Kik, P. G., and Atwater, H. A. (2002). Observation of coupled plasmon-polariton modes in Au nanoparticle chain waveguides of different lengths: Estimation of waveguide loss. *Appl. Phys. Lett.*, **87**, 1714–1716.

Maier, S. A., Kik, P. G., and Atwater, H. A. (2003*a*). Optical pulse propagation in metal nanoparticle chain waveguides. *Phys. Rev. B*, **67**, 205402–1–5.

Maier, S. A., Kik, P. G., Atwater, H. A., Meltzer, S., Hatrel, E., Koel, B. E., and Requicha, A. G. (2003*b*). Local detection of electromagnetic energy transport below the diffraction limit in metal nanoparticle plasmon waveguides. *Nature Mater.*, **2**, 229–232.

Mandelshtam, L. I. (1945). Group velocity in crystalline arrays. *Zh. Eksp. Teor. Fiz.*, **15**, 475–478. [*Complete Collected Works* (Akad. Nauk SSSR, Moscow, 1947) Vol. 2, p. 334; *Complete Collected Works* (Akad. Nauk SSSR, Moscow, 1950), Vol. 5, pp. 419, 465].

Mansuripur, M., Zakharian, A. R., and Moloney, J. V. (2007). Surface plasmon polaritons on metallic surfaces. *Opt. Photon. News*, **18**, 44–49.

Marcuvitz, N. (1956). Of field representations in terms of leaky modes or eigenmodes. *IRE Trans. Ant. Prop.*, **4**, 192–194.

Markos, P. and Soukoulis, C. M. (2008). *Wave propagation: from electrons to photonic crystals and left-handed materials.* Princeton University Press, Princeton.

Marques, R. and Freire, M. (2005). On the usefulness of split ring resonators for magnetic metamaterial design at infrared and optical frequencies. In *IEEE MELECON*, Benalmadena (Malaga), Spain, pp. 222–224.

Marques, R., Martel, J., Mesa, F., and Medina, F. (2002*a*). Left-handed-media simulation and transmission of em waves in subwavelength split-ring-resonator loaded metallic waveguides. *Phys. Rev. Lett.*, **89**, 183901–1–4.

Marques, R., Martel, J., Mesa, F., and Medina, F. (2002*b*). A new 2D isotropic left-handed metamaterial design: theory and experiment. *Microw. Opt. Technol. Lett.*, **36**, 405–408.

Marques, R., Martin, F., and Sorolla, M. (2008). *Metamaterials with negative parameters: Theory, design and microwave applications.* John Wiley and Sons Inc., New York.

Marques, R., Medina, F., and Rafii-El-Idrissi, R. (2002*c*). Role of bianisotropy in negative permeability and left-handed metamaterials. *Phys. Rev. B*, **65**, 144440–1–6.

Marques, R., Mesa, F., Martel, J., and Medina, F. (2003). Comparative analysis of edge- and broadside-coupled split ring resonators for metamaterial design - theory and experiments. *IEEE Trans. Ant. Prop.*, **51**, 2572–2581.

Martel, J., Marques, R., Falcone, F., Baena, J. D., Medina, F., Martin, F., and Sorolla, M. (2004). A new LC series element for compact bandpass filter design. *IEEE Microw. Wireless Comp. Lett.*, **14**, 210–212.

Martin, F., Bonache, J., Falcone, F., Sorolla, M., and Marques, R. (2003*a*). Split ring resonator-based left-handed coplanar waveguide. *Appl. Phys. Lett.*, **83**, 4652–465.

Martin, F., Falcone, F., Bonache, J., Marques, R., and Sorolla, M. (2003*b*). Miniature coplanar waveguide stop band filters based on multiple tuned split ring resonators. *IEEE Microw. Wireless Comp. Lett.*, **13**, 511–513.

Maslovski, S., Ikonen, P., Kolmakov, I., Tretyakov, S., and Kaunisto, M. (2005). Artificial magnetic materials based on the new magnetic particle: metasolenoid. *Progr. Electromagn. Res.*, **54**, 61–81.

Maslovski, S., Tretyakov, S., and Alitalo, P. (2004). Near-field enhancement and imaging in double planar polariton-resonant structures. *J. Appl. Phys.*, **96**, 1293–1300.

Maystre, D. and Enoch, S. (2004). Perfect lenses made with left-handed materials: Alice's mirror? *J. Opt. Soc. Am. A*, **21**, 122–131.

McAlister, A. J. and Stern, E. A. (1963). Plasma resonance absorption in thin metal films. *Phys. Rev.*, **132**, 1599 – 1602.

Mehring, M. and Freysoldt, F. (1980). A slotted tube resonator (STR) for pulsed ESR and ODMR experiments. *J. Phys. E. Sci. Instrum.*, **13**, 894–895.

Melville, D. O. S. and Blaikie, R. J. (2005). Super-resolution imaging through a planar silver layer. *Opt. Exp.*, **13**, 2127–2134.

Melville, D. O. S. and Blaikie, R. J. (2006). Experimental comparison of resolution and pattern fidelity in single- and double-layer planar lens lithography. *J. Opt. Soc. Am. B*, **23**, 461–467.

Melville, D. O. S. and Blaikie, R. J. (2007). Analysis and optimization of multilayer silver superlenses for near-field optical lithography. *Physica B*, **394**, 197–202.

Melville, D. O. S., Blaikie, R. J., and Alkaisi, M. M. (2006). A comparison of near-field lithography and planar lens lithography. *Current Appl. Phys.*, **6**, 415–418.

Melville, D. O. S., Blaikie, R. J., and Wolf, C. R. (2004). Submicron imaging with a planar silver lens. *Appl. Phys. Lett.*, **84**, 4403–4405.

Merlin, R. (2004). Analytical solution of the almost-perfect-lens problem. *Appl. Phys. Lett.*, **84**, 1290–1292.

Mesa, F., Freire, M. J., Marques, R., and Baena, J. D. (2005). 3D superresolution in metamaterial slab lenses: Experiment and theory. *Phys. Rev. B*, **72**, 235117–1–5.

Moharam, M. G., Grann, E. B., Pommet, D. A., and Gaylord, T. K. (1995). Formulation for stable and efficient implementation of the rigorous coupled-wave analysis of binary gratings. *J. Opt. Soc. Am. B*, **12**, 1068–1076.

Momo, F., Sotgin, A., and Zonta, R. (1983). On the design of a split ring resonator for ESR spectroscopy between 1 and 4 GHz. *J. Phys. E. Sci. Instrum.*, **16**, 43–46.

Moser, H. O., Casse, B. D. F., Wilhelmi, O., and Saw, B. T. (2005). Terahertz response of a microfabricated rod-split-ring-resonator electromagnetic metamaterial. *Phys. Rev. Lett.*, **94**, 063901–1–4.

Mossotti, O. F. (1850). Discussione analitica sull'influenza che l'azione di un mezzo dielettrico ha sulla distribuzione dell'elettricita alla superficie di piu corpi elettrici disseminati in esso. *Memorie di Matematica e di Fisica della Societa Italiana delle Scienze, Modena*, **XXIV**, Parte seconda, 49–74.

Nachman, A. I. (1988). Reconstructions from boundary measurements. *Ann. Math.*, **128**, 531–576.

Nefedov, I. S. and Tretyakov, S. A. (2005). On potential applications of metamaterials for the design of broadband phase shifters. *Microw. Opt. Technol. Lett.*, **45**, 98–103.

Notomi, M. (2000). Theory of light propagation in strongly modulated photonic crystals: Refraction-like behavior in the vicinity of the photonic band gap. *Phys. Rev. B*, **62**, 10696–10705.

O'Brien, S., McPeake, D., Ramakrishna, S. A., and Pendry, J. B. (2004). Near-infrared photonic band gaps and nonlinear effects in negative magnetic metamaterials. *Phys. Rev. B*, **69**, 241101–1–4.

O'Brien, S. and Pendry, J. B. (2002). Magnetic activity at infrared frequencies in structured photonic crystals. *J. Phys. Condens. Matter*, **14**, 6383–6394.

Oliner, A. A. and Tamir, T. (1962). Backward waves on isotropic plasma slabs. *J. Appl. Phys.*, **33**, 231–233.

Ono, A., Kato, J., and Kawata, S. (2005). Subwavelength optical imaging through a metallic nanorod array. *Phys. Rev. Lett.*, **95**, 267407–1–4.

Oseen, C. W. (1922). Die Einsteinsche Nadelstichstrahlung und die Maxwellschen Gleichungen. *Ann. d. Phys.*, **69**, 202–204.

Otto, A. (1968). Excitation of nonradiative surface plasma waves in silver by the method of frustrated total reflection. *Z. Phys.*, **216**, 398–410.

Oulton, R. F., Sorger, V. J., Genov, D. A., Pile, D. F. P., and Zhang, X. (2008). A hybrid plasmonic waveguide for subwavelength confinement and long-range propagation. *Nature Photonics*, **2**, 496–500.

Padilla, W. J. (2007). Group theoretical description of artificial magnetic metamaterials. *Opt. Exp.*, **15**, 1639–1646.

Paesler, M. and Mayer, P. (1996). *Near-field optics: Theory, instrumentation and applications.* John Wiley and Sons Inc., New York.

Panina, L. V., Grigorenko, A. N., and Makhnovskiy, D. P. (2002). Optomagnetic composite medium with conducting nanoelements. *Phys. Rev. B*, **66**, 155411–1–17.

Parimi, P. V., Lu, W. T., Vodo, P., Sokoloff, J., Derov, J. S., and Sridhar, S. (2004). Negative refraction and left-handed electromagnetism in microwave photonic crystals. *Phys. Rev. Lett.*, **92**, 127401–1–4.

Parimi, P. V., Lu, W. T., Vodo, P., and Sridhar, S. (2003). Photonic crystals: Imaging by flat lens using negative refraction. *Nature*, **426**, 404–1.

Park, S. Y. and Stroud, D. (2004). Surface-plasmon dispersion relations in chains of metallic nanoparticles: An exact quasistatic calculation. *Phys. Rev. B*, **69**, 125418–1–7.

Paul, J., Christopoulos, C., and Thomas, D.W.P. (2001). Time-domain modelling of negative refractive index material. *Electron. Lett.*, **37**, 912–913.

Pendry, J.B., Schurig, D., and Smith, D. R. (2006). Controlling electromagnetic fields. *Science*, **312**, 1780–1782.

Pendry, J. B. (2000). Negative refraction makes a perfect lens. *Phys. Rev. Lett.*, **85**, 3966–3969.

Pendry, J. B. (2001). Electromagnetic materials enter the negative age. *Phys. World*, **14**, 47.

Pendry, J. B. (2003*a*). Optics: positively negative. *Nature*, **423**, 22–23.

Pendry, J. B. (2003*b*). Perfect cylindrical lenses. *Opt. Exp.*, **11**, 755–760.

Pendry, J. B. (2004). A chiral route to negative refraction. *Science*, **306**, 1353–1355.

Pendry, J. B., Holden, A. J., Robbins, D. J., and Stewart, W. J. (1998). Low frequency plasmons in thin-wire structures. *J. Phys. Condens. Matter*, **10**, 4785–4809.

Pendry, J. B., Holden, A. J., Robbins, D. J., and Stewart, W. J. (1999). Magnetism from conductors and enhanced nonlinear phenomena. *IEEE Trans. Microw. Theory Tech.*, **47**, 2075–2084.

Pendry, J. B., Holden, A. J., Stewart, W. J., and Youngs, I. (1996). Extremely low frequency plasmons in metallic mesostructures. *Phys. Rev. Lett.*, **76**, 4773–4776.

Pendry, J. B. and Ramakrishna, S. A. (2002). Near field lenses in two dimensions. *J. Phys. Condens. Matter*, **14**, 8463–8479.

Pendry, J. B. and Ramakrishna, S. A. (2003). Refining the perfect lens. *Physica B - Condens. Matter*, **338**, 329–332.

Pfenninger, S., Forrer, J., and Schweiger, A. (1988). Bridged loop-gap resonator: A resonant structure for pulsed ESR transparent to high-frequency radiation. *Rev. Sci. Instrum.*, **59**, 752–760.

Podolskiy, V. and Narimanov, E. (2005). Near-sighted superlens. *Opt. Lett.*, **30**, 75–77.

Podolskiy, V., Sarychev, A. K., and Shalaev, V. M. (2002). Plasmon modes in metal nanowires and left-handed materials. *J. Nonlin. Opt. Phys. Mater.*, **11**, 65–74.

Podolskiy, V., Sarychev, A. K., and Shalaev, V. M. (2003). Plasmon modes and negative refraction in metal nanowire composites. *Opt. Exp.*, **11**, 735–745.

Pokrovsky, A. L. and Efros, A. L. (2002*a*). Diffraction in left-handed materials and theory of Veselago lens. arxiv/cond-mat/0202078.

Pokrovsky, A. L. and Efros, A. L. (2002*b*). Diffraction theory and focusing of light by left-handed materials. arxiv/cond-mat/0207534.

Pokrovsky, A. L. and Efros, A. L. (2002*c*). Electrodynamics of metallic photonic crystals and the problem of left-handed materials. *Phys. Rev. Lett.*, **89**, 093901–1–4.

Quidant, R., Girard, C., Weeber, J.-C., and Dereux, A. (2004). Tailoring the transmittance of integrated optical waveguide with short metallic nanoparticle chains. *Phys. Rev. B*, **69**, 085407–1–7.

Quinten, M., Leitner, A., Krenn, J. R., and Aussenegg, F. R. (1998). Electromagnetic energy transport via linear chains of silver nanoparticles. *Opt. Lett.*, **23**, 1331–1333.

Radkovskaya, A., Shamonin, M., Stevens, C. J., Faulkner, G., Edwards, D. J., Shamonina, E., and Solymar, L. (2005). Resonant frequencies of a combination of split rings: experimental, analytical and numerical study. *Microw. Opt. Technol. Lett.*, **46**, 473–476.

Radkovskaya, A., Sydoruk, O., Shamonin, M., Stevens, C. J., Faulkner, G., Edwards, D. J., Shamonina, E., and Solymar, L. (2007*a*). Experimental study of a bi-periodic magnetoinductive waveguide: comparison with theory. *IET Microw. Ant. Prop.*, **1**, 80–83.

Radkovskaya, A., Sydoruk, O., Shamonin, M., Stevens, C. J., Faulkner, G., Edwards, D. J., Shamonina, E., and Solymar, L. (2007*b*). Transmission properties of two shifted magnetoinductive waveguides. *Microw. Opt. Technol. Lett.*, **49**, 1054–1058.

Ramakrishna, S. A. and Pendry, J. B. (2002). The asymmetric lossy near-perfect lens. *J. Mod. Opt.*, **49**, 1747–1762.

Ramakrishna, S. A., Pendry, J. B., Wiltshire, M. C. K., and Stewart, W. J. (2003). Imaging the near field. *J. Mod. Opt.*, **50**, 1419–1430.

Ramo, S., Whinnery, J., and van Duzer, T. (1965). *Fields and waves in communication electronics.* Wiley, New York.

Ran, L., Huangfu, J., Chen, H., Li, Y., Zhang, X., Chen, K., and Kong, J. A. (2004). Microwave solid-state left-handed material with a broad bandwidth and an ultralow loss. *Phys. Rev. B*, **70**, 073102–1–3.

Rao, X. S. and Ong, C. K. (2003). Amplification of evanescent waves in a lossy left-handed material slab. *Phys. Rev. B*, **68**, 113103–1–4.

Rayleigh, Lord (1892). On the influence of obstacles arranged in rectangular order upon the properties of a medium. *Philos. Mag.*, **34**, 481–502.

Rockstuhl, C., Zentgraf, T., Guo, H., Liu, N., Etrich, C., Loa, I., Szassen, K., Kuhl, J., Lederer, F., and Giessen, H. (2006). Resonances of split-ring-resonator metamaterials in the near infrared. *Appl. Phys. B*, **84**, 219–227.

Rogla, J. L., Carbonell, J., and Boria, V. (2007). Study of equivalent circuits for open-ring and split-ring resonators in coplanar waveguide technology. *IET Microw. Ant. Prop.*, **1**, 170–176.

Rotman, W. (1962). Plasma simulation by artificial dielectrics and parallel-plate media. *IRE Trans. Ant. Prop.*, **10**, 82–95.

Ruppin, R. (2000). Surface polaritons of a left-handed medium. *Phys. Lett. A*, **277**, 61–64.

Ruppin, R. (2001). Surface polaritons of a left-handed material slab. *J. Phys. Condens. Matter*, **1**, 1811–1819.

Russell, P. St. J. (1984). Novel thick grating beam-squeezing device in Ta_2O_5 corrugated planar waveguide. *Electron. Lett.*, **20**, 72–73.

Saadoun, M. M. I. and Engheta, N. (1992). A reciprocal phase shifter using novel pseudochiral or omega medium. *Microw. Opt. Technol. Lett.*, **5**, 184–188.

Saadoun, M. M. I. and Engheta, N. (1994). Bianisotropic and bi-isotropic media and applications. *Progr. Electromagn. Res.*, **9**, 351–397.

Salandrino, A. and Engheta, N. (2006). Far-field subdiffraction optical microscopy using metamaterial crystals: Theory and simulations. *Phys. Rev. B*, **74**, 075103–1–5.

Sanada, A., Caloz, C., and Itoh, T. (2004). Characteristics of the composite right/left-handed transmission lines. *IEEE Microw. Wireless Comp. Lett.*, **14**, 68–70.

Sarid, D. (1981). Long-range surface-plasma waves on very thin metal films. *Phys. Rev. Lett.*, **47**, 1927–1930.

Sarychev, A. K., Shvets, G., and Shalaev, V. M. (2006). Magnetic plasmon resonance. *Phys. Rev. E*, **73**, 036609–1–10.

Sasaki, T., Iwata, K., Otsuka, H., and Itoh, I. (1995). Generation of homogeneous high magnetic fields within superconducting 'Swiss Roll'. *Cryogenics*, **35**, 339–343.

Sauviac, B., Simovski, C. R., and Tretyakov, S. A. (2004). Double split-ring resonators. analytical modeling and numerical simulations. *Electromagn.*, **24**, 317–338.

Scalora, M., Bloemer, M. J., Pethel, A. S., Dowling, J. P., Bowden, C. M., and Manka, A. S. (1998). Transparent, metallo-dielectric, one-dimensional, photonic band-gap structures. *J. Appl. Phys.*, **83**, 2377–2383.

Schneider, S. and Dullenkopf, P. (1977). Slotted tube resonator: a new NMR probe head at high observing frequencies. *Rev. Sci. Instrum.*, **48**, 68–73.

Schurig, D., Mock, J. J., Justice, B. J., Cummer, S. A., Pendry, J. B., Starr, A. F., and Smith, D. R. (2006). Metamaterial electromagnetic cloak at microwave frequencies. *Science*, **314**, 977–980.

Schurig, D. and Smith, D. R. (2004). Negative index lens aberrations. *Phys. Rev. E*, **70**, 065601–1–4.

Schurig, D. and Smith, D. R. (2005). Sub-diffraction imaging with compensating bilayers. *New J. Phys.*, **7**, 162–1–15.

Schuster, A. (1904). *An introduction to the theory of optics.* Edward Arnold, London.

Shadrivov, I. V., Zharov, A. A., and Kivshar, Y. S. (2003). Giant Goos-Hanchen effect at the reflection from left-handed metamaterials. *Appl. Phys. Lett.*, **83**, 2713–2715.

Shalaev, V. M., Cai, W., Chettiar, U. K., Yuan, H.-K., Sarychev, A. K., Drachev, V. P., and Kildishev, A. V. (2005). Negative index of refraction in optical metamaterials. *Opt. Lett.*, **30**, 3356–3358.

Shamonin, M., Shamonina, E., Kalinin, V. A., and Solymar, L. (2004). Properties of a metamaterial element: Analytical solutions and numerical simulations for a singly split double ring. *J. Appl. Phys.*, **95**, 3778–3784.

Shamonin, M., Shamonina, E., Kalinin, V. A., and Solymar, L. (2005). Resonant frequencies of a split-ring resonator: analytical solutions and numerical simulations. *Microw. Opt. Technol. Lett.*, **44**, 133–136.

Shamonina, E. (2008). Slow waves in magnetic metamaterials: history, fundamentals and applications. *phys. stat. sol. b*, **245**, 1471–1482.

Shamonina, E., Kalinin, V. A., Ringhofer, K. H., and Solymar, L. (2001). Imaging, compression and Poynting vector streamlines for negative permittivity materials. *Electron. Lett.*, **37**, 1243–1244.

Shamonina, E., Kalinin, V. A., Ringhofer, K. H., and Solymar, L. (2002*a*). Magneto-inductive waveguide. *Electron. Lett.*, **38**, 371–372.

Shamonina, E., Kalinin, V. A., Ringhofer, K. H., and Solymar, L. (2002*b*). Magnetoinductive waves in one, two, and three dimensions. *J. Appl. Phys.*, **92**, 6252–6261.

Shamonina, E. and Solymar, L. (2004). Magneto-inductive waves supported by metamaterial elements: components for a one-dimensional waveguide. *J. Phys. D: Appl. Phys.*, **37**, 362–367.

Shamonina, E. and Solymar, L. (2006). Properties of magnetically coupled metamaterial elements. *J. Magn. Magn. Mater.*, **300**, 38–43.

Shefer, J. (1963). Periodic cylinder arrays as transmission lines. *IEEE Trans. Microw. Theory Tech.*, **11**, 55–61.

Shelby, R. A., Smith, D. R., Nemat-Nasser, S. C., and Schultz, S. (2001*b*). Microwave transmission through a two-dimensional, isotropic, left-handed metamaterial. *Appl. Phys. Lett.*, **78**, 489–491.

Shelby, R. A., Smith, D. R., and Schultz, S. (2001*a*). Experimental verification of a negative index of refraction. *Science*, **292**, 77–79.

Shen, J. T., Catrysse, P. B., and Fan, S. (2005). Mechanism for designing metallic metamaterials with a high index of refraction. *Phys. Rev. Lett.*, **94**, 197401–1–48.

Shen, J. T. and Platzman, M. (2002). Near-field imaging with negative dielectric constant lenses. *Appl. Phys. Lett.*, **80**, 3286–3288.

Shen, Y. R. (1984). *Principles of nonlinear optics.* Wiley, New York.

Shvets, G. (2003). Applications of surface plasmon and phonon polaritons to developing left-handed materials and nano-lithography. *Proc. SPIE*, **5221**, 124–132.

Shvets, G. and Urzhumov, Y. (2004). Engineering the electromagnetic properties of periodic nanostructures using electrostatic resonances. *Phys. Rev. Lett.*, **93**, 243902–1–4.

Sihvola, A. (2007). Metamaterials in electromagnetics. *Metamaterials*, **1**, 2–11.

Silin, R. A. and Sazonov, V. P. (1966). *Slow wave structures.* Sovetskoe Radio, Moscow. in Russian.

Silveirinha, M. and Engheta, N. (2006). Tunnelling of electromagnetic energy through subwavelength channels and bends using ε-near-zero materials. *Phys. Rev. Lett.*, **97**, 157403–1–4.

Silveirinha, M. and Engheta, N. (2007). Design of matched zero-index metamaterials using nonmagnetic inclusions in epsilon-near-zero media. *Phys. Rev. B*, **75**, 075119–1–10.

Simonyi, K. (1963). *Foundations of electrical engineering.* Pergamon Press, Oxford.

Simovski, C. R. and He, S. (2003). Frequency range and explicit expressions for negative permittivity and permeability for an isotropic medium formed by a lattice of perfectly conducting omega particles. *Phys. Lett. A*, **311**, 254–264.

Simovski, C. R., Popov, M., and He, S. (2000*a*). Dielectric properties of a thin film consisting of a few layers of molecules or particles. *Phys. Rev. B*, **62**, 13718–1–13.

Simovski, C. R. and Tretyakov, S. A. (2007). Local constitutive parameters of metamaterials from an effective-medium perspective. *Phys. Rev. B*, **75**, 195111–1–10.

Simovski, C. R., Tretyakov, S. A., Sihvola, A. H., and Popov, M (2000*b*). On the surface effect in thin molecular or composite layers. *Eur. Phys. J.: Appl. Phys.*, **9**, 195–204.

Simovski, C. R., Viitanen, A. J., and Tretyakov, S. A. (2005). Resonator mode in chains of silver spheres and its possible application. *Phys. Rev. E*, **72**, 066606–1–10.

Sivukhin, D. V. (1957). The energy of electromagnetic waves in dispersive media. *Opt. Spectrosc.*, **3**, 308–312.

Smith, D. R., Kolinko, P., and Schurig, D. (2004*a*). Negative refraction in indefinite media. *J. Opt. Soc. Am. B*, **21**, 1032–1043.

Smith, D. R., Padilla, W. J., Vier, D. C., Nemat-Nasser, S. C., and Schultz, S. (2000). Composite medium with simultaneously negative permeability and permittivity. *Phys. Rev. Lett.*, **84**, 4184–4187.

Smith, D. R., Schultz, S., Markoš, P., and Soukoulis, C. M. (2002). Determination of effective permittivity and permeability of metamaterials from reflection and transmission coefficients. *Phys. Rev. B*, **65**, 195104–1–5.

Smith, D. R. and Schurig, D. (2003). Electromagnetic wave propagation in media with indefinite permittivity and permeability tensors. *Phys. Rev. Lett.*, **90**, 077405–1–4.

Smith, D. R., Schurig, D., Mock, J. J., Kolinko, P., and Rye, P. (2004*b*). Partial focusing of radiation by a slab of indefinite media. *Appl. Phys. Lett.*, **84**, 2244–2246.

Smith, D. R., Schurig, D., Rosenbluth, M., Schultz, S., Ramakrishna, S. A., and Pendry, J. B. (2003). Limitations on subdiffraction imaging with a negative refractive index slab. *Appl. Phys. Lett.*, **82**, 1506–1508.

Smith, D. R., Vier, D. C., Koschny, Th., and Soukoulis, C. M. (2005). Electromagnetic parameter retrieval from inhomogeneous Metamaterials. *Phys. Rev. E*, **71**, 036617–1–11.

Smolyaninov, I.I. (2003). Surface plasmon toy model of a rotating black hole. *New J. Phys.*, **5**, 147–1–8.

Smolyaninov, I.I. and Davis, C. C. (2004). Linear and nonlinear optics of surface-plasmon whispering-gallery modes. *Phys. Rev. B*, **69**, 205417–1–7.

Smolyaninov, I.I., Elliott, J., Zayats, A., and Davis, C. C. (2005). Far-field optical microscopy with a nanometer-scale resolution based on the in-plane mage magnification of the surface plasmon polaritons. *Phys. Rev. Lett.*, **94**, 057401–1–4.

Smolyaninov, I.I., Hung, Y.-J., and Davis, C. C. (2007). Magnifying superlens in the visible frequency range. *Science*, **315**, 1699–1701.

Solymar, L. (1984). *Lectures on electromagnetic theory.* Oxford University Press, Oxford.

Solymar, L. and Cook, D. J. (1981). *Volume holograms and volume gratings.* Academic Press, London.

Solymar, L. and Walsh, D. (2004). *Electrical properties of materials* (7th edn). Oxford University Press, Oxford.

Solymar, L., Webb, D. J., and Grunnet-Jepsen, A. (1996). *The physics and applications of photorefractive materials.* Clarendon Press, Oxford.

Solymar, L., Zhuromskyy, O., Sydoruk, O., Shamonina, E., Young, I. R., and Syms, R. R. A. (2006). Rotational resonance of magnetoinductive waves: Basic concept and application to nuclear magnetic resonance. *J. Appl. Phys.*, **99**, 123908–1–8.

Sondergaard, T. and Bozhevolnyi, S. (2007). Slow-plasmon resonant nanostructures: Scattering and field enhancements. *Phys. Rev. B*, **75**, 073402–1–4.

Soukoulis, C. M., Kafesaki, M., and Economou, E. N. (2006). Negative-index materials: New frontiers in optics. *Adv. Mater.*, **18**, 1941–1952.

Stratton, J. A. (1941). *Electromagnetic theory.* McGraw Hill, New York.

Svirko, Y., Zheludev, N., and Osipov, M. (2001). Layered chiral metallic microstructures with inductive coupling. *Appl. Phys. Lett.*, **58**, 498–500.

Sydoruk, O. (2007). *Tailoring the properties of metamaterials for linear and nonlinear applications.* Ph. D. thesis, University of Osnabrueck.

Sydoruk, O., Kalinin, V. A., and Shamonina, E. (2007*a*). Parametric amplification of magnetoinductive waves. *phys. stat. sol. b*, **244**, 1176–1180.

Sydoruk, O., Radkovskaya, A., Zhuromskyy, O., Shamonina, E., Shamonin, M., Stevens, C. J., Edwards, D. J., Faulkner, G., and Solymar, L. (2006). Tailoring the near-field guiding properties of magnetic metamaterials with two resonant elements per unit cell,. *Phys. Rev. B*, **73**, 224406–1–12.

Sydoruk, O., Shamonin, M., Radkovskaya, A., Zhuromskyy, O., Shamonina, E., Trautner, R., Stevens, C. J., Faulkner, G., Edwards, D. J., and Solymar, L. (2007*b*). A mechanism of subwavelength imaging with bi-layered magnetic metamaterials: theory and experiment. *J. Appl. Phys.*, **101**, 073903–1–8.

Sydoruk, O., Shamonina, E., and Solymar, L. (2007*c*). Parametric amplification in coupled magnetoinductive waveguides. *J. Phys. D*, **40**, 68796887.

Sydoruk, O., Shamonina, E., and Solymar, L. (2007*d*). Tayloring of the subwavelength focus. *Microw. Opt. Technol. Lett.*, **49**, 2228–2231.

Sydoruk, O., Zhuromskyy, O., Shamonina, E., and Solymar, L. (2005). Phonon-like dispersion curves for magnetoinductive waves. *Appl. Phys. Lett.*, **87**, 072501–1–3.

Syms, R. R. A. (1986). Perturbation analysis of nearly uniform coupled waveguide arrays. *Appl. Opt.*, **25**, 2988–2995.

Syms, R. R. A. (1987). Approximate solution of eigenmode problems for layered coupled waveguide arrays. *IEEE J. Quant. Electron.*, **23**, 525–532.

Syms, R. R. A. (2006). private communication.

Syms, R. R. A., Shamonina, E., Kalinin, V., and Solymar, L. (2005*a*). A theory of metamaterials based on periodically loaded transmission lines: Interaction between magnetoinductive and electromagnetic waves. *J. Appl. Phys.*, **97**, 064909–1–6.

Syms, R. R. A., Shamonina, E., and Solymar, L. (2005*b*). Positive and negative refraction of magnetoinductive waves in two dimensions. *Eur. Phys. J. B*, **46**, 301–308.

Syms, R. R. A., Shamonina, E., and Solymar, L. (2006*a*). Magneto-inductive waveguide devices. *IEE Proc. Microw. Ant. Prop.*, **153**, 111–121.

Syms, R. R. A., Solymar, L., and Shamonina, E. (2005*c*). Absorbing terminations for magneto-inductive waveguides. *IEE Proc. Microw. Ant. Prop.*, **152**, 77–81.

Syms, R. R. A., Solymar, L., and Young, I. R. (2008). Three-frequency parametric amplification in magneto-inductive ring resonators. *Metamaterials*, **2**, 122–134.

Syms, R. R. A., Sydoruk, O., Shamonina, E., and Solymar, L. (2007*a*). Higher order interactions in magnetoinductive waveguides. *Metamaterials*, **1**, 44–51.

Syms, R. R. A., Young, I. R., and Solymar, L. (2006*b*). Low-loss magnetoinductive waveguides. *J. Phys. D: Appl. Phys.*, **39**, 1945–1951.

Syms, R. R. A., Young, I. R., and Solymar, L. (2007*b*). Parametrically amplified magneto-inductive ring resonators. In *Proc. 1st Int.*

Congr. on Advanced Electromagnetic Materials in Microwaves and Optics (Metamaterials 2007), Rome, pp. 40–43.

Synge, E. H. (1928). A suggested method for extending microscpoe resolution into the ultra-microscopic region. *Philos. Mag.*, **6**, 356–362.

't Hooft, G. W. (2001). Comment on 'Negative refraction makes a perfect lens'. *Phys. Rev. Lett.*, **87**, 249701–1.

Tang, C. J. and Gao, L. (2004). Surface polaritons and imaging properties of a multi-layer structure containing negative-refractive-index materials. *J. Phys. Condens. Matter*, **16**, 4743–4751.

Tatartschuk, E. (2007). unpublished.

Tatartschuk, E. (2008). unpublished.

Taubner, T., Korobkin, D., Urzhumov, Y., Shvets, G., and Hillenbrand, R. (2006). Near-field microscopy through a SiC superlens. *Science*, **313**, 1595–1595.

Thompson, G. H. B. (1955). Unusual waveguide characteristics associated with the apparent negative permeability obtainable in ferrites. *Nature*, **175**, 1135–1136.

Thompson, G. H. B. (1963). Backward waves in longitudinally magnetized ferrite filled guides. Volume 6, Electromagnetic theory and antennas, E. C. Jordan, editor of *International series of monographs on electromagnetic waves.* Pergamon Press. Proc. Symp. Copenhagen, Denmark, June 1962.

Tilley, D. R. (1988). *Basic surface plasmon theory*, Volume Surface Plasmon-Polaritons of *IOP Short Meetings Series*, pp. 1–23. IOP.

Tonks, L. and Langmuir, I. (1929). Oscillations in ionised gases. *Phys. Rev.*, **33**, 195–211.

Toraldo di Francia, P. (1952). Nuovo pupille superresolventi. *Atti Fond. Girogio Ronchi*, **7**, 366–372.

Toraldo di Francia, P. (1953). Directivity, supergain and information. *Proc. IRE*, **3**, 25–26.

Tretyakov, S. (2007). On geometrical scaling of split-ring and double-bar resonators at optical frequencies. *Metamaterials*, **1**, 40–43.

Tretyakov, S., Nefedov, I., Sihvola, A., Maslovski, S., and Simovski, C. (2003). Waves and energy in chiral nihility. *J. Electromagn. Waves Appl.*, **17**, 695–706.

Tretyakov, S. A. (2003). *Analytical modeling in applied electromagnetics.* Artech House, Norwood, MA.

Tretyakov, S. A., Mariotte, F., Simovski, C. R., Kharina, T. G., and Heliot, J.-P. (1996). Analytical antenna model for chiral scatterers: comparison with numerical and experimental data. *IEEE Trans. Ant. Prop.*, **44**, 1006–1014.

Tsakmakidis, K. L., Boardman, A. D., and Hess, O. (2007). 'Trapped rainbow' storage of light in metamaterials. *Nature*, **450**, 397–401.

Tsakmakidis, K. L., Klaedtke, A., Aryal, D. P., Jamois, C., and Hess, O. (2006). Single-mode operation in the slow-light regime using oscillatory waves in generalized left-handed heterostructures. *Appl. Phys. Lett.*, **89**, 201103–1–3.

Uslenghi, P. L. E. (2007). Invisible metamaterial trenches. In *Proc. 1st Int. Congr. on Advanced Electromagnetic Materials in Microwaves and Optics (Metamaterials 2007)*, Rome, pp. 486–487.

Valanju, P. M., Walser, R. M., and Valanju, A. P. (2002). Wave refraction in negative-index media: always positive and very inhomogeneous. *Phys. Rev. Lett.*, **88**, 187401–1–4.

Valentine, J., Zhang, S., Zentgraf, T., Ulin Avila, E., Genov, D. A., Bartal, G., and Zhang, X. (2008). Three-dimensional optical metamaterial with a negative refractive index. *Nature*, doi:10.1038/nature07247, 1–5.

Veselago, V. G. (1968). The electrodynamics of substances with simultaneously negative values of ε and μ. *Sov. Phys. Usp.*, **10**, 509–514. (translated from *Usp. Fiz. Nauk* **92**, 517, 1967).

Wait, J. R. (1962). *Electromagnetic waves in stratified media.* MacMillan, New York.

Walter, C. H. (1965). *Traveling wave antennas.* McGraw Hill, New York.

Wang, S., Tang, C., Pan, T., and Gao, L. (2006). Effectively negatively reflective material made of negative-permittivity and negative-permeability bilayer. *Phys. Lett. A*, **351**, 391–397.

Webb, K. J. and Yang, M. (2006). Subwavelength imaging with a multilayer silver film structure. *Opt. Lett.*, **31**, 2130–2132.

Webb-Wood, G., Ghoshal, A., and Kik, P. G. (2006). In situ experimental study of a near-field lens at visible frequencies. *Appl. Phys. Lett.*, **89**, 193110–1–3.

Weber, W. H. and Ford, G. W. (2004). Propagation of optical excitations by dipolar interactions in metal nanoparticle chains. *Phys. Rev. B*, **70**, 125429–1–8.

Welford, K. (1988). *The method of attenuated total reflection*, Volume Surface plasmon-polaritons of *IOP Short Meetings Series*, pp. 25–78.

Welford, K. (1991). Surface plasmon-polaritons and their uses. *Opt. Quant. Electron.*, **23**, 1–27.

Wendler, L. and Haupt, R. (1986). Long-range surface plasmon-polaritons in asymmetric layer structures. *J. Appl. Phys.*, **59**, 3289–3291.

Whinnery, J. R. and Jamieson, H. W. (1944). Equivalent circuits for discontinuities in transmission lines. *Proc. IRE*, **32**, 98–114.

White, J., White, C. J., and Slocum, A. H. (2005). Octave-tunable miniature RF resonators. *Microw. Wireless Comp. Lett.*, **15**, 793–795.

Whitney, A. V., Elam, J. W., Zou, S., Zinovev, A. V., Stair, P. C., Schatz, G. C., and van Duyne, R. P. (2005). Localized surface plasmon resonance nanosensor: A high-resolution distance-dependence study using atomic layer deposition. *J. Phys. Chem. B*, **109**, 20522–20528.

Williams, J. M. (2001). Some problems with negative refraction. *Phys. Rev. Lett.*, **87**, 249703–1.

Wiltshire, M. C. K., Hajnal, J. V., Pendry, J. B., Edwards, D. J., and Stevens, C. J. (2003*a*). Metamaterial endoscope for magnetic field transfer: near field imaging with magnetic wires. *Opt. Exp.*, **11**, 709–715.

Wiltshire, M. C. K., Hajnal, J. V., Pendry, J. B., and Edwards, D. J. (2004*a*). RF field transmission through Swiss Rolls - an anisotropic magnetic metamaterial. In *Proc. 27th ESA Antenna Technology Workshop on Innovative Periodic Antennas: Electromagnetic Bandgap, Left-handed Materials, Fractal and Frequency Selective Surfaces*, Santiago de Compostela, Spain.

Wiltshire, M. C. K., Pendry, J. B., Young, I. R., Larkman, J., Gilderdale, D. J., and Hajnal, J. V. (2001). Microstructured magnetic materials for radio frequency operation in magnetic resonance imaging (MRI). *Science*, **291**, 849–851.

Wiltshire, M. C. K., Shamonina, E., Young, I. R., and Solymar, L. (2003*b*). Dispersion characteristics of magneto-inductive waves: comparison between theory and experiment. *Electron. Lett.*, **39**, 215–217.

Wiltshire, M. C. K., Shamonina, E., Young, I. R., and Solymar, L. (2004*b*). Experimental and theoretical study of magneto-inductive waves supported by one-dimensional arrays of 'Swiss Rolls'. *J. Appl. Phys.*, **95**, 4488–4493.

Wood, B., Pendry, J. B., and Tsai, D. P. (2006). Directed subwavelength imaging using a layered metal-dielectric system. *Phys. Rev. B*, **74**, 115116–1–8.

Xu, X., Quan, B., Gu, C., and Wang, L. (2006). Biansotropic response of microfabricated metamaterials in the terahertz region. *J. Opt. Soc. Am. B*, **23**, 1174–1180.

Yablonovitch, E. (1987). Inhibited spontaneous emission in solid-state physics and electronics. *Phys. Rev. Lett.*, **58**, 2059–2062.

Yang, F., Sambles, J. R., and Bradberry, G. W. (1991). Long-range surface modes supported by thin films. *Phys. Rev. B*, **44**, 5855–5872.

Yariv, A., Xu, Y., Lee, R. K., and Scherer, A. (1999). Coupled-resonator optical waveguide: a proposal and analysis. *Opt. Lett.*, **24**, 711–713.

Yeatman, E. (1996). Resolution and sensitivity in surface microscopy and sensing. *Biosens. Bioelectron.*, **6**, 635649.

Yen, T. J., Padilla, W. J., Fang, N., Vier, D. C., Smith, D. R., Pendry, J. B., Basov, D. N., and Zhang, X. (2004). Terahertz magnetic response from artificial materials. *Science*, **303**, 1494–1496.

Zayats, A. V., Smolyaninov, I. I., and Maradudin, A. A. (2005). Nano-optics of surface plasmon polaritons. *Phys. Rep.*, **408**, 131–314.

Zervas, M. N. (1991). Surface plasmon-polariton waves guided by thin metal films. *Opt. Lett.*, **16**, 720–722.

Zhang, S., Fan, W., Malloy, K. J., Brueck, S. R. J., Panoiu, N. C., and Osgood, R. M. (2005*a*). Near-infrared double-negative metamaterials. *Opt. Exp.*, **13**, 4922–4930.

Zhang, S., Fan, W., Minhas, B. K., Frauenglass, A., Malloy, K. J., Osgood, R. M., and Brueck, S. R. J. (2005*b*). Midinfrared resonance magnetic nanostructures exhibiting a negative permeability. *Phys. Rev. Lett.*, **94**, 037402–1–4.

Zhang, S., Fan, W., Panoiu, N. C., Malloy, K. J., Osgood, R. M., and Brueck, S. R. J. (2005*c*). Experimental determination of near-infrared negative-index metamaterials. *Phys. Rev. Lett.*, **95**, 137404–1–4.

Zhang, S., Fan, W., Panoiu, N. C., Malloy, K. J., Osgood, R. M., and Brueck, S. R. J. (2006). Optical negative-index bulk metamaterials consisting of 2D perforated metal-dielectric stacks. *Opt. Exp.*, **14**, 6778–6787.

Zhao, Y., Belov, P. A., and Hao, Y. (2006). Spatially dispersive finite-difference time-domain analysis of sub-wavelength imaging by the wire medium slabs. *Opt. Exp.*, **14**, 5154–5167.

Zhao, Y., Belov, P. A., and Hao, Y. (2007). Amplification of evanescent spatial harmonics and subwavelength imaging inside of a wire medium slab. In *Proc. 1st Int. Congr. on Advanced Electromagnetic Materials in Microwaves and Optics (Metamaterials 2007)*, Rome.

Zhou, J., Koschny, T., Kafesaki, M., Economou, E. N., Pendry, J. B., and Soukoulis, C. M. (2005*a*). Saturation of the magnetic response of split-ring-resonators at optical frequencies. *Phys. Rev. Lett.*, **95**, 223902–1–4.

Zhou, J., Zhang, L., Tuttle, G., Koschny, T., and Soukoulis, C. M. (2006). Negative index materials using simple short wire pairs. *Phys. Rev. B*, **73**, 041101(R)–1–4.

Zhou, L., Wen, W., Chan, C. T., and Sheng, P. (2005*b*). Electromagnetic wave tunneling through negative-permittivity media with high magnetic fields. *Phys. Rev. Lett.*, **94**, 243905–1–4.

Zhou, X. and Hu, G. (2007). Total transmission condition for photon tunnelling in a layered structure with metamaterials. *J. Opt. A: Pure Appl. Opt.*, **9**, 60–65.

Zhuromskyy, O. (2008). unpublished.

Zhuromskyy, O., Shamonina, E., and Solymar, L. (2005*a*). 2D metamaterials with hexagonal structure: spatial resonances and near field imaging. *Opt. Exp.*, **13**, 9299–9309.

Zhuromskyy, O., Shamonina, E., and Solymar, L. (2005*b*). Effect of radiation on dispersion of magneto-inductive waves in a metamaterial. *Proc. SPIE*, **5955**, 595506–1–6.

Ziolkowski, R. W. (2003*a*). Pulsed and CW Gaussian beam interactions with double negative metamaterial slabs. *Opt. Exp.*, **11**, 662–681.

Ziolkowski, R. W. (2003*b*). Pulsed and CW Gaussian beam interactions with double negative metamaterial slabs: Errata. *Opt. Exp.*, **11**, 1596–1597.

Ziolkowski, R. W. (2004). Propagation in and scattering from a matched metamaterial having a zero index of refraction. *Phys. Rev. E*, **78**, 046608–1–12.

Ziolkowski, R. W. and Heyman, E. (2001). Wave propagation in media having negative permittivity and permeability. *Phys. Rev. E*, **64**, 056625–1–15.

Index

Authors

Laszlo Solymar was born in 1930 in Budapest. He is Emeritus Professor of Applied Electromagnetism at the University of Oxford and Visiting Professor and Senior Research Fellow at Imperial College, London. He graduated from the Technical University of Budapest in 1952 and received the equivalent of a Ph.D in 1956 from the Hungarian Academy of Sciences. In 1956 he settled in England where he worked first in industry and later at the University of Oxford. He did research on antennas, microwaves, superconductors, holographic gratings, photorefractive materials, and metamaterials. He has held visiting professorships at the Universities of Paris, Copenhagen, Osnabrück, Berlin, Madrid and Budapest. He published 8 books and over 250 papers. He has been a Fellow of the Royal Society since 1995. He received the Faraday Medal of the IEE in 1992.

Ekaterina Shamonina was born in 1970 in Twer, Russia. She is Professor in Advanced Optical Technologies at the University of Erlangen-Nürnberg. She graduated in 1993 in Physics at the Moscow State University and received her doctorate in 1998 from the University of Osnabrück, Germany. She was a visiting scientist at the University of Campinas, Brazil in 1996 and 1998. In 2000 she was awarded the 7-year Emmy Noether Fellowship from the German Research Council (Deutsche Forschungsgemeinschaft). She spent the first leg of the fellowship, 2000-2002 at the University of Oxford. After further six months at Imperial College, London she returned to the University of Osnabrück where she built up a research group working on Metamaterials. She completed her habilitation in Theoretical Physics in 2006. Her main research areas apart from metamaterials have been amorphous semiconductors, photorefractive materials, antennas and plasmonics. She published over 70 research papers. She was awarded the Hertha-Sponer Prize 2006 of the German Physical Society.